AF471610

Lecture Notes in Physics

333

I. Appenzeller H. J. Habing
P. Léna (Eds.)

Evolution of Galaxies Astronomical Observations

Proceedings of the Astrophysics School I,
Organized by the European Astrophysics
Doctoral Network
at Les Houches, France, 5–16 September 1988

Springer-Verlag Berlin Heidelberg GmbH

Editors

I. Appenzeller
Universität Heidelberg and Landessternwarte
Königstuhl, D-6900 Heidelberg, FRG

H. J. Habing
Sterrewacht Leiden, Huyghens Lab.
Wassenaarseweg 78, NL-2300 RA Leiden, The Netherlands

P. Léna
Université de Paris VII and Observatoire de Meudon
F-92195 Meudon Cédex, France

DOI 10.1007/978-3-540-46194-4

Originally published by Springer-Verlag Berlin Heidelberg New York in 1989
MyCopy version of the original edition 1989

2158/3140-543210 – Printed on acid-free paper
www.springer.com/mycopy

Preface

Scientific progress in astronomy is heavily dependent upon increased scientific collaboration, a stronger mobility of scientists and an adequate use of large research facilities, available within Europe through multinational or bilateral cooperation.

These objectives led to the foundation in 1986 of the European Astrophysics Doctoral Network, which today federates 11 Departments of Astronomy in European Universities, all these Departments having a graduate program in Astrophysics.

The Network decided to organize each year a Summer School at a predoctoral level, gathering European PhD students at the beginning of their research period, placing them in interaction with the international community of scientists and among themselves, offering them a broad exposure to major fields of astronomy at an early stage of their own research in order to deepen their scientific education and enable them to gain maximum advantage from the possibilities offered by international and European collaboration, especially from the large observing capabilities provided by space– and ground–based European telescopes.

In 1988, this objective converged with the intention of the Fédération Francaise des Magistères de Physique of organizing each year, in the famous Les Houches School of Physics, a new set-up: Ecoles Pré-doctorales de Physique, aimed at PhD students in various fields of Physics, and somehow returning to the early "Les Houches" style. The merging of this objective with those of the Network led to a joint organization of the 1988 School, held in Les Houches.

The 1988 school dealt with two parallel topics : "The Origin, Structure and Evolution of Galaxies" and "Astronomical Observations: Methods and Tools".

These subjects, particularly active in Europe, were chosen to cover areas where considerable theoretical progress and fundamental discoveries are expected in the coming years, given the new observational tools Europe will have at its disposal: the Space Telescope, the Infrared Space Observatory, the Very Large Telescope, the IRAM radio-interferometer, the Hipparcos satellite, the southern millimetric telescope (SEST), the James Clerk Maxwell millimetric telescope in Hawaii, and many other instruments built either in bilateral cooperation or within the European Southern Observatory (ESO) or the European Space Agency (ESA).

This volume contains all lectures presented at the Astrophysics School I, with the exception of the course on "Instrumentation of Large Telescopes" by Sandro D'Odorico. Although an important part of the school programme, this lecture had to be omitted, as Professor D'Odorico's responsibilities in connection with initiating the ESO-VLT instrumentation programme unfortunately made it impossible for him to prepare a manuscript at the present time.

It became clear during the School that graduate studies in Europe are arranged with great diversity: some countries have, or can afford to have, many graduate courses, while others have none or, in some cases, lack the minimum geographic concentration of students needed for their organization. The opportunity of having specially prepared courses, understandable by students fulfilling the minimum requirement of a solid education in Physics, was therefore greatly appreciated. A careful planning of topics and their order made sure that these lectures were accessible to research students exposed to one year or less of research in Astrophysics.

We take here the opportunity to express our gratitude to Jean Heyvaerts, (the Network coordinator), the Conseil d'Administration de l'Ecole des Houches, the European bodies in Brussels (Erasmus Program) and Strasbourg (the European Council), the Fritz Thyssen Stiftung, the Ministries of Foreign Affairs (MAE) and of Research and Technology (MRT) of

France, the UK Science and Engineering Research Council, the Academie Suisse des Sciences Naturelles, the NWO of the Netherlands, and the JNICT of Portugal, which all supported this new venture both financially and morally.

The most important prerequisite for the success of the school was clearly the great enthusiasm of the 57 students (representing 15 different countries) who interacted continuously among themselves and with the lecturers. The spirit, as well as the languages, in the School were truly European, with, unfortunately, as yet no participants from Eastern Europe.

This publication was made possible thanks to the interest of Springer-Verlag, through Prof. W. Beiglböck.

Agnés Fave, Annie Glomot and Nicole Leblanc devoted a great deal of energy and smiling enthusiasm to the success of the School and the production of this volume.

Contents

PARTICIPANTS:

Appl, Stephan	Landessternwarte Heidelberg, FRG
Bernard, J.Philippe	LPSP Verrieres, France
Blommaert, Joris	Sterrewacht Leiden, NL
Breitfellner, Michel	Inst. für Astronomie Wien, A
Campos Aguilar, Ana	IAC Tenerife, Spain
Charlot, Stéphane	STSI Baltimore, USA
Cuddeford, Philip	SISSA Trieste, Italy
Davies, Jeremy R.	University Coll. Cardiff, UK
Deleuil, Magali	LAS Marseille, France
Dietrich, Matthias	Univ. Göttingen, FRG
Donati, J.-Francois	Observ. Paris-Meudon, France
Dougados, Catherine	Observ. Paris-Meudon, France
Dubath, Pierre	Observatoire de Genève, CH
Dutrey, Anne	Observatoire Toulouse, France
Eckert, Josef	Physik. Inst. Erlangen, FRG
Fruscione, Antonella	IAP Paris, France
Gallais, Pascal	Observ. Paris-Meudon, France
Gama, Filomena	Engineering Faculty Porto, Portugal
Garcia Burillo, Santiago	IRAM Grenoble, France
Garcia Gomez, Carlos	IAC Tenerife, Spain
Gourgoulhon, Eric	Univ. Paris 7 Meudon, France
Hunt, Leslie	CNR Firenze, Italy
Jablonka, Pascale	Univ. Paris 7 Meudon, France
Jenniskens, Peter	Leiden University, NL
Jorgensen, Inger	Copenhagen Univ. Obsv., DK
Kamphuis, Jurgen Jan	Kapteyn Lab. Groningen, NL
Kerschbaum, Franz	Inst. für Astronomie, A
Leeuwin, Francine	Univ. Paris 7 Meudon, France
Lehoucq, Roland	IAP Paris, France
Lima, Joao Jose	Fac. Ciencas Porto, Portugal
Liu, Ronghui	Cavendish Lab. Cambridge, UK
Loup, Cécile	CERMO Grenoble, France
Madejsky, Rainer	Landessternwarte Heidelberg, FRG
Maisack, Michael	Astron. Institut Tübingen, FRG
Mannucci, Filoppo	Instituto Astronomia Firenze, Italy
Martinez, Vicent	Univ. Valencia, Spain
Mauder, Wolfgang	Physik. Institut Erlangen, FRG
Moore, Benjamin	Durham University, UK
Morin, Stéphane	Observatoire Marseille, France
Pastor Server, Josefa	Univ. Barcelona, Spain
Pello Descayre, Roser	Univ. Barcelona, Spain
Peymirat, Christophe	CRPE St-Maur, France
Pisani, Armando	SISSA Trieste, Italy
Remy, Sophie	CEN Saclay, France
Rist, Claire	IRAM Grenoble, France
Salez, Morvan	ENS Paris, France
Sauvage, Marc	Univ. Paris 7 Meudon, France
Schertl, Dieter	Physik. Institut Erlangen, FRG
Stark, Ronald	Sterrewacht Leiden, NL

Udry, Stéphane	Observatoire de Genève
Valls-Gabaud, David	IAP Paris, France
Van den Broek, Albertus	University Amsterdam, NL
Vozikis, Christos	University Thessaloniki, Greece
Wozniak, Hervé	Observatoire Marseille, France
Xiluri, Kiriaki	Univ. Crete Iraklion, Greece
Yepes Alonso, Gustavo	University Madrid, Spain
Zwitter, Tomaz	SISSA Trieste, Italy

Galaxy Formation

Malcolm S. Longair
Royal Observatory, Blackford Hill, Edinburgh EH9 3HJ

1 Introduction

In this introductory course on galaxy formation, the emphasis is upon the basic physical processes needed to understand the vast literature which has grown up in recent years on this key topic of modern astrophysical cosmology. My intention is to provide you with a set of tools which you can then use to develop your own models of the ways in which galaxies may have come about. I will try to elucidate some of the trickier bits of the story but all the way through I will emphasise the simplicity of the ideas rather than their complexity.

The plan of the lecture course and the contents of the following chapters are as follows:

2. **The Basic Framework**. In this chapter we will look at the physics of the isotropic Hot Big Bang model and derive many ideas which will be crucial in the understanding of the problems of the origin of galaxies.
3. **The Evolution of Fluctuations in the Standard Hot Big Bang**. This is a key part of the story since the rules which come out of this study define the basic problems of galaxy formation.
4. **Dark Matter** Dark matter plays a central role in the most popular theories of galaxy formation and so we have to assess the evidence that it is present in the Universe in such quantities as could profoundly influence our view of how galaxies formed.
5. **Correlation Functions** Once we understand how structures might form, we have to make more detailed comparison of the theory with the observations and so look at the whole spectrum of structures which have formed in the Universe. One of the crucial confrontations between the theories and the observations concerns temperature fluctuation in the Microwave Background Radiation.
6. **The Post-Recombination Universe** To understand how to make real structures in the Universe, we have to follow the perturbations into their non-linear stages in the post-recombination epoch. In the non-linear stages of development, cooling processes may be more important than purely dynamical collapse in certain circumstances. We also need to look at observational evidence on the early evolution of real objects such as galaxies, quasars and radio sources.

A full set of references is given at the end of the lectures to more detailed texts on all these topics.

2 The Basic Framework

We begin by reviewing the framework which virtually all astrophysicists use to study the problems of galaxy formation. This is what has come to be known as the **Standard Hot Big Bang Model** of the Universe. We will study this picture quite carefully bringing out the features which are important in the subsequent discussion. The standard model has the great advantage of simplicity and most of the essential features can be understood by simple physical arguments. The standard model is based upon a few basic observations and assumptions and we begin by looking at the observational basis for the model. The two features of the large scale structure of the Universe which we need are its overall isotropy and the fact that it is in a state of uniform expansion at the present day. Let us look at the best modern evidence on these topics.

2.1 The Isotropy of the Universe

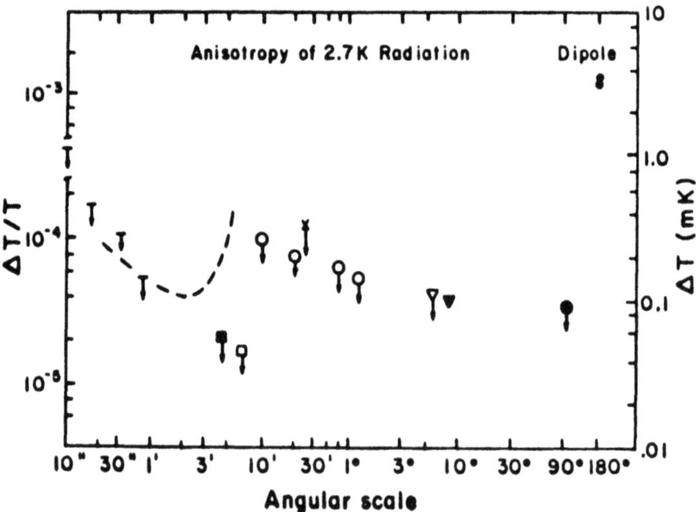

Fig. 1. Upper limits and measurements of the anisotropy of the Microwave Background Radiation (Wilkinson 1988).

The most direct evidence for the overall isotropy of the Universe comes from observations of the **Microwave Background Radiation.** This component of background radiation was discovered, more or less by chance, by Penzias and Wilson in 1965. We will show later that it is identified with the cool remnant of the hot early phases of the Universe but for the moment we are interested in it as a tool for measuring the large-scale isotropy of the Universe. The present limits to and measurements of its anisotropy on different angular scales are displayed in Fig. 1 (Wilkinson 1988). The global anisotropy of the radiation is dominated by the dipole term which has amplitude $\Delta T/T = (1.2 \pm 0.1) \times 10^{-3}$ with apex in the direction $\alpha = 11^{\mathrm{h}}\!.3 \pm 0^{\mathrm{h}}\!.16; \delta = -7^\circ\!.5 \pm 2^\circ\!.5$. This pure dipole term is attributed to the motion of the Earth at a velocity of about 350 km s^{-1} with respect to a frame of reference in which the Microwave Background Radiation would be 100% isotropic. Besides this dipole term, the upper limits to the temperature fluctuations all correspond to $\Delta T/T \leq 10^{-4}$. Until recently, only upper limits have been reported but Davies *et al.* (1987) have now claimed a detection of temperature fluctuations with rms deviation $\Delta T/T = 3.7 \times 10^{-5}$ on an angular scale of 8° at a wavelength of 3 cm. A key question, of course, is whether or not this signal is indeed a fluctuation in the radiation temperature of the Microwave Background Radiation or, say, due to diffuse Galactic emission or discrete sources. As Wilkinson remarks in his review, it will require only a modest improvement in sensitivity to measure correlated fluctuations at other wavelengths. The positive detection of fluctuations and the measurement of their spectral properties would open up a whole new range of astrophysical possibilities from the topology of the early Universe, through galaxy formation to the detection of extensive regions of hot electrons in the Universe (see, e.g. Melchiorri *et al.* 1986).

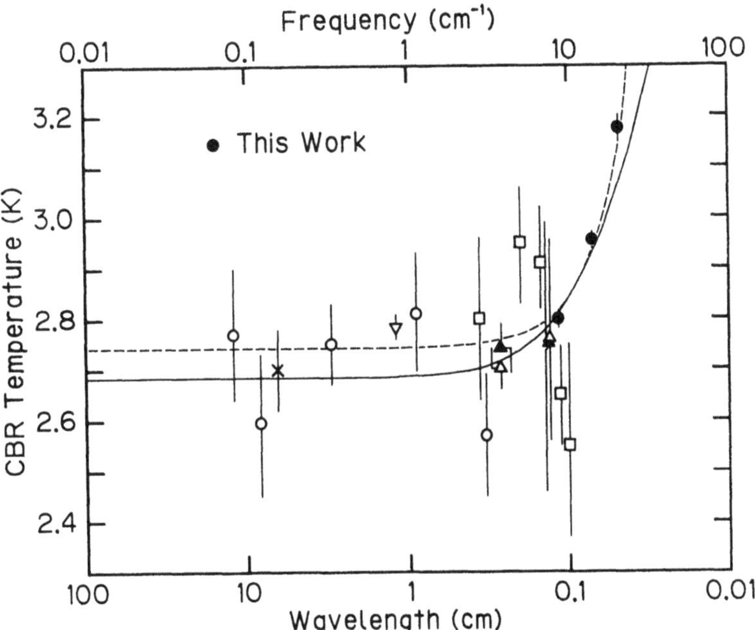

Fig. 2. Recent high precision measurements of the radiation temperature of the Microwave Background Radiation (from Matsumoto *et al.* 1988). Their submillimetre rocket observations are indicated by solid points. The solid line shows the effects of Compton scattering upon the spectrum of the background radiation. The dashed line shows a model in which there is a cool dust component in addition to a pure Planck spectrum.

Recent measurements of the spectrum the Microwave Background Radiation are shown in Fig. 2. There is good agreement at wavelengths longer than 1 mm. Wilkinson (1988) reports a mean temperature $T = 2.740 \pm 0.016$ K from these data. An observation of great interest is the recent Nagoya/Berkeley rocket experiment in which the sky background temperature beyond the peak of the blackbody curve was measured at wavelengths of 1.16, 0.71 and 0.48 mm (Matsumoto *et al.* 1988). The radiation temperature at 1.16 mm agrees with the longer wavelength observations but the values at the shorter wavelengths are significantly in excess of 2.74 K, values of 2.963 ± 0.017 K and 3.150 ± 0.026 K being measured at 0.71 and 0.48 mm respectively. The extra energy in the sub-millimetre region of the spectrum, over and above that of a 2.74 K blackbody, would amount to about 20% of the total energy in a blackbody radiation spectrum at 2.75 K.

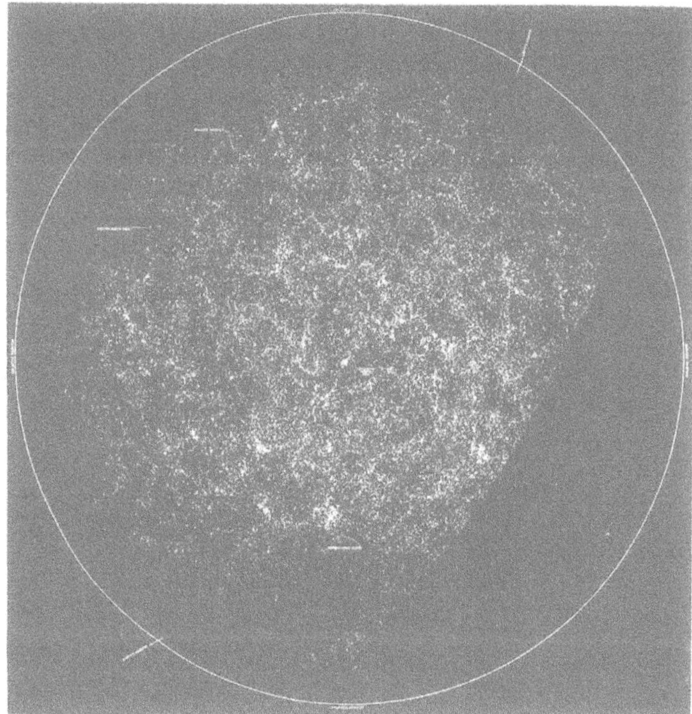

Fig. 3. The distribution of galaxies on the celestial sphere. This image was generated from the counts of galaxies over the whole northern sky and as far south as declination -20°by the Lick astronomers, C.D. Shane, C.A. Wirtanen and their colleagues. The picture is an equal area projection of the northern galactic hemisphere with the centre of the diagram representing the direction of the north galactic pole and the circumference of the circle corresponding to looking through the galactic equator. Over one million galaxies were counted in the Lick survey. The black segment to the lower right of the diagram is due to the lower declination limit of the survey. Towards the edges of the diagram the numbers of galaxies decrease because of interstellar obscuration by dust in the plane of the Galaxy. Towards the centre of the diagram, the picture is more or less unobscured and represents the true distribution of galaxies in the Universe, as seen projected onto the celestial sphere. The typical distances of most of the galaxies in this picture are between about 100 and 500 Mpc (from M. Seldner *et al.* 1977).

If the excess is real, it could be attributed to a number of possible effects. Cool, diffuse dust emission is one possibility and a model fit to the observations is indicated by the dashed line in Fig. 2. One intriguing possibility is that this dust emission might be associated with a population of stars formed very early in the history of galaxy formation, the so-called population III stars. Another possibility, illustrated by the solid line in Fig.

2, is Compton scattering of the blackbody background spectrum by hot electrons, a process described by Zeldovich and Sunyaev (1969). In what follows we will assume that the radiation can be described by a black-body spectrum with radiation temperature 2.75 K and the submillimetre excess, if confirmed, has some separate, possibly related, origin. The immediate implication of such a black-body spectrum is that at some stage in the evolution of the Universe the matter and radiation must have been in thermodynamic equilibrium at a single temperature.

The isotropy of the distribution of **galaxies** is a more complex issue. The nature of the problem is well illustrated by the famous picture of the distribution of galaxies in the northern galactic hemisphere produced by Peebles and his colleagues from the Lick counts of galaxies (Fig. 3 Seldner *et al.* 1977). In the picture the north galactic pole is at the centre and the plane of the Galaxy is represented by the circle surrounding the distribution of galaxies. The galaxies are shown in an equal area projection. One obvious feature is the decreasing surface density of galaxies towards the edge of the diagram. This is simply due to obscuration by interstellar dust. The regions towards the centre of the diagram should be relatively free from obscuration and provide a clear view of the distribution of galaxies on a large scale. The typical distance of the galaxies is about 100 to 500 Mpc. In the unobscured regions, it is apparent that the distribution of galaxies is far from uniform and homogeneous – there is prominent superclustering of galaxies, elongated stringy structures and apparent holes where there are low surface densities of galaxies. There has been considerable discussion about the reality of these features but it is now agreed that they are real properties of the observed distribution of galaxies.

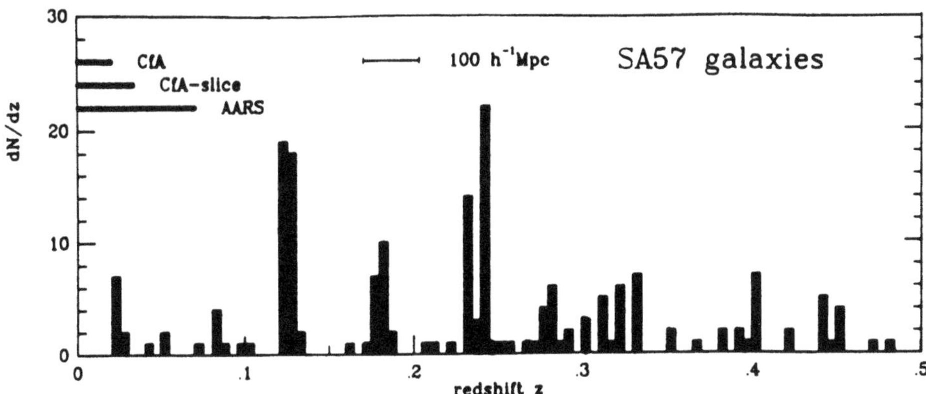

Fig. 4. The redshift distribution of a complete sample of faint galaxies in the deep optical survey of Koo and Kron (1988).

The most extensive database for defining these structures is the enormous Center for Astrophysics (CfA) Redshift Survey (Geller *et al.* 1987). This redshift catalogue of a large sample of galaxies selected from the Zwicky catalogue down to a limiting magnitude of about 15.5 already contains about 20,000 galaxies which represents about 70% of the final complete sample. In the areas of the survey for which the data are complete, there is very clear evidence for superclustering, voids and thin sheets of galaxies. These are the types of structures which must be explained by successful models of the origin of structure in the Universe. Despite the obvious irregularity of the distribution of galaxies, it can also be seen that, looked at on a large enough scale, one region of the Universe looks very much like another. This has been formally tested by Peebles and his colleagues. The conclusion is that, although the distribution of galaxies is inhomogeneous, the inhomogeneity decreases to a very low value on large enough physical scales. On the very largest scales, the galaxy distribution appears to be uniform. This is demonstrated formally using the two-point correlation function for galaxies which is described in more detail in Section 5.1

It is useful to have a physical picture of the large scale distribution of galaxies and this is provided by the remarkable analysis of Gott and his colleagues (1987). Using the data from the CfA Redshift survey, they have shown that the topology of the distribution of galaxies is **sponge-like** in the sense that the galaxies form the material of the sponge which is continuously connected and the holes also form a continuously connected network throughout the sponge. This type of topology has important implications for the assumptions made about the properties of the initial perturbations from which the large scale structure of the Universe formed.

The CfA Redshift survey is limited to the local region of space, corresponding to distances less that about 100 Mpc and even the Lick survey only probes the large scale structure to a distance of about 500 Mpc. The sponge-like structure seems to persist out to much larger distances. This has been demonstrated in the recent deep redshift surveys carried out by Ellis and his colleagues at the Anglo-Australian Telescope (Ellis 1987) and by Koo and Kron (1988) (Fig. 4). It is obvious that the galaxies are not randomly distributed but are clumped in redshift corresponding to superclustering and voids in the distribution of galaxies.

In view of the obvious irregularity of the distribution of galaxies, what evidence is there for the isotropy and homogeneity of objects on even larger scales? The best evidence comes from the distribution of radio sources contained in all-sky surveys. The 4C survey of radio sources shown in Fig. 5 is a good example of what is observed. In this diagram, which is drawn in an equal area projection, there is an obvious hole in the centre corresponding to the region of sky about the north celestial pole which was not included in the 4C survey. We now know that the redshifts of the objects contained in this survey, radio galaxies and quasars, are typically between about 1 and 3, meaning that they sample the largest scale structures in the Universe accessible to us. Webster (1977) showed that there is no evidence for anisotropy in the distribution of radio sources in the 4C, Parkes and Bologna catalogues within the statistical uncertainties of the numbers available. According to Webster, the statistical limits to the distribution of radio sources corresponds to $\Delta N/N \leq 0.03$ on a scale of 1 Gpc i.e. as a cube of side 1 Gpc is moved about the Universe, the fluctuations in the numbers of objects counted is less than this value. Although this figure is less impressive than the limits from the measurements of the Microwave Background Radiation, it refers to the large scale distribution of real astronomical objects observed **at roughly the present epoch.** In contrast, we will show that the limits obtained from the fluctuations (or lack of them) in the Microwave Background Radiation constrain the **temperature** fluctuations of the radiation at very much earlier epochs, corresponding to a **redshift of about 1000.**

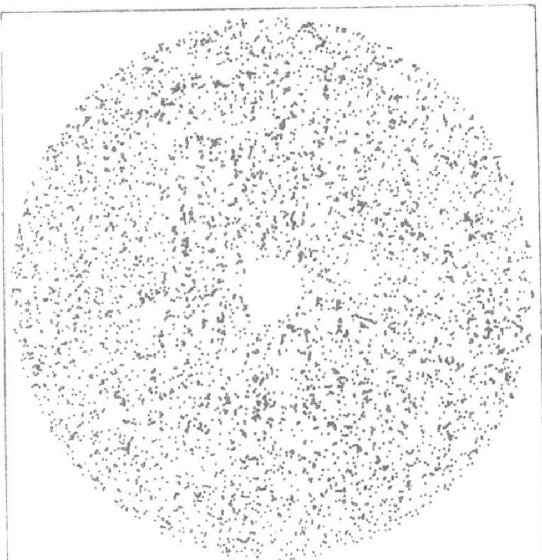

Fig. 5. The distribution of bright radio sources in the northern galactic hemisphere as derived from the 4C catalogue. There are over 3000 sources in this region of the survey. The diagram is plotted in an equal area projection and the circle surrounding the distribution corresponds to declination 0°. The hole in the centre of the diagram corresponds to the region north of $\delta = 80°$ which was not surveyed. (Data from Pilkington and Scott 1965 and Gower *et al.* 1967: the equal area projection is due to M. Seldner)

2.2 The Hubble Expansion

Hubble announced his discovery of the velocity-distance relation for galaxies in 1929. The modern version of Hubble's famous relation $v = H_0 r$, where v is the velocity of recession of the galaxy, r its distance from our Galaxy and H_0 is Hubble's constant, is provided by Sandage's classic determination of Hubble's relation for the brightest galaxies in clusters (Fig. 6, Sandage 1968, 1987, 1988). It is worth noting that nowadays this is not the only way in which the velocity-distance relation is defined. A good example is the use of radio galaxies selected from catalogues of bright radio sources. They display a tight redshift-magnitude relation which extends to redshifts of about 1.8 – we will return to this diagram later (Fig. 13, Lilly and Longair 1984). Even the quasars, which have a broad dispersion in intrinsic magnitude display a significant velocity-redshift relation out to large redshifts, provided attention is confined to radio loud objects (Wills and Lynds 1978, Wall and Peacock 1985). One topic to which we will return is the impact of **streaming** of galaxies upon the mean Hubble flow. These systematic motions of galaxies correspond to less than about 10% of the Hubble velocity and thus are insignificant compared with the intrinsic scatter in the redshift-magnitude relation.

It should be noted that these two facts, the overall isotropy of the Universe and Hubble's law are sufficient to show that the distribution of galaxies is expanding apart uniformly. This is illustrated by the simple diagram shown in Fig. 7 which shows the uniform expansion of a system of galaxies between two epochs. By fixing attention

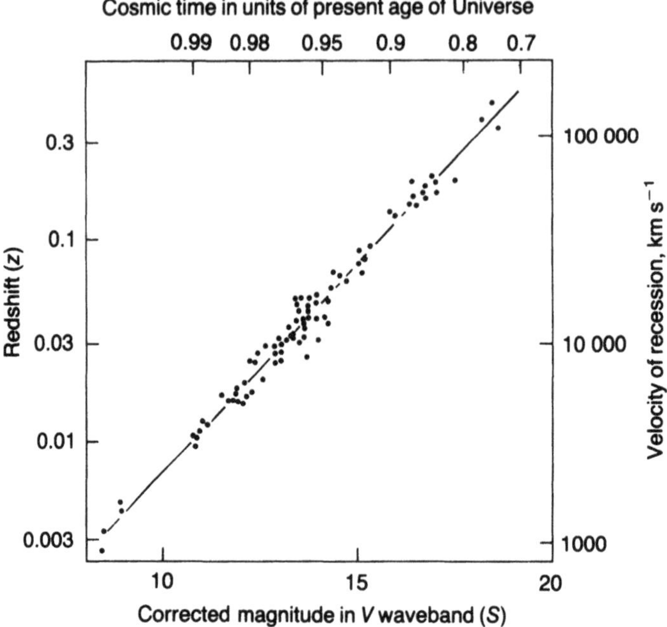

Fig. 6. A modern version of the Hubble diagram for the brightest galaxies in clusters (after Sandage 1968). In this logarithmic plot, the observed flux density S of the galaxy, expressed as a corrected V magnitude, is plotted against the redshift which is proportional to velocity v so long as $v \ll c$. The V magnitude is a measure of distance and the straight line shows the expected relation if the brightest galaxies in clusters all had the same intrinsic luminosity so that $S \propto z^{-2}$. This diagram indicates that, for this class of galaxy which can be readily observed to cosmological distances, velocity of recession is proportional to distance.

upon any galaxy in the array, it can be seen that an observer on each galaxy observes a velocity-distance relation. In other words, our first two facts strongly suggest that we should begin our study with isotropic, uniformly expanding cosmological models.

2.3 The Robertson-Walker Metric

The general metric for all world models consistent with the assumptions of isotropy and homogeneity can now be derived. We require one further assumption which is known as the **Cosmological Principle** – this is the statement that the observations we make are typical of what would be observed by any observer located anywhere in the Universe. In other words, any suitably selected observer would observe the same large scale features of the Universe at the present time as we do. Nowadays this is more than simply an assumption. Studies of the large scale distribution of galaxies show that, although it is inhomogeneous, the degree of inhomogeneity seems to be more or less the same as we look further and further away (see, e.g. Peebles 1980). I have no doubt but that we will soon have very good measures of the homogeneity of the distribution of galaxies extending to redshifts about 0.5 which is as far away as we can observe the Universe more or less as it is at the present epoch. The cosmological principle boils down to the statement that we are located at a typical point in the Universe and not at some highly privileged vantage point. We should note in passing that in some sense we are privileged observers because of the fact of our existence and because we are able to ask such questions! The discussion of the extent to which we are privileged would take us far beyond the scope of the present lectures but, for a thought-provoking read, I can thoroughly recommend **The Anthropic Cosmological Principle** by Barrow and Tipler. We will restrict attention to the strictly classical cosmological principle.

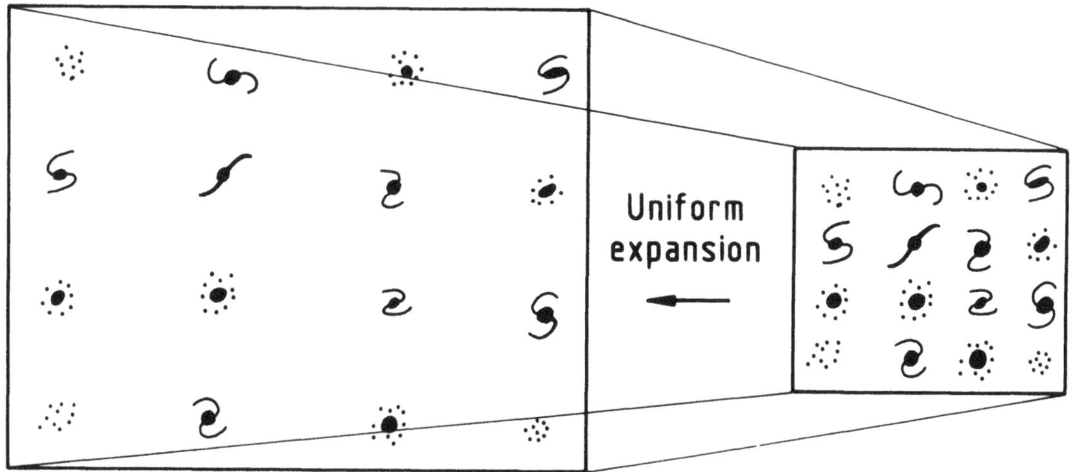

Fig. 7. Illustrating how the uniform expansion of a system of galaxies leads to a linear relation between their distances and their recession velocities. If attention is focussed upon any pair of galaxies, it will be observed that the further they are apart, the further they have to separate in order to preserve a uniform expansion.

It is a pleasant argument to derive, solely from the requirements of isotropy and homogeneity and that light be propagated according to the laws of special relativity, the highly restricted set of metrics which could describe the large-scale structure of space-time. These arguments are given in their simplest forms by Gunn (1978) and in Chapter 15 of my book **Theoretical Concepts in Physics**, hereafter referred to as TCP. I will use the notation of my book simply because I believe it provides the simplest intuitive approach to the full theory.

A requirement of suitable geometries is that they should all reduce locally to the Minkowski metric of special relativity

$$ds^2 = dt^2 - \frac{1}{c^2}(dr^2 + r^2 d\Phi^2) \tag{1}$$

This is no more than the standard metric of special relativity but written in spherical polar coordinates in which

$$d\Phi^2 = d\theta^2 + \sin^2\theta d\phi^2.$$

The requirement of isotropy reduces the possible geometries of space-time to general isotropic curved spaces for which the metric can be written

$$ds^2 = dt^2 - \frac{1}{c^2}\left[dx^2 + R_c^2 \sin^2\left(\frac{x}{R_c}\right)(d\theta^2 + \sin^2\theta d\phi^2)\right] \tag{2}$$

This is not perhaps the most familiar of forms for the metric of curved space-time but it has a simple geometric interpretation. In this form, the radius of curvature of two dimensional sections through the space is everywhere R_c and the curvature of the space $\kappa = 1/R_c^2$. Notice that I have changed the radial distance coordinate from r to x because I will want to give r a special meaning in a moment. It is well known that there are three possible values of the curvature κ and hence of the radius of curvature of the space R_c.

1. If κ is positive, R_c is real and the spatial sections of the space-time have spherical, closed geometry.
2. If κ is negative, R_c is imaginary and the spatial sections of the geometry are open and hyperbolic. In this case the $\sin(x/R_c)$ in equation (2) is replaced by $\sinh(x/R_c)$
3. If κ is zero, R_c is infinity and the spatial geometry is flat, open Euclidean space and the metric reduces to equation (1).

In perhaps the most popular form of the metric, these different cases correspond to geometries with $k = 1, -1$ and 0 respectively. This more popular form can be derived from the metric of equation (2) by a simple coordinate transformation (see TCP, page 320).

Now we can include in our metric the observation that the Universe is in a state of uniform expansion. Since we are dealing with uniformly expanding universes, this means that the relative separations of any two points in the Universe now, x_0, was smaller in the past by a factor R, i.e. $x = Rx_0$. By this simple substitution, we absorb the whole of the dynamics of the expansion of the Universe into the function $R(t)$ which is known as the **Scale Factor** of the Universe. We normalise the scale factor so that $R = 1$ at the present epoch $t = t_0$. Notice that when we refer to time we mean **cosmic time** which is defined to be proper time measured by a observer who moves in such a manner that the Universe always appears to be isotropic. It is then a simple exercise to show that the metric (2) becomes the famous **Robertson-Walker metric**

$$ds^2 = dt^2 - \frac{R^2(t)}{c^2}\left[dr^2 + \Re^2 \sin^2\left(\frac{r}{\Re}\right)(d\theta^2 + \sin^2\theta d\phi^2)\right] \tag{3}$$

There are several important remarks to be made about the variables and constants in the above metric. First of all, it will be noted that the dynamics and geometry of the models are defined by one function $R(t)$ and one constant $\Re$. As discussed above $R(t)$ describes the dynamics of the expansion. $\Re$ is the radius of curvature of space at the present epoch, i.e. $\Re = R_c(t_0)$, so that the curvature of space at the present epoch is $\kappa_0 = 1/\Re^2$. It is interesting to show from the Robertson-Walker metric that the curvature of space changes as the Universe expands as

$$R_c(t) = \Re R(t).$$

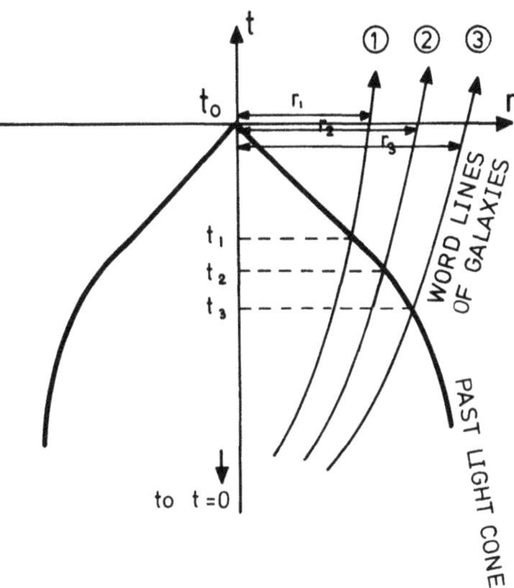

Fig. 8. A simple space-time diagram illustrating the definition of the radial comoving distance coordinate r. All observations of galaxies are made along our past light cone which is centred on the Earth. The trajectories of galaxies partaking in the expansion of the distribution of galaxies is shown.

Special attention should be paid to the meaning of the radial coordinate r. It is referred to as the **radial comoving distance coordinate** and is a measure of the distance the galaxy would have if its position were projected forward to the present epoch. This is illustrated in Fig. 8 which is a simple space-time diagram showing that we observe galaxies along our past light-cone. However, we have to define our distances and geometry in terms of their values at a particular reference epoch which we take to be the present. To obtain a picture of how you might imagine r to be measured, we could line up a whole army of comoving observers between the galaxy and ourselves and then tell them to measure the distance between each other at some prearranged epoch t_0. The sum of all the increments dr measured at the epoch t_0 is a measurement of r. Notice that in real cosmological problems, r is unmeasureable because, to find it, we have to project the position of the galaxy as we observe it in the past to the present epoch and this depends upon a knowledge of the dynamics of the Universe. Unfortunately we do not know this. The final point to be made about r is that it is a distance which is attached to the galaxy for all time. In fact, it is no more than a distance label. The variation of its distance from the Earth due to the expansion of the Universe is all absorbed in the variation of $R(t)$ with cosmic epoch.

It is important to emphasise that there is very little physics at all in this definition of the metric of isotropic space-times – in fact, only special relativity. As stated above, the physics is all built into the variation of $R(t)$ with cosmic epoch – if you wish to build your own model universes, you can do this by inventing theories which define $R(t)$.

One of the key aspects of observational cosmology is the relation between observable quantities and the intrinsic properties of the objects. This is an interesting and very important use of the Robertson-Walker metric and is indispensible for astrophysical cosmology. I will not derive the results here but I give simple derivations of the key results in Section 15.4 of TCP. We will only quote two of the most important results.

The first is the definition of **redshift** and its relation to $R(t)$. The redshift of a galaxy is so-called because it was discovered by the pioneers of observational cosmology that the spectral lines of galaxies were all shifted towards the red end of the optical spectrum relative to their rest wavelengths. If λ_{em} is the emitted wavelength of some spectral feature and λ_{obs} the wavelength with which it is observed, the redshift z of the object is defined to be

$$z = \frac{\lambda_{obs} - \lambda_{em}}{\lambda_{em}} \tag{4}$$

The usual interpretation of the redshift is as a velocity of recession and it is true that, provided the velocity of the galaxy is much less than the velocity of light, $v \ll c$, $v = cz$. However, there is a much deeper meaning of redshift in cosmology. You may show from the Robertson-Walker metric that in general the redshift is directly related to the scale factor of the Universe through the relation

$$R(t) = \frac{1}{1+z} \tag{5}$$

This result comes directly from the metric (3) (see TCP, Section 15.4.1, if necessary) and thus is independent of the physics of the expansion. What this relation tells us is that the redshift measures the size of the Universe, i.e. the physical separation between comoving test particles, when the radiation was emitted relative to its present size. For example, at a redshift $z = 3$, the galaxies were all closer together by a factor of 4 relative to their present separation. If we were also able to measure the cosmic time when the radiation was emitted from distant objects, we could derive directly the dynamics of the Universe. Unfortunately, the astrophysics of galaxies is too poorly understood at present to provide useful constraints, although we will give an interesting example in Section 6.4.4 of the use of this procedure as applied to large redshift radio galaxies.

The second point is that, using the Robertson-Walker metric, we can show that Hubble's constant H_0 is just the present rate of expansion of the Universe, i.e.

$$H_0 = \left(\frac{\dot{R}}{R}\right)_0 \tag{6}$$

The subscript 0 means the value of Hubble's constant at the present epoch. Note, however, that Hubble's constant may be defined at any epoch and in general changes with cosmic epoch. Thus, at any epoch $H = \dot{R}/R$.

The other parameter which we can define at the present epoch is the dimensionless **deceleration parameter**, q_0. This is simply the present deceleration of the Universe, $\ddot{R}$, written in the form

$$q_0 = -\left(\frac{\ddot{R}}{\dot{R}^2}\right)_0$$

2.4 The Dynamical Framework

We will consider the dynamics of world models in three parts. First of all we consider the standard dust models, then look at radiation-dominated models and finally consider briefly inflationary models. All the models begin with dynamical equations derived from Einstein's General Theory of Relativity. This is by far the best classical theory of gravity which we possess. This is not the place to go into the details of how well General Relativity has been tested experimentally but it is sufficient to say that it has survived the most precise experiments which have been made up till now. A particularly spectacular confirmation of many aspects of the theory has been the accurate prediction of the change in period of the binary pulsar system PSR 1913+16, which consists of a pair of neutron stars in a close binary orbit, due to the radiation of gravitational waves. This remarkable agreement between theory and observation enables wide classes of alternative theories to General Relativity to be excluded. An excellent summary of the current status of General Relativity and possible alternative theories of gravity is given by C.M. Will in his book **Theory and Experiment in Gravitational Physics.** We therefore have little hesitation in adopting General Relativity as the dynamical framework for our model universes.

The Einstein field equations can be written in the following form:

$$\ddot{R} = -\frac{4\pi GR}{3}\left(\rho + \frac{3p}{c^2}\right) + \left[\frac{1}{3}\Lambda R\right] \tag{7}$$

$$\dot{R}^2 = \frac{8\pi G\rho}{3}R^2 + \left[\frac{1}{3}\Lambda R^2\right] - \frac{c^2}{\Re^2} \tag{8}$$

These are the general equations for the dynamics of isotropic world models in which the density and pressure of the matter and radiation are ρ and p respectively. Notice that the pressure term in equation (7) is a relativistic correction so that the quantities in large round brackets represent the total inertial mass density. Unlike normal pressure forces which depend upon the gradient of the pressure and, for example, hold up stars, this pressure term depends linearly on the pressure and, since it contributes to the inertial mass, increases the gravitational force. $\Re$ is the radius of curvature of the geometry of the world model at the present epoch and so the last term of equation (8) is simply a constant of integration. I have included the famous **cosmological constant** Λ in equations (7) and (8) in large square brackets. This term has had a chequered history in that it was originally introduced by Einstein in order to produce static solutions of the field equations more than 10 years before it was discovered that the Universe is in fact non-static in the sense that it is expanding. As we will discuss below it has had a new lease of life with the development of models of an inflationary stage in the early history of the Universe.

2.4.1 The Standard Dust Models

This analysis is performed in all the standard text-books. By **dust**, we mean a pressureless fluid, $p = 0$. In addition, we set the cosmological constant $\Lambda = 0$. It is convenient to refer the density of matter to its value at the present epoch ρ_0. Because of conservation of mass, $\rho = \rho_0 R^{-3}$ and so the pair of equations reduces to the following simple form

$$\dot{R}^2 = \frac{8\pi G\rho_0}{3}R^{-1} - \frac{c^2}{\Re^2} \tag{9}$$

It is well known that a relation of this form can be derived using purely Newtonian dynamics. I show how this is done because we will use ideas implicit in this argument to understand some of the problems which arise in the theory of galaxy formation. This argument looks naive but it is in fact a very helpful exercise.

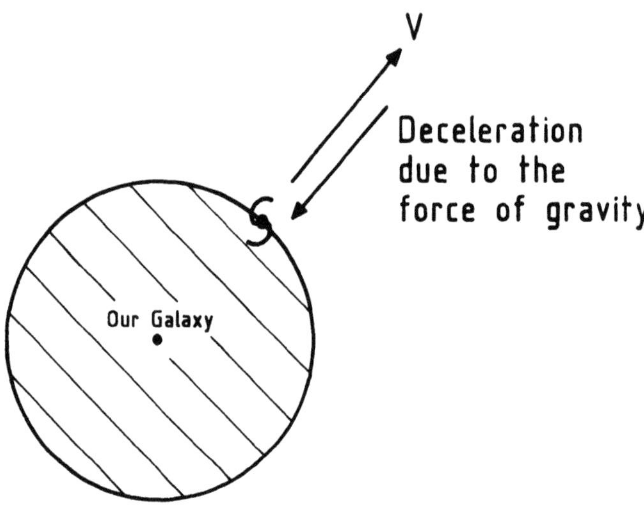

Fig. 9. Illustrating the dynamics of Newtonian world models.

We consider a galaxy at distance x from the Earth and ask what the deceleration of that galaxy is due to the attraction of matter inside the sphere of radius x centred on the Earth. By Gauss's theorem, because of the spherical symmetry of the distribution of matter within x, we can replace that mass by a point mass at the centre of the sphere and so the deceleration of the galaxy is

$$m\ddot{x} = -\frac{GMm}{x^2} = -\frac{4\pi x^3 \rho m}{3x^2}$$

Now, we make the substitutions as before – replace x by the comoving value x_0 using the scale factor R, $x = Rx_0$, and express the density in terms of its value at the present epoch, $\rho = \rho_0 R^{-3}$. Notice that the mass of the galaxy m cancels out on either side of the equation, showing that the deceleration refers to the dynamics of the Universe as a whole rather than to any particular galaxy. Therefore,

$$\ddot{R} = -\frac{4\pi G\rho_0}{3}\frac{1}{R^2}$$

which is identical to equation (7) for dust models with $\Lambda = 0$. Integrating this equation, we find

$$\dot{R}^2 = \frac{8\pi G\rho_0}{3}R^{-1} + \text{constant} \qquad (10)$$

This result is identical to equation (8) if we identify the constant with $-c^2/\Re^2$.

The above analysis brings out a number of important points about the world models of general relativity. First of all, note that, because of the assumption of isotropy, local physics is also global physics. This is why the Newtonian argument works. The same physics which defines the local behaviour of matter also defines its behaviour on the largest scales. For example, the curvature of space within one cubic metre is exactly the same as that on the scale of the Universe itself. A second point is to note that, although we might appear to have placed the Earth in a rather special position in Fig. 9, the observer located on the galaxy would perform exactly the same calculation to work out our deceleration relative to that galaxy. In other words, the Newtonian calculation applies for all observers who move in such a way that the Universe appears isotropic to them. Third, notice that at no point in the argument did we ask over what physical scale the calculation was to be valid. It is a remarkable fact that this calculation describes correctly the dynamics of the Universe on scales which are greater than the **horizon scale** which we take to be $r = ct$ i.e. the maximum distance between points which can be causally connected at the epoch t. The reason for this is again the same as for the first two points – local physics is also global physics and so, if the Universe were set up in such a way that it had uniform density on scales far exceeding the horizon scale, the dynamics on these very large scales would be exactly the same as the local dynamics. We will find this idea very helpful in understanding the evolution of small perturbations in Section 3.

The solutions of Einstein's field equations were discovered by A.A. Friedman in 1925, the year before his death in Leningrad. 1988 is the centenary of Friedman's birth and appropriately there have been celebrations this year in the USSR. A remarkable biography of Friedman by Tropp, Frenkel and Chernin has been published this year in the USSR which I can strongly recommend. The solutions of the equations are often appropriately referred to as the **Friedman models** of the Universe. It is convenient first of all to express the density of the world model in terms a **critical density** ρ_c which is defined to be $\rho_c = (3H_0^2/8\pi G)$ and then to refer the actual density of the model ρ to this value through a **density parameter** $\Omega = \rho/\rho_0$. Thus, the density parameter is given by

$$\Omega = \frac{8\pi G\rho}{3H_0^2} \tag{11}$$

The dynamical equation (8) therefore becomes

$$\dot{R}^2 = \frac{\Omega H_0^2}{R} - \frac{c^2}{\Re^2} \tag{12}$$

There are several important facts which can be deduced from this equation. If we set $t = t_0, R = 1$, i.e. their values at the present epoch, we find that

$$\Re = \frac{c/H_0}{(\Omega - 1)^{1/2}} \qquad \text{and} \qquad \kappa = \frac{(\Omega - 1)}{(c/H_0)^2} \tag{13}$$

This last result shows that there is a one-to-one relation between the density of the Universe and its spatial curvature, one of the most beautiful results of the Friedman

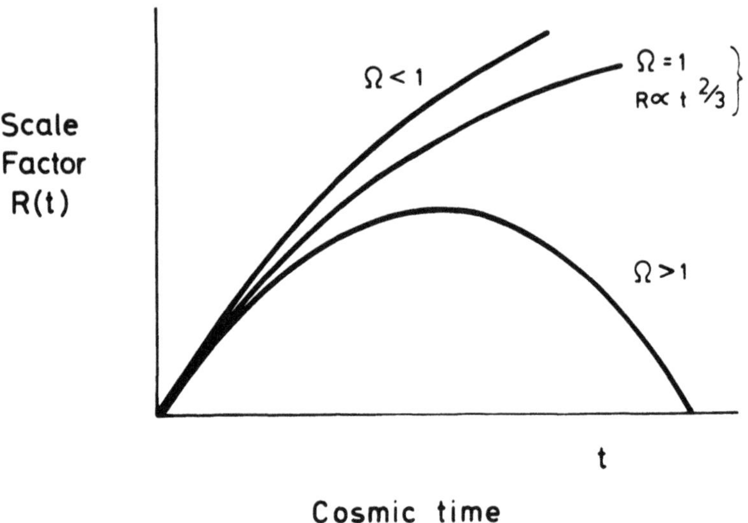

Fig. 10. The dynamics of the classical models of General Relativity. The models are parameterised by the density parameter Ω.

world models. The solutions of equation (12) are displayed in Fig. 10 which shows the well-known relation between the dynamics and geometry of the Friedman world models.

1. The models with $\Omega > 1$ have closed, spherical geometry and they collapse to an infinite density in a finite time;
2. The models with $\Omega < 1$ have open, hyperbolic geometries and expand forever. They would reach infinity with a finite velocity.
3. The model with $\Omega = 1$ is the critical model which separates the open from the closed models and the collapsing models from those which expand forever. This model is often referred to as the **Einstein-de Sitter model** or the **critical model.** The velocity of expansion tends to zero as R tends to infinity. It has a particularly simple variation of $R(t)$ with cosmic epoch,

$$R = \left(\frac{3}{2}H_0 t\right)^{\frac{2}{3}} \qquad \kappa = 0 \tag{14}$$

Another important result is the function $R(t)$ for the empty world model, $\Omega = 0$, $R(t) = H_0 t$, $\kappa = -(H_0/c)^2$. This model is sometimes referred to as the **Milne model.** It is an interesting exercise to show why it is that, in the completely empty world model, the global geometry of the Universe is hyperbolic. I give a derivation of this result in the Appendix to Chapter 15 of TCP.

Differentiating equation (12) with respect to time, or substituting into equation (7), we can show immediately that the present deceleration of the Universe q_0 is directly proportional to the density parameter Ω, $q_0 = \Omega/2$. Note that this result is only true if the cosmological constant Λ is zero. In general, we find

$$q_0 = \frac{\Omega}{2} - \frac{1}{3}\frac{\Lambda}{H_0^2}$$

In the same way, if $\Lambda \neq 0$, the relation between the curvature and the density parameter becomes

$$\kappa = \frac{(\Omega - 1)}{(c/H_0)^2} + \frac{1}{3}\frac{\Lambda}{c^2} \tag{15}$$

An important result for many aspects of cosmology is the relation between redshift z and cosmic time t. It is straightforward to show from equation (12) that

$$\frac{dz}{dt} = -H_0(1+z)^2(\Omega z + 1)^{\frac{1}{2}} \tag{16}$$

Cosmic time t measured from the big bang follows immediately by integration

$$t = \int_0^t dt = -\frac{1}{H_0}\int_\infty^z \frac{dz}{(1+z)^2(\Omega z + 1)^{1/2}} \tag{17}$$

It is a useful exercise to show that the present age of the Universe t_0 is H_0^{-1} if $\Omega = 0$ and $(2/3)H_0^{-1}$ if $\Omega = 1$.

Just as it is possible to define Hubble's constant at any epoch by $H = \dot{R}/R$, we can define a density parameter Ω at any epoch through the definition $\Omega = 8\pi G\rho/3H^2$. Since $\rho = \rho_0(1+z)^3$, it follows that

$$\Omega H^2 = \frac{8\pi G}{3}\rho_0(1+z)^3$$

It is a useful exercise to show that this relation can be rewritten

$$\left(1 - \frac{1}{\Omega}\right) = (1+z)^{-1}\left(1 - \frac{1}{\Omega_0}\right) \tag{18}$$

This is an important result because it shows that, whatever the value of Ω_0 now, because $(1+z)^{-1}$ becomes very small at large redshifts, Ω tends very closely to the value 1 in the distant past. There are two ways of looking at this result. On the one hand, it is very convenient that the dynamics of all world models tend to those of the Einstein-de Sitter model in the early stages of the dust filled models. On the other hand, we observe that it is remarkable that the Universe is within a factor of ten of the value $\Omega = 1$ at the present day. If the value of Ω were significantly different from 1 in the distant past, then it would be very widely different from 1 now as can be seen from equation (18). The fact that the curvature of space κ must be close to zero now results in what is often referred to as the **flatness problem.** The problem is that our Universe must have been very finely tuned indeed to the value $\Omega = 1$ in the distant past if we are to end up with a Universe with Ω close to 1 now. Some argue that it is so remarkable that our Universe is within a factor of ten of $\Omega = 1$ now, the only reasonable value the Universe can have is Ω precisely equal to 1. Proponents of the inflationary picture of the early Universe have a solution to this problem.

2.4.2 Radiation Dominated Universes

At the opposite extreme from dust-filled universes are those in which radiation contributes all the inertial mass. In this case, we cannot neglect the pressure term in equation (7). For a gas of photons, massless particles or a relativistic gas in the ultrarelativistic limit, $E \gg mc^2$, pressure is related to energy density by $p = \frac{1}{3}\varepsilon$ and the inertial mass density of the radiation ρ_{rad} is related to its energy density ε by $\varepsilon = \rho_{rad}c^2$. We can now work out simply how the energy density of radiation varies with redshift. If $N(h\nu)$ is the number density of photons of frequency ν, then the energy density of radiation is found by summing over all frequencies

$$\varepsilon = \sum_{\nu} h\nu N(h\nu)$$

Now the number density of photons varies as $N = N_0(1+z)^3$ and the energy of each photon changes with redshift by the usual redshift factor $\nu = \nu_0(1+z)$. Therefore, the variation of the energy density of radiation with epoch is

$$\varepsilon = \sum_{\nu_0} h\nu_0 N_0(h\nu_0)(1+z)^4$$

$$\varepsilon = \varepsilon_0(1+z)^4 = \varepsilon_0 R^{-4} \tag{19}$$

A case of particular interest is that of black-body radiation. The energy density is given by the Stefan-Boltzmann law, $\varepsilon = aT^4$ and its spectral energy density by the Planck law

$$\varepsilon(\nu) = \frac{8\pi h\nu^3}{c^3}\frac{1}{e^{h\nu/kT}-1}d\nu.$$

It immediately follows that for black body radiation the radiation temperature T_r varies with redshift as

$$T_r = T_0(1+z)$$

Correspondingly, the spectrum of the radiation changes as

$$\varepsilon(\nu_1)d\nu_1 = \frac{8\pi h\nu_1^3}{c^3}\left(e^{h\nu_1/kT_1}-1\right)^{-1}d\nu_1 = \frac{8\pi h\nu_0^3}{c^3}(e^{h\nu_0/kT_0}-1)^{-1}(1+z)^4 d\nu_0$$

Thus, it can be seen that upon redshifting, a black body spectrum preserves its form but the radiation temperature changes as $T_r(z) = T_r(0)(1+z)$ and the frequency of each photon as $\nu = \nu_0(1+z)$. Another way of looking at these results is in terms of the adiabatic expansion of a gas of photons. The adiabatic index, i.e. the ratio of specific heats γ, for radiation and a relativistic gas in the ultrarelativistic limit is $\gamma = 4/3$. It is a useful exercise to show that, in an adiabatic expansion, $T_r \propto V^{-\frac{1}{3}}$ which is exactly the same as the above result.

The variations of p and ρ with R are now substituted into equations (7) and (8). We find

$$\ddot{R} = \frac{8\pi G\varepsilon_0}{3c^2}\frac{1}{R^3} \tag{20}$$

$$\dot{R}^2 = \frac{8\pi G\varepsilon_0}{3c^2}\frac{1}{R^2} - \frac{c^2}{\Re^2} \tag{21}$$

We will show in a moment that the Universe becomes radiation-dominated at early epochs corresponding to values to $R \lesssim 10^{-3} - 10^{-4}$. At these early epochs we can neglect the constant term $c^2/\Re^2$ and then the integration of equation (21) is straightforward.

$$R = \left(\frac{32\pi G\varepsilon_0}{3c^2}\right)^{\frac{1}{4}} t^{\frac{1}{2}} \tag{22}$$

Thus, the dynamics of the radiation-dominated models are very simple, $R \propto t^{\frac{1}{2}}$, and depend only upon the total inertial mass density in relativistic or massless forms. Notice that we have to add all the contributions to ε at the relevant epochs.

2.4.3 Inflationary Models

These models have come into prominence as a result of the deepening understanding of elementary particle physics and its application to the early stages of the Hot Big Bang. These considerations lead to the possibility of physical processes which have some highly non-intuitive features. We can approach the dynamics of these models from two simple points of view.

First of all, we consider the dynamical equations according to Einstein's original prescription but keep in the cosmological constant Λ. Suppose the Universe is empty, $\rho = 0$. Then, equation (7) becomes

$$\ddot{R} = \frac{1}{3}\Lambda R \tag{23}$$

As Zeldovich has remarked, this equation shows that the cosmological constant describes the **repulsive effect of a vacuum** – any test particle introduced into the vacuum acquires an acceleration simply by virtue of being located there. According to classical physics, there is no simple physical picture for this process but it does indicate the type of physics one is forced to think about if the cosmological constant is included in Einstein's field equations.

In the second case, we consider some of the recent developments in the theory of elementary particles. The key development for cosmology has resulted from the introduction of the Higgs field into the theory of the weak interactions involving the W and Z bosons. The Higgs field is introduced in order to eliminate high order singularities in the theory and it has the property of being a **scalar** field. These fields have properties quite unlike those of vector fields, such as electromagnetism, or tensor fields, such as General Relativity, in that they can result in a negative pressure equation of state $p = -\rho c^2$. This may be thought of as a **tension**, the opposite of a pressure, associated with the energy density ρc^2. Suppose the volume V contains an internal energy E. Then, on expanding, the work done is pdV which is derived from the internal energy E so that the total internal energy in the volume $V + dV$ becomes $E - pdV = E + \rho c^2 dV$ i.e. the effect of the negative pressure equation of state is that the energy density remains constant during the expansion! A naive way of thinking about this result is that as the vacuum expands there is more of it and there is therefore more, not less, vacuum energy during the expansion. If we substitute the above negative pressure equation of state into equation (7), we obtain the result

$$\ddot{R} = \frac{8\pi GR}{3}\rho \tag{24}$$

where ρ is the constant vacuum energy density. It can be seen that this result is formally identical to equation (23) and gives a physical basis for the apparently strange nature of the cosmological constant. An introduction to many of these new physical ideas aimed at astronomers and astrophysicists is given by Zeldovich (1986).

The solutions of equation (23) are exponentially growing solutions. Let us start from the first integral of the equation as embodied in equation (8) with $\rho = 0$. The dynamical equation for R therefore becomes

$$\dot{R}^2 = \frac{1}{3}\Lambda R^2 - \frac{c^2}{\Re^2} \tag{25}$$

The solution of this equation is

$$R = \frac{c}{H}\sinh Ht \tag{26}$$

with $H = \sqrt{\Lambda/3}$. It is evident that for large values of t, R grows exponentially with cosmic time. The benefit of expressing the solution in the form of expression (25) is that it brings out clearly the point that when the exponential growth of R takes place, the curvature term $c^2/\Re^2$ becomes negligibly small compared with the size of R. Remember that $\Re$ is a constant. This means that exponential growth results in a spatial geometry which is arbitrarily close to flat, Euclidean space. After this period of exponential growth, we have to arrange that the dynamics transform into the Universe as we know it at some early epoch. This **inflationary model** of the early Universe has a number of attractions. First of all, it produces very naturally a world model with flat space, which, when we transfer over to the Universe we know, will have $\Omega = 1$. Second, the exponential growth of the scale factor means that regions which were originally close together, and hence causally connected, separate by enormous factors beyond the horizon scale during the exponential expansion phase. This could account for the large scale isotropy and homogeneity of the present Universe on scales which apparently could not have been causally connected in the early Universe. These are intriguing ideas and appear to offer a possible explanation for two of the fundamental problems of modern cosmology, the flatness problem and that of accounting for the overall isotropy and homogeneity of the Universe. The proponents of the inflationary scenario also believe that it can explain the origin of the fluctuations from which galaxies form and the absence of magnetic monopoles in the Universe now.

2.5 The Determination of Cosmological Parameters - the Matter and Radiation Content of the Universe

2.5.1 Hubble's Constant

The value of Hubble's constant H_0 remains controversial. The problem is not that the slope of the redshift-magnitude relation is ill-defined but that it is difficult to calibrate the relation, in other words, to find methods of determining the distances to galaxies which are independent of their redshifts. The values quoted in the literature lie in the range roughly $50 \lesssim H_0 \lesssim 100$ km s^{-1} Mpc^{-1}. The classical calibration procedures used by Sandage and Tammann result in values close to the lower end of this range whilst the infrared Tully-Fisher method of distance calibration tends to give values towards the upper end of the range (see e.g. Tammann 1987, Aaronson 1987). This difficult problem is reviewed in the book **The Cosmic Distance Ladder** by Michael Rowan-Robinson (1985). He has recently reassessed his conclusions on the basis of data obtained between 1984 and 1988 (Rowan-Robinson 1989). My own preference is for values close to 50 km $s^{-1}Mpc^{-1}$ but this is because I have looked at other data, in particular, estimates of the age of our Galaxy from studies of globular clusters.

In view of the uncertainty about the exact value of Hubble's constant, it is conventional to write it in the following form, $H_0 = 100h$ km $s^{-1}Mpc^{-1} = 3.24 \times 10^{-18}h\ s^{-1}$ or

$$h = \left(\frac{H_0}{100 \text{ km s}^{-1}\text{Mpc}^{-1}}\right) \tag{27}$$

Thus the value of h probably lies between about 0.5 and 1. It is common practice to use a standard value of $H_0 = 100$ km s^{-1} Mpc^{-1} and to include h in the various expressions to show explicitly the sensitivity of the answers to the precise value of Hubble's constant. A particularly common use of h is in describing the density of matter in the Universe ρ relative to the critical density $\rho_c = 3H_0^2/8\pi G$. The value of ρ_c is $1.88 \times 10^{-26}\ h^2$ kg m^{-3} and hence the density of matter is $\rho = \Omega\rho_c = 1.88 \times 10^{-26}\ \Omega h^2$ kg m^{-3} which is the reason why the combination Ωh^2 appears in many of the cosmological formulae.

2.5.2 The Age of the Universe

The oldest stellar systems we know of in our Galaxy are the globular clusters. Their Hertzsprung-Russell diagrams have been the subject of intensive study by specialists in the theory of stellar structure and evolution with a view to establishing their ages as accurately as possible. For the oldest clusters, ages of between 13 and 20×10^9 years are found corresponding to acceptable ranges of H_0 of $50 \lesssim H_0 \lesssim 77$ km s^{-1} Mpc^{-1} if $\Omega = 0$ or $33 \lesssim H_0 \lesssim 50$ km s^{-1} Mpc^{-1} if $\Omega = 1$. It is important to recall that the ages of the globular clusters are based upon present understanding of the theory of stellar structure and evolution. It therefore depends upon the theory being correct and we should bear in mind that it cannot yet explain the low flux of neutrinos emitted by the Sun. There must therefore be some unease about giving this argument too much weight. The age of the Universe can also be estimated from radioactive dating. These estimates result in ages of our Galaxy of about 10^{10} years (Fowler 1987).

2.5.3 The Deceleration Parameter

We have already derived the relation between the deceleration parameter q_0 and the density parameter Ω. It is important to recognise that these are separately measurable quantities and thus provide a test of the validity of General Relativity on the largest scales accessible to us since , if $\Lambda = 0$, General Relativity requires $2q_0 = \Omega$. Alternatively, the comparison of q_0 with Ω provides a measure of Λ, the cosmological constant, from equation (15). I will describe the determination of the value of q_0 using the infrared magnitude-redshift relation for radio galaxies to illustrate the problems which arise.

The basic idea is to find some standard properties of galaxies which can be observed nearby and far away. Because the geometry of space depends on Ω and because of the different relations between comoving coordinate distance and redshift, a standard galaxy of fixed luminosity has a redshift-magnitude relation which depends upon the value of

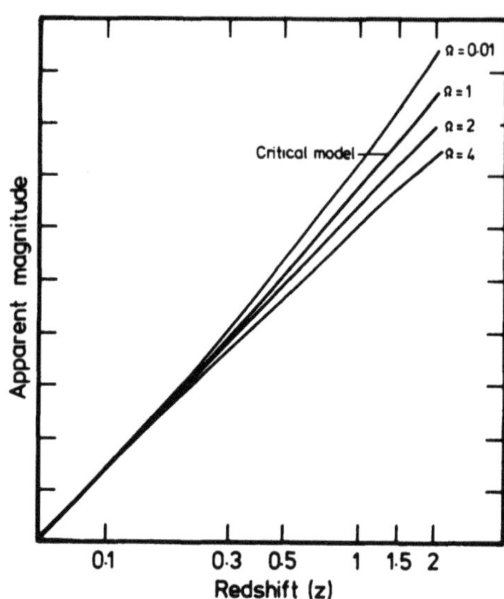

Fig. 11. The redshift-apparent magnitude relation for a source of fixed luminosity with a power-law spectrum, $I_\nu \propto \nu^{-1}$, according to the Friedman world models as described by the density parameter Ω.

Ω (see, e.g. TCP Chapter 15). An example of these differences for a source with a power-law energy spectrum, $I_\nu \propto \nu^{-1}$, is shown in Fig. 11. Thus, if we are able to find standard objects, we might hope to determine the deceleration parameter from such a relation. The problem illustrated by Fig. 11 is that, in order to observe a significant difference between the world models, we have to observe the objects at large redshifts, $z \approx 1$, but this also means that we have to observe them not as they are now but as they were in the distant past. This is illustrated by the cosmic time-redshift relation shown in Fig. 12 in which the relations are displayed for Friedman world models with $\Omega = 0$ and $\Omega = 1$. It can be seen that an object with redshift $z = 1$ emitted its radiation when the Universe was certainly less than half its present age and so there is no reason to expect the standard objects to have the same properties as they have at the present epoch. We therefore have to understand the **astrophysical evolution** of the objects used in this type of cosmological test if we are to obtain a convincing result. For example, if we use galaxies to perform this test, we have make appropriate corrections for the evolution of their stellar content.

The radio galaxies are important because they are the only stellar systems available in reasonable numbers at redshifts greater than one for which studies of their stellar populations can be made. It turns out that the strong radio sources appear to be only associated with very massive galaxies which have a small dispersion in their intrinsic luminosities. The other interesting point is the use of the near infrared waveband for these studies. First, the typical spectrum of a giant elliptical galaxy peaks at about 1 μm and therefore, when observed at large redshifts, most of the energy is shifted into the infrared waveband, $1 - 2$ μm. Thus, it is relatively easier to detect very distant radio galaxies at 2 μm as compared with the optical waveband. The second point is astrophysical in that the stars which contribute most of the light in the infrared waveband are stars belonging to the cool red giant branch of the Hertzsprung-Russell diagram. These stars originated from the old population of the galaxy, i.e. stars with masses about that of the Sun. Therefore, when the integrated light of the galaxy in the infrared waveband is measured, the evolution of the stellar population of the galaxy is averaged over cosmological timescales. This contrasts with what is observed in the optical region of the spectrum in which much of the light can be contributed by young stellar populations which are still undergoing their main sequence evolution. For example, the optical light of a galaxy can be strongly influenced by bursts of star formation occurring throughout the life of the galaxy whereas the infrared observations sample the majority old stellar population.

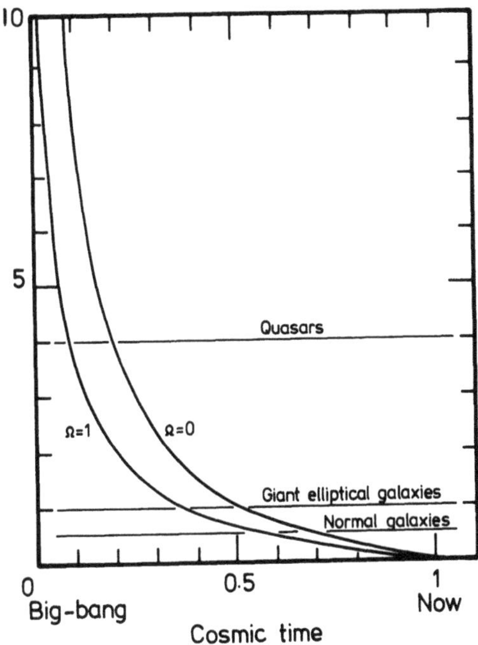

Fig. 12. The relation between cosmic time and redshift for Friedman world models with $\Omega = 0$ and 1. The horizontal lines indicate the redshifts to which different classes of object can be observed more or less at the limits of current technology.

These expectations are borne out in practice – the redshift-magnitude relation in the K waveband (i.e. at 2.2 μm) is very tight out to redshifts of 1.5 or more (Fig. 13) whereas the R magnitude-redshift relation shows a much wider dispersion in observed magnitudes at redshifts greater than about 0.5. This confirms the theoretical expectations which are illustrated by the various models discussed by Spinrad (1987). The K magnitude-redshift relation supplemented by optical-to-infrared colours and optical spectroscopy of the galaxies provide a picture in which the galaxies have been undergoing passive evolution superimposed upon which there have been bursts of star formation, possibly associated with the events which gave rise to the radio sources. By passive evolution, we mean the underlying evolution of the primordial stellar populations which takes place as various classes of star evolve off the main sequence and become red giants. It turns out that the expected evolution of the K–luminosity of a giant elliptical galaxy can be worked out in a remarkably model-independent way because the red giant branches for stars with mass roughly that of the Sun are remarkably similar. On very general grounds, it is expected that the galaxies should be about 1 magnitude brighter at a redshift of 1 as compared with their luminosities at the present day (see Section 6.4.4). We have used these models in conjunction with our observed K-magnitude-redshift relation to solve for q_0 (Lilly and Longair 1984). The results are not particularly impressive, the estimated value of q_0 probably lying in the range $0.1 \lesssim q_0 \lesssim 0.9$. The important point is that the evolutionary changes expected can now be observed and, with large enough statistics, it may be possible to obtain improved estimates of this evolution and possibly of q_0. I have told this story in some detail so that the difficulties of estimating Ω from the standard approach can be appreciated.

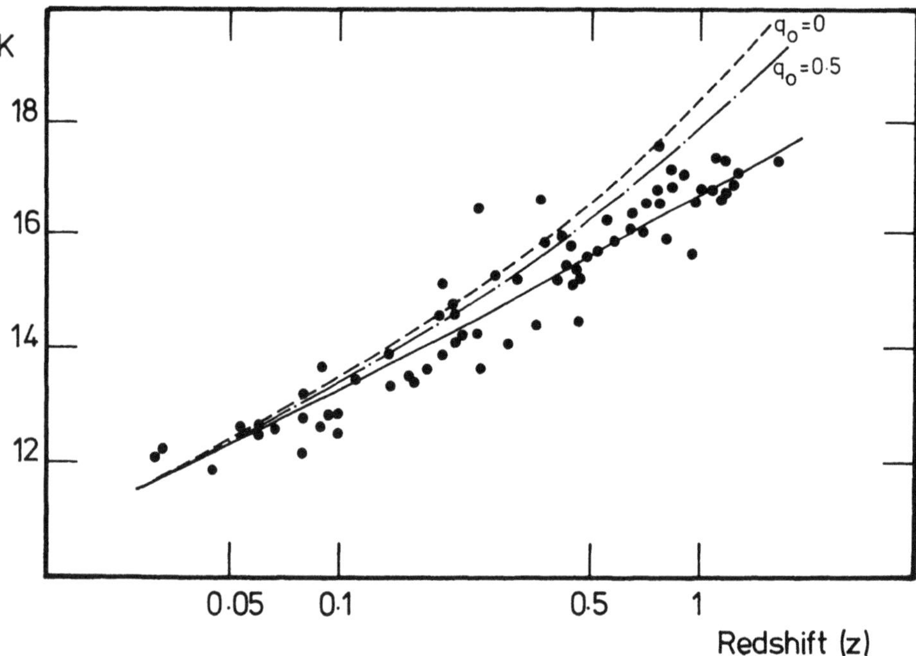

Fig. 13. The redshift-K-magnitude relation for a complete sample of radio galaxies selected from the 3CR catalogue of radio sources (Lilly and Longair 1984). The expectation of uniform world models with $\Omega = 0$ and 1 ($q_0 = 0$ and $\frac{1}{2}$ respectively) are shown as dotted lines as well as models incorporating corrections for the evolution of the stellar populations of the galaxies (solid line). This topic is discussed in more detail in Section 6.4.4.

In my view, all the evidence on distant galaxies and quasars is consistent with values of q_0 lying in the range 0 to 1 but they do not provide a good estimate of where about in this range it might lie.

2.5.4 The Density Parameter

There are several arguments which lead to good lower limits to the amount of gravitating matter in the Universe

1. The matter contained in the visible parts of galaxies can be found by evaluating average values of the mass-to-light ratios for different types of galaxy and then, knowing the average luminosity per unit volume due to galaxies, the average density of visible matter in the Universe can be found. A number of independent estimates agree that this amounts to $\Omega_{gal} \approx 0.02$.

2. To the visible matter we have to add the **dark** or **hidden matter** which is found to be present in the outer regions of giant spiral galaxies and in rich clusters of galaxies. The dark matter is inferred to be present because the dynamics of these systems show that there must be much more mass present in these systems than is contained in the visible parts of galaxies. The typical mass-to-light ratios found in these regions correspond to masses about ten times the mass contained in the visible parts of galaxies. When account is taken of the dark or hidden matter, the total mass density increases by about a factor of ten, i.e. to $\Omega \approx 0.2$.

3. On even larger scales, estimates of the mass density in the general field can be found from what is known as the **cosmic virial theorem**. In this procedure, the random velocities of galaxies in the Universe are compared with the varying component of the gravitational acceleration due to the large scale structure in the distribution of galaxies. As in the other methods described above, the mass density is found by comparing the kinetic energy of the system with its gravitational potential energy. The cosmic virial theorem applied to galaxies selected from the general field has suggested values for the density parameter of about 0.2 to 0.3.

4. A similar argument involves studies of the infall of galaxies into the local supercluster of galaxies. The local supercluster is an extensive region, roughly centred upon the Virgo cluster of galaxies, in which the galaxy density exceeds the density in the general field by a factor of about 2. Therefore, galaxies in the vicinity of the supercluster should feel a gravitational acceleration towards it, thus providing a measure of the mean density of gravitating matter within the system. This method has also resulted in values of Ω about 0.2 to 0.3.

5. A similar procedure has involved interpretation of the dipole anisotropy in the distribution of the Microwave Background Radiation. According to observations made with the IRAS satellite, there is a dipole anisotropy in the distribution of galaxies which are strong far infrared emitters which is similar that of the Microwave Background Radiation (Yahil *et al.* 1986). Although there is some controversy about the exact nature of this dipole component (see e.g. Clowes *et al.* 1987), even if it is assumed to exist, there are complications in relating the distribution of these galaxies to the velocity of the Local Group's motion through the frame of reference in which the Microwave Background Radiation is 100% isotropic. If it is assumed that matter is distributed overall like the distribution of IRAS galaxies, the inferred value of Ω is about 0.83. It is well known, however, that the IRAS galaxies are mostly spiral and irregular galaxies which avoid dense associations of galaxies such as rich clusters. Strauss and Davis (1987) have made corrections to the spatial distribution

of galaxies to take account of this effect and find that the inferred value of Ω drops to 0.39.

The last three arguments depend upon the assumption that the dark matter is distributed like the visible matter in galaxies. There is a basic problem about this assumption in that the velocities induced by large density perturbations depend upon the **density contrast** $\Delta\rho/\rho$ between the background and the discrete system and not upon the absolute value of the density ρ. A typical formula for the infall velocity u of test particles into a supercluster is

$$u \propto H_0 r \Omega^{0.6}(\Delta\rho/\rho) \tag{28}$$

(see e.g. Section 3.4 and Gunn (1978)). Thus, if, in addition to the observed distribution of galaxies, there existed a uniform background distribution of dark matter, this would have the effect of decreasing the value of $\Delta\rho/\rho$ and the net result would be that, for a given observed velocity of infall, u, a larger value of Ω would be inferred. Thus, one can contrive models in which these dynamical estimates would be consistent with $\Omega = 1$. The agreement is, however, obtained at the expense of assuming that there is a difference in the spatial distribution of the visible and the dark matter i.e. there is biasing, a topic to which we will return in Section 5.3.

Finally, as we will show in Section 2.7, the mean density in baryons is constrained by the production of the light elements in the early stages of the hot Big Bang. This results in a limit $\Omega_{bar} \leq 0.05h^{-2}$– otherwise less than the observed abundance of deuterium is created primordially.

The upshot of all of this is that I believe it is correct to say that there is no direct observational evidence that the value of Ω is greater than about 0.2 to 0.3. From the purely observational point of view, all the matter in the Universe could be in the form of ordinary baryonic matter if Hubble's constant is about 50 km s^{-1} Mpc^{-1}. On the other hand, there is also no evidence against the postulate that the actual value of the density parameter Ω is 1 and that most of the matter in the Universe is in some as yet undetermined non-baryonic form. The one clear requirement of such matter is that it should not be distributed like the visible matter. In other words, if you require the Universe to have $\Omega = 1$, you are placed in the unhappy position that you have to put most of it where it can least readily be detected, which is an interesting philosophical position, to say the least!

2.5.5 Streaming

The latest manifestation of the problem of determining the mean amount of gravitating matter in the Universe has been that of streaming. There is now a considerable amount of evidence which suggests that there exist significant streaming velocities of galaxies present in the Universe. As emphasised above, these streaming velocities are small systematic perturbations on the smooth Hubble flow.

1. First, there is the velocity of the Local Group of galaxies relative to the frame of reference in which the Microwave Background Radiation would be 100% isotropic. This corresponds to a velocity of the Local Group of about 600 km s^{-1}.
2. The oldest evidence on streaming is that described by Rubin, Ford and Rubin (1973) who noted that a shell of spiral galaxies with recession velocities in the range 3500

to 6500 km s^{-1} appeared to show a net streaming velocity of about 800 km s^{-1}. Their observations have been confirmed by infrared photometric distance estimates by Collins *et al.* (1986).

3. A separate demonstration of streaming has been carried out by Staveley-Smith (1985) who used the Tully-Fisher relation to work out the distances and streaming velocities of spiral galaxies. Again, a net streaming velocity is found but in a different direction from both the Rubin and Ford velocity vector and that of the apex of the dipole term in the Microwave Background Radiation.
4. Finally, a group of workers (Burstein *et al.* 1986, 1987, Lynden-Bell *et al* 1988) have used distance indicators for elliptical galaxies to show that there is some form of streaming towards the Centaurus supercluster in the Southern Hemisphere. The exact nature of these motions is not clear. Originally, it was considered to be a general streaming motion but more recently these authors believe that these velocities are caused by a **Great Attractor** in the general direction of the Centaurus supercluster.

The observational problem boils down to two questions – if the streaming velocities exist, what are their magnitudes and over what physical scales do coherent streaming motions persist? At the moment the picture is confused but the definition of the detailed velocity field of galaxies has rapidly become an important area which will eventually give us better information about the large scale distribution of mass in the Universe.

2.5.6 The Cosmological Constant

At the present epoch there is no observational evidence which demands that the cosmological constant Λ should be non-zero. Interest in models with non-zero cosmological constant has been rekindled by theories of the inflationary behaviour of the early Universe. Although there is no evidence that $\Lambda \neq 0$, equally there is no definite evidence which shows that Λ is identically zero. Limits to the value of Λ can be found from the broad requirement that q_0 probably lies in the range 0.05 to 1 and Ω in the range 0.1 to 2, i.e. $(\frac{1}{3}\Lambda/H_0^2) \lesssim 1$.

2.5.7 The Radiation Content of the Universe

The energy density in radiation is much better defined than that in the matter content of the Universe. Isotropic background radiation has been detected in the radio, microwave, X-ray and γ-ray wavebands (see e.g. Longair and Sunyaev 1971). In terms of energy density, by far the greatest contributor to the isotropic background is the Microwave Background Radiation which has energy density 4.33×10^{-14} J m^{-3} if $T_{rad} = 2.75$ K. Even in those wavebands for which there are only upper limits at the moment, the far infrared, infrared, optical and ultraviolet backgrounds, the energy density of the background radiation must be considerably less than this value. The corresponding density parameter for radiation is $\Omega_{rad} \approx 10^{-4}$. This is the amount of inertial mass in the radiation and thus is negligible compared with the estimates of the mass density in matter at the present epoch, $\Omega \sim 0.1 - 1$. We can therefore safely assume that our Universe is matter-dominated at the present epoch. Another useful figure is the number density of photons of the Microwave Background Radiation which is about 4.2×10^8 m^{-3} if $T_{rad} = 2.75$ K.

2.6 The Thermal History of the Universe

We now put together the above results to derive the thermal evolution of the Hot Big Bang. First, we have to estimate the major contributors to the inertial mass density of the Universe and this is found by comparing the inertial mass densities in matter and radiation.

$$\frac{\rho_{rad}}{\rho_{matter}} = \frac{aT^4(z)}{\Omega\rho_c(1+z)^3c^2} = \frac{2.6\times10^{-5}(1+z)}{\Omega h^2} \tag{29}$$

Thus, as discussed in Section 2.5.7, the Universe is expected to be **matter-dominated** at redshifts $z \lesssim 4\times10^4\Omega h^2$ and the dynamics are described by the standard Friedman models of Section 2.4.1, $R \propto t^{2/3}$ provided $\Omega z \gg 1$. At redshifts $z \gtrsim 4\times10^4\Omega h^2$, the Universe is **radiation-dominated** and then the dynamics are described by the solutions of Section 2.4.2, $R \propto t^{1/2}$.

Another important number is the present photon to baryon ratio. Assuming $T_{rad} =$ 2.75 K,

$$\frac{N_\gamma}{N_B} = \frac{3.75\times10^7}{\Omega h^2} \tag{30}$$

If photons are neither created nor destroyed during the expansion of the Universe, this number is an invariant. This ratio is a measure of the factor by which the photons outnumber the baryons in the Universe at the present and is also proportional to the specific entropy per baryon during the radiation-dominated phases of the expansion.

As shown in Section 2.4.2, the spectrum of the Microwave Background Radiation preserves its black body form during the expansion of the Universe but is shifted to higher photon energies and radiation temperatures by a factor $(1+z)$, $T(z) = 2.75(1+z)$ K, and the total energy density increases by a factor $(1+z)^4$. We can therefore identify certain epochs which are of special significance in the temperature history of the Universe.

2.6.1 The Epoch of Recombination

At a redshift $z \approx 1500$, the radiation temperature of the Microwave Background Radiation $T_r \approx 4000$ K and then there are sufficient photons with energies $h\nu \geq 13.6$ eV in the tail of the Planck distribution to ionise all the neutral hydrogen in the intergalactic medium. It may at first appear strange that the temperature is not closer to 150 000 K at which temperature $\langle h\nu\rangle = kT = 13.6$ eV for the ionisation of neutral hydrogen. The important points to remember are that the photons far outnumber the baryons in the intergalactic medium and there is a broad range of photon energies present in the Planck distribution. It only needs about one in 10^8 of the photons present to have energy greater than 13.6 eV to have as many ionising photons as hydrogen atoms. This effect of ionisation or excitation occurring at somewhat lower temperatures than would be predicted by simply equating $h\nu$ to kT appears in a number of astronomical problems and is due to exactly the same feature of broad equilibrium distributions such as the Planck and Maxwell distributions - examples include the photoionisation of the regions of ionised hydrogen, the temperature at which nuclear burning is initiated in the cores of stars and the temperature at which dissociation of light nuclei by background thermal photons takes place in the early Universe.

The result is that at redshifts $z \geq 1500$, the intergalactic gas is an ionised plasma and for this reason the redshift $z_r = 1500$ is referred to as the **epoch of recombination.**

The hydrogen is fully ionised and at earlier epochs, $z \approx 6000$, the helium is fully ionised as well. The most important result is that the Universe becomes opaque to Thomson scattering. This is the simplest of the scattering processes which impede the propagation of photons from their sources to the Earth through an ionised plasma. The photons are simply scattered without loss of energy by free electrons. It is useful to work out the optical depth of the intergalactic gas to Thomson scattering. It is simplest to write this in the form

$$d\tau_T = \sigma_T N_e(z) c \frac{dt}{dz} dz \tag{31}$$

where σ_T is the Thomson scattering cross-section $\sigma_T = 6.665 \times 10^{-29}$ m^2. Detailed calculations of the ionisation state of the intergalactic gas with redshift are discussed in Section 5.4 where it is shown that the optical depth of the intergalactic gas becomes unity at a redshift very close to 1000. Let us evaluate this integral in the limit of large redshifts, assuming that the Universe is matter-dominated so that the cosmic-time redshift relation can be written $dt/dz = -H_0 \Omega^{\frac{1}{2}} z^{5/2}$. Then,

$$\tau_T = \frac{2}{3} \frac{c}{H_0} \frac{\sigma_T \rho_c \Omega^{\frac{1}{2}}}{m_p} [z^{3/2} - z_0^{3/2}] = 0.04(\Omega h^2)^{\frac{1}{2}} [z^{3/2} - z_0^{3/2}] \tag{32}$$

It can be seen that the optical depth to Thomson scattering becomes very large as soon as the intergalactic hydrogen becomes fully ionised. The immediate result is that the Universe beyond a redshift of about 1000 becomes unobservable because any photons originating from larger redshifts are scattered many times before they are propagated to the Earth and consequently all the information they carry about their origin is rapidly lost. The net result is that there is a **photon barrier** at a redshift of 1000 beyond which we cannot obtain information directly using photons. We will return to the process of recombination and the variation of the optical depth to Thomson scattering in Section 5.4 because this is a crucial topic in evaluating the observability of fluctuations in the Microwave Background Radiation. If there is no further scattering of the photons of the background radiation, the redshift of about 1000 becomes the **last scattering surface** and therefore it is the fluctuations imprinted on the radiation at this epoch which determine the fluctuations in the radiation temperature of the background radiation.

2.6.2 The Epoch of Equality of Matter and Radiation Inertial Mass Densities

At a redshift $z = 4 \times 10^4 \Omega h^2$, the matter and radiation make equal contributions to the inertial mass density and at larger redshifts the Universe is radiation dominated. The difference in the variation of the scale factor with cosmic epoch has already been discussed. There are two other important changes. First, after the intergalactic gas recombines, and specifically at redshifts $z \leq 100$, there is negligible coupling between the matter and the photons of the Microwave Background Radiation because all the matter is neutral. This statement would of course be incorrect if the intergalactic medium were ionised at some later epoch. At redshifts greater than 1000, however, there is no ambiguity about the fact that the intergalactic hydrogen is ionised and the matter and radiation have a very large optical depth for Thomson scattering as shown by equation (32).

If the matter and radiation were thermally uncoupled, they would cool independently, the hot gas having ratio of adiabatic indices $\gamma = 5/3$ and the radiation $\gamma = 4/3$. It is simple to show that these result in adiabatic cooling rates which depend upon the scale factor R as $T_m \propto R^{-2}$ and $T_r \propto R^{-1}$ respectively. We would therefore expect the matter to cool much more rapidly than the radiation. This is not the case, however, because the matter and radiation are coupled together by Compton scattering. In particular, there are sufficient Thomson scatterings of the electrons by the photons of the background radiation that the Compton effect becomes important in maintaining the matter temperature at the same value as that of the radiation.

The exchange of energy between photons and electrons is an enormous subject and has been treated by Weymann (1965), Sunyaev and Zeldovich (1980) and Pozdnyakov *et al* (1983). The equation for the rate of exchange of energy between a thermal radiation field at radiation temperature T_r and a plasma at temperature T_e interacting solely by Compton scattering has been derived by Weymann (1965).

$$\frac{d\varepsilon_r}{dt} = 4N_e\sigma_T c\varepsilon_r\left(\frac{kT_e - kT_{rad}}{m_e c^2}\right)$$

where ε_r is the energy density of radiation. This equation expresses the fact that, if the electrons are hotter than the radiation, the radiation is heated up by the matter and, contrariwise, if the radiation is hotter than the matter, the matter is heated by the radiation. The astrophysical difference between the two cases arises because of the enormous difference in the number densities of the photons and the electrons $N_\gamma/N_e = 3.75 \times 10^7(\Omega h^2)^{-1}$. Let us look at this difference from the point of view of the optical depths for the interaction of an electron with the radiation field and of a photon with the electrons of the intergalactic gas. In the first case, the optical depth for interaction of an electron with the radiation field is $\tau_e = \sigma_T c N_\gamma t$ whereas that of the photon with the electrons is $\tau_\gamma = \sigma_T c N_e t$ where σ_T is the Thomson cross-section and t is the age of the Universe. This means that it is much more difficult to modify the spectrum of the photons as opposed to the energy distribution of the electrons because in the time any one photon is scattered by an electron, the electron has been scattered many times by the photons. Another way of expressing this is to say that the heat capacity of the radiation is very much greater than that of the matter.

We consider two uses of these formulae. In the first, we consider the heating of the electrons by the Compton scattering of the photons of the Microwave Background Radiation. The collision time between electrons, protons and atoms is always much shorter than the age of the Universe and hence, when energy is transferred from the radiation field to the electrons, it is rapidly communicated to the matter as a whole. This is the process by which the matter and radiation are maintained at the same temperature in the early Universe. Following Peebles (1968), let us work out the redshift to which Compton scattering can maintain the matter and radiation at the same temperature. At epoch t, the total energy transfer to the matter is $(d\varepsilon_r/dt)t$. When this is of the same order as the energy density in the radiation field, no more energy can be transferred and the heating ceases. At this point $(T_{rad} - T_e)/T_{rad} \approx 1$ and hence the condition becomes $4T_{rad}N_e\sigma_T ctk/m_e c^2 \approx 1$. We can write $N_e = 11x\Omega h^2(1+z)^3$ m^{-3}where x is the degree of ionisation of the intergalactic gas. The thermal contact between the photons and the plasma is very strong up to the epoch of recombination and it is the heating after recombination which is of interest. We need the variation of x with redshift following recombination to work out the redshift at which the heating ceases to maintain the matter and radiation at roughly the same temperature. Inspecting the tables for the ionisation of the intergalactic (see, e.g. Peebles 1968), we find that the decoupling occurs at a redshift $z \approx 100$ when $x \approx 10^{-5}$. Thus, the Compton scattering process maintains the matter and radiation at the same temperature even at epochs well after the epoch of recombination.

In the second case, we study the necessary condition for significant distortions of the spectrum of the Microwave Background Radiation to take place. If, by some process, the electrons are heated to a temperature greater than the radiation temperature and if no photons are created, the spectrum of the radiation is distorted from its black-body form by Compton scattering. The interaction of the hot electrons with the photons results in an average frequency change of $\Delta\nu/\nu \approx kT_e/m_e c^2$. Thus, to obtain a significant change in the energy of the photon, $\Delta\nu/\nu \approx 1$, we require the Compton optical depth

$$\tau_C = \int \left(\frac{kT_e}{m_e c^2}\right) \sigma_T c N_e dt \tag{33}$$

to be one or greater. If we take $T_e = T_r(1+z)$ K, we find that $\tau = 1$ at a redshift $z = 2\times10^4(\Omega h^2)^{-1/5}$. Thus, unless the temperature of the electrons is raised to temperatures very much greater than $T_r(1+z)$, significant distortions of the spectrum of the Microwave Background Radiation are expected to originate at redshifts $z \sim 10^4$. Sunyaev and Zeldovich (1980) have surveyed the different types of distortion which would result from large injections of thermal energy into the intergalactic gas at large redshifts.

The second important effect is that the speed of sound changes rapidly with redshift at about this epoch. All sound speeds are roughly the square root of the ratio of total energy density to total mass density. More precisely, the speed of sound c_s is given by

$$c_s^2 = \left(\frac{\partial p}{\partial \rho}\right)_S$$

where the subscript S means "at constant entropy" i.e. an adiabatic sound speed. The complication is that, from the epoch when the energy densities of matter and radiation are equal to beyond the epoch of recombination and the subsequent neutral phase, the dominant contributors to p and ρ change dramatically as the Universe changes from being radiation-dominated to matter-dominated, the coupling between the matter and the radiation becomes weaker and finally the plasma recombines at redshifts of about 1000.

We can write the expression for the sound speed as follows:

$$c_s^2 = \left(\frac{\partial p}{\partial \rho}\right)_S$$

So long as the matter and radiation are closely coupled, this can be written

$$c_s^2 = \frac{\left(\frac{\partial p}{\partial T}\right)_{rad}}{\left(\frac{\partial \rho}{\partial T}\right)_{rad} + \left(\frac{\partial \rho}{\partial T}\right)_{mat}} \tag{34}$$

where the partial derivatives are taken at constant entropy. It is a useful exercise to show that this reduces to the following result:

$$c_s^2 = \frac{c^2}{3}\frac{4\rho_{rad}}{4\rho_{rad} + 3\rho_{mat}} \tag{35}$$

Thus, in the radiation-dominated phases, $z \gtrsim 4 \times 10^4 \Omega h^2$, the speed of sound tends to the relativistic sound speed, $c_s = c/\sqrt{3}$. However, at smaller redshifts, the sound speed decreases as the contribution of the inertial mass density in the matter becomes more important. After recombination, the sound speed is just the thermal sound speed of the matter which, because of the close coupling between the matter and the radiation, has temperature $T \approx 4000$ K at $z = 1500$.

2.6.3 Early epochs

We can now extrapolate the Hot Model back to much earlier epochs. First, we can extrapolate back to redshifts $z \approx 10^8$ when the radiation temperature increases to about $T_r = 3 \times 10^8$ K. These temperatures are sufficiently high for the background photons to have γ-ray energies. At this high temperature, the photons are energetic enough to dissociate light nuclei such as helium and deuterium. At earlier epochs, all nuclei are dissociated. We will study the process of primordial nucleosynthesis of the light elements in the next sub-section.

At a slightly greater redshift, $z \approx 10^9$, electron-positron pair production from the thermal background radiation becomes feasible and at a slightly earlier epoch the opacity of the Universe for weak interactions becomes unity. The Universe is flooded with electron-positron pairs, roughly one pair for every pair of photons present in the universe now.

We can extrapolate even further back in time to $z \approx 10^{12}$ when the temperature of the background radiation is sufficiently high for baryon-antibaryon pair production to take place from the thermal background. Just as in the case of the epoch of electron-positron pair production, the Universe is flooded with baryons and antibaryons, roughly one pair for every pairs of photons present in the Universe now.

We can carry on this process of extrapolation back into the mists of the early Universe as far as we believe we understand high energy particle physics. I show schematically in Fig. 14 the thermal history which comes out of the Hot Model. How far back one is prepared to extrapolate is largely a matter of taste. The most ambitious theorists have no hesitation in extrapolating back to the very earliest Planck eras, $t \approx 10^{-44}$ s when the relevant physics is certainly very different from the physics of the Universe from redshifts of about 10^{12} to the present day. I will say very little about the very earliest phases but refer the brave and ambitious to the texts at the end of this chapter.

2.7 Nucleosynthesis in the Early Universe

There are two good reasons for investigating this process in a little detail. First of all, primordial nucleosynthesis provides one of the most important constraints upon the density parameter of matter in the form of baryons and this is a key part of our story. A second reason is that it provides an example of the decoupling processes which may be important for other types of unknown weakly interacting particles. We will find a qualitatively similar example when we study possible forms of the dark matter.

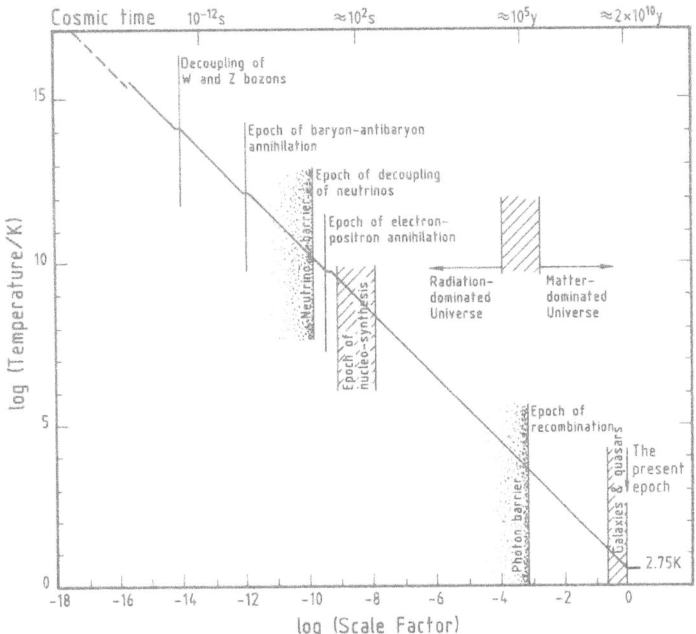

Fig. 14. The thermal history of the radiation temperature of the Microwave Background Radiation according to the standard Hot Big Bang. The radiation temperature decreases as $T_r \propto R^{-1}$ except for abrupt jumps as different particle-antiparticle pairs annihilate at $kT \approx mc^2$. Various important epochs in the standard model are indicated. An approximate time scale is indicated along the top of the diagram. The neutrino and photon barriers are indicated. In the standard model, the Universe is optically thick to neutrinos and photons prior to these epochs.

The basic physics is as follows. Consider a particle of mass m at very high temperatures such that its total energy is much greater than its rest mass energy, $kT \gg mc^2$. If the timescales of the interactions which maintain this species in thermal equilibrium with all the other species present at temperature T are shorter than the age of the Universe at that epoch, statistical mechanics tells us that the equilibrium number densities of the particle and its antiparticle are

$$N = \bar{N} = \frac{4\pi g}{h^3}\int_0^\infty \frac{p^2\,dp}{e^{E/kT} \pm 1} \tag{36}$$

where g is the statistical weight of the particle, p is its momentum and the $\pm$ sign depends upon whether the particles are fermions (+) or bosons (−). It will be recalled that photons are massless bosons for which $g = 2$, nucleons and antinucleons are fermions for which g =2 and neutrinos are fermions which possess the helicity and hence $g = 1$. As a result, the following equilibrium number densities N and energy densities ε are found:

$$g = 2 \qquad N_\gamma = 0.244\left(\frac{2\pi kT}{hc}\right)^3 \mathrm{cm}^{-3} \qquad \varepsilon = aT^4 \qquad \text{Photons}$$

$$g = 2 \qquad N^+ = N^- = 0.183\left(\frac{2\pi kT}{hc}\right)^3 \mathrm{cm}^{-3} \qquad \varepsilon = \frac{7}{8}aT^4 \qquad \text{Nucleons, Antinucleons}$$

$$g = 1 \qquad N = 0.091\left(\frac{2\pi kT}{hc}\right)^3 \mathrm{cm}^{-3} \qquad \varepsilon = \frac{7}{16}aT^4 \qquad \text{Electron \& Muon Neutrinos}$$

To find the total energy density, we have to add all the equilibrium energy densities together i.e.

$$\text{Total Energy Density} = \varepsilon = \chi(T)aT^4 \tag{37}$$

When the particles become non-relativistic, $kT \ll mc^2$, but the thermal abundance of the species are still maintained by interactions between the particles, the non-relativistic limit of the integral (36) should be taken. In this case, we find

$$N = g\left(\frac{mkT}{h^2}\right)^{\frac{3}{2}} e^{-mc^2/kT} \tag{38}$$

Let us look at the decoupling of protons and neutrons in the early Universe. We consider the case in which the neutrons and protons are non-relativistic, $kT \ll mc^2$, but their equilibrium abundances are maintained by the electron-neutrino interactions

$$e^+ + n \rightleftharpoons p + \bar{\nu}_e \qquad \nu_e + n \rightleftharpoons p + e^- \tag{39}$$

For the neutrons and protons the values of g are the same and so the relative abundance of neutrons to protons is

$$\left[\frac{n}{p}\right] = \exp\left(-\frac{\Delta mc^2}{kT}\right) \tag{40}$$

This abundance ratio freezes out when the neutrino interactions can no longer maintain the equilibrium abundances of neutrons and protons. The condition for "freezing out" is that the timescale for the weak interactions becomes greater than the age of the Universe. The timescale for the weak interactions is $t_{weak} = (\sigma N c)^{-1}$ where σ is the weak interaction cross-section which is proportional to the square of the energy $\sigma \propto E^2$. N is the number density of nucleons which decreases as the Universe expands as R^{-3}. Since $R \propto T^{-1}$ and $E \propto T$, it follows that the weak interaction cross section decreases as $t_{weak} \propto T^{-5}$.

This timescale has to be compared with the timescale of the expansion of the Universe which is given by equation (22). We recall that we now have to include all the contributors to the energy density of the Universe and so we have to use equation (37) for the relation between energy density and cosmic time. We therefore find

$$\varepsilon = \chi(T)aT^4 = \frac{3c^2}{32\pi G}t^{-2}$$

$$t \propto T^{-2}$$

It can be seen that the time scale of the weak interactions decreases much more rapidly with temperature than does the expansion time scale. Decoupling takes place when $t = t_{weak}$. Substituting $\sigma = 3 \times 10^{-49}(E/m_e c^2)^2$ m^2 into the above formula, we find that decoupling takes place at an energy $kT \approx 1$ MeV. Since the difference in rest masses of the neutron and proton corresponds to $\Delta mc^2 = 1.28$ MeV, substituting into equation (40), we find that, at a temperature $kT = 1$ MeV when the Universe is only 1 s old, the neutron fraction is

$$\left[\frac{n}{n+p}\right] = 0.21$$

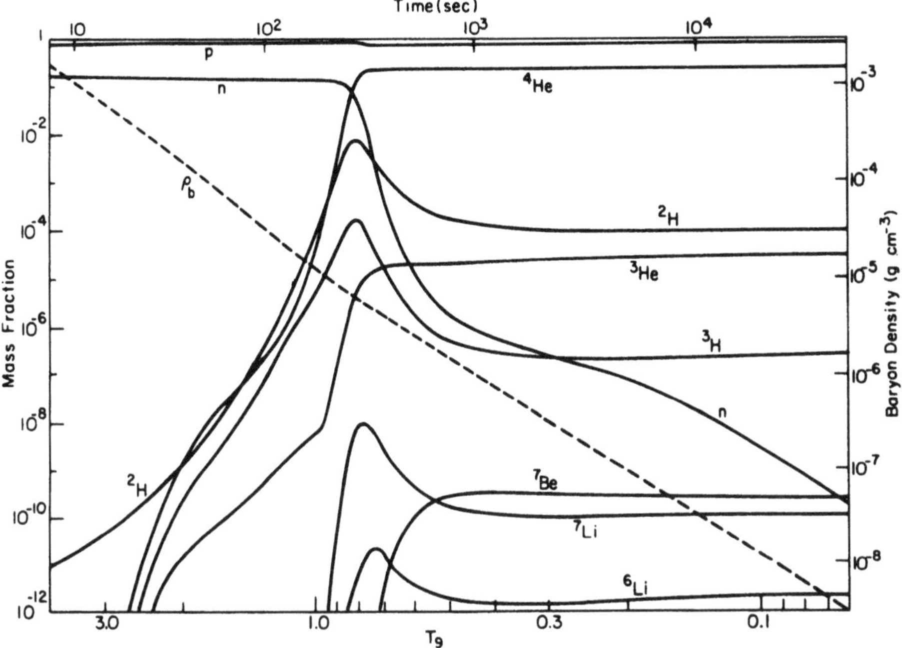

Fig. 15. An example of the time and temperature evolution of the abundances of different light elements in the standard Hot Model of the Universe from detailed computer calculations by Dr. Robert Wagoner (1973). Before about 10 s from the origin of the model, no significant synthesis of the light elements takes place because deuterium ^{2}H is destroyed by hard γ-rays in the high energy tail of the black-body spectrum. As the temperature decreases, more and more of the deuterium survives and the synthesis of heavier light elements becomes possible through the reactions involved in the p-p chain

$$\mathrm{p+n \rightarrow D \quad D+D \rightarrow {}^3He+n \quad {}^3He+n \rightarrow T+p \quad D+D \rightarrow T+p \quad T+D \rightarrow {}^4He+n}$$

Notice that the synthesis of elements such as D, ^{3}He, ^{4}He, ^{7}Li and ^{7}Be is completed after about 15 m.

The neutron fraction does not decrease according to equation (40) after this epoch but only very slowly because the reactions (34) can no longer maintain the equilibrium abundances. The detailed calculations by Peebles (1966) quoted by Weinberg (1972) show that after 300 s the neutron fraction has fallen to 0.123. It is at this epoch that the bulk of the formation of the light elements takes place as shown in Fig. 15 (Wagoner 1973). In the nuclear reactions, almost all the neutrons are combined with protons to form ^{4}He nuclei so that for every pair of neutrons a helium nucleus is formed. The predicted helium to hydrogen mass ratio is therefore just twice the neutron fraction

$$\left[\frac{^4\mathrm{He}}{\mathrm{H}}\right] \approx 0.25$$

The detailed evolution of the light elements during the epoch of nucleosynthesis illustrated in Fig. 15 is the result of detailed calculations by Wagoner (1973). It turns out that in addition to ^{4}He which is always produced with an abundance of about 24 to 25%, there are traces of the light elements deuterium (D), helium-3 (^{3}He) and lithium-7 (^{7}Li).

These are quite remarkable results. It has always been a great problem to understand why the abundance of helium is so high wherever it can be observed in the Universe. Its chemical abundance always appears to be greater than about 24% and it has been very difficult to account for this value by stellar nucleosynthesis. The problem is that in stellar nucleosynthesis the helium produced is rapidly converted into heavier elements. In addition, it has always been a mystery where the deuterium in the Universe could have been synthesised. It is a very fragile nucleus and is destroyed rather than created in stellar interiors. The same argument applies to the isotope of helium, ^{3}He. It is remarkable that it is precisely these elements which are synthesised in the early stages of the Hot Big Bang. The reason is simple. In stellar interiors, nucleosynthesis takes place in roughly thermodynamic equilibrium over very long timescales whereas in the early stages of the Hot Big Bang the "explosive" nucleosynthesis is all over in a few minutes. The distinction is between stationary and non-stationary nucleosynthesis.

Notice that the physics which determines the abundance of ^{4}He is different from the synthesis of the other light elements. It can be observed from the above analysis that the synthesis of ^{4}He is essentially thermodynamic, in that it is fixed by the ratio of neutrons to protons when the neutrinos decouple from the nuclear reactions which maintain equilibrium between the protons and neutrons. In other words, the ^{4}He abundance is a measure of the **temperature** of the Universe at the epoch of decoupling of the neutrinos. On the other hand, the abundances of the other light elements are entirely determined by how far through the p-p chain the reactions can proceed before the temperature falls below that at which nucleosynthesis can take place. Thus, in high density Universes, there is time for essentially all the neutrons to combine into deuterium nuclei which then combine to form ^{4}He nuclei. On the other hand, if the matter density is low, there is not time for all the intermediate stages in the synthesis of helium to be completed and the result is a much higher abundance of deuterium and ^{3}He. Thus, the abundances of the deuterium and ^{3}He are measures of the **density** of the Universe. This has been quantified by Wagoner's calculations which are displayed in Fig 16. It can be seen that for the standard Hot Big Bang, the ^{4}He abundance is remarkably insensitive to the present mass density in the Universe, in contrast to that of the other light elements.

The deuterium and ^{3}He abundances provide strong constraints upon the present baryon density in the Universe. It is found that the deuterium abundance relative to hydrogen is always about $[D/H] \approx 1.5 \times 10^{-5}$. Therefore, since we only know of ways of destroying deuterium rather than creating it, this figure provides a firm lower limit to the amount of deuterium which should be produced by primordial nucleosynthesis. In turn, this sets an upper limit to the present baryon density of the Universe. The figure which results from the most recent analyses is

$$\Omega h^2 \lesssim 0.05 \tag{41}$$

Thus, even adopting a small value of $h = 0.5$, it is apparent that this argument strongly suggests that baryonic matter cannot close the Universe. This will prove to be a key part of the story of the problems of galaxy formation.

To repeat a point we have made earlier, this same type of decoupling process is used in calculating the abundances of massive neutrino-like particles which might have been present in equilibrium in the early Universe. It turns out that particles with mass about $1 - 10$ GeV could have been present in the early Universe and that, making reasonable assumptions about the cross-sections for interaction of these particles and their antiparticles, sufficient of them could survive to close the Universe now. We will return to this topic later.

Another point of interest is that, as a by-product of the above analysis, we have derived the epoch at which the Universe becomes opaque to neutrinos. Just as there is a barrier for photons at a redshift of about 1000, so there is a **neutrino barrier** at an energy $kT \approx 1$ MeV. This means that, if it were possible to undertake neutrino astronomy, we would expect the background neutrinos to be last scattered at the epoch corresponding to $kT = 1$ MeV.

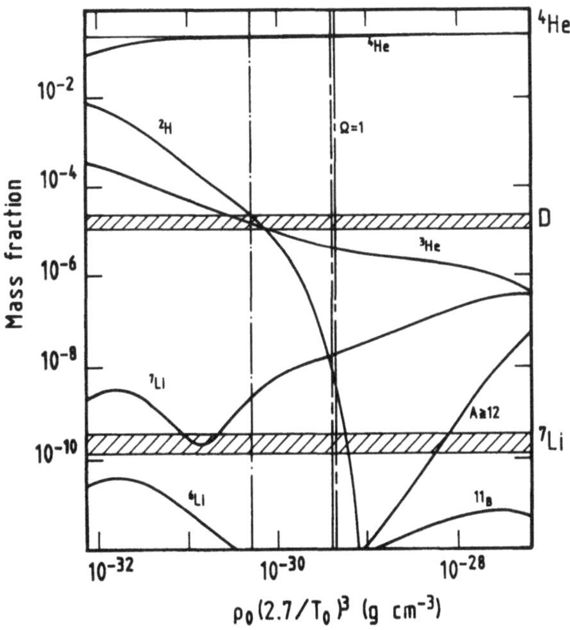

Fig. 16. The predicted primordial abundances of the light elements compared with their observed abundances. The present density of the Universe is shown along the abscissa. The observed abundances are in reasonable agreement with models which have $\Omega h^2 \lesssim 0.05$. This value is significantly smaller than the closure value $\Omega = 1$ (after Audouze 1987).

There is an interesting piece of physics associated with the evaluation of the thermal temperature of this neutrino background. We have shown above that the weak interactions maintain the equilibrium abundances of the neutrons and protons until $kT \approx 1$ MeV. Prior to this epoch, the photons, neutrinos, electrons and their antiparticles are the only species which are relativistic and they all have the same thermal temperature. However, at an energy $kT \approx 0.5$ MeV, the electrons and positrons annihilate creating γ-ray photons. These high energy photons are rapidly thermalised by Compton scattering and so the thermal temperature of the radiation becomes greater than that of the neutrinos. The expansion is adiabatic and so the thermal temperatures of the neutrinos and photons can be worked out assuming all the entropy of the electrons and positrons is transferred to the radiation. The net result is that, if the radiation temperature is 2.75 K at the present epoch, the temperature of the neutrinos is expected to be $(4/11)^{1/3}$ of this value, i.e. $T_\nu = 1.96$ K. This would be the temperature of the neutrino background which should be detectable and which was last scattered at the epoch when $kT \approx 1$ MeV i.e. $t \approx 1$ s.

There is one other important point about the influence of the neutrinos upon the dynamics of the Universe. We notice that the dynamics during the radiation-dominated phase is determined by the total inertial mass density of massless particles and this should include the neutrinos. In fact, throughout the radiation dominated phase from $kT \approx 1$ MeV to the epoch of equality of the energy densities of massless and cold matter, the appropriate value of χ is 1.7. This has the effect of changing slightly the dynamically important epochs described in Section 2.6.

2.8 Comment

This introduction has proved to be much longer than might have been expected. The reason is that many of the conclusions about the problems of galaxy formation are intimately related to features of the standard Hot Model. In addition, we have been able to develop a number of very useful tools which will clarify many of the pieces of physics needed in the succeeding Sections. Everything which we have said so far is no more than the background against which we now wish to tackle the theory of the formation of galaxies.

3 The Evolution of Fluctuations in the Standard Hot Big Bang

Galaxies are complex systems but the aim of the cosmologist is not to explain all their detailed features. It is the job of the astrophysicist to explain the detailed astrophysics of galaxies and how they evolve. The aims of the cosmologist who studies the processes of galaxy formation are much more modest. The goal is to explain how it is that large scale structures can form in the expanding Universe in the sense that regions of overdensity $\delta\rho/\rho$ can reach amplitude 1 from initial conditions which must have been remarkably isotropic and homogeneous. Once the initial perturbations have grown in amplitude to $\delta\rho/\rho \approx 1$, the growth of the perturbation becomes non-linear and it rapidly evolves towards a bound structure during which star formation and other astrophysical phenomena lead to galaxies as we know them. The cosmologist's objective is therefore to account for the initial conditions necessary for the formation of galaxies and other large scale structures in the Universe. In its simplest form, the cosmologist therefore seeks to explain how fluctuations can grow to amplitude $\delta\rho/\rho \approx 1$ in the expanding Universe. This may appear to be a rather modest goal but it turns out to create one of the most difficult problems of modern cosmology. Indeed, this is yet one more problem for which we are forced to investigate seriously processes in the very early Universe.

We can make a convincing case that structures such as galaxies, clusters and other large scale structures must have formed relatively late in the Universe. We can deduce this from the typical mean densities of these objects now. Roughly speaking, the density contrasts $\delta\rho/\rho$ for galaxies, clusters of galaxies and superclusters are $\sim 10^6$, 1000 and a few respectively. Since the density of matter in the Universe changes as $(1+z)^3$, it follows that galaxies could not have separated out as discrete objects at redshifts greater than about 100, and the corresponding redshifts for clusters and superclusters are $z \sim 10$ and 1 respectively. We conclude that the galaxies and larger scale structures must have separated out from the expanding gas at redshifts less than 100 which is well into the matter dominated phase of the Hot Big Bang. This is in itself an important conclusion since it means that galaxies as we know them were not formed in the inaccessible remote past but in the redshift range which should in principle be observable. We will also find that this observation is consistent with theoretical arguments about the epochs when large scale baryon perturbations could have begun to collapse in the conventional Hot Big Bang. These conclusions provides the motivation for studying how small density perturbations grow in the expanding Universe.

3.1 The Non-relativistic Wave Equation for the Growth of Small Perturbations in the Expanding Universe

The analysis which follows is one of the classics of theoretical astrophysics. The origin of the problem of the growth of small perturbations under gravity dates back to the work of Jeans in the first decade of this century and then to a classic paper by Lifshitz in 1946.

The problem gets off to a very bad start. Let us first write down the standard equations of gas dynamics for a fluid under gravity. These consist of three partial differential equations which describe the conservation of mass, the equation of motion for an element of the fluid and the equation for the gravitational potential in the presence of a density distribution ρ. These are:

$$\textit{Equation of Continuity}: \frac{\partial \rho}{\partial t} + \nabla.(\rho \mathbf{v}) = 0 \tag{42}$$

$$\textit{Equation of Motion}: \frac{\partial \mathbf{v}}{\partial t} + (\mathbf{v}.\nabla)\mathbf{v} = -\frac{1}{\rho}\nabla p - \nabla\phi \tag{43}$$

$$\textit{Gravitational Potential}: \nabla^2\phi = 4\pi G\rho \tag{44}$$

Let us recall the meaning of these equations. They describe the dynamics of a fluid of density ρ and pressure p in which the velocity distribution is $\mathbf{v}$. The gravitational potential ϕ at any point is given by Poisson's equation (44) in terms of the density distribution ρ. It is important to remember exactly what the partial derivatives mean. In equations (42), (43) and (44), the partial derivatives describe the variation of the quantities **at a fixed point in space.** These coordinates are often referred to as **Eulerian coordinates.** There is another way of writing the equations of fluid dynamics in which the motion of a particular fluid element is followed. These are known as **Lagrangian coordinates.** Derivatives which follow the fluid element are written as total derivatives d/dt and it is straightforward to show that

$$\frac{d}{dt} = \frac{\partial}{\partial t} + (\mathbf{v}.\nabla) \tag{45}$$

This is a well-known result and the proof may be found in TCP, Appendix to Chapter 5. Notice also the operator $(\mathbf{v}.\nabla)$. There is no ambiguity when this operator is used with a scalar quantity. When it operates upon a vector quantity, it means that the derivative $v_x\partial/\partial x + v_y\partial/\partial y + v_z\partial/\partial z$ should be taken for each component of the vector.

The equations of motion can therefore be written in Lagrangian form in which we follow the behaviour of one element of the fluid.

$$\frac{d\rho}{dt} = -\rho\nabla.\mathbf{v} \tag{46}$$

$$\frac{d\mathbf{v}}{dt} = -\frac{1}{\rho}\nabla p - \nabla\phi \tag{47}$$

$$\nabla^2\phi = 4\pi G\rho \tag{48}$$

Notice that, for the cosmological problem we are analysing, we can think of the equations (46), (47) and (48) as being written in comoving form - i.e. the behaviour of a particular element of the expanding Universe is followed rather than what would be observed if one sat at a fixed point in space and watched the Universe expand past it. Notice that in deriving (46) we have used the identity $\nabla.(\rho\mathbf{v}) = \rho\nabla.\mathbf{v} + \mathbf{v}.\nabla\rho$.

It is standard practice now to establish the zero order solution for the unperturbed medium i.e. a uniform state in which ρ and p are the same everywhere and $\mathbf{v} = 0$. Unfortunately this solution does not exist. Equation (46) shows that if everything is uniform and the velocity is zero, we only obtain solutions if $\rho = 0$! This is a bit of a problem since it means that there is no static solution with finite density and pressure. This is a worry for those who insist upon mathematical rigour. We will find a way of circumventing this problem later.

Fortunately, we want to treat the growth of fluctuations in an expanding medium and this eliminates this particular problem. We are interested in the zero order solutions for the velocity $\mathbf{v}$, the density ρ, the pressure p and the gravitational potential ϕ. The zero order solutions are $\mathbf{v}_0$, ρ_0, p_0 and ϕ_0 and these satisfy the above equations (46), (47) and (48).

$$\frac{d\rho_0}{dt} = -\rho_0\nabla.\mathbf{v}_0 \tag{49}$$

$$\frac{d\mathbf{v}_0}{dt} = -\frac{1}{\rho_0}\nabla p_0 - \nabla\phi_0 \tag{50}$$

$$\nabla^2\phi_0 = 4\pi G\rho_0 \tag{51}$$

The next step is to write down the equations with first order perturbations so that we write

$$\mathbf{v} = \mathbf{v}_0 + \delta\mathbf{v} \quad \rho = \rho_0 + \delta\rho \quad p = p_0 + \delta p \quad \phi = \phi_0 + \delta\phi \tag{52}$$

These are substituted into equations (46), (47) and (48). The equations are expanded to first order in small quantities and then equations (49), (50) and (51) are subtracted from each of them in turn. From the subtraction of (49) from (46), we find

$$\frac{d}{dt}\left(\frac{\delta\rho}{\rho_0}\right) = -\nabla.\delta\mathbf{v} \tag{53}$$

The quantity $\Delta = \delta\rho/\rho_0$ is often referred to as the **density contrast** and it is the growth of Δ with cosmic epoch which is the subject of the present exercise.

To make progress with equation (47), we first note the following expansion for $\delta\mathbf{v}$:

$$\frac{d(\mathbf{v}_0+\delta\mathbf{v})}{dt}=\frac{\partial\mathbf{v}_0}{\partial t}+(\mathbf{v}_0.\nabla)\mathbf{v}_0+\frac{d(\delta\mathbf{v})}{dt}+(\delta\mathbf{v}.\nabla)\mathbf{v}_0 \tag{54}$$

This relation is found by expanding $d\mathbf{v}/dt$ to first order in small quantities using equation (45). In expanding the right hand side of equation (50), we assume that the initial state is **homogeneous** and **isotropic** so that $\nabla p_0 = 0$ and $\nabla\rho_0 = 0$. We then find when we subtract equation (50) from (54),

$$\frac{d(\delta\mathbf{v})}{dt}+(\delta\mathbf{v}.\nabla)\mathbf{v}_0=-\frac{1}{\rho_0}\nabla\delta p-\nabla\delta\phi \tag{55}$$

The third equation results from the subtraction of equation (51) from equation (48). Because of the linearity of Poisson's equation (48), we find

$$\nabla^2\delta\phi=4\pi G\delta\rho \tag{56}$$

Equations (53), (55) and (56) are the key differential equations in the present analysis.

We now look at the case of perturbations in the expanding Universe. It is convenient to write the distances in terms of comoving coordinates by writing $\mathbf{x} = R(t)\mathbf{r}$ where $\mathbf{r}$ is comoving coordinate distance and $R(t)$ is the scale factor. We can therefore write

$$\delta\mathbf{x}=\delta(R(t)\mathbf{r})=\mathbf{r}\delta R(t)+R(t)\delta\mathbf{r}$$

Therefore the velocity can be written

$$\mathbf{v}=\delta\mathbf{x}/\delta t=\frac{dR}{dt}\mathbf{r}+R(t)\frac{d\mathbf{r}}{dt}$$

Thus, we can identify $\mathbf{v}_0$ with the Hubble expansion term $(dR/dt)\mathbf{r}$ and the perturbation to the Hubble flow $\delta\mathbf{v}$ with the term $R(t)(d\mathbf{r}/dt)$. It is convenient to write the perturbed velocity as $R(t)\mathbf{u}$ so that $\mathbf{u}$ is the perturbed comoving velocity. Equation (55) therefore becomes

$$\frac{d}{dt}(R\mathbf{u})+(R\mathbf{u}.\nabla)\dot{R}\mathbf{r}_0=-\frac{1}{\rho_0}\nabla\delta p-\nabla\delta\phi \tag{57}$$

It will prove to be a convenience to write the derivatives with respect to the comoving coordinate $\mathbf{r}$ rather than $\mathbf{x}$ so that $d/dx = (1/R)d/dr$. I will write the differentials with respect to comoving coordinates as ∇_c. Therefore, since $(R\mathbf{u}.\nabla)\dot{R}\mathbf{r} = \mathbf{u}\dot{R}$, equation (57) becomes

$$\frac{d\mathbf{u}}{dt}+2\left(\frac{\dot{R}}{R}\right)\mathbf{u}=-\frac{1}{\rho_0 R^2}\nabla_c\delta p-\frac{1}{R^2}\nabla_c\delta\phi \tag{58}$$

Now, let us consider adiabatic perturbations in which the perturbations in pressure and density are related to the adiabatic sound speed c_s^2 by $\delta p/\delta\rho = c_s^2$. Thus, δp can be replaced by $c_s^2\delta\rho$ in equation (58). We now combine equations (53) and (58) by taking the divergence (in comoving coordinates) of equation (58) and the time derivative of equation (53).

$$\nabla_c.\dot{\mathbf{u}} + 2\left(\frac{\dot{R}}{R}\right)\nabla_c.\mathbf{u} = -\frac{c_s^2}{\rho_0 R^2}\nabla_c^2(\delta\rho) - \frac{1}{R^2}\nabla_c^2(\delta\phi) \tag{59}$$

$$\frac{d^2}{dt^2}\left(\frac{\delta\rho}{\rho}\right) = -\nabla_c.\dot{\mathbf{u}}$$

Therefore

$$\frac{d^2\Delta}{dt^2} + 2\left(\frac{\dot{R}}{R}\right)\frac{d\Delta}{dt} = \frac{c_s^2}{\rho_0 R^2}\nabla_c^2\delta\rho + 4\pi G\delta\rho \tag{60}$$

We now seek wave solutions for Δ of the form $\Delta \propto \exp i(\mathbf{k}_c.\mathbf{r} - \omega t)$ and hence derive a wave equation for Δ.

$$\frac{d^2\Delta}{dt^2} + 2\left(\frac{\dot{R}}{R}\right)\frac{d\Delta}{dt} = \Delta(4\pi G\rho_0 - k^2 c_s^2) \tag{61}$$

where $\mathbf{k}_c$ is the wavevector in comoving coordinates. The proper wavevector $\mathbf{k}$ is related to $\mathbf{k}_c$ by $\mathbf{k}_c = R\mathbf{k}$. Equation (61) is the result we have been seeking and from it follow a number of important conclusions. I make no apology for deriving equation (61) in somewhat boring detail because it is as important as any equation in astrophysical cosmology.

3.2 The Jeans' Instability

Let us first of all return to the problem originally studied by Jeans. We obtain the differential equation for gravitational collapse in a static medium by setting $\dot{R} = 0$ in equation (61). Then for waves of the form $\Delta = \Delta_0 \exp i(\mathbf{k}.\mathbf{r} - \omega t)$, the dispersion relation is

$$\omega^2 = c_s^2 k^2 - 4\pi G\rho_0 \tag{62}$$

It is intriguing that this relation was first derived by Jeans in 1902. The corresponding equation in the electrostatic case was derived by Langmuir in the 1920s and describes the dispersion relation for longitudinal plasma oscillations

$$\omega^2 = c_s^2 k^2 + \frac{N_e e^2}{m_e \epsilon_0}$$

where N_e is the electron density and m_e is the mass of the electron. The formal similarity of the physics may be appreciated from comparison of the attractive gravitational acceleration in a region of mass density ρ_0 and the repulsive electrostatic acceleration in a region of electron charge density $N_e e$. The equivalence of $-G\rho_0$ and $N_e e^2/4\pi\epsilon_0 m_e$ is apparent.

The dispersion relation (62) describes oscillations or instability depending upon the sign of its right-hand side.

(a) If $c_s^2 k^2 > 4\pi G\rho_0$, the right-hand side is positive and the perturbations are oscillatory, i.e. they are sound waves in which the pressure gradient is sufficient to provide support for the region. Writing the inequality in terms of wavelength, stable oscillations are found for wavelengths less than the critical **Jeans' wavelength** λ_J

$$\lambda_J = \frac{2\pi}{k_J} = c_s\left(\frac{\pi}{G\rho}\right)^{\frac{1}{2}} \tag{63}$$

(b) If $c_s^2k^2 < 4\pi G\rho_0$, the right-hand side is negative, corresponding to unstable modes. The solutions can be written

$$\Delta = \Delta_0 \exp(\Gamma t + i\mathbf{k}.\mathbf{r})$$

where

$$\Gamma = \pm\left[4\pi G\rho_0\left(1 - \frac{\lambda_J^2}{\lambda^2}\right)\right]^{\frac{1}{2}}$$

Notice that the positive solution corresponds to exponentially growing modes. For wavelengths much greater than the Jeans' wavelength, the growth rate Γ becomes $\pm(4\pi G\rho_0)^{\frac{1}{2}}$. Thus, the characteristic growth time for the instability is

$$\tau = \Gamma^{-1} = (4\pi G\rho_0)^{-\frac{1}{2}} \approx (G\rho_0)^{-\frac{1}{2}}$$

This is the famous **Jeans' Instability** and the time scale τ is the typical collapse time for a region of density ρ_0. Notice that the expression for the Jeans' length is just the distance a sound wave travels in a collapse time.

The physics of this result is very simple. The instability is driven by the self-gravity of the region and the tendency to collapse is resisted by the internal pressure gradient. We can easily derive the Jeans' instability criterion by considering the pressure support of a region of internal pressure p, internal densityρ and radius r. The equation for hydrostatic support for the region can be written

$$\frac{dp}{dr} = -\frac{G\rho M(< r)}{r^2}$$

To order of magnitude, $dp/dr \approx -p/r$ and $M \approx \rho r^3$. Therefore, since $c_s^2 \approx p/\rho$, the critical scale is $r \approx c_s/\sqrt{G\rho}$. Thus, the Jeans' length is the scale which is just stable against gravitational collapse. If the region were any larger, the gravitational forces would overwhelm the internal pressure gradients and the region would collapse under gravity. This classical Jeans' instability is almost certainly of central importance for the processes of star formation in galaxies.

3.3 The Jeans' Instability in an Expanding Medium

We now return to the full version of equation (61).

$$\frac{d^2\Delta}{dt^2} + 2\left(\frac{\dot{R}}{R}\right)\frac{d\Delta}{dt} = \Delta(4\pi G\rho - k^2c_s^2) \tag{61}$$

The second term $2(\dot{R}/R)(d\Delta/dt)$ modifies the classical Jeans' analysis in crucial ways. It is apparent from the right-hand side of equation (61) that the Jeans' instability criterion applies in this case also but the growth rate is significantly modified. Let us work out the growth rate of the instability in the long wavelength limit $\lambda \gg \lambda_J$ in which case we can neglect the pressure term $c_s^2k^2$. We therefore have to solve the equation

$$\frac{d^2\Delta}{dt^2} + 2\left(\frac{\dot{R}}{R}\right)\frac{d\Delta}{dt} = 4\pi G\rho_0\Delta \tag{64}$$

Rather than deriving the general solution, let us consider the special cases $\Omega = 1$ and $\Omega = 0$ for which the scale factor–cosmic time relations are $R = (\frac{3}{2}H_0t)^{\frac{2}{3}}$ and $R = H_0t$ respectively.

1. $\Omega = 1$ In this case,

$$4\pi G\rho = \frac{2}{3t^2} \qquad \text{and} \qquad \frac{\dot{R}}{R} = \frac{2}{3t}$$

Therefore

$$\frac{d^2\Delta}{dt^2} + \frac{4}{3t}\frac{d\Delta}{dt} - \frac{2}{3t^2}\Delta = 0 \tag{65}$$

Because of the power-law dependence upon t, we seek power-law solutions of the form $\Delta = at^n$. Substituting into equation (65), we find

$$n(n-1) + \frac{4}{3}n - \frac{2}{3} = 0$$

which has solutions $n = \frac{2}{3}$ and $n = -1$. The latter solution corresponds to a decaying mode. The $n = \frac{2}{3}$ solution corresponds to the growing mode we are seeking $\Delta \propto t^{\frac{2}{3}} \propto R = (1+z)^{-1}$. This is the key result

$$\frac{\delta\rho}{\rho} \propto (1+z)^{-1} \tag{66}$$

In contrast to the exponential growth found in the static case, the growth of the perturbation in the case of the expanding Universe is **algebraic**. This is the origin of the problems of forming galaxies by gravitational collapse.

2. $\Omega = 0$ In this case,

$$\rho = 0 \qquad \text{and} \qquad \frac{\dot{R}}{R} = \frac{1}{t}$$

and hence

$$\frac{d^2\Delta}{dt^2} + \frac{2}{t}\frac{d\Delta}{dt} = 0 \tag{67}$$

Again, seeking power-law solutions of the form $\Delta = at^n$, we find $n = 0$ and $n = -1$ i.e. in this case there is a decaying mode and one of constant amplitude $\Delta =$ constant.

These simple results describe the evolution of small amplitude perturbations, $\delta\rho/\rho \ll 1$. In the early stages of the matter-dominated phase, the dynamics of the world models approximate to the Einstein-de Sitter model, $R \propto t^{2/3}$ and so the amplitude of the density contrast grows linearly with R. In the late stages, when the Universe may approximate to the $\Omega = 0$ model, the amplitude of the perturbations grow very slowly and in the limit $\Omega = 0$ does not grow at all. This last result is not particularly surprising since if $\Omega = 0$ there is no gravitational driving force to make the perturbation grow!

The physical reason for this behaviour can be understood from consideration of the dynamics of the Friedman world models. We demonstrated in Section 2 how the dynamics of these models could be understood in terms of a simple Newtonian model. We can model the development of a spherical perturbation in the expanding Universe by embedding a spherical region of density $\rho + \delta\rho$ in an otherwise uniform Universe of density ρ. Using the same logic as in Section 2.4, the spherical region behaves like a Universe of slightly higher density. We can therefore use equation (12) for which it is straightforward to derive a parametric solution

$$R = a(1 - \cos\theta) \qquad t = b(\theta - \sin\theta) \tag{68}$$

where

$$a = \frac{\Omega}{2(\Omega - 1)} \qquad \text{and} \qquad b = \frac{\Omega}{2H_0(\Omega - 1)^{\frac{3}{2}}}$$

The trick is now to look at solutions for small values of θ, corresponding to early epochs of the matter-dominated phase. Expanding to first order in θ, $\cos\theta = 1 - \frac{1}{2}\theta^2$, $\sin\theta = \theta - \frac{1}{6}\theta^3$, we find the solution

$$R = \Omega^{\frac{1}{3}}\left(\frac{3H_0 t}{2}\right)^{\frac{2}{3}} \tag{69}$$

This solution corresponds to the conclusion derived from equation (18) that, in the early stages, the dynamics of the world models tend towards those of the Einstein-de Sitter model, $\Omega = 1$, i.e. $R \propto t^{\frac{2}{3}}$, but with a different constant of proportionality.

Now let us look at a region of slightly greater density embedded within the background model. To derive this behaviour, we expand the expressions for R and t to higher order in θ, $\cos\theta = 1 - \frac{1}{2}\theta^2 + \frac{1}{24}\theta^4 \ldots$, $\sin\theta = \theta - \frac{1}{6}\theta^3 + \frac{1}{120}\theta^5 \ldots$ This solution follows in exactly the same manner as equation (69)

$$R = \Omega^{\frac{1}{3}}\left(\frac{3H_0 t}{2}\right)^{\frac{2}{3}}\left[1 - \frac{1}{20}\left(\frac{6t}{b}\right)^{\frac{2}{3}}\right] \tag{70}$$

From this, we can immediately write down an expression for the evolution of the density of the model with cosmic epoch

$$\rho(R) = \rho_0 R^{-3}\left[1 + \frac{3}{5}\frac{(\Omega - 1)}{\Omega}R\right] \tag{71}$$

Notice that if $\Omega = 1$, there is no growth of the perturbation. The density perturbation may be considered to be a mini-Universe of slightly higher density embedded in the $\Omega = 1$ model. Therefore, taking the density contrast to be the difference between the model with $\Omega > 1$ and the critical model $\Omega = 1$, we find

$$\frac{\delta\rho}{\rho} = \frac{\rho(R) - \rho_0(R)}{\rho_0(R)}$$

Therefore

$$\frac{\delta\rho}{\rho} = \frac{3}{5}\frac{(\Omega - 1)}{\Omega}R \tag{72}$$

This result shows why the perturbation grows only linearly with cosmic epoch. The growth corresponds to the slow divergence between the variation of the scale factors with cosmic epoch of the model with $\Omega = 1$ and one with slightly greater density.

This model has another very great merit in that it demonstrates clearly that this law of growth of the perturbations applies to fluctuations on any physical scale, including those of wavelength greater than the scale of the horizon, $r > ct$. This follows from the same reasoning which we used in discussing the global dynamics of the Universe in Section 2.4.1. If a perturbation is set up on a scale greater than the horizon, it behaves just like a closed Universe and the amplitude of the fluctuation grows according to $\delta\rho/\rho \propto R$.

3.4 The Evolution of Peculiar Velocities in the Expanding Universe

The development of velocity perturbations in the expanding Universe can be derived from equation (58). Let us look at the case in which we can neglect the pressure gradients so that the velocity perturbations are driven by the potential gradient $\delta\phi$.

$$\frac{d\mathbf{u}}{dt} + 2\left(\frac{\dot{R}}{R}\right)\mathbf{u} = -\frac{1}{R^2}\nabla_c\delta\phi \tag{73}$$

Let us divide the velocity perturbations into those parallel and perpendicular to the gravitational potential gradient, $\mathbf{u} = \mathbf{u}_{\parallel} + \mathbf{u}_{\perp}$ where $\mathbf{u}_{\parallel}$ is parallel to $\nabla_c\delta\phi$. The component associated with $\mathbf{u}_{\parallel}$ is often referred to as potential motion since it is driven by the potential gradient. On the other hand, the perpendicular component $\mathbf{u}_{\perp}$ is not driven by potential gradients and corresponds to vortex or rotational velocities. We will consider the development of the velocity perturbations under the influence of the growing modes.

Rotational Velocities First of all, let us consider the rotational component $\mathbf{u}_{\perp}$. Equation (73) reduces to

$$\frac{d\mathbf{u}_{\perp}}{dt} + 2\left(\frac{\dot{R}}{R}\right)\mathbf{u}_{\perp} = 0 \tag{74}$$

The solution of this equation is straightforward $\mathbf{u}_{\perp} \propto R^{-2}$. We recall that $\mathbf{u}_{\perp}$ is a comoving perturbed velocity and the proper velocity is $\mathbf{v}_{\perp} = R\mathbf{u}_{\perp} \propto R^{-1}$. Thus, the rotational velocities decay as the Universe expands. This is no more than the conservation of angular momentum in an expanding medium, mvr = constant. This poses a grave problem for models of galaxy formation which involve primordial turbulence. The rotational turbulent velocities decay and there must be further sources of turbulent energy if the rotational velocities are to be maintained.

Potential Motions The development of potential motions is most directly derived from equation (53)

$$\frac{d\Delta}{dt} = -\nabla.\delta\mathbf{v} \tag{53}$$

The peculiar velocity $\mathbf{v}_{\parallel}$ is parallel to the wave vector of the perturbation $\Delta = \Delta_0 \exp i(\mathbf{k}.\mathbf{x} - \omega t) = \Delta_0 \exp i(\mathbf{k}_c.\mathbf{r} - \omega t)$. It can be seen that the divergence of the peculiar velocity is just proportional to the rate of growth of the density contrast. Using comoving derivatives, equation (53) can be rewritten

$$\frac{d\Delta}{dt} = -\frac{1}{R}\nabla_c.(R\mathbf{u}) = -i\mathbf{k}_c.\mathbf{u} \tag{75}$$

i.e.

$$|v_{\parallel}| = \frac{R}{k_c}\frac{d\Delta}{dt}$$

Notice that we have written this expression in terms of the comoving wave vector k_c which means that this expression describes the evolution of a particular perturbation with cosmic epoch. Let us consider separately the cases $\Omega = 1$ and $\Omega = 0$.

(1) $\Omega = 1$ As shown above, in this case $\Delta = \Delta_0(t/t_0)^{2/3}$ and $R = (3H_0t/2)^{2/3}$ and hence

$$|\delta v_{\parallel}| = |Ru| = \frac{H_0 R^{\frac{1}{2}}}{k}\left(\frac{\delta\rho}{\rho}\right)_0 = \frac{H_0}{k}\left(\frac{\delta\rho}{\rho}\right)_0 (1+z)^{-\frac{1}{2}} \tag{76}$$

where $(\delta\rho/\rho)_0$ is the density contrast at the present epoch. This calculation shows how the peculiar velocities grow with cosmic time in the critical model, $\delta v \propto t^{\frac{1}{3}}$. In addition, it can be seen that the peculiar velocities are driven by both the amplitude of the perturbation and its scale. Equation (76) shows that, if $\delta\rho/\rho$ is the same on all scales, the peculiar velocities are driven by the smallest values of k, i.e. by the perturbations on the largest physical scales. Thus, local peculiar velocities can be driven by density perturbations on the very largest scales which is an important result for understanding the origin of the peculiar motion of the Galaxy with respect to the frame of reference in which the Microwave Background Radiation is 100% isotropic and of the various types of large-scale streaming velocities reported recently.

(2) $\Omega = \mathbf{0}$ In this case, it is simplest to proceed from equation (58) in which there is no driving term in the equation

$$\frac{d\mathbf{u}}{dt} + 2\left(\frac{\dot{R}}{R}\right)\mathbf{u} = 0$$

The solution is the same as that for $u_{\perp}$ given above. i.e. $v_{\parallel} \propto R^{-1}$. i.e. the peculiar velocities decay with time.

These solutions provide us with the general rules for the evolution of peculiar velocities in the expanding Universe. So long as $\Omega z \gg 1$, the velocities driven by potential gradients grow as $t^{\frac{1}{3}}$ but at redshifts $\Omega z \ll 1$, the velocities decrease. For a given value of Ω, there is a redshift at which the peculiar velocities of galaxies selected randomly from the general field has a maximum value and, if this could be measured, an estimate of Ω would be obtained.

3.5 The Relativistic Case

We have to investigate next the case of a relativistic gas because, in the early stages of the radiation-dominated phase of the Hot Big Bang, the primordial perturbations are in a radiation-dominated plasma for which the relativistic equation of state $p = \frac{1}{3}\varepsilon$ is applicable. We therefore require the relativistic generalisations of equations (46), (47) and (48). Equation (46), the equation of continuity, becomes an equation describing the conservation of energy. There is no simple way of demonstrating this except by using the general energy-momentum tensor for a fully relativistic gas. The equation of energy conservation becomes

$$\frac{d}{dt}(\rho + \frac{p}{c^2}) = \frac{\dot{p}}{c^2} - \left(\rho + \frac{p}{c^2}\right)(\nabla.\mathbf{v}) \tag{77}$$

Substituting $p = \frac{1}{3}\rho c^2$ into equation (77), we derive the relativistic continuity equation

$$\frac{d\rho}{dt} = -\frac{4}{3}\rho(\nabla.\mathbf{v}) \tag{78}$$

The differential equation for the gravitational potential ϕ becomes

$$\nabla^2\phi = 4\pi G(\rho + \frac{3p}{c^2}) \tag{79}$$

For a fully relativistic gas, this becomes

$$\nabla^2\phi = 8\pi G\rho$$

Finally, the expression for the acceleration of an element of the fluid in the gravitational potential ϕ remains the same as before

$$\frac{d\mathbf{v}}{dt} = -\nabla\phi \tag{80}$$

where the pressure gradient term has been neglected. The net result is that the equation for the evolution of the perturbations for a relativistic gas are formally of exactly the same form as for the non-relativistic case but with slightly different constants. Therefore, an analysis essentially identical to the one we have carried out in Sections 3.1 to 3.4 applies in the relativistic case as well.

Going through the same analysis as before, neglecting the pressure gradient terms, we find the following differential equation for the growth of the instability

$$\frac{d^2\Delta}{dt^2} + 2\left(\frac{\dot{R}}{R}\right)\frac{d\Delta}{dt} - \frac{32\pi G\rho}{3}\Delta = 0 \tag{81}$$

This equation is formally identical to equation (64). Using exactly the same approach as in Section 3.3, we seek solutions of the form $\Delta = at^n$, recalling that in the radiation-dominated phases, the scale factor–cosmic time relation is given by equation (22) in which $R \propto t^{\frac{1}{2}}$. Going through precisely the same procedure, we find solutions $n = \pm 1$. The growing solution corresponds to

$$\Delta \propto t \propto R^2 \propto (1+z)^{-2} \tag{82}$$

Thus, once again, the unstable modes grow algebraically with cosmic time. It will be noted again that nowhere does the analysis describe the scale of the perturbation relative to the horizon scale.

3.6 The Evolution of Adiabatic Fluctuations in the Standard Hot Big Bang

We now have all the information we need to discuss the simplest case, that of the evolution of adiabatic perturbations in the standard Hot Big Bang model. We need the following information:

(1) **The Jeans' length** is the maximum scale for stable fluctuations at any epoch and is given by the distance a sound wave can travel in a collapse time at that epoch.

(2) **The horizon scale** is the maximum distance over which information can be communicated at a particular cosmic epoch t and hence is just $r_H \approx ct$.

(3) **The growth rates** of the unstable models are algebraic with epoch. In the matter-dominated phase, the perturbation grows as R so long as $\Omega z \gg 1$. The growth is much slower at smaller redshifts and becomes zero in the limit $\Omega = 0$. In the early radiation-dominated phases, the growth rate is algebraic with $\Delta \propto R^2$.

Let us use these general rules to study the evolution of perturbations of different masses in the standard model. Although there is some ambiguity about how to relate the wavelength λ_J to the mass of the object which ultimately forms from it, we will use for illustrative purposes the concept of the **Jeans' mass** which is the mass of baryons contained within a region of radius λ_J, $M_J = (4\pi\lambda_J^3/3)\rho_B$. The expectation is that this is roughly the ultimate mass of the object which forms from a perturbation of this scale.

Let us consider first of all the radiation-dominated phases. The mass density in baryons is $\rho_B = 1.88 \times 10^{-26}\Omega_B h^2(1+z)^3$ kg m^{-3} where Ω_B is the density parameter in baryons at the present epoch. The Jeans' length in the radiation-dominated phase is

$$\lambda_J = \frac{c}{\sqrt{3}}\left(\frac{\pi}{G\rho}\right)^{\frac{1}{2}}$$

where ρ is the total mass density including both photons and neutrinos i.e. $\rho = 4.81 \times 10^{-31}\chi(1+z)^4$ kg m^{-3}, recalling that $\chi = 1.7$ when the neutrinos are taken into account. Therefore, the Jeans' mass in the early stages of the radiation dominated phase, $z \gg 4 \times 10^4 \Omega h^2$, is

$$M_J = 2.8 \times 10^{30} z^{-3}\Omega_B h^2 M_\odot \tag{83}$$

There are several important conclusions which can be derived from this result. The first is that the Jeans' mass grows as $M_J \propto R^3$ as the Universe expands. Thus, the Jeans' mass is one solar mass $M_\odot$ at a redshift $z = 10^{10}$ and increases to the mass of a large galaxy $M = 10^{11}M_\odot$ at redshift $z = 3 \times 10^6$. The second conclusion follows from a comparison of the Jeans' length with the horizon scale $r_H = ct$. Using equation (22), the horizon scale can be written

$$r_H = ct = c\left(\frac{3}{32\pi G\rho}\right)^{\frac{1}{2}} \qquad \text{and} \qquad \lambda_J = c\left(\frac{\pi}{3G\rho}\right)^{\frac{1}{2}}$$

It is apparent that, in the radiation dominated phase, the Jeans' length is of the same order as the horizon scale.

The physical meaning of these results is clear. If we consider a perturbation of galactic mass, say $M = 10^{11}M_\odot$, in the early stages of the radiation-dominated phase, its scale far exceeds the horizon scale and hence the amplitude of the perturbation grows as R^{-2}. At a redshift $z \approx 3 \times 10^6$, the perturbation enters the horizon and, at the same time, the Jeans' length becomes larger than the scale of the perturbation. The perturbation is therefore stable against gravitational collapse and becomes a sound wave which oscillates at constant amplitude. As long as the Jeans' length remains greater than the scale of the perturbation, the perturbation does not grow in amplitude.

The variation of the Jeans' mass with redshift is shown in Fig. 17. The variation of the sound speed with redshift has been included in these calculations. At the epoch of equality of the rest mass energies in matter and radiation, the sound speed becomes less than the relativistic sound speed $c/\sqrt{3}$ according to equation (35). It is an interesting question whether or not there exists this regime prior to the epoch of recombination. According to the analysis of Section 2.6, if Ω were as low as 0.1 and $h = 0.5$, the epoch of equality of the matter and radiation energy densities would occur about a redshift of 1000, the epoch of recombination. In this case, there would be a precipitous drop in the appropriate sound speed from $c/\sqrt{3}$ to the thermal sound speed of a gas at 4000 K. If the values of Ω and h are greater than these values, this intermediate regime would exist and

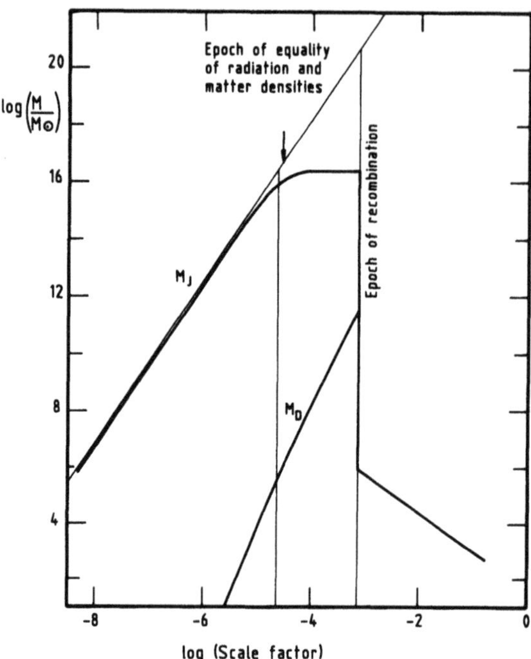

Fig. 17. The evolution of the Jeans' mass with scale factor. Also shown is the evolution of the mass scales which are damped by photon diffusion. The dependences of various scale lengths upon scale factor are shown.

the appropriate sound speed is $c_s = c(4\rho_{rad}/9\rho_B)^{\frac{1}{2}}$ and the mass density variation with scale factor is $\rho \propto R^{-3}$. A simple calculation shows that the Jeans' mass is independent of scale factor (or redshift) until the epoch of recombination.

The next crucial epoch is the epoch of recombination when the primordial plasma recombines and there is an abrupt drop in the sound speed. The pressure within the perturbation is no longer provided by the radiation but by the internal thermal pressure. Because of the close coupling between the matter and radiation even at the redshift of 1500, the matter and radiation temperatures are more or less the same. This means that the appropriate sound speed is the adiabatic sound speed for a gas at temperature 4000 K. The sound speed is therefore $c_s = (5kT/3m_H)^{\frac{1}{2}}$ and the Jeans' mass

$$M_J = \frac{4\pi}{3}\left(\frac{\pi c_s}{G}\right)^{\frac{3}{2}} \rho_B^{-\frac{1}{2}} = 10^6(\Omega h^2)^{-\frac{1}{2}} M_\odot \tag{84}$$

Thus, the Jeans' mass drops abruptly to masses much less than typical galactic masses. This means that all perturbations with mass greater than about 10^6 $M_\odot$ grow according to $\Delta \propto R$ until $\Omega z \sim 1$. It is intriguing to note that the Jeans' mass immediately following recombination corresponds roughly to the mass of globular clusters which are known to be the oldest stellar systems in our Galaxy.

The evolution of the Jeans' mass following recombination depends strongly upon the subsequent thermal history of the gas. If the gas continued to cool adiabatically as the Universe expands, $T \propto R^{-2}$, the Jeans' mass would decrease as $M_J \propto R^{-1.5}$. However, it is unlikely that the gas cools in this simple fashion. We know that the intergalactic gas is very highly ionised at epochs corresponding to $z \sim 4$ from the absence of Lyman-α

absorption in the spectra of distant quasars and so the intergalactic gas must have been strongly heated at some epoch between $z = 1500$ and $z = 4$.

3.7 Dissipation Processes in the Pre-Recombination Phases of the Hot Big Bang

To complete our discussion of the physics of adiabatic baryonic fluctuations in the standard model, dissipative processes in the radiation-dominated phase have to be considered. As we have discussed above, the matter and radiation are closely coupled in the radiation dominated phases but the coupling is not perfect and the radiation can diffuse out of the the fluctuations. Since the radiation provides all the pressure support for the perturbation, the net result is that the perturbation is damped out if the radiation can diffuse out of the fluctuation. This process was first described by Silk (1968) and is often referred to as **Silk damping.** The important process is photon diffusion from a region of scale r. We consider the process of Thomson scattering and use a simple diffusion approximation. A more complete discussion of this process is given by Weinberg (1972).

The following simple arguments present the essence of the full calculation. At any epoch there is a photon mean free path for scattering by electrons $\lambda = (N_e\sigma_T)^{-1}$ where $\sigma_T = 6.665 \times 10^{-29}$ m^2. We recall that the protons and electrons are closely coupled electrostatically as in a fully ionised plasma and so the photons are also closely coupled to the protons as well. We therefore ask how far the photons can diffuse in the cosmic time scale at epoch t. There are several ways of looking at this process. One can either think in terms of the diffusion coefficient for photons D which, according to kinetic theory, is related to the mean free path λ by $D = \frac{1}{3}\lambda c$. Then, the scale over which the photons can diffuse is $r \sim (Dt)^{\frac{1}{2}}$ i.e. $r \approx (\frac{1}{3}\lambda ct)^{\frac{1}{2}}$ where t is cosmic time. Another way of looking at this same result is in terms of the total distance travelled in the diffusion process. The distance travelled from the point of origin by the photon is $r \approx N^{\frac{1}{2}}\lambda$ where N is the number of scatterings. We can now work out the mass of baryons contained within this scale r as a function of cosmic epoch.

In the early pre-recombination phase, $z > 4 \times 10^4 \Omega h^2$, the Universe is radiation-dominated and, from equation (22), $t = (3/32\pi G\rho)^{-\frac{1}{2}} = 2.3 \times 10^{19}(1+z)^{-2}$ s. The number density of electrons varies as $N_e = \rho(1+z)^3/m_p = 11\Omega_B h^2(1+z)^3$ m^{-3} where m_p is the mass of the proton. Thus the damping mass, sometimes referred to as the **Silk mass,** is

$$M_D = \frac{4\pi}{3} r^3 \rho_B = 2 \times 10^{26} (\Omega_B h^2)^{-\frac{1}{2}} (1+z)^{-\frac{9}{2}}\ M_\odot \tag{85}$$

In the redshift range just prior to recombination, the Universe may become matter dominated if there exists a period during which $4 \times 10^4 \Omega h^2 > z > 1500$. In this case, the cosmic time–redshift relation, equation (69), becomes $t = \frac{2}{3H_0}(1+z)^{-\frac{3}{2}}\Omega^{-\frac{1}{2}}$. Repeating the above analysis for this case, we find

$$M_D = 1.8 \times 10^{23} (\Omega_B h^2)^{-\frac{5}{4}} (1+z)^{-\frac{15}{4}}\ M_\odot \tag{86}$$

The key question is what range of masses survives to the epoch of recombination $z = 1500$. If $\Omega h^2 = 1$, the Universe is matter-dominated throughout the redshift range 4×10^4 to 1500 and then using equation (86), we find $M_D = 2 \times 10^{11} M_\odot$; if $h = 0.5$,

$M_D = 10^{12} M_\odot$. In the opposite extreme, if the Universe is of low density, $\Omega_B = 0.1$ and $h = 0.5$, the Universe remains radiation-dominated to $z = 1500$ and then according to equation (85), the damping mass is $M_D = 6 \times 10^{12} M_\odot$. A more detailed treatment of the damping process is given by Weinberg (1972) with essentially the same results.

The upshot of these calculations is that all perturbations with masses $M \lesssim 10^{12} M_\odot$ are damped out because of the diffusion of photons relative to the matter perturbations. According to the strict theory of adiabatic baryonic perturbations, only massive perturbations on the scales of massive galaxies and greater survive into the post-recombination eras. The perturbations which would have resulted in stars, star clusters and even normal galaxies such as our own are damped to exponentially small values of Δ and it is assumed that these structures must form by the process of fragmentation of the larger scale structures which survive to $z < 1500$. The evolution of the damping mass M_D with cosmic epoch is included schematically in Fig. 17.

3.8 Baryonic Pancake Theory

We can now put together these ideas to produce the standard baryonic pancake theory of the origin of galaxies. The considerations of Section 3.6 show that, in the standard baryonic fluctuation picture, only large scale perturbations with $M \gtrsim 10^{12}\ M_\odot$ survive to the epoch of recombination. All the fluctuations on smaller mass scales are damped very effectively by photon diffusion. This was the theory which was developed in detail in the 1970s, principally by Zeldovich and his colleagues (see e.g. Doroshkevich *et al.* 1974).

We use the results developed in Section 3.3 for the subsequent evolution of the baryonic perturbations. All perturbations with masses greater than about $10^6 M_\odot$ grow as $\Delta \propto (1+z)^{-1}$ until the epoch at which $\Omega z \sim 1$. In the pure baryonic picture, the density parameter in baryons Ω_B must be less than $0.05h^{-2}$ because of the constraints provided by primordial nucleosynthesis (Section 2.7) and so, even in the case $h = 0.5$, the perturbations only grow very slowly at redshifts $z \lesssim 5$. We know that galaxies certainly exist at a redshift of 1 and it is likely that they had already formed by a redshift of about $2-3$. Therefore the fluctuations must have developed to large amplitude by this epoch, i.e. $\Delta \sim 1$. Since $\Delta \propto (1+z)^{-1}$, the amplitude of the perturbations at the epoch of recombination must have been $\Delta \gtrsim 3 \times 10^{-3}$. In the pre-recombination epochs, this fluctuation is a sound wave which oscillates with constant amplitude and, as we demonstrated in Section 3.6, this is also the amplitude with which it enters the horizon at earlier epochs. This argument illustrates one of the fundamental problems with all theories of galaxy formation which is that the amplitude of the fluctuations when they enter the horizon cannot be of infinitesimal amplitude. Their amplitudes are finite and far exceed statistical fluctuations. This statement will have to be refined for the other theories we will discuss later. It has to be assumed that these finite-sized fluctuations on scales which are greater than the horizon scale in the early Universe are produced by elementary processes in the early Universe or that they are part of the initial conditions from which the Universe evolved. There is some prospect of explaining their amplitude in the inflationary picture of the early Universe.

The baryonic pancake scenario results from consideration of the ultimate development of the large scale perturbations when they become non-linear. The structures which survive on the scale of clusters and superclusters of galaxies are unlikely to be perfectly spherical and, in a simple approximation, they can be approximated by ellipsoids with three unequal axes. It can be shown that such ellipsoids collapse most rapidly along the shortest axis with the result that flattened structures like pancakes form. The density becomes large in the plane of the pancake and the infalling matter is heated to a high temperature as the matter collapses into the pancake. It is assumed that galaxies form by fragmentation or thermal instabilities (see Section 6.2) within the pancakes.

This theory has a number of successes but also a number of serious problems. Among the successes we can include the fact that it accounts for the large-scale structure of the Universe rather naturally. The pancakes create flattened, stringy structures which form late in the Universe. The fact that quasar activity is at a maximum at epochs $z \sim 2-3$ might be explained by the fact that most galaxy formation only occurred at these epochs.

The problems are, however, rather great. The most important of these are:

1. The fluctuations in the matter density at the epoch of recombination have to be rather large. We discuss the theory of the origin of temperature fluctuations in the Microwave Background Radiation later but we can give a simple indication of the nature of the problem. The fluctuations which survive to recombination are adiabatic and so there are temperature fluctuations corresponding to the density and pressure perturbations. The adiabatic perturbations result in temperature fluctuations

$$\left(\frac{\delta T}{T}\right) \approx \frac{1}{3}\left(\frac{\delta \rho}{\rho}\right) \tag{87}$$

Thus, even in the most favourable scenario, it can be seen that temperature fluctuations $\delta T/T \gtrsim 10^{-3}$ are expected and this exceeds by at least an order of magnitude the upper limits to the temperature fluctuations on the scale of clusters of galaxies at the epoch of recombination. There is a way out of this problem and that is to assume that the intergalactic gas is fully ionised at redshifts less than 1500 so that there is a large optical depth to Thomson scattering which would damp the amplitude of the temperature fluctuations. However, in this theory in which everything forms late in the Universe, there are no discrete sources which could release energy at these early epochs.
2. A second problem is that even at redshifts $z \sim 4$ there must have been considerable amounts of star formation. The evidence for this comes from the element abundances in the most distant quasars which are not very different from those which we observe nearby. Therefore, there must have been considerable amounts of star formation in order to create the heavy elements we see in these distant systems. Another piece of information which suggests that the star formation must have happened quite early is the observation of a giant branch in the most distant radio galaxies (Lilly 1988, 1989). These observations strongly suggest that there is already an old population of stars present in these systems and thus, according to this theory, would push the epoch of formation of the large scale structures further into the past. This simply exacerbates the problems of temperature fluctuations in the Microwave Background Radiation.

3.9 Concluding Remarks

It may seem remarkable that I have spent so much time developing a theory which ends up in serious conflict with the observations. The important point about the above analysis is that we have had to develop many of the tools which are necessary for the formulation of any theory of galaxy formation and we will return to them again and again in the subsequent Sections. It is also important to appreciate that this theory is the best one can do within the conventional scenario in which one assumes that all the matter in the Universe is in the form of baryonic matter. The fact that there are serious conflicts with the observations is one strong motivation for taking seriously those models in which we do not know exactly what form most of the mass in the Universe is at the present time and this is the topic which we have to address now. My own view is that it is a pity that the simplest "dull man's" view of the origin of structure in the Universe runs into these difficulties since the model would have a much more solid physical foundation if we did not have to introduce new physics. The other side of the coin is however perhaps even more exciting in that we are forced to take seriously new pieces of physics which may well lead to new understanding of fundamental physical processes. This would be yet one more example of astronomical problems leading to fundamental physical processes which cannot yet be studied in the laboratory. This is the story we take up now.

4 Dark Matter and Galaxy Formation

4.1 Introduction

Let us recall the motivation for taking dark matter seriously in the context of galaxy formation. It is very difficult to avoid the conclusion that the fluctuations in the Microwave Background Radiation should be much greater than those observed if the simple baryonic adiabatic theory were correct. We will show how models in which the dominant mass contribution is in the form of non-baryonic dark matter can overcome this particular problem. The second strong statement which can be made is that if the Universe is in fact of critical density, $\Omega = 1$, the value which comes naturally out of the inflationary model for the early Universe, then there must be non-baryonic dark matter present because of the constraints which come from primordial nucleosynthesis. It is reasonable to assume that this dominant contribution of the dark matter would also solve the problem of the nature of the dark haloes in massive galaxies and the dark matter problem in clusters of galaxies.

Before embarking upon this story, it is salutary to remember that there is no positive evidence for non-baryonic dark matter in any of the tests for Ω described in Section 2.5.4. I believe that all the data can be reconciled with values of $\Omega \sim 0.1 - 0.2$ and that this mass could all be baryonic. It is intriguing that we are driven to take seriously the possibility that the present Universe is dominated by non-baryonic dark matter because of the difficulties in forming galaxies and because of the theoretical attraction of the inflationary picture of the very early Universe.

4.2 Forms of Dark Matter

There are many possibilities for the dark matter which must be present in the outer regions of large galaxies, in clusters of galaxies and other large scale systems. The fundamental problem is that we are limited by observation in the types of dark matter which can be easily detected. There are even many forms of ordinary baryonic matter which would be very difficult to detect, let alone the more speculative forms of dark matter such as non-baryonic matter or black holes.

Forms of **Baryonic Matter** which are difficult to detect are those which are very weak emitters of electromagnetic radiation. For example, very low mass stars with $M \leq 0.05 M_\odot$ which are of too low mass to burn hydrogen into helium, are very faint objects and have proved very difficult to detect. Some of these objects, which are often referred to as **brown dwarfs**, may now have been detected (Hawkins and Bessell 1988) but their number density is poorly known. By the same token, low mass solid bodies such as planets, asteroids and other small lumps of rock are extremely difficult to detect. One amusing example which is often quoted is that the dark matter could all be in the form of standard bricks (or copies of the Astrophysical Journal!) and, even if they were sufficiently common to make $\Omega = 1$, they would be extremely difficult to detect by either their emission or absorption properties. It is quite possible that the dark matter in the outer regions of galaxies and in clusters of galaxies is in the form of low mass stars since the inferred mass-to-light ratios in these systems are about a factor of 5 to 10 less than that necessary to close the Universe. We will exclude baryonic dark matter from the rest of the discussion because of the constraint $\Omega_B \lesssim 0.05 h^{-2}$ which comes from primordial nucleosynthesis.

Another possible candidate for the dark matter is **Black Holes**. Nowadays there are useful limits to the number density of black holes in certain ranges of masses. These come from studies of the number of gravitational lenses observed among large samples of extragalactic radio sources and from the absence of gravitational lensing effects by stellar mass black holes in the haloes of galaxies (Canizares 1987, Hewitt *et al.* 1987). The limits for massive black holes, $M \sim 10^{10} - 10^{12} M_\odot$ correspond roughly to $\Omega \lesssim 1$ and similar limits are found for solar mass black holes. At the moment it cannot be excluded that the dark matter might consist of a very large population of low mass black holes but these would have to be produced by a rather special initial perturbation spectrum in the very early Universe before the epoch of primordial nucleosynthesis. To produce primordial black holes, the fluctuations would have to be exceed $\Delta = 1$ on scales greater than the horizon. The fact that black holes of mass less than about 10^{12} kg evaporate by Hawking radiation on a cosmological timescale sets a firm lower limit to the possible masses of mini-black holes which could contribute to the dark matter at the present epoch (Hawking 1975).

The most fashionable form of dark matter is **Non-Baryonic Dark Matter**. One of the attractions of these ideas is that they can be related to the types of particles which may exist according to current theories of elementary particles. There are many possibilities and I will only mention the most popular suggestions.

1. The smallest mass candidates are the **axions** which have rest mass energies about $10^{-2} - 10^{-5}$ eV. If they exist, they must have been born when the thermal temperature of the Universe was about 10^{12} K but they never acquire thermal velocities as they are never in equilibrium. Thus, they remain "cold" and behave in a manner similar to the very massive particles discussed in (3).

2. Another possibility is that the three known types of neutrino have finite rest mass. The most interesting possibility is that the rest mass energy of the neutrino lies in the range 10 – 30 eV. Laboratory experiments have provided firm upper limits to the rest mass of the electron antineutrino of the order of 20 to 30 eV (see e.g. Perkins 1987). There was a great deal of excitement when the Soviet group of Lyubimov *et al* (1980) reported that a finite rest mass had been measured of about 30 eV. The reason for this is that this is almost exactly the value needed to close the Universe with the known types of neutrino. The number density of neutrinos of a single type in thermal equilibrium at temperature T is $N = 0.091 \times 10^4 (kT/\hbar c)^3$ m^{-3}. If there are N_ν neutrino types present all with rest mass m_ν, the present mass density of neutrinos in the Universe would be $\rho_\nu = N N_\nu m_\nu$. If this is to equal the critical density $\rho_c = 1.88 \times 10^{-26} h^2$ kg m^{-3}, the necessary rest mass energy of the neutrino is $m_\nu = 184 h^2/N_\nu$ eV. Since we know that there are at least four neutrino species present with this number density, the electron neutrinos and anti-neutrinos, the muon neutrinos and anti-neutrinos (and we should include the pair of tau neutrinos as well), $N_\nu \geq 4$ and hence the necessary rest mass of the neutrino is $46h^2$ eV. Since h lies in the range 0.5 to 1 with some preference for the lower end of this range, it follows that if the neutrino rest mass were about 10 to 20 eV, the neutrinos could close the Universe. This range of neutrino masses would just be consistent with the remarkable observations of the distribution of arrival times of the neutrinos associated with the supernova explosion in the Large Magellanic Cloud which occurred in February 1987.
3. A third possibility is that the dark matter is in some form of more massive ultra-weakly interacting particle with mass, say, 1 keV or in the range 1 to 10 GeV. This might be the gravitino, the supersymmetric partner of the graviton, or the photino, the supersymmetric partner of the photon, or possibly some from of massive neutrino-like particle as yet unknown. The interest of the high energy particle physicists in these arguments is obvious since the cosmological arena may prove to be the only way in which these types of particles can be readily studied. It is only proper to point out that the possible existence of these types of unknown particles are theoretical extrapolations quite far beyond the range of energies which have been explored experimentally but these ideas are sufficiently compelling on theoretical grounds that many particle theorists are looking seriously at the possibility that cosmological studies will prove to be important in constraining theories of elementary processes at ultra-high energies.

There are useful limits which can be set astrophysically to the possible number densities of these different forms of dark matter. Let us look at the case of massive ultraweakly interacting particles and massive neutrinos. Useful limits to their masses can be found from the possibility that they account for the dark matter in the outer regions of giant galaxies and in clusters of galaxies. If these are the types of particles which bind these systems, we can find limits to their masses from the fact that they are collisionless fermions and therefore there are constraints on the phase space density of these particles which translates into a limit upon their masses since for a given momentum, only a finite number of particles within a given volume is allowed.

Let us perform a simple sum which indicates the nature of the calculation. Massive neutrino-like particles are fermions and consequently are subject to the Fermi Exclusion Principle according to which there is a maximum number of available states in phase space for a given maximum momentum p_0. The elementary phase volume is h^3 and consequently, recalling that there can be two particles of opposite spin per state, the number of particles with momenta in the range p to $p + dp$ is $dN = 2(4\pi p^2 dp)/h^3$ per unit volume. Integrating, the total number of particles in volume V is

$$N = \frac{8\pi V}{3h^3} p_0^3$$

Since there may be more than one neutrino species present, we can multiply this number by N_ν. Now, we are interested in bound gravitating systems such as galaxies and clusters of galaxies which are subject to the **Virial Theorem** according to which the kinetic energy of the particles which make up the system must be equal to half of its gravitational potential energy. If σ is the root-mean-square velocity dispersion of the objects which bind the system, we can write $\frac{1}{2}M\sigma^2 = \frac{1}{2}GM^2/R$ where M is the mass of the system and R is a suitably defined radius. We can therefore write $\sigma^2 = GM/R$. Now, the neutrino-like particles are to bind the system and therefore the total mass of the system is $M = NN_\nu m_\nu$ where m_ν is the rest mass of the neutrino. The typical Fermi momentum of the particles p_0 is just $m_\nu \sigma$ and therefore we find the following relation between observable quantities and the rest mass of the neutrino

$$m_\nu^4 \gtrsim \frac{9\sqrt{2}\pi\hbar^3}{8N_\nu G\sigma R^2} \tag{88}$$

Let us insert different values for the velocity dispersions and radii of the systems in which there is known to be a dark matter problem.

In **clusters of galaxies**, typical values are $\sigma = 1000$ km s^{-1} and $R = 1$ Mpc. In this case, if there is only one neutrino species, $N_\nu = 1$, we find $m_\nu \gtrsim 1.5$ eV. If there were six neutrino species $N_\nu = 6$, we would find $m_\nu \gtrsim 0.9$ eV.

For **giant galaxies**, for which $\sigma = 300$ km s^{-1} and $R = 1$ kpc, $m_\nu \gtrsim 65$ eV if $N_\nu = 1$ and $m_\nu \gtrsim 40$ eV if $N_\nu = 6$. For **small galaxies** for which $\sigma = 100$ km s^{-1}, the corresponding figures are $m_\nu \gtrsim 80$ eV and $m_\nu \gtrsim 50$ eV respectively.

These are useful limits to the masses of neutrino-like particles which could form the dark haloes. For example, it is clear that standard neutrinos with rest masses $m_\nu \sim 10-20$ eV could bind clusters of galaxies but they cannot bind the haloes of giant or small galaxies. Thus, we conclude that, if standard neutrinos with rest mass $m_\nu \sim 10-20$ eV are indeed the form of dark matter which closes the Universe, some other form of dark matter must be present in the haloes of galaxies. Although perhaps less attractive than a model in which there is only one type of dark matter, it is quite possible, for example, that some form of baryonic dark matter could bind the haloes of galaxies.

Another important conclusion is that, if the masses of the particles were much greater than 10 - 20 eV and they are as common as the neutrinos and photons, which is expected in a simple picture of the Hot Big Bang model with massive ultraweakly interacting particles present, the present density of the Universe would far exceed the critical mass density $\Omega = 1$. Therefore there would have to be some suppression mechanism to ensure that if $m \approx 1$ keV or $1-10$ GeV, these particles are much less common than the photons and electron neutrinos at the present day.

There are at least two possibilities. In the first, which we can call late decoupling, the decoupling of the particles from the equilibrium early phases occurs when $kT \approx mc^2$ and then the abundances of the particles decrease exponentially as $N = N_0 \exp -(mc^2/kT)$ until the time scales of the reactions which maintain the coupling exceed the age of the Universe. It can be seen that this argument bears a distinct family resemblance to the arguments used in the case of primordial nucleosynthesis (Section 2.7). Putting in reasonable values for the cross-sections thought to be appropriate for these unknown species, the theorists have suggested that the number densities of the massive neutrinos can be suppressed by the appropriate factor if $m \approx 1 - 10$ GeV (see e.g. Barrow 1983). Similar arguments have been made about the number density of relict photinos which are expected to have masses and number densities consistent with these values.

A second possibility is that the unknown particles decouple very early. For example, suppose the dark matter consists of gravitinos. Then, in the very early Universe indeed, they were in thermal equilibrium with everything else. Since we know that there are at least 35 elementary entities such as quarks of all colours, gauge bosons and so on, it can be seen that since all the other things eventually decay into photons and neutrinos, one could account for the relatively low number density of gravitinos simply by stating that they appear in their equilibrium abundances relative to everything else which was present in the very early Universe. Thus, to order of magnitude, a suppression factor of about 100 could be explained in this way so that gravitinos with masses $m \sim 1$ keV are a possible form for the dark matter.

Is this all looking rather far-fetched? It must be agreed that one is extrapolating known particle physics far beyond what has been definitely established by laboratory experiments. However, there are two reasons why I believe we have to take these ideas seriously. Our simplest baryonic model of galaxy formation failed and this strongly suggests that there is some essential ingredient missing from the simple baryonic picture. The above suggestions are the best that is on offer from the high energy particle physicists and they bear more than a passing resemblance to the types of particles required by modern theories of elementary particles.

A second reason is that there are real possibilities of testing some of the astronomical predictions for certain types of dark matter experimentally. Massive particles with masses $m \sim 1$ GeV with velocities appropriate to those of the galactic halo could be detected in laboratory recoil experiments involving cryogenically cooled detectors. In these types of detector, the kinetic energy of the particle is absorbed by the lattice of a very pure semiconductor material and the very small temperature rise in the sample is measured. It has been shown that, if the halo of our galaxy were bound by 1 GeV particles, there would be a significant detection rates of events in such a cryogenic detector. This is a particularly intriguing class of experiment and, if it were to produce a positive result, would have a very profound impact upon the theory of elementary particles. It is interesting to note that the 1 GeV particles have the attraction that they could be not only the particles which bind the halo of our Galaxy but also close the Universe.

4.3 Instabilities in the Presence of Dark Matter

It is conventional to consider three types of non-baryonic dark matter according to the rest masses of the species. The terms **hot, warm** and **cold dark matter** are used to describe particles with rest masses about 10 eV, 1 keV and 1 GeV respectively. The terms refer to the velocity dispersions of the material now. The species remain relativistic until $mc^2 \lesssim kT$ and therefore the least massive particles decouple latest in the Universe. They therefore have the greatest thermal velocity dispersion now. From the comparison $kT = mc^2$, it can be seen that the hot, warm and cold dark matter species were relativistic at redshifts $z \sim 4 \times 10^4$, 4×10^6 and 4×10^{12} respectively. Thus, the cold dark matter is expected to be very cold at the present epoch with essential zero velocity dispersion. Notice that the cold dark matter comes most directly from the decoupling of particles in the early Universe.

The key result concerns the coupling of fluctuations in the dark matter and in the baryons. The important point is that the ordinary matter and radiation are completely decoupled from the dark matter except through their mutual gravitational influence. Let us write down again the expression for the development of the gravitational instability when the internal pressure of the fluctuations can be neglected, i.e. equations (61) and (81), which we can write

$$\ddot{\Delta} + 2\left(\frac{\dot{R}}{R}\right)\dot{\Delta} = A\rho\Delta \tag{89}$$

where $A = 4\pi G$ in the matter dominated case and $A = 32\pi G/3$ in the radiation dominated case. The following points should be noted. First, in the radiation dominated case, this equation applies to fluctuations on scales greater than the horizon scale. If cold dark matter is the dominant form of matter, its dynamical role is much less than that of the radiation and thus the dominant gravitational perturbations are associated with the standard adiabatic fluctuations in the closely coupled radiation dominated plasma. The second point is that, after the epoch of recombination, most of the inertial mass is in the dark matter and therefore the evolution of these perturbations dominates the development of the baryonic perturbations. The third point is that, for all the dark matter perturbations, the non-baryonic particles are collisionless and hence there is no internal pressure to support the fluctuations.

Let us write the density contrast in the baryons and the dark matter as Δ_B and Δ_D respectively. We consider first the epochs immediately after recombination. We have to solve the coupled equations

$$\ddot{\Delta}_B + 2\left(\frac{\dot{R}}{R}\right)\dot{\Delta}_B = A\rho_B\Delta_B + A\rho_D\Delta_D \tag{90}$$

$$\ddot{\Delta}_D + 2\left(\frac{\dot{R}}{R}\right)\dot{\Delta}_D = A\rho_B\Delta_B + A\rho_D\Delta_D \tag{91}$$

Rather than find the general solution, let us find the solution for the case in which the dark matter has $\Omega = 1$ and baryon density is negligible compared with that of the dark matter. Then equation (91) reduces to equation (89) for which we have already found the solution $\Delta_D = BR$ where B is a constant. Therefore, the equation for the evolution of the baryon perturbations becomes

$$\ddot{\Delta}_B + 2\left(\frac{\dot{R}}{R}\right)\dot{\Delta}_B = 4\pi G\rho_D BR$$

Since the background model is the critical model for which $R = (3H_0t/2)^{2/3}$ and $3H_0^2 = 8\pi G\rho_D$, this equation simplifies to

$$R^{\frac{3}{2}}\frac{d}{dR}\left(R^{-\frac{1}{2}}\frac{d\Delta}{dR}\right)+2\frac{d\Delta}{dR}=\frac{3}{2}B \tag{92}$$

We find that the solution, $\Delta = B(R - R_0)$, satisfies equation (92). This is a rather pleasant solution because it has the property that at the epoch corresponding to $R = R_0$, the amplitude of the baryon perturbations is zero.

This result has the following significance. Suppose that, at some redshift z_0, the amplitude of the baryon fluctuations is very small, i.e. very much less than that of the perturbations in the dark matter. The above result shows how the amplitude of the baryon perturbation develops subsequently under the influence of the dark matter perturbations. In terms of redshift we can write

$$\Delta_B = \Delta_D\left(1-\frac{z}{z_0}\right) \tag{93}$$

Thus, it can be seen that the amplitude of the perturbations in the baryons grows rapidly to the same amplitude as that of the dark matter perturbations. To put it crudely, the baryons fall into the dark matter perturbations and, within a factor of two in redshift, have amplitude fluctuations half that of the dark matter perturbations.

The same result is found in the early development of the perturbations when the dark matter and baryonic perturbations have scales greater than the horizon. Most of the inertial mass is in the radiation and so the development of the perturbation in the dark matter is closely tied to those in the radiation-dominated plasma.

Important differences occur when the perturbations enter the horizon. The baryonic perturbations are stabilised because the Jeans' length is of the same order as the horizon scale. Therefore, the amplitudes of the baryonic perturbations remain more or less exactly the same as when they entered the horizon right up to the epoch of recombination when the decoupling of the matter and radiation takes place. So long as the radiation dominated plasma is the principal source of inertia, the dark matter perturbations are also stabilised and do not grow in amplitude. After the epoch of equality of the energy densities in the dark matter and the radiation, however, the dark matter perturbations grow independent of those in the radiation dominated plasma. We see now why the above calculation is of considerable importance. The baryon perturbations are stabilised from the redshift at which they enter the horizon to the epoch of recombination but the amplitude of the perturbations in the dark matter grows from z_{eq} to the epoch of recombination. Therefore, the relative amplitudes of the fluctuations in the dark matter and the baryons is roughly $\Delta_B/\Delta_D \approx 1500/z_{eq}$ i.e. the baryon perturbations are of much smaller amplitude than those in the dark matter at the epoch of recombination. Perturbations on scales larger than those which come through the horizon at redshift z_{eq} have relatively smaller differences between Δ_D and Δ_B. In the limit in which the perturbations come through the horizon at the epoch of recombination, the amplitudes of the fluctuations are of the same order of magnitude. As soon as the matter and radiation decouple, the amplitude of the perturbations in the baryonic matter rapidly grows to the same amplitude as that in the dark matter as demonstrated by equation (93). As shown above, the amplitude of the perturbations in the baryons has grown to values close to that in the dark matter by a redshift a few times smaller than the recombination redshift. Thus, even if the fluctuations in the matter were completely washed out by damping processes, the presence of fluctuations in the dark matter ensures that baryon fluctuations are regenerated after recombination.

4.4 The Evolution of Hot and Cold Dark Matter Perturbations

4.4.1 Hot Dark Matter

Let us consider first the case of Hot Dark Matter. For the sake of definiteness, I will assume that the rest mass of the neutrino is 30 eV which means that they have $m_\nu c^2 = kT$ at a redshift $z = 1.26 \times 10^5$. This means that during the processes of decoupling of the neutrinos and nucleosynthesis, the neutrinos were fully relativistic and none of the predictions of the standard model are affected.

The key process in this neutrino picture is the process of **free streaming** which occurs as soon as the relativistic neutrinos enter the horizon. In all the models, it is assumed that the perturbations are set up on scales much greater than the horizon. Therefore, although the particles are collisionless, they cannot escape from perturbations on scales larger than the horizon since that is as far as they can travel in the available cosmic time. However, as soon as they come through the horizon, if the neutrinos are relativistic, they can stream freely out of the perturbation. This process of free streaming means that the neutrino perturbations are damped out as soon as they enter the horizon, provided the neutrinos are relativistic. In fact, it is only after the neutrinos become non-relativistic that they no longer escape freely from the perturbations. Thus, the only masses which can survive are those on very large scales. We can make a simple estimate of the range of masses which survive by working out the mass contained within the horizon when the neutrinos become non-relativistic. This mass corresponds to $M_\nu = \frac{4\pi}{3} r_H^3 \rho_\nu$ where ρ_ν is the mass density in neutrinos at the epoch when they become non-relativistic. It is not quite clear in this simple presentation exactly what one means by the neutrinos becoming non-relativistic. Detailed calculations show that the particles may be considered non-relativistic by a redshift $z = 3 \times 10^4 (m_\nu / 30 \text{ eV})$ and the damping mass is

$$M_\nu = 4 \times 10^{15} \left(\frac{m_\nu}{30 \text{ eV}} \right)^{-2} M_\odot \tag{94}$$

This means that all smaller masses are damped out by the free streaming of the neutrinos.

The subsequent evolution of the fluctuations is straightforward. These perturbations begin to grow at the redshift z_{eq}. In the case of the relict neutrinos, this redshift is of the same order as the redshift at which the neutrinos become non-relativistic. The reason for this is the fact that, according to the canonical Hot Big Bang, the energy density in the neutrinos is more or less the same as the energy density in the photons. At the epoch when the neutrinos become non-relativistic, their inertial mass no longer decreases as the Universe expands in contrast to the case of the photons which continue to decrease in energy as R^{-1}. The neutrino perturbations grow from redshift z_{eq}, although fluctuations with masses $M \leq M_\nu$ given by equation (96) are damped out by free streaming. In parallel with this development, adiabatic baryon fluctuations within the horizon may be damped by Silk damping. After the epoch of recombination, the baryons fall into these perturbations and attain the same amplitude as the neutrino perturbations. It will be noticed that this picture looks very like the standard adiabatic picture in which only the largest scale structures are formed. It is assumed that the subsequent behaviour of the perturbations is not dissimilar from the standard adiabatic picture in which smaller scale structures have to form by fragmentation of the large scale structures. The model has all the attendant advantages and disadvantages of the adiabatic model. The objects are formed late in the Universe with the added advantage that the dark haloes of clusters of galaxies could be formed out of the neutrinos. It is difficult in this picture to account for the early formation of galaxies.

4.4.2 Cold Dark Matter

For definiteness, we consider the particles to have mass 1 GeV. The mass within the horizon when the particles became non-relativistic is therefore very small, $M \ll M_{\odot}$ and consequently a very wide range of masses survives. Once again, the growth of the cold dark matter perturbations begins after z_{eq} and the baryonic matter falls into the growing perturbations after the epoch of recombination.

There are a number of advantages in this model. First of all, unlike the case of the Hot Dark Matter picture, perturbations on essentially all mass scales survive to the epoch of recombination. According to the standard picture, the baryons fall into these perturbations and it is out of these fluctuations that real objects form. It is important that in this model globular cluster sized systems can begin to collapse immediately after recombination. Indeed, there is no reason why most of the baryonic mass in the Universe cannot now begin to collapse and begin the processes of star formation soon after the epoch of recombination. Once discrete objects have formed, they then begin the process of hierarchical clustering under the gravitational influence of the initial fluctuation spectrum which extends to the largest masses. The important difference as compared with the adiabatic model is that discrete objects can form early and therefore dissipative gas dynamical processes which are crucial in forming thin pancakes on the scale of clusters and superclusters of galaxies do not work. Rather, the formation of large scale structure must take place through a process of clustering under gravity.

4.4.3 Conclusion

We have now made considerable progress towards the development of a consistent picture of galaxy formation. Whilst the Cold Dark Matter picture appears to present no difficulties at this stage and is the preferred model of many, we will keep all the models in play since they all have distinctive features which can explain particular features of the present state of the Universe. We have now to turn our attention to the spectrum of the fluctuations which, so far, we have only discussed in the most general terms.

5 Correlation Functions and the Spectrum of the Initial Fluctuations

To make a quantitative comparison between the theories and the observed distribution of galaxies, we have to put in a spectrum for the fluctuations from which galaxies and larger scale structures are to form. The best way of describing the large scale structure is in terms of **Correlation Functions** which it is hoped can be related to the spectrum of initial fluctuations which came out of the early stages of the Hot Big Bang.

5.1 The Two-point Correlation Function for Galaxies

The simplest tool which comes directly from observation is the **two-point correlation function.** It is defined to be the function $\xi(r)$ which describes the excess probability over random of finding a galaxy at distance r from a chosen galaxy. The simplest definition is to write the number of galaxies in the volume element dV at distance r in the form

$$dN(r) = n_0[1 + \xi(r)]dV \tag{94}$$

where n_0 is the average background number density of galaxies. The reason the function $\xi(r)$ is called a correlation function is that it can also be written in terms of the probability of finding pairs of galaxies separated by r,

$$dN_{pair} = n_0^2[1 + \xi(r)]dV_1 dV_2$$

This function can also be directly related to the density contrast $\Delta(x)$. We can write $\rho = \rho_0[1 + \Delta(x)]$. Therefore the number of pairs separated by distance r is just

$$dN_{pair}(\mathbf{r}) = \rho(\mathbf{x})dV_1\rho(\mathbf{x} + \mathbf{r})dV_2$$

Therefore

$$dN_{pair}(\mathbf{r}) = \rho_0^2 dV_1 dV_2[1 + \Delta(\mathbf{x})][1 + \Delta(\mathbf{x} + \mathbf{r})]$$

Now, when we take averages over a large volume, the average value of Δ is zero by definition and therefore the two-point correlation function is just

$$dN_{pair}(r) = \rho_0^2 dV_1 dV_2[1 + \langle\Delta(\mathbf{x})\Delta(\mathbf{x} + \mathbf{r})\rangle]$$

This shows explicitly the relation between the density perturbations and the two point correlation function which is an observable

$$\xi(r) = \langle\Delta(\mathbf{x})\Delta(\mathbf{x} + \mathbf{r})\rangle \tag{95}$$

The determination of the two-point correlation function for galaxies has been one of the most active area of observational cosmology for the last 15 years. It was greatly stimulated by the work of Peebles and his coworkers who devoted a large amount of effort to the determination of $\xi(\mathbf{r})$ for all the large samples of galaxies available to them (see Peebles 1980). These studies are difficult because the determination of the correlation function is very sensitive to the completeness and reliability of the galaxy catalogues and large samples of galaxies are needed so that the correlation functions can be determined with good statistical significance. An example of a recent determination of the two point correlation function for galaxies is shown in Fig. 18. This function describes the clustering properties of galaxies on physical scales from about 200 kpc to 20 Mpc. The function is often described by a power-law correlation function of the form

$$\xi(r) = \left(\frac{r}{r_0}\right)^{-\gamma} \tag{96}$$

in which the scale $r_0 = 5h^{-1}$ Mpc and the exponent $\gamma = 1.8$. Notice that throughout this discussion, we will work in terms of the three dimensional correlation function although much of the data is derived from angular two-point correlation functions $\omega(\theta)$ which are described on the celestial sphere. For a power-law correlation function, $\omega(\theta) \propto \theta^{-(\gamma-1)}$

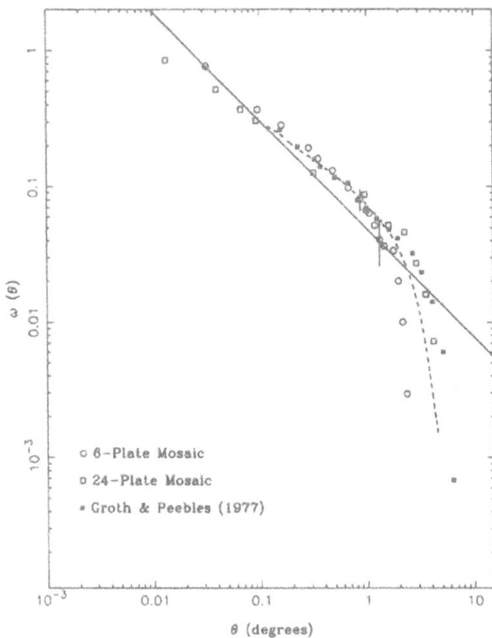

Fig. 18. The angular two-point correlation function $\omega(\theta)$ for galaxies determined from a number of recent surveys of large complete samples of galaxies (Collins *et al.* 1988).

Equation (96) describes the excess number density of galaxies at distance r relative to the average density and it gives a quantitative description of the clustering properties of galaxies on a wide range of scales. It should be noted that it is a rather broad-brush approach to the description of clustering in the Universe in that it is a spherically symmetric function so that it cannot describe more than the basic clumping of galaxies. Higher order correlation functions, such as the three and four point correlation functions take more of this information into account (see Peebles 1980). There are several notes to be made about the correlation functions.

1. Fig. 18 shows that the correlation function for galaxies is quite smooth. There are no obvious preferred scales, say, on the scale of the rich clusters of galaxies. Thus, it appears that structure on a very wide range of scales must be present in the initial perturbation spectrum from which the structure formed.
2. There is a characteristic scale $r_0 = 5h^{-1}$ Mpc which defines the distance at which the density of galaxies is greater than that of the background by a factor of two. This may be interpreted roughly as a measure of the scale to which the perturbations have become non-linear in the sense that all structure on smaller scales has $\xi(r) > 1$. This cannot be the whole story, however, since the sponge-like structure extends to scales much greater than $5h^{-1}$ Mpc.
3. There may be some structure in the correlation function as suggested by the dashed line. Some determinations of the function show a broad bump at about $10h^{-1}$ Mpc.
4. Of particular interest is the behaviour of the function on large physical scales. The

determinations of the two-point function show that the function falls off more rapidly than a power-law on large scales and, in some determinations, the function $\xi(r)$ becomes negative on scales greater than $20h^{-1}$ Mpc. Notice that, on these very large scales, the amplitude of the two-point correlation function is very much less than one. A problem concerns the clustering of objects on very large scales. It has been found by Bahcall and her coworkers (see e.g. Bahcall 1988) that Abell clusters, the richest giant clusters of galaxies, are correlated with a characteristic clustering scale $r_0 \approx (15 - 25)h^{-1}$ Mpc. The exact relation between this type of clustering of clusters and the two-point correlation function for galaxies is not at all clear. If this result is confirmed, it is difficult to reconcile it with a simple picture of the evolution of the correlation function. It should be noted that there are other phenomena which may have to be explained on these very large scales. There is now some evidence for the clustering of quasars on large scales with $r_0 = (10 - 20)h^{-1}$ Mpc (Shanks *et al.* 1987, Miller and Mitchell 1989) and we have yet to account for the streaming velocities of galaxies.

5.2 The Perturbation Spectrum

We have spent a lot of time on the evolution of the perturbations with time and so now we have to treat more formally their distribution in space, i.e. the function $\Delta(r)$. The standard procedure is to take the Fourier transform of the density distribution so that the amplitude of the perturbations with different wavelengths λ, or wavevectors $\mathbf{k} = (\frac{2\pi}{\lambda})\mathbf{i}_k$, can be found. Since we are dealing with a three-dimensional distribution of galaxies, we have to take a three-dimensional Fourier transform. However, since we are using an isotropic two-point correlation function, there are many simplifying features of the analysis. First of all, we define the Fourier transform pair for $\Delta(x)$

$$\Delta(x) = \frac{V}{(2\pi)^3} \int \Delta_k e^{-i\mathbf{k}.\mathbf{x}} d^3k \tag{97}$$

$$\Delta_k = \frac{1}{V} \int \Delta(x) e^{i\mathbf{k}.\mathbf{x}} d^3x \tag{98}$$

To relate the statistical averages of $\Delta(x)$ and Δ_k, we use Parseval's theorem which provides a relation between the integrals of the squares of $\Delta(x)$ and its Fourier transform Δ_k

$$\frac{1}{V} \int \Delta^2(x) d^3x = \frac{V}{(2\pi)^3} \int |\Delta_k|^2 d^3k \tag{99}$$

The quantity on the left hand side of equation (99) is just the mean square amplitude of the fluctuation per unit volume and therefore we can write

$$\langle \Delta^2 \rangle = \frac{V}{(2\pi)^3} \int |\Delta_k|^2 d^3k \tag{100}$$

Since we are dealing with the isotropic two-point correlation function, the element of k-space can be written $d^3k = 4\pi k^2 dk$.

The final step is to relate $\langle \Delta^2 \rangle$ to the correlation function through equation (95). It is simplest to begin with a Fourier series and then transform the series summation into a Fourier integral. Thus, we begin by writing $\Delta(x)$ as

$$\Delta(x) = \sum \Delta_k e^{-\mathbf{k}.\mathbf{x}}$$

$\Delta(x)$ is a real function and therefore we can find $\xi(r)$ by writing $\xi(x) = \langle \Delta(x)\Delta^*(x)\rangle$ where $\Delta^*(x)$ is the complex conjugate of $\Delta(x)$. Taking the average value of the product of $\Delta(\mathbf{x})$ and $\Delta(\mathbf{x}+\mathbf{r})$ in this way we find

$$\xi(r) = \left\langle \sum_k \sum_{k'} \Delta_k \Delta^*_{k'} e^{-i(\mathbf{k}-\mathbf{k}').\mathbf{x}} e^{i\mathbf{k}'.\mathbf{r}} \right\rangle$$

Now, when we take the cross terms in this summation, they all vanish except for those for which $\mathbf{k} = \mathbf{k}'$ and then we find

$$\xi(r) = \sum |\Delta_k|^2 e^{i\mathbf{k}.\mathbf{r}}$$

We now convert this simple relation into a Fourier integral

$$\xi(r) = \frac{V}{(2\pi)^3} \int |\Delta_k|^2 e^{i\mathbf{k}.\mathbf{r}} d^3 k \tag{100}$$

Finally, we note that $\xi(r)$ is a real function and therefore we are only interested in the integral over the real part of $e^{i\mathbf{k}.\mathbf{r}}$ i.e. the integral over $\cos \mathbf{k}.\mathbf{r} = \cos(kr\cos\theta)$. We have to integrate over an isotropic distribution of angles θ i.e. integrate $\cos(kr\cos\theta)$ over $\frac{1}{2}\sin\theta d\theta$. Performing this integral, we obtain the final answer

$$\xi(r) = \frac{V}{(2\pi)^3} \int |\Delta_k|^2 \frac{\sin kr}{kr} 4\pi k^2 dk \tag{101}$$

This is the relation between the two-point correlation function $\xi(r)$ and the power spectrum of the fluctuations $|\Delta_k|^2$ we have been seeking.

It is to be hoped that it will eventually be possible to predict the initial power spectrum $|\Delta_k|^2$ from theory. The observations suggest that the spectrum is broad with no preferred scales and it is natural therefore to work with power spectra of power-law form

$$|\Delta_k|^2 \propto k^n \tag{102}$$

According to equation (101), this means that the correlation function $\xi(r)$ should have the form $\xi(r) \propto \int k^{(n+2)} dk$. If we now consider a volume of space $V \sim r^3$, then because the functions $\sin kr/kr$ decreases rapidly to zero when $kr > 1$, we can integrate k from 0 to $k_{max} \approx 1/r$. Taking the integral, we find

$$\xi(r) \propto r^{-(n+3)} \tag{103}$$

Since the mass of the fluctuation is just proportional to r^3, this result can also be written $\xi \propto M^{-\frac{(n+3)}{3}}$. Finally, to relate ξ to the root-mean-square density fluctuation Δ, we have to take the square root of ξ, i.e.

$$\Delta = \frac{\delta\rho}{\rho} = \langle \Delta^2 \rangle^{\frac{1}{2}} \propto M^{-\frac{(n+3)}{6}} \tag{104}$$

The above analysis is useful because it shows the functional relations between the various ways of describing the density perturbations. A power spectrum $|\Delta_k|^2$ of power-law form k^n corresponds to a two-point correlation function $\xi(r) \propto r^{-(n+3)} \propto M^{-\frac{(n+3)}{3}}$

and to a spectrum of density perturbations $\Delta \propto M^{-\frac{(n+3)}{6}}$. A number of deductions follow from these results.

1. It is clear that so long as $n > -3$, the spectrum decreases to large mass scales so that the Universe is isotropic and homogeneous on the very largest scales.

2. A Poisson noise spectrum has equal power on all scales and hence $n = 0$. This results in a density perturbation spectrum $\Delta \propto M^{-\frac{1}{2}}$ and to a correlation function $\xi \propto M^{-1}$.

3. There is one case of special interest, $n = 1$, for which $\Delta \propto M^{-\frac{2}{3}}$ and $\xi \propto r^{-4} \propto M^{-\frac{4}{3}}$. This spectrum has the property that the root-mean-square density contrasts Δ on all scales are the same when they come through the horizon. In the Cold Dark Matter picture, the perturbations in the dark matter grow from the epoch z_{eq} to the present. They have constant amplitude from the redshift z_{eq} back to the epoch when they came through the horizon. The rate of growth of the perturbation before the perturbations enter the horizon is $\Delta(z) \propto R^2$. Therefore, if the initial fluctuation spectrum was $\Delta \propto M^{-\beta}$, at any epoch before the fluctuation came through the horizon its amplitude would be proportional to $M^{-\beta}R^2$. The perturbation grows until the mass comes within the horizon. Now, the mass of cold dark matter within the horizon at any redshift is just $(ct)^3 \rho_{dark}(t_0) R^{-3}$ where $\rho_{dark}(t_0)$ is the density of dark matter at the present epoch t_0. Since $t \propto R^2$ in the radiation-dominated phase, the mass within the horizon in dark matter varies as R^3. Therefore, the perturbation grows until that redshift at which this mass comes through the horizon which is proportional to $M^{\frac{1}{3}}$. The spectrum of mass perturbations once the perturbations come through the horizon is thus proportional to $M^{-\beta+\frac{2}{3}}$. This shows that, if $\beta = \frac{2}{3}$, the fluctuations are of the same amplitude on different mass scales when they come through the horizon. This rather special value, $n = 1$, is known as the **Zeldovich spectrum** (Zeldovich 1972). It is intriguing that, if $n = 1$, the Universe is a **fractal** in the sense that every perturbation which comes through the horizon has the same amplitude. In other words, as the Universe expands, we always find perturbations of the same amplitude coming through the horizon. Zeldovich has given some physical arguments why the value $n = 1$ might arise in the early Universe. In the inflationary picture of the early Universe, there is some reason to suppose that fluctuations with the Zeldovich spectrum and roughly the correct amplitude may be generated from thermal fluctuations during the inflationary expansion. Another intriguing point is that we can work out explicitly how large the amplitude of these fluctuations have to be when they come through the horizon. The epoch of equality of the inertial mass densities in the matter and radiation occurs at a redshift $4 \times 10^4 h^2$ if $\Omega = 1$. Now, the dark matter perturbations grow as $(1+z)^{-1}$ from that epoch to the present so long as they remain linear. We know that galaxies must have been created by a redshift of about 4 and therefore the amplitude of the perturbations when they came through the horizon must have been about $\Delta \approx 10^{-4}$. These are certainly not infinitesimal perturbations.

4. Another way of looking at the problem is to invert it and ask what value of n would give the observed exponent in the two-point correlation function. Since $\xi \propto r^{-(n+3)}$,

the preferred value would appear to be $n = -1.2$. The problem with this approach is that it assumes that the observed two-point correlation function represents the initial fluctuation spectrum. Unfortunately, it cannot be assumed that non-linear effects have not modified considerably the spectrum from its initial form. This is shown rather dramatically by the n-body simulations of the growth of structure in the Universe by clustering under gravity. In physical terms, this may be understood in terms of the fact that on the scales over which the two-point correlation function has been well measured, relaxation under gravity is likely to be well under way and there has been time for systems on the scale of, say, 1 Mpc to have virialised.

5.3 The Evolution of the Initial Perturbation Spectrum

We have already discussed some of effects which modify significantly the initial perturbation spectrum, such as Silk Damping. We now want to put all these effects together to understand how the initial perturbation spectrum evolves to form the structures we observe in the Universe today. Let us consider three separate cases.

We have already dealt with the case of **Adiabatic Baryonic Perturbations** in some detail. In this case there is no dark matter present. Silk damping eliminates all the small scale perturbations. One interesting phenomenon which is specific to this model is that, after the perturbations come within the horizon and they begin to oscillate as sound waves, if we observe the spectrum at any subsequent time, there will also be oscillations in the mass spectrum of the perturbations since the oscillations on different mass scales begin to grow at different times as they come through the horizon. A second point is that the sound waves are cooled adiabatically as the Universe expands resulting in a gradual decay in the amplitude of the perturbations. These phenomena are illustrated schematically in Fig. 19.

In the case of the **Hot Dark Matter** picture, we have already described the influence of the free streaming of the neutrinos upon the mass spectrum of the perturbations. Only the most massive perturbations survive (Fig. 19). The picture resembles the classical adiabatic picture after recombination with the baryons falling into the large scale neutrino perturbations.

In the case of the **Cold Dark Matter** picture, we have to look in a little more detail at the spectrum of perturbations observed at any epoch prior to the redshift z_{eq}. On scales larger than the horizon, the mass spectrum preserves its initial form $\Delta \propto M^{-\frac{(n+3)}{6}}$ and these continue to grow as R^2 until this mass scale enters the horizon, i.e. $\Delta \propto M^{-\frac{(n+3)}{6}} R^2$ However, the perturbations stop growing when they enter the horizon. As shown above, the scale factor R at which a perturbation of mass M enters the horizon is proportional to $M^{\frac{1}{3}}$ and so the density perturbation spectrum is

$$\Delta \propto M^{-\frac{(n+3)}{6}-\frac{2}{3}} = M^{-\frac{(n-1)}{6}} \tag{104}$$

If we look at this result in terms of the power-spectrum of the perturbations, masses greater than the horizon mass have power-spectrum $|\Delta_k|^2 \propto k^n$ whereas masses within the horizon have a flatter spectrum, $|\Delta_k|^2 \propto k^{(n-4)}$. Since the cold dark matter perturbations became non-relativistic very early in the Universe, this perturbation spectrum is maintained to very low masses. This spectrum is illustrated in Fig. 19.

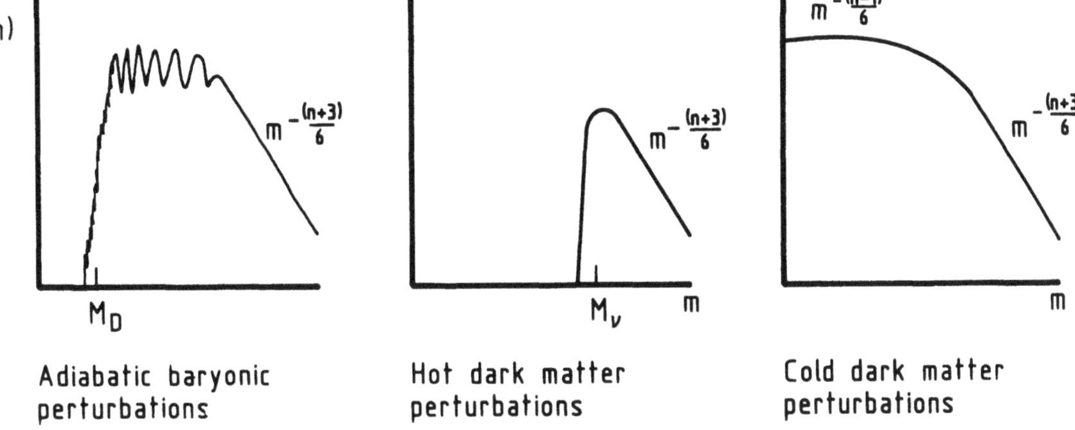

Fig. 19. The forms of the density contrast Δ as a function of mass for an initial fluctuation power spectrum of the form $|\Delta_k|^2 \propto k^n$ by the epoch of recombination for the case of (i) adiabatic baryonic fluctuations, (ii) hot dark matter in the form of neutrinos with rest mass 30 eV and (iii) cold dark matter.

These models have been the subject of detailed computer simulations to determine how well they can reproduce the observed large scale structure in the Universe. We have already mentioned the problems which arise in the case of the standard adiabatic baryonic perturbation picture. The Hot and Cold Dark Matter pictures result in quite different pictures of how galaxies came about and of the origin of the large scale structure of the Universe. Fig. 20 is a sample of the results of computer simulations of the Hot and Cold Dark Matter models carried out by Frenk (1986).

In the case of the **Hot Dark Matter** model it can be seen that the model is very effective in producing flattened structure like pancakes. In this picture, the baryonic matter forms pancakes within the large neutrino haloes and their evolution is similar to the adiabatic picture from that point on. It can be seen that the model is too effective in producing flattened, stringy structures. Essentially everything collapses into the thin pancakes and filaments and the observed Universe is not as highly structured as this. In addition, galaxies must form rather late in this picture because it is only the most massive structures which survive to the recombination epoch. This means that it is difficult to produce stars and galaxies which are younger than the structures on the scale of $\sim 4\times10^{15} M_{\odot}$. Everything must have formed rather late in the Universe in this picture.

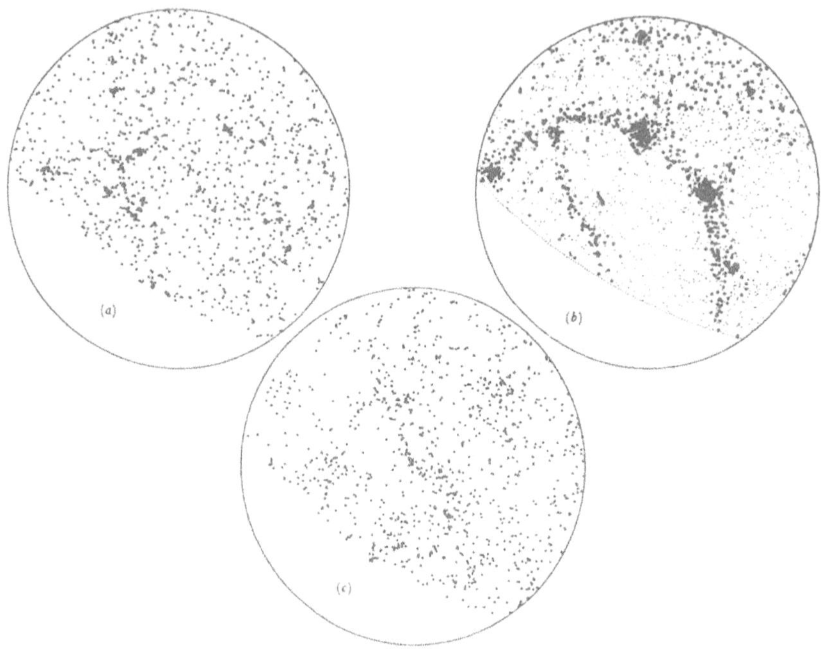

Fig. 20. Recent simulations of the expectations of (a) the Cold Dark Matter and (b) the Hot Dark Matter models of the origin of the large scale structure of the Universe compared with the observations (c) (Frenk 1986). The unbiased cold dark matter model does not produce sufficient large scale structure in the form of voids and filaments of galaxies whereas the unbiased hot dark matter model produces too much structure.

In the case of the **Cold Dark Matter** picture, masses on all scales can begin to collapse soon after recombination and star clusters and the first generations of stars can be old in this picture. The process of formation of the large scale structure is different in this case since it forms by gravitational clustering under the influence of the initial power spectrum of the perturbations. Large scale systems like galaxies and clusters of galaxies are assembled from their component parts by dynamical processes which can be simulated by computer modelling. Fig. 20 shows that structure indeed develops but is not as pronounced as in the observed Universe. This is because it is difficult to produce elongated structures by gravitational clustering which tends to make more symmetrical structures than the sheets and filaments of galaxies found in the Universe. However, one of the successes of the Cold Dark Matter picture is that it is can account for the observed two-point correlation function assuming $n = 1$ and that the phases of the waves which make up Δ_k are random (Frenk 1986).

Evidently, neither model gives a particularly good fit to the observations and this has led to the idea that there may well be bias mechanisms which lead to the preferential formation of galaxies in certain regions of space rather than others. **Biasing** has become a major growth area in astrophysical cosmology. In the case of the Cold Dark Matter picture, the biasing must be such that it results in galaxies being formed preferentially in regions of high density and not in the regions in between. In the case of the Hot

Dark Matter picture, some **Anti-biasing** mechanism is needed so that all the formation of galaxies is not concentrated in the sheets and filaments which form too readily in this model.

Is there any evidence for biasing in the Universe? Many studies are underway to find out if there are regions in the Universe where there is a preference for galaxies forming in one region rather than another. A good example is the case of structures such as the Coma Cluster of galaxies. This cluster has been studied kinematically in greater detail than any other cluster and so the estimates of its mass and mass-to-light ratio are well known. There is a dark matter problem which amounts to about a factor of 10 but the mass-to-light ratio found when all the dark matter is included corresponds to a factor of only one third of the value necessary to close the Universe. If the Universe really has the critical density, $\Omega = 1$, this means that there must be biasing by a factor of three towards the formation of galaxies in the region of the Coma cluster as opposed to the general field.

Many possible biasing and anti-biasing mechanisms have been discussed by Dekel and Rees (1987) and by Dekel (1987). As they emphasise, perhaps the most important requirements are reliable observational estimates of how much biasing (or anti-biasing) really occurs in the Universe. I refer readers to their papers for a survey of the many possible bias mechanisms and how these ideas can be tested by observations of the nearby Universe. I mention some of these ideas to give a flavour of the astrophysical questions to be addressed.

1. An interesting example discussed by Kaiser (1986) is the influence of large scale perturbations upon a region in which the fluctuations from which galaxies are forming are already well developed. In this situation, the large-scale density perturbation enhances the density perturbations associated with the collapsing galaxies and so can lead to enhanced galaxy formation in this region. The study of the density peaks in the random density fluctuation fields is an important aspect of the detailed evolution of the perturbation spectrum (see e.g. Peacock and Heavens 1985).
2. Galactic explosions may sweep away the gas from the vicinity of a galaxy, possibly having a positive or negative effect upon biasing. A violent explosion can remove the gas from the vicinity of the galaxy and make it too hot for further galaxy formation to occur in its vicinity. On the other hand, the swept-up gas may be greatly increased in density at the interface between the hot expanding gas and the intergalactic gas. By analogy with the case of galactic supernova remnants in which star formation can be stimulated by the passage of a shock wave, the same process on a galactic scale may stimulate the formation of new galaxies (Ostriker and Cowie 1981). One possibility is that this process might thicken the pancakes which form in the Hot Dark Matter picture.
3. The gas in the voids between superclusters may be so hot that galaxies cannot form in these regions.

It is evident from these examples that the questions are largely astrophysical. In my view, the most important discrepancy is between the total mass in the largest systems for which we can make reasonable estimates of the mass density and the critical model for which $\Omega = 1$. I find it striking that all the determinations of the amount of gravitating matter in rich clusters of galaxies, in the vicinity of superclusters and the cosmic virial theorem all suggest that the matter they contain, including the dark matter, is insufficient to close the Universe. Therefore, if $\Omega = 1$, most of the dark matter must be located between the largest systems for which mass estimates can be readily made.

5.4 Fluctuations in the Microwave Background Radiation

We have mentioned on several occasions the critical importance for cosmology of the limits to the temperature fluctuations in the Microwave Background Radiation. This is a complex topic and I will do no more than indicate the steps necessary in the full calculation. We have established the picture of the growth of the gravitational instabilities in the above sections and now we ask how these impact the Microwave Background Radiation. Of crucial importance in this calculation is the ionisation state of the intergalactic gas through the epoch of recombination. The estimate of the optical depth of the intergalactic gas due to Thomson scattering given by equation (32) shows how rapidly it grows to very large values at redshifts greater the redshift of recombination. Thus, temperature fluctuations which originate at redshifts greater than the redshift of recombination are damped out by scattering and the fluctuations we observe originate in a rather narrow redshift range about that at which the optical depth of the intergalactic gas is unity. This is why the precise determination of the ionisation history of the intergalactic gas is so important.

The problem, first discussed by Zeldovich, Kurt and Sunyaev (1968) and Peebles (1968), is a well known one in that during the recombination process, the photons released in recombination of hydrogen atoms are sufficiently energetic to ionise another hydrogen atom and thus there is no direct way of destroying the photons liberated in the recombination process. The answer is that Lyman-α photons are destroyed by the two-photon process in which two photons are liberated from the 2s state of hydrogen in a rare quadrupole transition. The spontaneous transition probability for this process is 0.1 sec but it turns out to be the dominant process which determines the rate of recombination of the intergalactic gas.

Detailed calculations of the degree of ionisation through the critical redshift range have been carried out by Jones and Wyse (1985) who find a very strong dependence of the fractional ionisation x upon redshift

$$x = 2.4 \times 10^{-3} \frac{(\Omega h^2)^{\frac{1}{2}}}{\Omega_B h^2} \left(\frac{z}{1000}\right)^{12.75} \tag{106}$$

In this formula Ω is the density parameter for the Universe as a whole at the present epoch and Ω_B is the present density parameter for the baryons. Using the same formalism which led to equation (26), we can find a remarkably simple expression for the optical depth of the intergalactic gas through these crucial epochs

$$\tau = 0.37 \left(\frac{z}{1000}\right)^{14.25} \tag{107}$$

Because of the enormously strong dependence upon redshift, the optical depth at which the intergalactic gas is unity is always very close to 1070, independent of the exact value of Ω and h. We can now work out the range of redshifts at which the photons of the Microwave Background Radiation were last scattered. This probability distribution is given by

$$dp/d\tau = e^{-\tau} d\tau/dz$$

This probability distribution is well fitted by a gaussian distribution with mean redshift 1070 and standard deviation $\sigma = 80$. This result formalises the statement that the last scattering of the photons did not take place at a single redshift but that half of the photons of the Microwave Background Radiation were last scattered between redshifts 1010 and 1130.

The implications of these results are important in working out the predicted temperature fluctuation spectrum. First of all, let us work out the physical scale corresponding to the thickness of the last scattering layer. The formula for the element of comoving distance at redshift z can be derived from equation (16)

$$dr = cdt(1+z) = \frac{cdz}{H_0(1+z)(\Omega z+1)^{\frac{1}{2}}}$$

Notice that this is the distance projected to the present epoch. Taking the approximation for large redshifts, we find $dr = cdz/H_0 z^{\frac{3}{2}} \Omega^{\frac{1}{2}}$ Thus, the redshift interval of 80 at a redshift of 1070 corresponds to a scale of $7(\Omega h^2)^{-\frac{1}{2}}$ Mpc at the present day. Thus, fluctuations on scales smaller than this at the present day will tend to be washed out by the superposition of perturbations at the decoupling epoch. The typical mass contained within this scale is $M \approx 10^{14}(\Omega h^2)^{1/2} M_\odot$ and consequently we expect the temperature fluctuations associated with smaller mass scales to be depressed by smearing. The angular scale of these fluctuations as observed at the present epoch is $4\Omega^{\frac{1}{2}}$ arcmin.

The problem is now to convert the density fluctuations and their associated velocities in the redshift range $1150 \lesssim z \lesssim 1000$ into temperature fluctuations. There are three processes which are of prime importance.

1. The **Adiabatic Perturbations** result in temperature perturbations $\delta T/T \approx \frac{1}{3}\delta\rho/\rho$ and therefore it is expected that there will be hot spots imprinted on the Microwave Background Radiation with amplitudes corresponding to the density perturbations. Even in this simplest case there are complications. A full calculation has to take into account the fact that in the hotter regions, recombination takes place later and in addition the full dynamics of recombination and the decoupling of matter and radiation should be included. The net result is that on angular scales much greater than 4 arcmin, the temperature fluctuations are roughly $\frac{1}{3}\delta\rho/\rho$.
2. A second effect which is important on large physical scales is the **Sachs-Wolfe Effect**. This is the gravitational redshift effect associated with the fact that the photons of the Microwave Background Radiation have to pass through regions of varying gravitational potential ϕ and is important on large physical scales.
3. A third effect is the **Doppler Scattering** of the photons by the growing perturbations. We showed in Section 3.1 that there must be peculiar velocities associated with the growth of the perturbations as soon as they start to grow

$$\frac{d\Delta}{dt} = -\nabla.\delta\mathbf{v} \tag{53}$$

and the photons undergo first order Doppler scatterings with these collapsing clouds.

The full calculation involves taking into account all these effects for a given power spectrum of fluctuations modified by the various effects discussed in Section 5.3. Integrals have to be taken over the forms of power-spectrum shown in Fig. 19 and then normalised so that the amplitude of the density perturbations Δ produce galaxies by the present epoch. The following results are typical of those which come out of the detailed computations (from Peacock, private communication).

1. For the case of Adiabatic Baryonic fluctuations, the predicted values of the root-mean-square fluctuation in the radiation temperature of the Microwave Background Radiation are $\delta T/T = 10^{-3.5}$ if $\Omega = 1, h = \frac{1}{2}$ and $10^{-2.8}$ if $\Omega = 0.1, h = 0.5$. These results can be understood in terms of the simple rules about the evolution of the perturbations with cosmic epoch. The fluctuations are expected to be much greater in the case of the low density model, $\Omega = 0.1$, because there is less time for the fluctuations to grow before $\Omega z \sim 1$.
2. For the case of Cold Dark Matter fluctuations, the corresponding temperature fluctuations are $\delta T/T = 10^{-4.8}$ if $\Omega = 1, h = 0.75$ and $10^{-4.0}$ if $\Omega = 0.2, h = 0.75$. A number of factors contribute to the much smaller values of $\delta T/T$ expected in this model, in particular, the fact that a very much wider spectrum of values of k contribute to the density fluctuation Δ.

These results are to be compared with the observed limits to fluctuations in the Microwave Background Radiation which correspond to $\delta T/T \lesssim 5 \times 10^{-5}$ on the scale of 4.5 arcmin and a possible measurement of $\delta T/T \approx 4 \times 10^{-5}$ on an angular scale of 8°. As expected, it can be seen that the adiabatic baryonic picture is in serious conflict with the observations. It can only be rescued if there is further heating and ionisation of the intergalactic gas after the epoch of recombination and there is no obvious source of energy to achieve this in this model. The cold dark matter picture is in satisfactory agreement with the observations.

5.5 Concluding Remarks

The net result of this discussion is that the one model which seems to be capable of satisfying the many constraints upon physical models of galaxy formation is the Cold Dark Matter theory. It should be emphasised that there are ways of reconciling the other theories with the observations but they require further, more or less arbitrary, additions to the straight-forward theory to ensure consistency with the observations. The aim of this section has been to provide the reader with the tools for constructing alternative theories of galaxy formation.

As an example of the type of theory to which we have paid little attention and to which the above considerations may be relevant, there might be **Isothermal Perturbations** present in the Hot Big Bang. These are simply density perturbations in the initial conditions which are not accompanied by corresponding pressure perturbations in the radiation dominated plasma. There is no obvious origin for such perturbations in the standard picture but if they did exist, they behave very like cold dark matter perturbations except that they are frozen into the background radiation field and so do not grow after they enter the horizon until after the epoch of recombination. This means that they are essentially stationary at the epoch of recombination and so they cause much smaller temperature fluctuations in the Microwave Background Radiation. Similar calculations to those given above for isothermal perturbations show that $\delta T/T \approx 10^{-4.2}$ if $\Omega = 1, h = 0.75$ and $\delta T/T \approx 10^{-4.6}$ if $\Omega = 0.1, h = 0.5$. This would be consistent with observation. Another way of describing these types of fluctuations is in terms of **isocurvature fluctuations.**

The reason for giving this example is to demonstrate that there are many variations which can be invented given the basic rules described above. It might be that a combination of many different types of perturbation contribute to the final scenario. All the models involve to a lesser or greater extent arbitrary initial conditions from which they evolve. It is generally agreed that the inflationary picture leading to a Cold Dark Matter dominated Universe is probably the simplest of the theories which is consistent with most of the observations. If this is indeed the case, there are obvious implications for elementary particle physics.

6 The Post-Recombination Universe

As we come closer and closer to the present epoch, the questions become much more astrophysical and concern the non-linear development of the initial perturbations. In this chapter we will look at some aspects of the physical processes which are likely to be important during the post-recombination epochs.

6.1 The Non-linear Collapse of Density Perturbations

This is a huge subject and one which has to be tackled if we wish to be able to make realistic comparison between the theories and the observations. We have already discussed qualitatively the non-linear collapse of pancakes. Let us look at some of the simple exact results for spherical collapse which act as a paradigm for what one would like to be able to do in the general case.

One calculation which can be carried out exactly is the collapse of a spherical region. The dynamics are exactly the same as those of a closed Universe with $\Omega > 1$ i.e. they describe the cycloidal variation of the scale factor R as given by the pair of equations (68)

$$R = a(1 - \cos\theta) \qquad t = b(\theta - \sin\theta) \tag{68}$$

where, in the cosmological case, $a = \Omega/[2(\Omega - 1)]$ and $b = \Omega/[2H_0(\Omega - 1)^{\frac{3}{2}}]$. It is apparent from equation (68) that the expansion reaches its maximum value when $\theta = \pi$ and collapses back to infinite density when $\theta = 2\pi$. There are some interesting results which come out of this model. For example, when the perturbation stops expanding, $\dot{R} = 0$ at $\theta = \pi$, the scale factor of the perturbation is $R_p = 2a = \Omega/(\Omega - 1)$. This occurs at time $t = \pi b = \pi\Omega/[2H_0(\Omega - 1)^{\frac{3}{2}}]$. Therefore, the density in the perturbation relative to that of the background density, which we take to be the critical $\Omega = 1$ model, is $\rho_p/\rho_0 = R^3/R_p^3 = 9\pi^2/16 = 5.55$. This means that, by the time the comoving sphere has stopped expanding, its density is already 5.55 times the background density.

The perturbation collapses to infinite density at time $t = 2\pi b$ which is twice the time the perturbation takes to reach maximum expansion. Since $R \propto t^{\frac{2}{3}}$, it follows that the relation between the redshift of maximum expansion z_{max} and the redshift of collapse z_{form} is

$$\frac{1+z_{form}}{1+z_{max}} = 2^{-\frac{2}{3}}$$

This means that collapse occurs very rapidly, within a factor of two in redshift, once the perturbation starts to collapse. For example, if $z_{max} = 10, z_{form} = 6$.

This is not a very useful example because the perturbation collapses to a black hole. In more realistic case, the perturbation will not be spherical but ellipsoidal in the lowest order approximation (see e.g. Peacock and Heavens (1985) for a discussion of the expected ellipticity distributions) and it collapses most rapidly along its shortest axis, as in the classical pancake picture. Another important point is that it is likely that the collapsing cloud fragments into subcomponents and the process of **violent relaxation** may take place in which the subunits reach a dynamical equilibrium under the influence of large scale potential gradients in the cloud. This may be the process by which clusters of galaxies reach a dynamical equilibrium. Such a state must satisfy the Virial Theorem according to which the kinetic energy of the system is just half its (negative) gravitational potential energy. At z_{max} all the energy of the system is potential energy $\Omega_p = -GM^2/R$. If the system collapses to half this radius, its gravitational potential energy becomes $-GM^2/(R/2)$ and, by conservation of energy, the kinetic energy acquired is $\frac{1}{2}Mv^2 = -GM^2/R - [-GM^2/(R/2)] = GM^2/R$ i.e the kinetic energy is half of the negative gravitational energy. This simple calculation shows that to reach a dynamical equilibrium, the radius of the perturbation has to decrease by a further factor of 2 and hence the density increase by a further factor of 8.

Thus, the final density of the bound object relative to the background density is at least a factor of 5.55×8 times the background density. For illustration let us see what this means for the redshifts when galaxies and other large scale systems separated out of the expanding background. The above factor is the minimum enhancement in the density of the bound object relative to the background. The density contrast may be much greater if other dissipative processes played a role in determining the final stable configuration. We can therefore state with some confidence that the density of the virialised object is at least 8×5.55 times the background density. We can write this as

$$\rho_{vir} \gtrsim 5.55 \times 8 \times \frac{3\Omega H_0^2}{8\pi G}(1+z_{max})^3 \tag{108}$$

Now, we can make a rough estimate of ρ_{vir} from the Virial Theorem. If M is the mass of the system and v^2 its velocity dispersion, the condition that the kinetic energy be half the gravitational potential energy means that

$$\frac{1}{2}Mv^2 = \frac{1}{2}\frac{GM^2}{R}$$

where R is some suitably defined radius and hence

$$\rho_{vir} \approx \frac{v^6}{\frac{4\pi}{3}G^3M^2} \tag{109}$$

Inserting this value into equation (108), we can find a limit to the redshift of formation of the object which we can write in the form

$$(1 + z_{max}) \lesssim 0.62\left(\frac{v}{100 \text{ kms}^{-1}}\right)^2 \left(\frac{M}{10^{12} M_\odot}\right)^{-\frac{2}{3}} (\Omega h^2)^{-\frac{1}{3}} \tag{110}$$

What this sum amounts to is an improved version of the simple calculation presented at the beginning of Section 3 concerning the relative densities of objects and the background density. Let us put in some representative figures. For **galaxies** having $v \sim 300$ km s^{-1} and $M \sim 10^{11} M_\odot$, the redshift of formation must be less than about 25. For **clusters of galaxies** for which $v \sim 1000$ km s^{-1} and $M \sim 10^{15} M_\odot$, the redshift of formation cannot be much greater than 1. This means that on average, the clusters of galaxies must have formed in the relatively recent past. The origin of this result can be understood from the fact that the relative density of a rich cluster to the background density is probably about 100 and once we take account of the above enhancement factors, the clusters must have virialised relatively recently.

These simple calculations illustrate the limitations of the simple linear theory for the formation of structure in the Universe.

6.2 The Role of Dissipation

By dissipation, we mean energy loss by radiation, the result being to remove thermal energy from the system. So far we have mostly been considering the development of perturbations under the influence of gravity alone. In a number of circumstances, however, once the gas within the system is stabilised by thermal pressure, the loss of energy by radiation is an effective way of decreasing the internal pressure, thus enabling the region to contract to preserve pressure equilibrium. If the radiation process is very effective in removing pressure support from the system, this can result in what is known as a **thermal instability.**

There are two important ways in which dissipation plays an important part in the story. The first is that we have to make stars in our protogalaxies. In the case of star formation in our own Galaxy, we know that the stars form within dusty regions. The sequence of events is that a region of a cool dust cloud becomes unstable, either through the standard Jeans' instability described in Section 3.2 or because of some external influence such as the passage of the gas cloud through a spiral arm or perhaps by compression of the gas by the blast wave of a supernova remnant. The cloud collapses and loses its internal binding energy by radiation. This process continues until the cloud becomes optically thick to radiation. The loss of energy is then mediated by the dust grains in the contracting gas cloud. The cool dust grains are heated to temperatures of about 100 K and consequently they reradiate away the binding energy of the protostar at far infrared wavelengths at which the collapsing cloud is transparent. In the case of the very first generations of stars there is an obvious problem with this mechanism in that there are no heavy elements present in the primordial gas out which dust grains could be formed. For these first generation stars, the star formation process presumably has to take place in a gas of essentially pure molecular hydrogen. This suggests that the process of formation of the first generation of stars may well be different from the star formation we observe at the present day in our Galaxy. Once the first generation of massive stars has formed, it can be shown that the fraction of heavy elements in the interstellar gas builds up quickly and it is out of this enriched gas that dust and the subsequent generations of stars form.

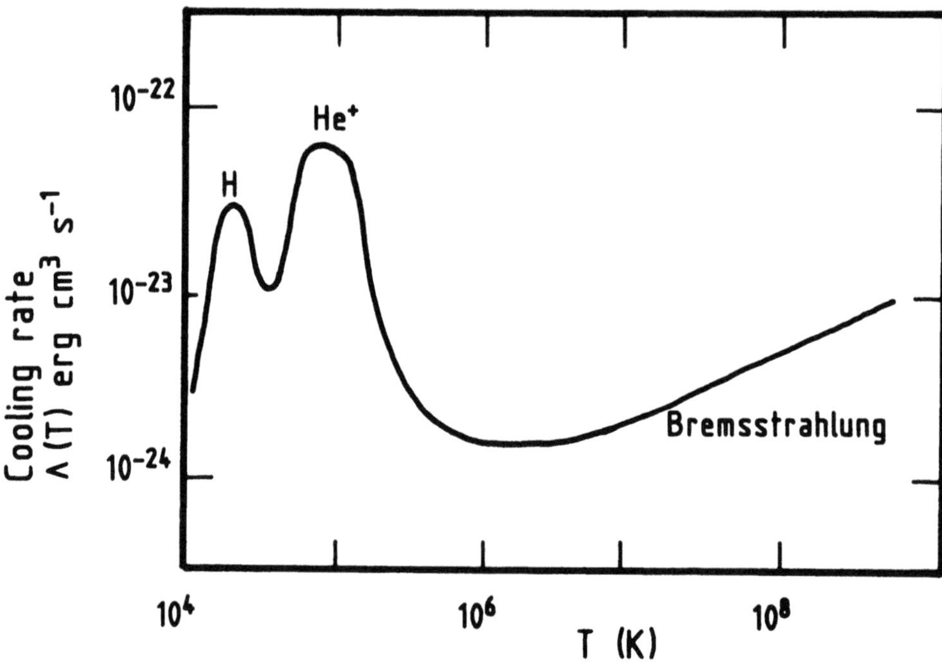

Fig. 21. A schematic diagram showing the cooling rate per unit volume for a hydrogen-helium plasma by radiation as a function of temperature.

It is obvious that dissipative processes play a dominant role in the formation of stars and this naturally leads to the question of whether or not similar processes might be important in the formation of larger scale systems. These processes were elegantly described by Rees and Ostriker (1977) whose presentation we will follow. The key diagram is the loss rate by radiation of a hydrogen-helium plasma as a function of temperature. A simplified version of this function is shown in Fig. 21, the cooling rate being given in a form such that the loss rate per unit volume is given by $dE/dt = -N^2\Lambda(T)$ where N is the number density of hydrogen ions. At high temperatures the dominant loss mechanism is bremsstrahlung whilst at lower temperatures the main loss mechanisms are free-bound transitions of hydrogen and ionised helium. Therefore the cooling time of the plasma is simply the time it takes the plasma to lose all its thermal energy $t_{cool} \approx \frac{3}{2}NkT/N^2\Lambda(T)$. This time scale should be compared with the time-scale for collapse which is given by $t_{dyn} \approx (G\rho)^{-\frac{1}{2}} \propto N^{-\frac{1}{2}}$. The significance of these timescales is best appreciated by inspecting the locus of the equality $t_{cool} = t_{dyn}$ in a temperature-number density diagram (Fig. 22).

It can be seen that that locus $t_{cool} = t_{dyn}$ is a mapping of the cooling curve of the hydrogen-helium plasma loss-rate into the $T - N$ plane. Inside this locus, the cooling time is shorter than the collapse time and so it is expected that dissipative processes are more important than dynamical processes. It is straightforward to draw lines of constant mass on Fig. 22. It can be seen that the range of masses which lie within the curve correspond to $10^6 \lesssim M/M_{\odot} \lesssim 10^{12}$. This is the most important conclusion of this analysis. The fact that the masses lie so naturally in the range of observed galaxy masses suggests that the typical masses of galaxies may not be wholly determined by the initial fluctuation spectrum but by more astrophysical processes as well. The other lines show the loci corresponding to the radiation loss times being equal to the age of the Universe and to perturbations having such low density that they do not collapse gravitationally in 10^{10} years.

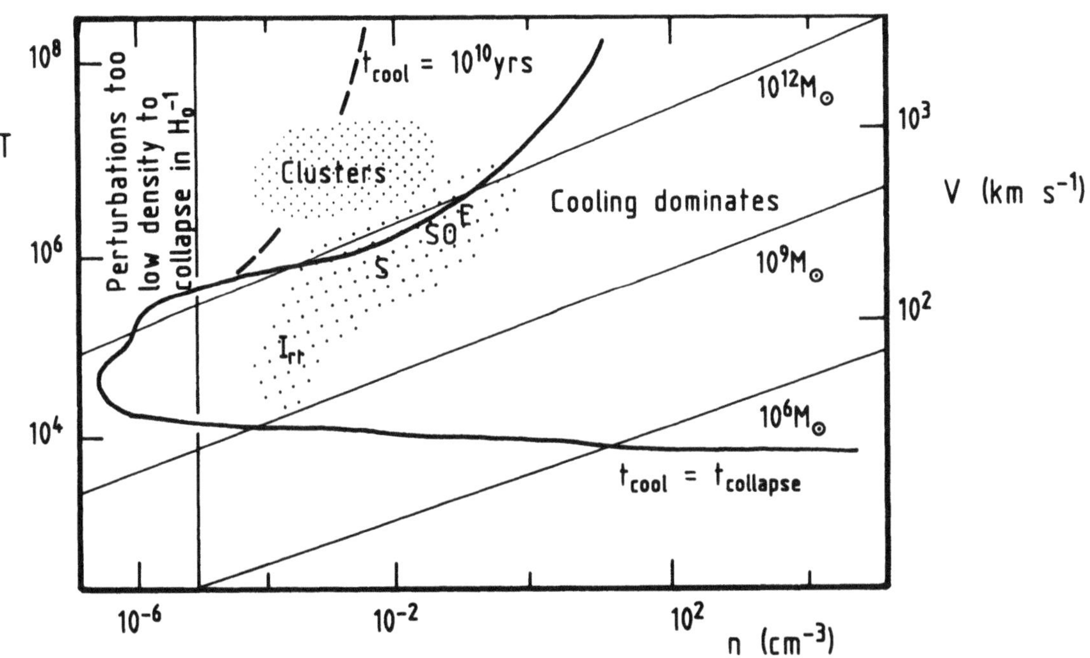

Fig. 22. A temperature-number density diagram showing the locus defined by the condition that the collapse time of a region t_{dyn} should be the same as its cooling time by radiation t_{cool} (after Rees and Ostriker (1977)). Also shown are lines of constant mass, a cooling time of 10^{10} years and the density at which the perturbations are of such low density that they do not collapse in the age of the Universe. Superimposed upon these lines is an indication of the typical number densities and equivalent temperatures for different types of galaxies and clusters of galaxies (after Blumenthal *et al.* 1984).

This diagram can be used astrophysically in the following way. For any theory of the origin of the large scale structure of the Universe, one works out at each stage the density and temperature of the gas in the structures as they are forming. Fig. 22 can then be used to determine whether or not the cloud is unstable to thermal collapse. A good example of this is found in the various forms of the pancake theory. When the gas collapses into pancakes, the matter falls onto a singular plane and, as a result, a shock wave passes out through the infalling matter and heats it to a high temperature. In this model, galaxies form by thermal instabilities in the heated gas. This is the means by which objects the mass of galaxies are formed within the pancake picture. It will be recalled that in these models all small scale structure is damped out either by Silk damping or by the free streaming of neutrinos. To put it another way, if the gas is heated above 10^4 K, there is no stable region for masses in the range 10^9 to $10^{12} M_\odot$.

The second thing one can do is to plot the observed location of galaxies on the temperature-number density diagram. This is also shown in Fig. 22. In this case the temperature is not used but rather an effective temperature associated with the velocity dispersion of the stars in the galaxy $\frac{1}{2}kT_{eff} \approx \frac{1}{2}mv^2$. It can be seen that the irregular galaxies (Irr) fall well within the cooling locus and the spirals (S), S0 and elliptical (E) galaxies all appear to lie close to the critical line. On the other hand, the clusters of galaxies lie outside the cooling locus. This suggests that cooling may well have been an important factor in the formation of galaxies but that, in the case of the clusters, their properties probably reflect the initial perturbation spectrum.

6.3 Primaeval Galaxies

It is obvious from the above considerations that the understanding of the processes of galaxy formation would be much improved if galaxies could be observed in the process of formation. The considerations of the Section 6.2 suggest that galaxy formation may well have occurred in the redshift range $2 \lesssim z_{form} \lesssim 25$. It is therefore interesting to work out whether galaxies in the process of forming their first generations of stars are observable. This topic has become a major growth area in astrophysical cosmology. The aim of these calculations is to predict the evolutionary behaviour of the populations of stars in the galaxy as it evolves, one condition being that the galaxies should end up resembling galaxies as we know them at the present epoch.

The essential ingredients of the **galaxy population models** are the initial mass function of the stars and a grid of models for the evolution of stars of different masses (see e.g. Rocca-Volmerange and Guiderdoni 1987). For very young galaxies most of the luminosity is associated with massive stars on the main sequence. The results of these modelling exercises are clearly sensitive to the input parameters but a useful approximation described by Baron and White (1987) is that a star formation rate of $1 M_\odot$ per year results in a luminosity of $2.2 \times 10^9 L_\odot$ provided the initial mass function is similar to the standard mass function. The net result is that a galaxy of mass M is expected to have an initial luminosity of

$$L = 2.2 \times 10^{12} \left(\frac{M}{10^{11} M_\odot}\right) \left(\frac{10^8 \mathrm{y}}{t_{form}}\right) L_\odot \tag{111}$$

where t_{form} is the time during which the initial burst of star formation takes place.

The initial luminosities of galaxies in the different models of galaxy formation described above can be estimated using this result. In a scenario in which galaxies form rapidly on a collapse time scale, $t_{coll} \approx 10^8$ years, galaxies such as our own, for which $M \approx 10^{11} M_\odot$, would be very luminous objects, $L \approx 10^{12} L_\odot$. This may be compared with the typical luminosity of galaxies, as characterised by the luminosity L^* in the standard Schechter luminosity function of galaxies

$$dN(L/L^*) = \phi^* \left(\frac{L}{L^*}\right)^{-1.25} \exp\left(-\frac{L}{L^*}\right) \frac{dL}{L^*} \tag{112}$$

Recent determinations show that $L^* \approx 1.6 \times 10^{10} h^{-2} L_\odot$. Therefore, in the picture of rapid galaxy formation, the primaeval galaxies would be at least a factor of 10 more luminous than the typical galaxy observed at the present epoch. This type of picture might be applicable to the Hot Dark Matter model. These galaxies would be observable at redshifts $z \approx 5$ or greater.

In the Cold Dark Matter picture the galaxies can be assembled over a much longer period out of smaller mass structures. Therefore, at any epoch, the luminosities of the galaxies are expected to be much smaller than those in the rapid galaxy formation picture. For example, if the galaxy builds up its stellar populations over a cosmological time-scale, the age of the galaxy at a redshift of 1 would be about $5 \times 10^9 h^{-1}$ years and therefore the mean luminosity of the galaxy, according to equation (111), would be only $4.4 \times 10^{10} h^{-1} L_\odot$ which is not so different from L^*. It would be difficult to detect these galaxies at very large redshifts.

What would the spectra of primaeval galaxies look like? In a simple picture they may be thought of as gigantic versions of the regions of ionised hydrogen in which star formation is taking place in our own Galaxy. Their spectra would therefore be flat with an abrupt cut of at the Lyman limit. There would be strong narrow emission lines associated with the excitation of the ambient gas by the hot young stars. The whole spectrum would be redshifted by a factor of $(1+z)$ to longer wavelengths. Thus, if the epoch when the first generation of stars formed occurred at a redshift of, say, 5, most of the radiation would be emitted in the red and infrared regions of the spectrum. If the epoch of formation was 10, that the primaeval galaxies should be searched for in the infrared region of the spectrum at $\lambda \gtrsim 1\mu$m. Searches have been carried out in the optical waveband and no convincing candidates for primaeval galaxies have been found. These searches are now possible in the infrared wavebands with the development of sensitive infrared cameras. These observations will soon be able to set firm limits to the length of the period during which the bulk of the star formation activity in galaxies took place.

It is interesting that some of the large redshift radio galaxies have properties which strongly resemble what might be expected of a primaeval galaxy. Spinrad and his colleagues (1987) have made beautiful observations of the most distant radio galaxies and they appear to have distinctly different properties from those observed at redshifts $z \lesssim 1$. Most of the strong radio galaxies at a redshift of about 2 appear to be associated with large clouds of ionised hydrogen which are intense emitters of Lyman$-\alpha$ radiation. These ionised gas clouds extend to scales of up to about 100 kpc about the radio galaxy and there is evidence for large amounts of star formation in these galaxies from observations of their optical-infrared colours. In a number of cases there is good evidence that the galaxies are undergoing mergers. The exact status of these galaxies in the scheme of galactic evolution is not clear but it seems beyond question that, in this class of highly luminous galaxy, active star formation is proceeding at redshifts $z \approx 2$ and that the stars are forming out of the huge gas clouds associated with the radio galaxies (see Section 6.4.5).

6.4 The Evolution of Galaxies and Quasars with Cosmic Epoch

An important development over recent years has been the capability to study selected samples of galaxies and quasars at cosmological epochs significantly earlier than the present. These observations have provided convincing evidence for the astrophysical evolution of different classes of object with cosmological epoch and hence a number of important clues about the early evolution of galaxies and quasars. It is useful to recall the relation between cosmic time and redshift (Fig. 12). Marked on that diagram are the redshifts to which different classes of object can be observed more or less at the limits of present capabilities. Normal galaxies such as our own can only be studied out to redshifts of about 0.5. The giant galaxies such as the brightest galaxies in clusters and the radio galaxies can be observed to redshifts of 1 and greater if they have prominent emission lines in their spectra. The most distant radio galaxies known have redshifts up to about 4 meaning that they emitted their light when the Universe was about one fifth of its present age. The quasars span a similar range in redshift, the largest redshift so far measured being 4.43. It is not at all unexpected that the properties of objects observed at redshifts of 1 and greater should differ significantly from those observed nearby and this is indeed found to be the case from many separate pieces of evidence. For all classes of object which we can observe at redshifts of about 0.5 and greater, there are significant changes in their properties relative to nearby objects.

6.4.1 Normal Galaxies

It is right that we should begin with the majority galactic population of the Universe, the normal galaxies. Counts of galaxies now extend to very faint magnitudes and there is good agreement among the observers about the nature of the number counts. In his review, Ellis (1987) has shown that counts made in the blue waveband show a significant excess of faint galaxies as compared with the expectation of uniform world models in which the comoving numbers and spectra of galaxies do not change with cosmic epoch (Fig. 23). The excess of faint galaxies appears to be smaller in surveys carried out in the red waveband. A second important point is that there appears to be a real scatter in the counts. At least part of this scatter must arise because of the clumping of galaxies along the line of sight illustrated in Fig. 4, i.e. the number of galaxies fluctuates depending upon the line of sight which one chooses through the spongy large scale distribution of galaxies. The third point, which is likely to account for the first result, is that spectroscopic surveys of faint galaxies have shown that they are significantly bluer than expected. Typically these galaxies have redshifts about 0.5 and the effect can be interpreted as the result of enhanced star formation activity at a redshift of 0.5 as compared with zero redshift.

A similar effect is found in the properties of galaxies which are members of rich clusters. Butcher and Oemler (1978, 1984) first noted that, with increasing redshift, the fraction of blue galaxies increases. This effect has been much discussed and, although it is possible to find clusters which show little blue excess at moderate redshifts, there is no question but that there appear to be more blue galaxies at large redshifts. The cause of this is not clear. In some cases it may be due to the presence of active galaxies but the most likely explanation is that there is again more star formation activity at large redshifts. In the case of the rich clusters of galaxies, it is not surprising to see these effects because, as discussed in Section 6.1 above, for all but the very richest relaxed clusters, there have not been more than a few crossing times during the life of the cluster by a redshift of 0.5 and therefore some traces of the formation process might be expected.

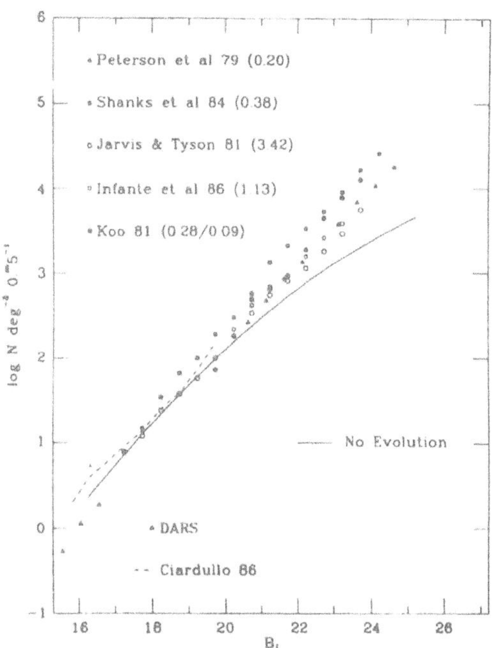

Fig. 23. Recent counts of faint galaxies in the B_J waveband compared with the expectations of a uniform world model (Ellis 1987).

6.4.2 Extragalactic Radio Sources

The counts of radio sources and the V/V_{max} test have long provided convincing evidence for strong changes in the evolutionary history of the radio source population. There have been significant advances in recent years as more complete samples of sources have been identified and the fraction of these sources for which redshifts have been measured has increased thanks largely to the heroic efforts of Spinrad and his colleagues (1987). Another important realisation has been the importance of the infrared waveband $1-2\mu m$ for identifications and photometry of radio sources as discussed in Section 2.5.3. The remarkable result of these developments is that, for most of the samples of bright sources, it is now possible to identify all the radio sources either in the optical or infrared wavebands.

All the radio, infrared, optical and redshift data have been used to construct models of the evolution of the luminosity function of radio sources with cosmic epoch. The free-form modelling techniques of Peacock (1985) have been extended by Dunlop (1987) to include all the new data. A sample of his results is shown in Fig. 24 in which the evolution of the radio luminosity function of the radio sources with redshift is displayed. The models for both steep and flat spectrum radio sources show the same rapid increase in their comoving space densities with increasing redshift followed by a rapid decline at the largest redshifts. Another important development has been that, with the increased numbers of identifications and redshifts, it is possible to determine the evolution of the comoving space density of sources directly. Excellent agreement between these observations and the free-form models has been found by Dunlop (1987).

Another interesting recent result which Dunlop (1987) has derived from these data is the ability of pure luminosity evolution of the strong radio source population to account

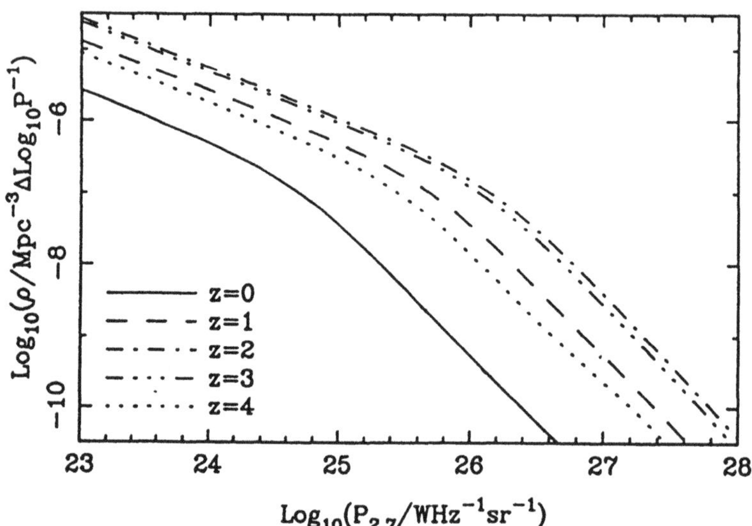

Fig. 24. Changes in the form of the radio luminosity function per unit comoving volume with redshift. This model can account for all the present observations and is described in terms of luminosity-evolution in which the luminosity function of the strong radio sources is shifted to higher radio luminosities at early epochs. The maximum luminosity occurs at redshifts $z \sim 2-3$. At larger redshifts, the average luminosity decreases from the maximum value (Dunlop 1987).

for the observed changes in the form of the luminosity function, as previously suggested for the radio-quiet quasars (see Section 6.4.3). The radio luminosity function can be split into two components, one consisting of intrinsically weak radio sources associated with normal and Seyfert galaxies and the other consisting of the powerful extended and compact radio sources which typify the strong radio source phenomenon. Shifting the strong radio source luminosity function purely in radio luminosity whilst the normal galaxy function remains unchanged can account very well for all the observations. The sense of these changes is indicated in Fig. 24. Interpreted literally, the evolution of the radio luminosity function suggests that the sources were at their most luminous at a redshift of about 2 - 3. Running the clocks forward, the mean radio luminosity of the sources increased to a maximum value when the Universe was about 25% of its present age and it subsequently declined to its present value. The interpretation of this behaviour requires much more study of the astrophysics of strong radio sources which is still not understood in predictive quantitative detail.

One way in which these results impinge directly upon the formation of galaxies is that it is now well-established that the strong radio sources are only associated with the most luminous galaxies. It is not understood why this occurs but the enhanced level of radio source activity at $z \sim 2-3$ might be associated with the fact that the most massive galaxies are only assembled rather late in the Universe. This has a natural interpretation in either the Cold or Hot Dark Matter pictures. In the former, the clustering of small-scale structures may only reach the size of giant galaxies at a redshift of about 3; in the latter, the pancakes only collapse at late epochs.

6.4.3 Radio Quiet Quasars

Similar evolutionary behaviour over cosmological time scales has been established for the radio quiet quasars. They display a steep number count and the changes of the optical luminosity function with cosmic epoch has been derived by Boyle and his colleagues (Boyle *et al.*, 1987). At zero redshift, the luminosity function of the overall quasar population joins smoothly onto the luminosity function of Seyfert galaxies, an important but natural continuity of the properties of these classes of object. With increasing redshift, there are many more quasars of a given luminosity than would be expected. The range of redshifts over which the analysis of Boyle *et al.* can be carried out is limited to $0 < z < 2$ as a result of their colour selection criteria for radio-quiet quasars. Over this redshift range, the change in the luminosity function is very similar to that of the radio luminosity function for radio sources shown in Fig. 24.

Systematic surveys have been made to determine complete samples of larger redshift quasars by a number of authors (Schmidt, Schneider and Gunn, 1987, Warren *et al.* 1987). It is relatively easy to find large numbers of quasars with redshifts up to about 3 but it has proved very difficult to discover quasars with redshifts greater than 4. Suddenly, in mid-1986, the breakthrough occurred and now 6 quasars with redshifts greater than 4 are known. These have been found in systematic surveys but there has also been an element of chance. The second largest redshift quasar was found accidentally on a long slit spectrogram while another object was being observed. These data provide important information about the large redshift behaviour of the luminosity function of quasars. Despite the discovery of quasars with redshifts greater than 4, there still appears to be a significant deficit of large redshift objects. The data suggest that there may well be a decline in the large redshift behaviour of the quasar population similar to that established for the radio sources.

6.4.4 The Stellar Populations of Distant Radio Galaxies

The problem with studying the evolutionary behaviour of radio sources and quasars is that their astrophysics is not secure enough for detailed studies of the implications of the observed evolution to be made. In contrast, there are good reasons to suppose that the evolution of the stellar content of these galaxies may be more susceptible to quantitative analysis. The reason radio galaxies are so important is that they are the only stellar systems available in reasonable numbers at redshifts greater than one for which studies of their stellar populations can be made. This programme would not have been possible without the spectra of these very faint objects which can be as faint as $R = 24.5$ and Spinrad's perseverance has been well rewarded by the discovery that these objects have strong narrow emission lines which makes the redshift determination possible. Another reason why the radio galaxies are so useful is that they form complete statistical samples with well defined selection criteria with the result that, although there might be astrophysical biases within the sample, because the samples are completely identified, the selection criteria are well defined.

The other key point is the importance of the near infrared waveband for these studies. First, the typical spectrum of a giant elliptical galaxy peaks at about 1μm and therefore, when observed at large redshifts, most of the energy is shifted into the infrared waveband, $1 - 2\mu$m. Thus, it is relatively easier to detect very distant radio galaxies at 2μm as compared with the optical waveband. The second point is astrophysical and that is that the stars which contribute most of the light in the infrared waveband are stars belonging to the cool red giant branch. These stars are derived from the oldest stellar populations of the galaxy. Therefore, when the integrated light of the galaxy is measured, the evolution of the stellar population of the galaxy is averaged over cosmological timescales. This contrasts strongly with what is observed in the optical region of the spectrum in which much of the light can be contributed by young stellar populations which are still undergoing their main sequence evolution. Thus, the optical light of a galaxy can be strongly influenced by bursts of star formation occurring throughout the life of the galaxy whereas the infrared observations sample the majority old stellar population.

These expectations are borne out in practice - the redshift-magnitude relation in the K waveband (2.2μm) is very tight out to redshifts of 1.5 or more (Fig. 13) whereas the R magnitude-redshift relation shows a wider dispersion in the observed magnitudes at redshifts greater than about 0.5. This confirms the theoretical expectations which are illustrated by the various models discussed by Spinrad (1987). The K magnitude - redshift relation supplemented by optical-to-infrared colours and optical spectroscopy of the galaxies provide a picture in which the galaxies have been undergoing passive evolution superimposed upon which there have been bursts of star formation, possibly associated with the events which gave rise to the radio sources. By passive evolution, we mean the underlying evolution of the primordial stellar populations which occur as various classes of star evolve off the main sequence and become red giants. It turns out that the expected evolution of the K-luminosity of a giant elliptical galaxy can be worked out in a remarkably model independent way because the red giant branches for stars with mass roughly that of the Sun are remarkably similar. On very general grounds, it is expected that the galaxies should be about 1 magnitude brighter at a redshift of 1 as compared with their luminosities at the present day. Let us demonstrate how this comes about.

The $K-$luminosity of the galaxy is proportional to the number of red giant stars N_{RG}. If τ_{RG} is the average lifetime of stars on the giant branch, then

$$N_{RG} = \left(\frac{dN}{dM}\right)_{MS}\left(\frac{dM_{to}}{dt}\right)\tau_{RG}$$

where $N(M)$ is the initial mass function of stars on the main sequence and M_{to} is the main sequence termination point at which stars move off the main sequence onto the giant branch. If we make the assumption that the mass-luminosity relation is of power-law form, $L \propto M^{\gamma}$, the lifetime of the star on the main sequence t_{to} is proportional to the ratio of the available energy, which is proportional to the mass of the star, divided by the energy loss rate which is just the luminosity of the star L, i.e. $t_{to} \propto M^{-(\gamma-1)}$ If the initial mass function is of power-law form, $dN/dM \propto M^{-x}$, then the luminosity of the red giant branch is

$$L \propto N_{RG} \propto t^{-\left(\frac{\gamma-x}{\gamma-1}\right)}$$

For stars with $M \sim M_{\odot}$, appropriate values are $\gamma = 4, x = 1$ and hence $L \propto t^{-1}$. At a redshift $z = 1, t = t_0/2$ if $\Omega = 0$ and $t = \frac{2}{3}t_0$ if $\Omega = 1$. Therefore, the galaxy should be about a factor of 2 to 3 more luminous at a redshift of 1 as compared with the present epoch, i.e. about a magnitude brighter. Obviously, this simple argument can be made much more precise by more detailed modelling.

As discussed earlier in Section 2.5.4, we have used these models in conjunction with our observed K–magnitude - redshift relation to solve for q_0 (Lilly and Longair 1984). The value of q_0 probably lies somewhere in the range $0.1 < q_0 < 0.9$. From my perspective, the important point is that the types of evolutionary change which must have taken place can now be observed and, with large enough statistics, it should be possible to obtain improved estimates of this evolution and possibly of q_0.

A fascinating example of how observations of this type can provide important information related to galaxy formation is the recent observation of the very distant radio galaxy 0902+34 which has redshift 3.4 (Lilly 1988). In addition to measuring its redshift, Lilly obtained photometry at optical and infrared wavelengths which showed clearly the presence of a giant branch in the broad-band galaxy spectrum. Similar spectra have been found by him in other very faint radio galaxies (Lilly 1989). By fitting galaxy population models to these observations, he was able to estimate the age of the stellar population of the galaxy as 2×10^9 years. If the population were any younger, the galaxy would be much brighter in the optical relative to the infrared waveband. Let us adopt as long time-scales as possible. If we assume $\Omega = 0, h = 0.5$, the Universe was only 4.5×10^9 years old at a redshift of 3.4 and therefore the stellar population must have formed when the Universe was only 2.5×10^9 years old, i.e. at a redshift of 7. In other words, the stellar population must have formed when the Universe was only about one-tenth of its present age. In principle, this line of reasoning can be used to constrain the parameters of cosmological models. For example, if $\Omega > 0, h = 1$, the Universe would be less than 2.25×10^9 years old at $z = 3.4$ and there is barely time for the stellar population to have formed. This analysis is indicative of the type of information which can be derived from observations of the stellar populations of very distant galaxies.

6.4.5 The Lyman-α Galaxies

The success of Spinrad and his colleagues in measuring the redshifts of these very distant radio galaxies has been due to the fact that they possess strong, narrow emission lines in their spectra. These are ideal for studying the astrophysics of galaxies because the strong narrow lines enable the redshift to be found without contaminating the underlying stellar continuum radiation. Among the most remarkable discoveries of this programme has been that of **Lyman–α galaxies** which have such large redshifts, $z > 1.6$, that the Lyman–α line is redshifted into the accessible optical waveband (Spinrad 1987, Djorgovski 1987). These galaxies have enormously powerful Lyman–α emission lines and, in addition, the Lyman–α emission does not come from the nuclear regions of the galaxies but from a region which can extend well beyond the confines of the galaxy. In some cases, the extended Lyman–α emission has the same extent as the blue continuum radiation which is inferred to be associated with active star formation. These results seem to provide a consistent picture in which there is extensive star formation occurring over the bulk of the galaxy and possibly outside its normal confines.

One of the most remarkable results of these studies has been that the radio structure of the double sources seems to be aligned with the major axis of the Lyman–α cloud which introduces a remarkable new feature into the relation between the optical and radio activity in these objects (McCarthy *et al.* 1987). Is the star formation stimulated by the formation of the radio source? Does the infall of material into the galaxy define the axis along which the radio jets are ejected?

Whilst pursuing these studies, the unidentified sources 3C 326.1 and 3C 294 were observed and, to the observers' surprise, they were found to be strong Lyman$-\alpha$ sources, with redshifts 1.825 and 1.779 respectively. These **Giant Lyman$-\alpha$ Clouds** are very extended and have large internal velocity dispersions. The ionising spectrum appears to be soft. Exactly what these objects are is a matter of the greatest interest. One interpretation is that they are galaxies in the process of formation. It should be a relatively easy matter with the new generation of infrared array cameras to find the galaxies which are presumably underlying these Giant Clouds. These are the first really strong candidates for any class of galaxy at an early stage in their evolution.

7 Conclusion

The present survey is a gentle introduction to many of the most important developments in our understanding of the formation of galaxies. It should be noted that there are many topics which I have not covered but which should form part of a more complete survey. Among topics which merit proper discussion I would include the re-ionisation of the intergalactic gas, the nature of the absorption line systems observed in distant quasars and their role in cosmology, the origin of the angular momentum of galaxies and many other theoretical topics. The books listed below cover many of these topics.

The survey also highlights areas which are ripe for exploration by the new observing facilities. It is unnecessary to list all the observations one would like to make of direct relevance to the problems of galaxy formation. What is striking about the subject is that many questions have arisen which can be answered by observation if the time is made available for their study by the present and next generation of telescopes. Many key programmes which are now at the very limit of what is possible should become quite straightforward with the new generation of telescopes. Prime among these facilities will be the **Hubble Space Telescope**, the **Infrared Space Observatory**, the **Advanced X-ray Astronomy Facility** and the coming generation of **Very Large Telescopes**. It will be intriguing to compare what are currently seen to be the major issues of galaxy formation with the position in ten years time.

Acknowledgements

I am particularly grateful to Alan Heavens, John Peacock and Sergei Shandarin for reading parts of this survey and offering very helpful comments. John Peacock very kindly allowed me to use his lecture notes in preparing some of the material for these lectures. Some of the text is based upon joint work with Rashid Sunyaev which dates from 1975–1980. I am very grateful for his contribution to clarifying my own ideas. I should emphasise that the responsibility for the contents of the written version of these lecture notes and for any misrepresentations or expressions of personal opinion is entirely my own.

The following books and reviews are recommended for many more details of the topics discussed in the lecture notes.

References

BOOKS AND REVIEWS

1. Observational Cosmology, J.E. Gunn, M.S. Longair and M.J. Rees. 8th Advanced Course, Swiss Society of Astronomy and Astrophysics, Saas-Fee, Geneva Observatory, 1978.
2. The Large Scale Structure of the Universe, P.J.E. Peebles, Princeton University Press, 1980.
3. The Cosmic Distance Ladder: Distance and Time in the Universe, M. Rowan-Robinson, Freeman, New York, 1985.
4. Theory and Experiment in Gravitational Physics, C.M. Will, Cambridge University Press, 1981.
5. Theoretical Concepts in Physics, M.S. Longair, Cambridge University Press, 1984.
6. Gravitation and Cosmology, S. Weinberg, John Wiley and Sons, 1972.
7. A.A. Friedman, E.A. Tropp, B.Ya. Frenkel and A.D. Chernin, Nauka, Moscow, 1988.
8. The Anthropic Cosmological Principle, J.D. Barrow and F.J. Tipler, Oxford University Press, 1986.
9. The Formation of Galaxies, G. Efstathiou and J. Silk, Fundamentals of Cosmic Physics, 9, 1 - 138, 1983.
10. The Formation of Galaxies and Large Scale Structure with Cold Dark Matter, G.R. Blumenthal, S.M. Faber, J.R. Primak and M.J. Rees, Nature, 311, 517.
11. Observational Cosmology, IAU Symposium No. 124. (eds. A. Hewitt, G. Burbidge and Fang Li Zhi), D. Reidel and Co., Dordrecht, 1987.
12. Nearly Normal Galaxies: from the Planck Time to the Present, (ed. S.M. Faber), Springer Verlag, New York, 1987.
13. Large Scale Structures of the Universe, IAU Symposium No. 130 (eds. J. Audouze, M-C. Pelletan and A.S. Szalay), Kluwer Academic Publishers, Dordrecht, 1988.
14. High Redshift and Primeval Galaxies, (eds. J. Bergeron, D. Kunth, B. Rocca-Volmerange and J. Tran Thanh Van), Edition Frontieres, 1987.

LITERATURE REFERENCES

15. Aaronson, M., 1987. IAU Symposium No. 124, op. cit., 187.
16. Audouze, J., 1987. IAU Symposium No 124, op. cit., 89.
17. Bahcall, N.A., 1988. Ann. Rev. Astr. Astrophys., 26, 631.
18. Baron, E. and White, S.D.M., 1987. Astrophys. J., 322, 585.
19. Barrow, J.D., 1983. Fund. Cosmic Phys., 8, 83.
20. Boyle, B.J., Shanks, T., Fong, R. and Peterson, B.A., 1987. IAU Symposium No. 124, op. cit., 643.
21. Burstein, D., Davies, R.L., Dressler, A., Faber, S.M., Lynden-Bell, D., Terlevich, R.J. and Wegner, G.A. 1986. Galaxy Distances and Deviations from Hubble Expansion, (eds. B. Madore and R.B. Tully), 123, D. Reidel and Co., Boston.
22. Burstein, D., Davies, R.L., Dressler, A., Faber, S.M., Stone, R.P.S., Lynden-Bell, D., Terlevich, R.J. and Wegner, G.A., 1987. Astrophys. J. Suppl., 64, 601.
23. Butcher, H. and Oemler, A., 1978, Astrophys. J., 219, 18.
24. Butcher, H. and Oemler, A., 1984. Nature, 310, 31.
25. Canizares, C., 1987. IAU Symposium No. 124, op cit., 729
26. Clowes, R.G., Savage, A., Wang, G., Leggett, S.K., MacGillivray, H.T. and Wolstencroft, R.D., 1987. Mon. Not. R. astr. Soc., 229, 27P.
27. Collins, C.A., Heydon-Dumbleton, N.H. and MacGillivray, H.T., 1988. Mon. Not. R. astr. Soc., (in press).
28. Collins, C.A., Joseph, R.D. and Robertson, N.A., 1986. Nature, 320, 506.
29. Davies, R.D., Lasenby, A.N., Watson, R.A., Daintree, E.J., Hopkins, J., Beckman, J., Sanchez-Almeida, J. and Rebolo, R., 1987. Nature, 326, 462.
30. Dekel, A., 1987. Comments Astrophys. 11, 235.
31. Dekel, A. and Rees, M.J., 1987. Nature, 326, 455.
32. Doroshkevich, A.G., Sunyaev, R.A. and Zeldovich, Ya.B., 1974. Confrontation of Cosmological Theories with Observational Data, IAU Symposium No. 63, (ed. M.S. Longair), 213, D. Reidel and Co., Dordrecht.

33. Dunlop, J.S., 1987. Ph.D. Dissertation, University of Edinburgh.
34. Dunlop, J.S. and Longair, M.S., 1987. High Redshift and Primeval Galaxies, op. cit., 93.
35. Djorgovski, S., 1988. Towards Understanding Galaxies at Large Redshifts, (eds. A. Renzini and R. Kron), D. Reidel and Co., Dordrecht,259.
36. Ellis, R., 1987. IAU Symposium No. 124, op. cit., 367.
37. Fowler, W.A., 1987. Q. Jl. R. astr. Soc., 28, 87.
38. Frenk, C.S., 1986. Phil. Trans. R. Soc. Lond. A 330, 517.
39. Geller, M.J., Huchra, J.P. and de Lapparent, V., 1987. IAU Symposium No. 124, op. cit., 301.
40. Gott, J.R., 1987. IAU Symposium No. 124, op. cit., 433.
41. Gower, J.F.R., Scott, P.F. and Wills, D., Mem. R. astr. Soc., 71, 49.
42. Hawking, S.W., 1975. Comm. Math. Phys., 43, 199 and Quantum Gravity: An Oxford Symposium eds. C.J. Isham, R. Penrose and D.W. Sciama, 219, Oxford University Press.
43. Hawkins, M.R.S. and Bessell, M.S., 1988. Mon. Not. R. astr. Soc., 234, 177.
44. Hewitt, J.N., Turner, E.L., Burke, B.F., Lawrence, C.R., Bennett, C.L., Langston., G.I. and Gunn, J.E., 1987. IAU Symposium No. 124, op cit., 747.
45. Jones, B.J.T. and Wyse, R.F.G., 1985. Astron. Astrophys., 149, 144.
46. Kaiser, N., 1984. Astrophys. J., 284, L9.
47. Koo, D.C. and Kron, R.G., 1988. Towards Understanding Galaxies at Large Redshifts, (eds. A. Renzini and R.G. Kron), D. Reidel and Co., Dordrecht, 209.
48. Lifshitz, E.M., 1946 J. Phys., USSR Academy of Sciences, 10, 116.
49. Lilly, S.J., 1988. Astrophys. J., 333, 161.
50. Lilly, S.J., 1989. Astrophys. J., (in press).
51. Lilly, S.J. and Longair, M.S., 1984. Mon. Not. R. astr. Soc., 211, 833.
52. Longair, M.S. and Sunyaev, R.A., Uspekhi Fiz. Nauk, 105,41 (English translation: Soviet Physics Uspekhi, 14, 569).
53. Lynden-Bell, D., Faber, S.M., Burstein, D., Davies, R.L., Dressler, A., Terlevich, R.J. and Wegner, G., 1988. Astrophys. J., 325, 19.
54. Lyubimov, V.A., Novikov, E.G., Nozik, V.Z., Tretyakov, E.F. and Kosik, V.S., 1980. Phys. Lett., 94B, 266.
55. Matsumoto, T., Hayakawa, S., Matsuo, H., Murakami, H., Sato, S., Lange, A.E. and Richards, P.L., 1988. Astrophys. J., 329, 567.
56. McCarthy, P.J., van Breugel, W., Spinrad, H. and Djorgovski, S., 1987. Astrophys. J., 321, L29.
57. Melchiorri, F., Dall'Oglio, G., De Bernardis, P., Mandolesi, N., Masi, S., Moreno, G., Olivo, B. and Pucacco, G., 1986. Space-borne Sub-Millimetre Astronomy Mission, (ed. N. Longdon), ESA SP-260, 33.
58. Miley, G., 1987. IAU Symposium No. 124, op. cit., 267.
59. Miller, L. and Mitchell, P., 1988. Mon. Not. R. astr. Soc., (in press).
60. Ostriker, J.P. and Cowie, L.L., 1981. Astrophys. J., 243, L127.
61. Peacock, J.A., 1985. Mon. Not. R. astr. Soc., 217, 601.
62. Peacock. J.A. and Heavens, A., 1985. Mon. Not. R. astr. Soc., 217, 805.
63. Peebles, P.J.E., 1966. Astrophys. J., 146, 542
64. Peebles, P.J.E., 1968. Astrophys. J., 153, 1.
65. Perkins, D.H., 1987. Introduction to High Energy Physics, 1987. Addison-Wesley Publishing Company.
66. Pilkington, J.D.H. and Scott, P.F., 1965. Mem. R. astr. Soc., 69, 183.
67. Pozdnyakov, L.A., Sobol, I.M. and Sunyaev, R.A., 1983. Sov. Sci. Rev. E, Astrophys. Sp. Phys. Reviews, 2,
68. Rees, M.J. and Ostriker, J.P., 1977 Mon. Not. R. astr. Soc., 179, 541.
69. Rocca-Volmerange, B and Guiderdoni, B., 1987. High Redshift and Primeval Galaxies, op. cit., 239
70. Rowan-Robinson, M., 1989. Space Science Rev., (in press).
71. Rubin, V.C., Ford, W.K. and Rubin, J.S., 1976. Astrophys. J., 183, L111.
72. Sandage, A.R., 1968. Observatory, 88, 91.
73. Sandage, A.R., 1987. IAU Symposium No. 124, op.cit., 1.
74. Sandage, A.R., 1988. Ann. Rev. astr. astrophys., 26, 561.
75. Schmidt, M., Schneider, D.P. and Gunn, J.E., 1987. Astrophys. J., 321, L7.
76. Seldner, M., Siebars, B., Groth, E. and Peebles, P.J.E., 1977. Astron. J., 82, 249.
77. Shanks, T., Boyle, B.J.,Fong, R. and Peterson, B.A., 1987. Mon. Not. R. astr. Soc., 227, 739.
78. Silk, J., 1968. Nature, 215, 1155
79. Spinrad, H., 1987. IAU Symposium No. 124, op. cit., 129.
80. Spinrad, H., 1987. High Redshift and Primeval Galaxies, op. cit., 59.
81. Staveley-Smith, L., 1985. Ph.D. Dissertation, University of Manchester.

82. Strauss, M.A. and Davis, M., 1988. IAU Symposium No. 130, op. cit., 191.
83. Sunyaev, R.A. and Zeldovich, Ya.B., 1980. Ann. Rev. astr. astrophys., 18, 537.
84. Tammann, G.A., 1987. IAU Symposium No. 124, op. cit., 151.
85. Wagoner, R.V., 1973. Astrophys. J., 179, 343.
86. Wall, J.V. and Peacock, J.A., 1985. Mon. Not. R. astr. Soc., 216, 173.
87. Warren, S.J., Hewett, P.C., Osmer, P.S. and Irwin, M.J., 1987. Nature, 330, 453.
88. Webster, A., 1977. Radio Astronomy and Cosmology, IAU Symposium No 74, (ed D.L. Jauncey), 75, D. Reidel and Co., Dordrecht.
89. Weymann, R.J., 1965. Phys. Fluids, 8, 2112.
90. Wilkinson, D.T., 1988. IAU Symposium No 130, op cit., 7.
91. Wills, D. and Lynds, R., 1978. Astrophys. J. Suppl., 31, 143.
92. Yahil, A., Walker, D. and Rowan-Robinson, M., 1986. Astrophys. J., 301, L1.
93. Zeldovich, Ya. B., 1972. Mon. Not. R. astr. Soc.,160, 1P.
94. Zeldovich, Ya.B., 1986. Sov. Sci. Rev. E., Astrophys. Sp. Phys. Reviews, 5, 1.
95. Zeldovich, Ya. B., Kurt, V.G. and Sunyaev, R.A., 1968. Zh. Eksp. Teor. Fiz., 55, 278. (English translation: Soviet Physics - JETP, 28, 146.
96. Zeldovich, Ya.B. and Sunyaev, R.A., 1969. Astrophys. Sp. Phys., 4, 301.

Stellar Dynamics

James Binney
Department of Theoretical Physics, Oxford University

1 Introduction

These lectures fall into three parts: orbits, equilibrium models and perturbation theory. The lectures on orbits and equilibrium models follow closely §§3.1–3.5 and 4.1–4.5 of the book *Galactic Dynamics*,[1] reworked to fit into four lectures by omission of the least vital or most technical material, and abbreviation of some derivations; more detail on most topics will be found in GD. The last two lectures introduce the student to perturbative galactic dynamics.

This last topic has been for too long the exclusive preserve of a small coterie of professional dynamicists. In practice physics proceeds by first solving exactly a small number of idealized problems, and then perturbing these solutions into approximate models of systems of real physical interest. Hence nearly all field theory is based on perturbation of the harmonic oscillator, much of statistical mechanics is concerned with perturbed ideal gases, and so forth. Hence the failure over the years of stellar dynamicists to emphasize adequately the methods of Hamiltonian perturbation theory may in some measure account for the disappointingly slow progress of our understanding of galaxies compared, for example, with the explosive growth in our knowledge of atomic and subatomic structure that has been achieved over the same period. I hope that this situation may be in some small measure ameliorated by my devoting a third of an introductory course in stellar dynamics to perturbation theory. Since perturbation theory is most conveniently formulated in terms of action-angle variables, the material selected from GD for the first four lectures gives relatively more space to these coordinates than does the book itself.

2 The Continuum Approximation

The stellar components of galaxies appear to be rather smooth. This suggests that we model their gravitational force-fields by those generated by continuous density distributions $\rho(\mathbf{x})$ equivalent to their actual star densities. Clearly this is only an approximation to the truth, and our first task must be to estimate the time during which a stellar orbit in such a smooth mass distribution yields a decent approximation to the true motion.

---

1 *Galactic Dynamics*, J. Binney & S. Tremaine, Princeton University Press, 1987 – hereafter "GD".

We consider first an infinite homogeneous system of identical stars, mass m and number density n. A test star moves at velocity $\mathbf{v}$ with respect to the other stars, which we imagine to be stationary. If we replace this system's discrete mass density with its smoothed value $\rho = nm$, the gravitational force vanishes everywhere and the test star's orbit is a straight line. Actually the test star is pulled now this way, now that as it passes each background star. Since the star's velocity is approximately constant we can estimate impulse the star receives perpendicular to its unperturbed path each time it passes a background star by integrating the gravitational force between the two stars along the unperturbed orbit. It is straightforward to show (see GD §4.0) that during a passage at impact parameter b the test star acquires transverse velocity satisfying

$$|\delta\mathbf{v}_\perp| \simeq \frac{2Gm}{bv}. \tag{1}$$

Since the background stars are randomly distributed in space, these velocity changes add incoherently. After time t the cumulative perturbation $|\Delta\mathbf{v}_\perp|_t$ satisfies

$$\begin{aligned}|\Delta\mathbf{v}_\perp|_t^2 = \sum|\delta\mathbf{v}_\perp|^2 &= \frac{4G^2m^2t}{v^2}\int\frac{nv2\pi b\mathrm{d}b}{b^2}\\ &= \frac{8\pi G^2m^2nt}{v}\int\frac{\mathrm{d}b}{b}.\end{aligned} \tag{2}$$

The integral over impact parameter b diverges at small b because we have integrated the forces along the unperturbed trajectory and this procedure becomes meaningless for impact parameters smaller than that, $b_{\min} \equiv Gm/v^2$, associated with scattering through a non-negligible angle. This difficulty can be entirely eliminated by employing exact Keplerian trajectories—see §7.1 of GD for details. For our order-of-magnitude calculation it suffices to simply trucate the integral at $b_{\min}$. The divergence of the integral in equation (2) at large b is more interesting. Clearly, in a real system the largest impact parameter $b_{\max}$ cannot exceed the characteristic size R of the system. Indeed, one might think that b should not exceed the mean interparticle separation $n^{-1/3}$. Careful examination of this question leads to the conclusion that $b_{\max} \simeq R$ (Spitzer 1987, §2.1). Defining the **Coulomb logarithm** by

$$\ln\Lambda \equiv \ln(b_{\max}/b_{\min}) \simeq \frac{Rv^2}{Gm}, \tag{3}$$

we conclude that $|\Delta\mathbf{v}_\perp|_t^2 \simeq v^2$ after a time

$$t_{\text{relax}} = \frac{v^3}{8\pi(Gm)^2n\ln\Lambda}. \tag{4}$$

According to the virial theorem (§4.3 of GD),

$$v^2 \simeq \frac{GmN}{R}, \quad \text{where} \quad N \equiv \tfrac{4}{3}\pi nR^3.$$

Thus the ratio of the **relaxation time** t_{relax} to the crossing time $t_{\text{cross}} \equiv R/v$ is

$$\frac{t_{\text{relax}}}{t_{\text{cross}}} = \frac{v^3N\frac{4}{3}\pi R^3}{8\pi(v^2R)^2\ln\Lambda}\frac{v}{R} = \frac{N}{6\ln\Lambda}. \tag{5}$$

Typically $\ln\Lambda$ lies in the range $5 \lesssim \ln\Lambda \lesssim 25$. So in systems in which N exceeds a few hundred, the continuum approximation should yield orbits that are valid for several crossing times. $N \gtrsim 10^5$ in a globular cluster and $N \gtrsim 10^{10}$ in a galaxy.

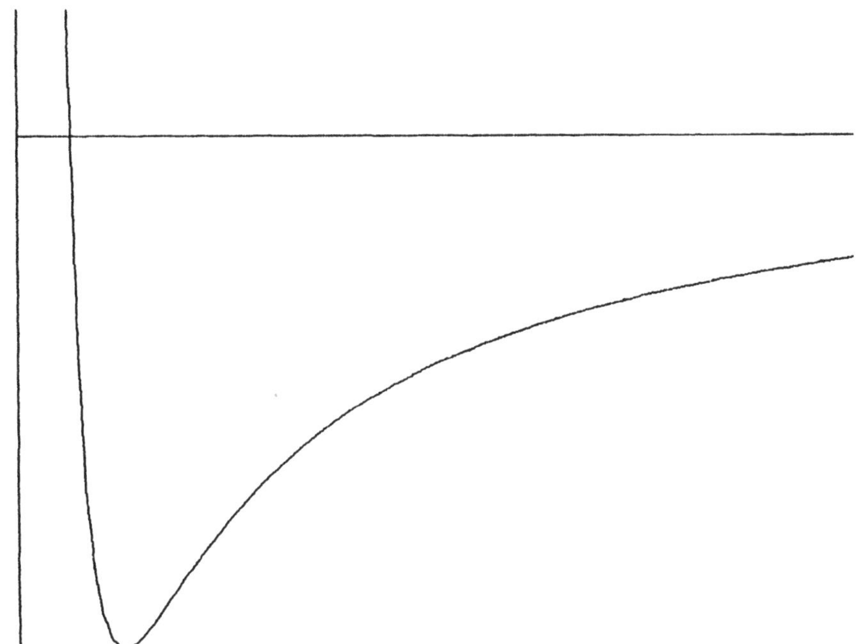

Fig. 1. The effective potential $\Phi_{\rm eff}$ is very anharmonic. Here $\Phi_{\rm eff}$ is plotted for the case $\Phi \propto 1/r$.

3 Orbits

3.1 Orbits in spherical potentials

A review of the structure of orbits in spherical potentials serves to introduce the important concept of an isolating integral.

The equation of motion is

$$\ddot{\mathbf{r}} = F(r)\hat{\mathbf{e}}_r. \tag{6}$$

Dotting through by $\mathbf{r}$ we show that the angular momentum vector $\mathbf{L} \equiv \mathbf{r} \times \mathbf{v}$ is conserved. Hence the motion occurs in a plane. Let (r, ψ) be polar coordinates in the orbital plane. In terms of these coordinates the equations of motion read

$$\begin{aligned} \ddot{r} - r\dot{\psi}^2 &= F(r) \\ 2\dot{r}\dot{\psi} + r\ddot{\psi} = 0 \;&\Rightarrow\; r^2\dot{\psi} = L. \end{aligned} \tag{7}$$

Eliminating ψ from the first equation, we obtain an equation of one-dimensional motion:

$$\ddot{r} - L^2/r^3 = F(r). \tag{8}$$

Figure 1 is a plot of the effective potential

$$\Phi_{\rm eff} \equiv \Phi(r) + \frac{L^2}{2r^2}, \quad \text{where} \quad \Phi(r) \equiv \int_r^\infty F(r)\,\mathrm{d}r, \tag{9}$$

which governs this motion. From the asymmetry of this potential about its minimum we see that we are dealing with a thoroughly anharmonic oscillator.

The oscillator's energy equation is

$$\tfrac{1}{2}\dot{r}^2 + \Phi_{\rm eff} = \tfrac{1}{2}\dot{r}^2 + \Phi + \frac{L^2}{2r^2} \equiv E \quad \text{(a constant)}. \tag{10}$$

Since $\frac{1}{2}L^2/r^2$ is just the kinetic energy of the orbiting particle's tengential motion, E is simultaneously the energy of the anharmonic oscillator and of the underlying orbit.

If we eliminate t rather than ψ between equations (7), we obtain an equation that determines the orbit's shape. This equation is most conveniently written in terms of the variable $u \equiv 1/r$:

$$\frac{\mathrm{d}^2u}{\mathrm{d}\psi^2} + u = -\frac{F(1/u)}{L^2u^2}, \quad \text{where} \quad u \equiv \frac{1}{r}. \tag{11}$$

As an example of the use of this equation, consider the case of Keplerian motion. Then $F(r) = -GM/r^2$ and equation (11) becomes

$$\frac{\mathrm{d}^2u}{\mathrm{d}\psi^2} + u = \frac{GM}{L^2} \quad \Rightarrow \quad u = \frac{GM}{L^2}[1 + e\cos(\psi - \psi_0)]. \tag{12}$$

Here the **eccentricity** e and **semi-major axis** a are constants. According to equation (12), u is periodic in ψ with period 2π, and thus the orbit is closed. In fact, the orbits are elliptical with the centre of attraction at one focus. Physically, the particle's trajectory closes on itself because the period T_r of the anharmonic oscillator that describes the radial motion is exactly equal to the time required for the particle to orbit once around the centre. In terms of angular frequencies we express this circumstance by the equation $2\pi/T_r \equiv \omega_r = \omega_\psi$.

Consideration of motion in the harmonic potential

$$\Phi = \Phi_0 + \tfrac{1}{2}\Omega^2r^2 \tag{13}$$

shows that in general $\omega_r \neq \omega_\psi$. The equations of motion of the Cartesian coordinates now decouple: $\ddot{x} = -\Omega^2x$ etc, and have solutions

$$x = X\cos(\Omega t + \phi_x), \qquad y = Y\sin(\Omega t + \phi_y), \tag{14}$$

where X, Y and the ϕ_i are arbitrary constants. It follows that the orbits are ellipses centred on the centre of attraction. The particle completes two radial oscillations in the time taken to revolve once around the centre. In terms of frequencies $\omega_r = 2\omega_\psi$.

3.2 Integrals

We define an integral to be a function $I(\mathbf{x}, \mathbf{v})$ of the phase-space coordinates which is such that $\left.\frac{\mathrm{d}I}{\mathrm{d}t}\right|_{\rm orbit} = 0$. (Notice that we do not allow time to appear explicitly in I.) L_x, L_y, L_z and E are all integrals, as is any function of any number of integrals. Thus in a strict sense there are infinitely many different integrals. However, $I \equiv L_x + L_y$ is not independent of L_x and L_y since we can predict its value as soon as we know L_x and L_y. When I speak of the "number of integrals" I mean the number of integrals in the largest set of mutually independent integrals.

Since each independent integral imposes through I = constant a fresh constraint on the phase-space coordinates, we expect the number of integrals to determine the dimensionality of the orbit. Kepler orbits are one-dimensional, which suggets that these orbits admit five integrals. We know of only four; L_x, L_y, L_z and E. To find the fifth consider equation (12) rewritten as

$$r = \frac{a(1-e^2)}{1+e\cos(\psi-\psi_0)} \quad \text{where} \quad a \equiv \frac{L^2}{GM(1-e^2)} = -\frac{GM}{2E}. \tag{15}$$

Clearly the semi-major axis $a(E)$ and eccentricity $e(E,L)$ are integrals. Solving for ψ_0 we find that it is also an integral:

$$\psi_0(\mathbf{x},\mathbf{v}) = \psi - \arccos\left\{\frac{1}{e}\left[\frac{a}{r}(1-e^2)-1\right]\right\}. \tag{16}$$

As the case of Keplerian motion displays, the number of integrals cannot be smaller than $N \equiv (6 -$ dimensionality of orbit). However, the number of integrals can exceed N as the case of motion in the potential $\Phi = -GM\left(\frac{1}{r}+\frac{r_0}{r^2}\right)$ shows. Here r_0 is a constant. If you integrate orbits in this potential you will find they look like the orbit of Figure 2. The orbit fills a two-dimensional portion of real space. Furthermore, at each point in the orbit's annulus just two velocity vectors are allowed. Hence the orbit occupies a two-dimensional region of phase space also. Yet it is easy to show that these orbits still have a fifth integral ψ_0. Equation (11) now reads

$$\frac{\mathrm{d}^2u}{\mathrm{d}\psi^2} + \left(1-\frac{2GMr_0}{L^2}\right)u = \frac{GM}{L^2}. \tag{17}$$

As for Keplerian motion, this is the equation of a harmonic oscillator, but the frequency is no longer 2π. Solving for u yields

$$u = \frac{GM}{L^2}\left[K^2 + e\cos\left(\frac{\psi-\psi_0}{K}\right)\right], \quad \text{where} \quad K \equiv 1\Big/\sqrt{1-\frac{2GMr_0}{L^2}}. \tag{18}$$

From this it follows that ψ_0 is an integral. To see why it does not reduce the orbit's dimensionality, suppose we are given $\mathbf{L}$, E, ψ_0 and r and seek to find the corresponding value(s) of ψ. We have

$$\psi = \psi_0 + K\arccos\left\{\frac{1}{e}\left[\frac{a}{r}(1-e^2)-K^2\right]\right\}. \tag{19}$$

But the we can always add $2m\pi$ to the value of the arccos in this equation, and when we do so we add $2mK\pi$ to ψ. But K will almost always be irrational. So by adding $2m\pi$ to the arccos with sufficiently large m we can make ($\psi \bmod 2\pi$) approach any given value as closely as we please. Hence ψ_0 imposes no useful constraint on the particle's motion. Such an integral is said to be **non-isolating**. Only isolating integrals reduce the dimensionality of an orbit.

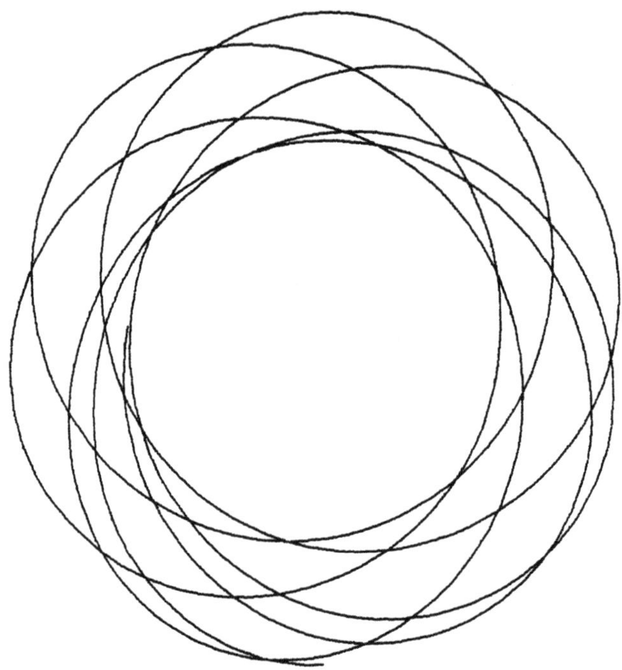

Fig. 2. Most orbits in central potentials form rosettes rather than closed curves.

3.3 Axisymmetric Potentials

Let (R, ϕ, z) be a system of cylindrical polar coordinates. Then writing $\mathbf{r} = R\hat{\mathbf{e}}_R + z\hat{\mathbf{e}}_z$, the equations of motion in an axisymmetric potential become

$$\frac{\mathrm{d}}{\mathrm{d}t}(R^2\dot{\phi}) = 0 \quad \Rightarrow \quad R^2\dot{\phi} = L_z, \text{ a constant,} \tag{20}$$

and

$$\left.\begin{aligned} \ddot{R} &= -\frac{\partial \Phi_{\text{eff}}}{\partial R} \\ \ddot{z} &= -\frac{\partial \Phi_{\text{eff}}}{\partial z} \end{aligned}\right. \quad \text{where} \quad \Phi_{\text{eff}} \equiv \Phi + \frac{L_z^2}{2R^2}. \tag{21}$$

Figure 3 shows isopotential contours of Φ_{eff} for the case of the potential

$$\Phi = \tfrac{1}{2}v_0^2 \ln\left(R^2 + \frac{z^2}{q^2}\right) \tag{22}$$

which generates a constant circular velocity. Φ_{eff} has a minimum at the radius of the circular orbit with angular momentum L_z.

Equations (21) reduce the orbit to motion in the meridional plane (R, z). The orbit's energy,

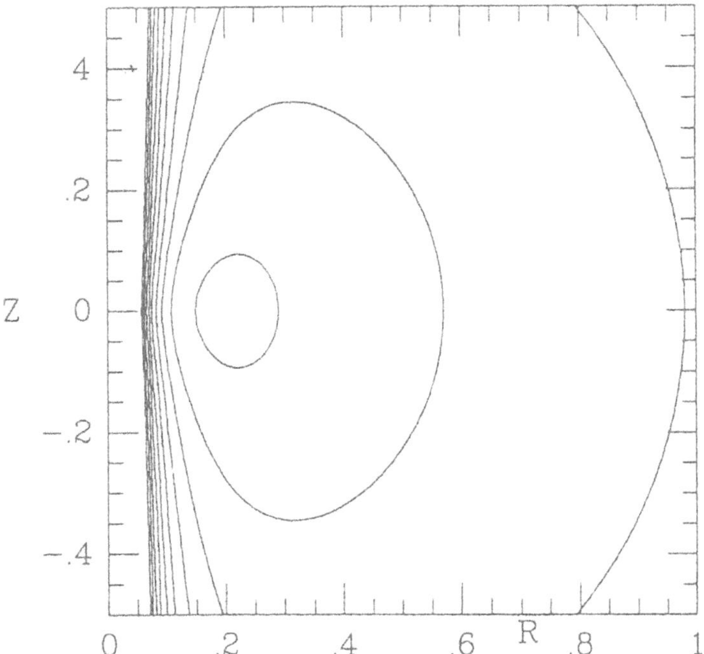

Fig. 3. Isopotential contours of the $\Phi_{\rm eff}$ defined by equations (21) and (22) with $L_z = 0.2$ and $q = 0.9$.

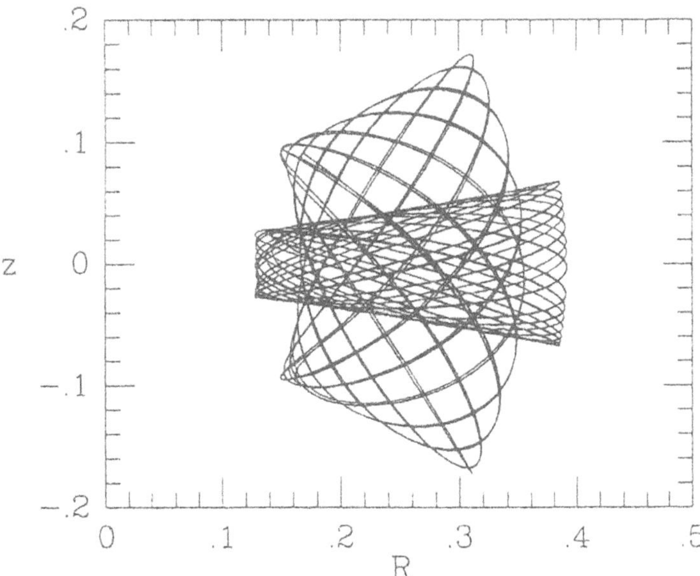

Fig. 4. Orbits in the effective potential Fig. 3 are constrained by an effective integral in addition to E; the two orbits have the same energy.

$$E = \tfrac{1}{2}(\dot{R}^2 + \dot{z}^2) + \Phi_{\rm eff} \tag{23}$$

is the only analytically available integral of this two-dimensional motion. However, Figure 4 clearly shows that orbits are constrained by some additional integral. What is it? As is often the case in physics, one can give a simple answer in two extreme cases.

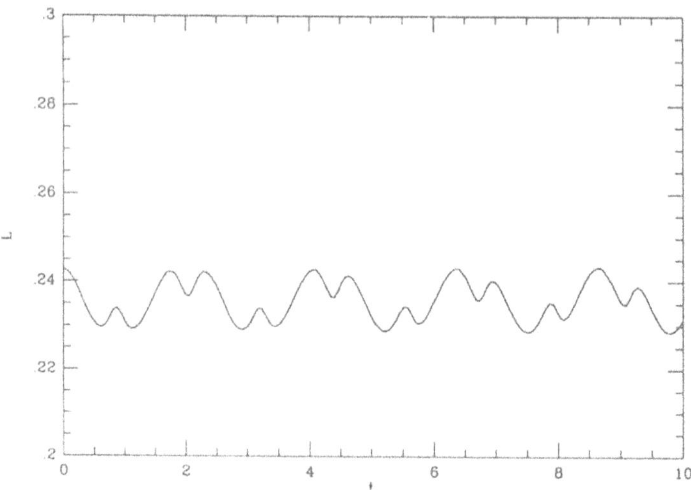

Fig. 5. The magnitude of the angular momentum vector of a star orbiting in a flattened potential does not vary greatly, though the vector's direction does. Here $L(t)$ is plotted for the taller of the orbits of Fig. 4

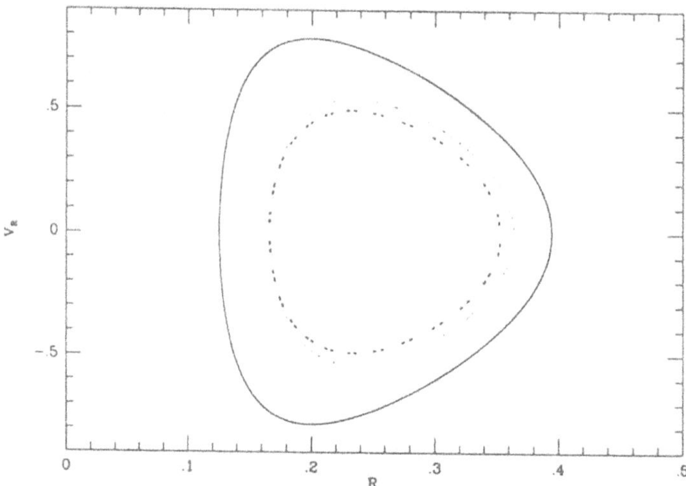

Fig. 6. Consequents in the $(R, \dot{R})$ surface of section of the orbit of Figs 4 and 5. If L were exactly constant the consequents would lie on the dotted curve. The full curve is the contour $\frac{1}{2}v_R^2 + \Phi_{\text{eff}}(R, 0) = E$.

3.3.1 Nearly spherical Φ

In a spherical potential the orbit would be confined to a plane. L/L_z would determine the inclination of this plane to the z-axis (where $L \equiv |\mathbf{L}|$), and if $L_c(E)$ is the angular momentum of a circular orbit of energy E, L/L_c would determine the width of the annulus formed by the orbit in its plane. In a flattened potential, the vector $\mathbf{L}$ is not conserved, but its magnitude nearly is (Figure 5). The orbit becomes a slightly elliptical rosette in a precessing plane perpendicular to the instantaneous value of **L**. The inclination of this plane to the z-axis is approximately constant as it slowly rotates about the axis. Figure 6 illustrates the truth of these statements in terms of a device called a **surface of section** introduced by Poincaré.

To make a surface of section for a two-dimensional orbit, we choose a coordinate condition, for example $z = 0$, and note down the value of the other coordinate, say x, and its conjugate momentum v_x every time the condition $z = 0$ is satisfied as we numerically integrate the orbit. The pair of points (x, v_x) thus obtained is called a **consequent**. The surface of section is simply a plot of each consequent in the (x, v_x) plane.

The condition $z = 0$ constitutes one restriction on the four phase-space coordinates of a two-dimensional orbit, and the equation $H(\mathbf{x}, \mathbf{v}) = E$ (where H is the Hamiltonian) is a further restriction. Hence if an orbit is not restricted by a further integral, it is free to explore a fully two-dimensional subset of the surface of section. However, if some additional integral $I(\mathbf{x}, \mathbf{v})$ is conserved, there is a third restriction on the four phase-space coordinates and the consequents should form a one-dimensional set, i.e. they should lie on a curve—a so-called **invariant curve**. Figure 6 shows that the consequents of the orbits shown in Figure 4 do indeed lie on a curve. Furthermore, it is easy to derive the curve the consequents would follow if the third restriction were $L =$ constant. This curve is also shown in Figure 6, and the consequents do indeed lie close to this curve.

3.3.2 The epicycle approximation

Suppose we develop $\Phi_{\rm eff}$ in a Taylor series about its minimum at the radius R_g of the circular orbit of angular momentum L_z:

$$\Phi_{\rm eff} = {\rm const} + \tfrac{1}{2}\kappa^2 x^2 + \tfrac{1}{2}\nu^2 z^2 + \cdots, \tag{24a}$$

where

$$x \equiv R - R_g, \quad \kappa^2 \equiv \frac{\partial^2 \Phi_{\rm eff}}{\partial R^2}\bigg|_{(R_g,0)}, \quad \nu^2 \equiv \frac{\partial^2 \Phi_{\rm eff}}{\partial z^2}\bigg|_{(R_g,0)}. \tag{24b}$$

Then the x and z motions decouple and we obtain two energy integrals

$$\begin{aligned} x &= X\cos(\kappa t + \psi_0) \qquad & z &= Z\cos(\nu t + \zeta), \\ E_R &\equiv \tfrac{1}{2}[v_R^2 + \kappa^2 (R - R_g)^2] \qquad & E_z &\equiv \tfrac{1}{2}[v_z^2 + \nu^2 z^2]. \end{aligned} \tag{25}$$

(Here X, Z, ψ_0 and ζ are constants.) It is easy to show that these integrals would cause invariant curves in the (R, v_R) surface of section to be ellipses centred on $(R_g, 0)$. The approximation (25) to an orbit is known as the **epicycle approximation**. This approximation is often employed, so it pays dividends to look a bit more closely at it and the quantities it introduces.

The frequency κ defined by (24b) is called the **epicycle frequency**. It is interesting to relate it to the circular frequency Ω. We have[2]

$$\Omega^2 \equiv \frac{v_c^2}{R^2} = \frac{1}{R}\frac{\partial \Phi}{\partial R} = \frac{1}{R}\frac{\partial \Phi_{\rm eff}}{\partial R} + \frac{L_z^2}{R^4}, \tag{26}$$

so

[2] I shall always reserve Ω for the angular frequency and v_c for the speed of a circular orbit.

$$\kappa^2 = \frac{\partial(R\Omega^2)}{\partial R} + \frac{3L_z^2}{R^4} = R\frac{\partial\Omega^2}{\partial R} + 4\Omega^2. \tag{27}$$

Since in galaxies Ω invariably decreases with increasing R, but never faster than it does for Keplerian motion ($\Omega \propto R^{-3/2}$), it follows that

$$\Omega \le \kappa \le 2\Omega. \tag{28}$$

I introduced the epicycle approximation in the context of motion in the meridional plane. But through conservation of $L_z = R^2\dot{\phi}$ it makes perfectly definite predictions for the ϕ-motion. Let

$$y \equiv R_g[\phi - (\phi_0 + \Omega t)] \tag{29}$$

be the tangential displacement of the star from the position it would occupy on a circular orbit of the same L_z (the "guiding centre"). Then expanding the constant $L_z = R^2\dot{\phi}$ to first order in the displacements from the guiding centre, we obtain

$$\phi = \phi_0 + \Omega t - \frac{2\Omega X}{\kappa R_g}\sin(\kappa t + \psi_0).$$

Hence

$$y = -Y\sin(\kappa t + \psi_0) \quad \text{where} \quad \frac{Y}{X} = \frac{2\Omega}{\kappa} \equiv \gamma \ge 1. \tag{30}$$

Thus the epicycles are elongated tangentially.

Finally, it is useful to employ this last result to eliminate $(R - R_g)$ from E_R. Let $\mathbf{v}'$ denote the star's velocity with respect to the galactic centre in an inertial frame and $\mathbf{v} \equiv \mathbf{v}' - \mathbf{v}_{\rm LSR}$ be the velocity with respect to the Local Standard of Rest (i.e. the circular speed at radius R). Then

$$\begin{aligned} v_\phi(R_0) &\equiv v'_\phi(R_0) - v_c(R_0) = R_0(\dot{\phi} - \Omega_0) \\ &= R_0(\dot{\phi} - \Omega_g + \Omega_g - \Omega_0) \\ &\simeq R_0\left[(\dot{\phi} - \Omega_g) - \left(\frac{\mathrm{d}\Omega}{\mathrm{d}R}\right)_{R_0} x\right]. \end{aligned} \tag{31}$$

Replacing $\dot{\phi}$ by $\Omega_g(1 - 2x/R_g)$ this becomes

$$\begin{aligned} v_\phi &\simeq -R_0 x\left[\frac{2\Omega_g}{R_g} + \left(\frac{\mathrm{d}\Omega}{\mathrm{d}R}\right)_{R_0}\right] \simeq -x\left[2\Omega_0 + R_0\left(\frac{\mathrm{d}\Omega}{\mathrm{d}R}\right)_{R_0}\right] \\ &\equiv 2Bx. \end{aligned} \tag{32}$$

Thus $E_R \simeq \frac{1}{2}[v_R^2 + \gamma^2 v_\phi^2]$ where $\gamma = \frac{1}{2}\kappa/B = 2\Omega_g/\kappa$. In this form E_R looks like the kinetic energy of peculiar motion, excepting that in general $\gamma \ne 1$. We shall find that important consequences follow from $\gamma \ne 1$.

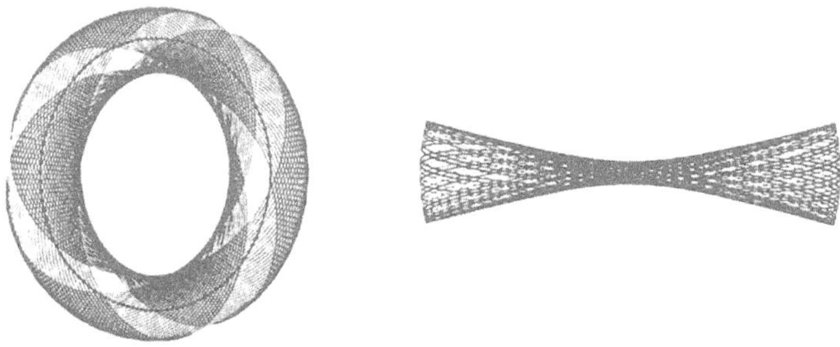

Fig. 7. Two orbits of the same energy in the barred potential (33a).

3.4 Non-Axisymmetric Potentials

Many, perhaps most, galaxies are barred. In general the major axis of the galaxy's potential rotates with respect to inertial space. Motion in such potentials is very complex. So here I shall restrict the discussion to non-rotating potentials; our results will apply sufficiently close to the centre of a rotating potential. A further important restriction will be to motion confined to the galaxy's equatorial plane.

Consider the toy potential

$$\Phi_L(x,y) = \tfrac{1}{2}v_0^2 \ln(R_c^2 + x^2 + y^2/q^2) \quad \text{where } v_0,\ R_c,\ q \text{ are constants.} \tag{33a}$$

This has a harmonic core

$$\Phi_L \simeq \frac{v_0^2}{2R_c^2}\left(x^2 + \frac{y^2}{q^2}\right) + \text{constant} \quad \text{for} \quad x, y \ll R_c, \tag{33b}$$

and at large radii the central force falls off as $1/R$ as in a galaxy with a flat rotation curve.

Figure 7 shows two orbits in (33a). Most orbits resemble one or other of them. Those with holes at their centres are called **loop** or **tube orbits**, and the others are known as **box orbits**. At each point on a tube orbit only two velocity vectors are possible, while box orbits generally permit only four velocity vectors at a given point in space. Thus these orbits appear to occupy two-dimensional surfaces in phase space. The surface of section ($y = 0,\ \dot{y} > 0$) for motion in (33a) that is shown in Figure 8(a) confirms this conclusion. What integral other than the Hamiltonian confines the orbits?

Figure 8(b) shows what the invariant curves in Figure 8(a) would look like if the sought-after integral were either (i) L_z or (ii) E_x for the harmonic approximation (33b) to (33a). Evidently L_z is a very crude approximation to the additional integral of loop orbits. But we clearly need better approximations to the integrals of loop and box orbits.

3.4.1 Stäckel potentials

So far we have encountered two cases in which we have an analytic expression for the additional integral of two-dimensional orbits; the harmonic oscillator and an axisymmetric potential. Careful examination of how these integrals arise will lead us to analytic models of box and loop orbits.

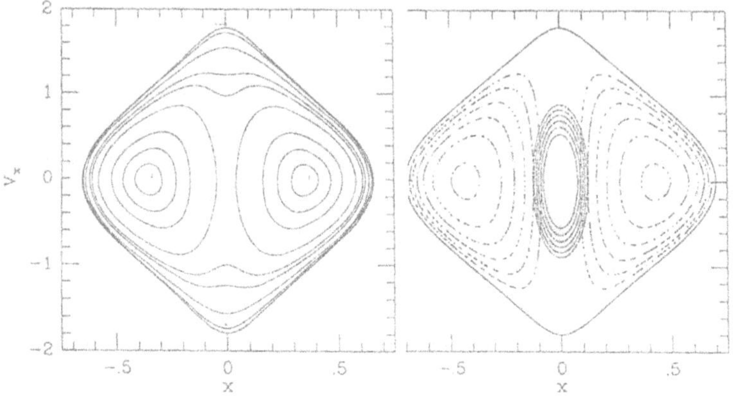

Fig. 8. (a) The $(x, \dot{x})$ surface of section for motion in (33a) at the energy of the orbits shown in Fig. 7. (b) The invariant curves that would arise if the additional integral were either L_z (dashed curves) or E_x (dotted curves).

Harmonic-oscillator orbits are bounded by the curves $x =$ constant and $y =$ constant on which $v_x(x) = 0$ and $v_y(y) = 0$. Similarly, orbits in an axisymmetric potential are bounded by curves $\phi =$ constant on which $v_r(r) = 0$.

In the case of the harmonic oscillator we obtain an expression for v_x as a function of x alone by writing the Hamiltonian as a function $H(x, y, v_x, v_y)$ of the coordinates that run parallel to the orbit boundaries and their conjugate momenta, and then noticing that this is the sum of a piece that depends on (x, v_x) alone and a piece that depends on (y, v_y) alone. In the case of an axisymmetric potential we obtain an expression $v_r(r)$ by noticing that $r^2 H = r^2[\frac{1}{2}v_r^2 + \Phi(r)] + \frac{1}{2}p_\phi^2$ is a sum of terms that depend either on (r, v_r) only or on (ϕ, p_ϕ) only.[3]

These observations suggest that we seek a coordinate system (u, v) such that:

(i) $\left.\begin{array}{l} u = \text{constant} \\ v = \text{constant} \end{array}\right\}$ form boundaries of box/loop orbits;

(ii) $H(u, v, p_u, p_v) \times f(u, v)$ breaks down into sum $F_u(u, p_u) + F_v(v, p_v)$ of terms that involve only one pair of conjugate variables each.

Figure 9 shows that condition (i) is approximately satisfied by the coordinates which are defined by

$$x = \Delta \sinh u \cos v, \quad y = \Delta \cosh u \sin v. \tag{34}$$

In these coordinates the Lagrangian reads

$$\mathcal{L} = \tfrac{1}{2}\Delta^2 (\sinh^2 u + \cos^2 v)(\dot{u}^2 + \dot{v}^2) - \Phi. \tag{35}$$

---

[3] The momentum p_ϕ conjugate to ϕ is just L_z.

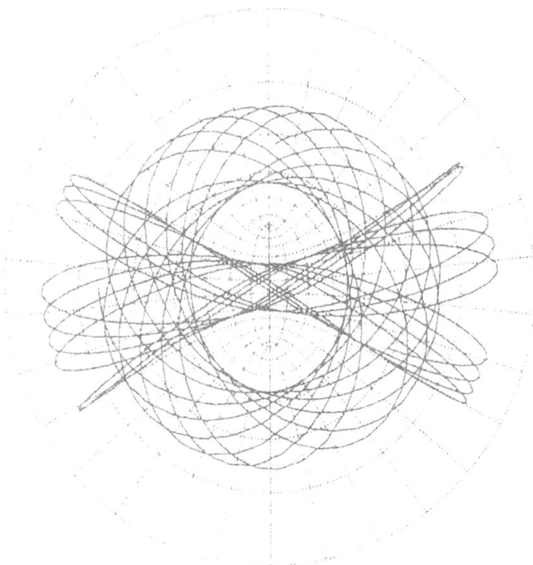

Fig. 9. Box and loop orbits in a non-rotating barred potential are approximately bounded by the hyperbolæ and ellipses of a system of confocal elliptical coordinates.

The momenta are therefore

$$p_u \equiv \frac{\partial \mathcal{L}}{\partial \dot{u}} = \Delta^2 (\sinh^2 u + \cos^2 v)\dot{u}, \quad p_v \equiv \frac{\partial \mathcal{L}}{\partial \dot{v}} = \Delta^2 (\sinh^2 u + \cos^2 v)\dot{v}, \tag{36}$$

and thus the Hamiltonian is

$$\begin{aligned} H(u, v, p_u, p_v) &= p_u \dot{u} + p_v \dot{v} - \mathcal{L} \\ &= \tfrac{1}{2}\Delta^2 (\sinh^2 u + \cos^2 v)(\dot{u}^2 + \dot{v}^2) + \Phi \\ &= \frac{p_u^2 + p_v^2}{2\Delta^2(\sinh^2 u + \cos^2 v)} + \Phi. \end{aligned} \tag{37}$$

Our strategy is to multiply H through by some function $f(u, v)$ and try to split the result into bits depending on (u, p_u) and (v, p_v). From the last line of (37) it follows that $f = \sinh^2 u + \cos^2 v$. Furthermore, if the resulting product is to split as desired, Φ must be of the form

$$\Phi(u, v) = \frac{U(u) - V(v)}{\sinh^2 u + \cos^2 v}. \tag{38}$$

Then $(\sinh^2 u + \cos^2 v) \times (H = E)$ breaks into

$$E \sinh^2 u - \frac{p_u^2}{2\Delta^2} - U(u) \equiv I_2 = \frac{p_v^2}{2\Delta^2} - V(v) - E \cos^2 v. \tag{39}$$

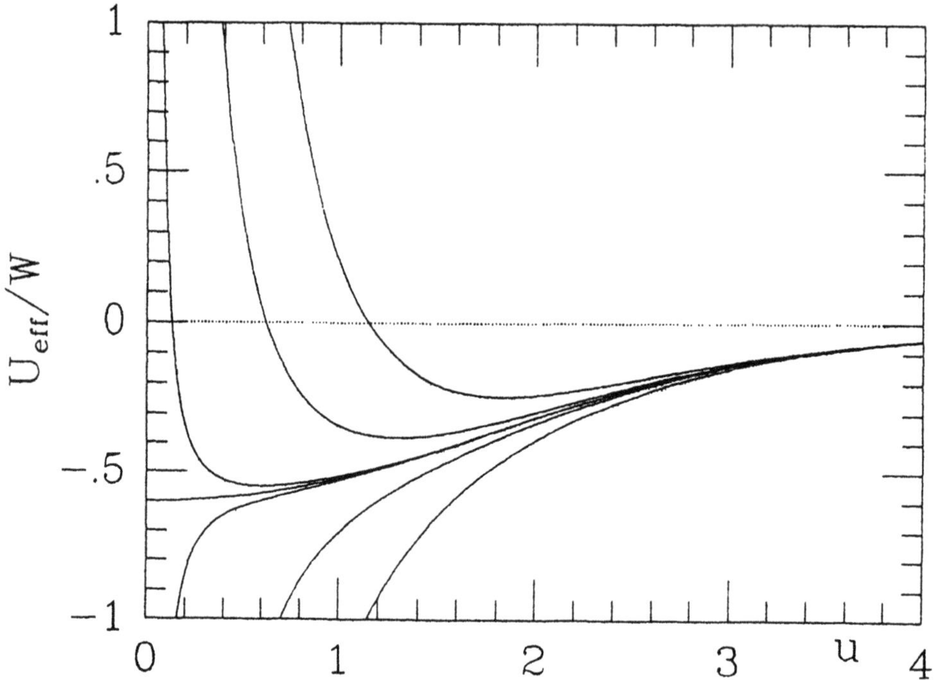

Fig. 10. Plots of the effective potential defined by (40) for several values of the integral I_2; from top to bottom $I_2 = 1, 0.25, 0.01, 0, -0.01, -0.25, -1$.

Since the quantity I_2 defined by this equation depends on neither (v, p_v) (from the equivalence at left), nor on (u, p_u) (from the equality at right), it must be a constant. In other words, it is our desired additional integral.

We obtain expressions $p_u(u)$ and $p_v(v)$ by rearranging (39):

$$\begin{aligned} p_u &= \pm\sqrt{2}\Delta \sinh u \sqrt{E - U_{\rm eff}}, \quad \text{where} \quad U_{\rm eff}(u) \equiv \frac{I_2 + U(u)}{\sinh^2 u}, \\ p_v &= \pm\sqrt{2}\Delta \cos v \sqrt{E - V_{\rm eff}}\ , \quad \text{where} \quad V_{\rm eff}(v) \equiv -\frac{I_2 + V(v)}{\cos^2 v}. \end{aligned} \tag{40}$$

To proceed further we must make a definite choice of Φ. Figure 10 shows curves of constant $U_{\rm eff}$ for several values of I_2 and the potential of a **perfect ellipsoid**; this is a body in which the density is of the form

$$\rho(\mathbf{x}) = \frac{\text{constant}}{1 + m^2}, \quad \text{where} \quad m^2(\mathbf{x}) \equiv \frac{x^2}{a^2} + \frac{y^2}{b^2} + \frac{z^2}{c^2}, \tag{41}$$

and (a, b, c) are constants specifying the ellipsoid's shape. A full discussion of perfect ellipsoids will be found in de Zeeuw (1985). When $a > b = c$ the body is cigar-shaped and its potential is given by (38) with

$$\begin{aligned} U(u) &= -W \sinh u \arctan\left(\frac{\Delta \sinh u}{c}\right) \\ V(v) &= W \cos v \,\text{arctanh}\left(\frac{\Delta \cos v}{c}\right) \end{aligned} \qquad (W \text{ a constant}). \tag{42}$$

According to Figure 10, when $I_2 > 0$, $U_{\rm eff} \to +\infty$ as $u \to 0$. Hence with $I_2 > 0$ the equation $0 = p_u^2 = 2\Delta^2(E - U_{\rm eff}) \sinh u$ has two roots, at $u_{\rm min}$ and $u_{\rm max}$. The only allowed values of u are those in $[u_{\rm min}, u_{\rm max}]$ since outside this range p_u is imaginary.

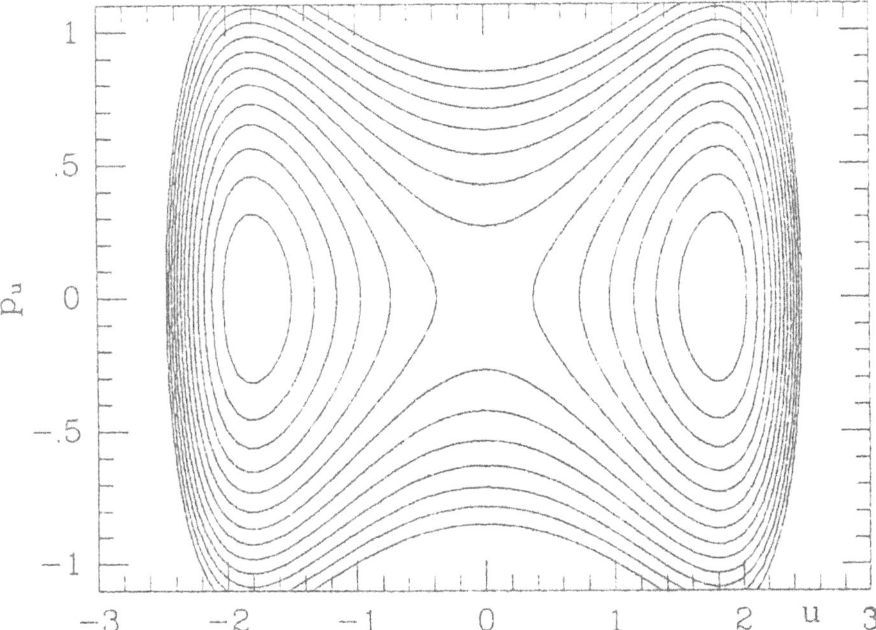

Fig. 11. Each curve shows the relationship $p_u(u)$ defined by (40) for a different value of I_2. The outer curves are for $I_2 < 0$.

The bounding curves $u = u_{\text{min}}$ etc are of course ellipses, which suggests that orbits with $I_2 > 0$ may be loop orbits.

When $I_2 \leq 0$, $0 = p_u^2 = 2\Delta^2(E - U_{\text{eff}}) \sinh u$ has but one root, u_{max}, and all $u \leq u_{\text{max}}$ are allowed.

A similar analysis of the equation $p_v^2 = 0$ shows that when $I_2 \geq 0$ all values of v are allowed, but that when $I_2 < 0$ the only allowed values of v lie in $[-v_{\text{max}}, v_{\text{max}}]$. Hence orbits which are forbidden to enter within the ellipse $u = u_{\text{min}} > 0$ circulate freely in the annulus bounded by this ellipse and $u = u_{\text{max}}$, while orbits that can penetrate to the centre are not permitted to wander further from the x-axis than the hyperbolae $v = \pm v_{\text{max}}$. We recognize the former orbits as loops, and the latter as boxes.

When an angular coordinate such as v takes all possible values along an orbit, the orbit is said to **circulate** in that coordinate. On the other hand, when the coordinate's range is restricted, the orbit is said to **librate** in the coordinate. Thus loop orbits circulate in v, while box orbits librate around $v = 0$.

What do these orbits look like in phase space? We know that they form two-surfaces $(u, v, p_u[u], p_v[v])$, where u and v may be regarded as parameters free to vary within the limits we have just derived. Let's cut such a surface through with the surface $v =$ constant, $p_v =$ constant. No matter what values we choose for these constants we always see the same figure, namely the curve $p_u(u)$. Figure 11 shows the forms of this figure for different values of I_2. When $I_2 > 0$ (loop orbit), the curve $p_u(u)$ encircles either a point on the positive u-axis or the negative u-axis. When $I_2 < 0$ (box orbit), $p_u(u)$ encircles the origin. In either case $p_u(u)$ is topologically equivalent to[4] a circle. A similar exercise shows that when we cut the orbital surface by ($u =$ constant, $v =$ constant), the resulting curve $p_v(v)$ is always topologically equivalent to a circle.[5] A mathematician

[4] That is, "continuously deformable into".

[5] One must bear in mind that v is an angular coordinate.

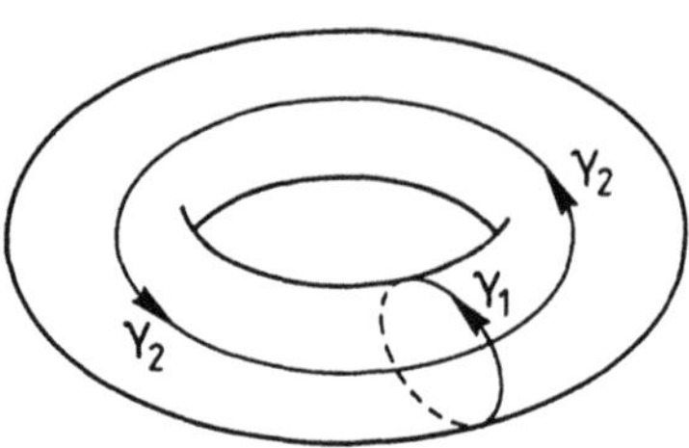

Fig. 12. An orbital torus and the paths γ_i defining the torus's two action integrals.

would express these results by saying that the orbital surface is a two-torus, since the latter is defined to be the product of two circles.

How shall we label these tori? The obvious labels are E and I_2, but much better labels are the **actions** defined by

$$J_u \equiv \frac{1}{\pi}\int_{u_{\min}}^{u_{\max}} p_u(u)\,\mathrm{d}u, \quad J_v \equiv \frac{1}{\pi}\int_{v_{\min}}^{v_{\max}} p_v(v)\,\mathrm{d}v. \tag{43}$$

J_u is just $(2\pi)^{-1}$ times the area enclosed by the orbit's curve in the (u,p_u) plane and similarly for J_v. However, it is better to think of the actions in terms of the orbital torus; one can show that

$$J_i = \frac{1}{2\pi}\oint_{\gamma_i} \mathbf{v}\cdot\mathrm{d}\mathbf{x}, \tag{44}$$

where $i = u, v$ and γ_u is any path on the torus that can be continuously deformed into the torus's curve in the (u,p_u) plane, and similarly for γ_v. Figure 12 illustrates this state of affairs. Since these line integrals can be converted into surface integrals in the usual way, I shall sometimes refer to the J_i as "the torus's cross-sections".

The actions are superior to (E, I_2) as labels of the tori for two reasons:

(i) they are **adiabatic invariants**;

(ii) there exist **angle variables** $\boldsymbol{\theta}$ conjugate to the J_i.

By saying they are adiabatic invariants, I mean the following. If we deform the potential Φ in which the star orbits, we shall move the star to a new orbital torus. However, if the timescale on which we deform Φ is long compared to the characteristic orbital time, the new torus will yield the same values for the integrals (44) as did the original torus. This is a very handy property in certain problems—for example studies of the evolution of globular clusters and galactic nuclei.

Given any point in phase space we can in principle find the torus that passes through that point and evaluate its cross-sections J_i. So we can regard the actions as functions $J_i(\mathbf{x},\mathbf{v})$ of the phase-space coordinates. Since these functions are constant along each orbit (i.e. they are integrals), it would be nice to use them as phase-space coordinates in their own right. Let (θ_1, θ_2) be the two additional coordinates required to make up a complete set $(\mathbf{J}, \boldsymbol{\theta})$ of four phase-space coordinates. Evidently the curves θ_i = constant define a grid on each torus. Hence if we follow a curve of, say, $\{\mathbf{J} = \text{constant}, \theta_2 = \text{constant}\}$, we must eventually wrap around the torus and return to our starting point. Let $\theta_1^{(\max)}$ be the value of θ_1 at which we return to the point from which we departed with $\theta_1 = 0$. Then the Cartesian phase-space coordinates must be periodic functions of θ_1 with period $\theta_1^{(\max)}$. It would be handy to scale θ_1 such that $\theta_1^{(\max)} = 2\pi$. Similarly we'd like to define θ_2 such that $(\mathbf{x}, \mathbf{v})$ is periodic in θ_2 with period 2π. Finally, we'd like to require that the set $(\mathbf{J}, \boldsymbol{\theta})$ be canonical—that is, one of those privileged systems in which Poincaré inariants and Poisson brackets take on especially simple forms.[6] Can these requirements be simultaneously satisfied? Miraculously, the answer is "yes". However, if we had sought to label the tori with (E, I_2) or practically any other labels, the answer would have been a firm "no".

3.4.2 Quasiperidic motion

Figure 11 qualitatively resembles the surface of section of Figure 8(a), which suggests that the invariant curves of the latter figure are also cross-sections through orbital tori. This is indeed the case. In fact one can show (Arnold 1978) that whenever a two-dimensional orbit has at least two isolating integrals, the orbit must lie on a torus. Equation (44) can be used to define two independent actions for each such torus and canonically conjugate angle variables θ_i can then be constructed exactly as one does in the case of a Stäckel potential. Again we find that all physical quantities are periodic in θ_i with period 2π. The Hamiltonian, being constant along each orbit, is a function $H(\mathbf{J})$ of $\mathbf{J}$ alone.

The evolution along an orbit of any phase-space coordinate w may be written in terms of the Poisson bracket $[.,.]$ as $\dot{w} = [w, H]$. In any canonical coordinate system $(\mathbf{p}, \mathbf{q})$ the Poisson bracket $[f, g]$ is evaluated as

$$[f, g] = \frac{\partial f}{\partial \mathbf{q}} \cdot \frac{\partial g}{\partial \mathbf{p}} - \frac{\partial f}{\partial \mathbf{p}} \cdot \frac{\partial g}{\partial \mathbf{q}}. \tag{45}$$

Thus in $(\mathbf{J}, \boldsymbol{\theta})$ coordinates the equations of motion are $\dot{\mathbf{J}} = 0$ and

$$\dot{\boldsymbol{\theta}} = [\boldsymbol{\theta}, H] = \frac{\partial H(\mathbf{J})}{\partial \mathbf{J}} \equiv \boldsymbol{\omega}(\mathbf{J}), \text{a constant characteristic of the orbit.} \tag{46}$$

Since $\mathbf{J}$ is constant, we may immediately integrate this equation to $\boldsymbol{\theta}(t) = \boldsymbol{\theta}(0) + \boldsymbol{\omega} t$; the angle variables increase linearly in time.

Since the Cartesian phase-space coordinates are periodic in the θ_i they may be expanded in Fourier series

$$\mathbf{x}(\mathbf{J}, \boldsymbol{\theta}) = \sum_{\mathbf{n}} \mathbf{X_n}(\mathbf{J}) e^{i\mathbf{n}\cdot\boldsymbol{\theta}} \quad \Rightarrow \quad \mathbf{x}(t) = \sum_{\mathbf{n}} \widetilde{\mathbf{X}}_{\mathbf{n}}(\mathbf{J}) e^{i\mathbf{n}\cdot\boldsymbol{\omega} t}, \tag{47}$$

where $\mathbf{n}$ is a vector with integer components and $\widetilde{\mathbf{X}}_{\mathbf{n}} \equiv \mathbf{X_n} e^{i\boldsymbol{\theta}(0)}$. Thus if we Fourier transform the time series $x_i(t)$ obtained by numerically integrating orbits such as those of Figure 7, we shall obtain a line spectrum in which the frequencies of all lines are integer combinations of two fundamental frequencies ω_i. Motion of this kind is called **quasiperiodic.**[7] By identifying the integer combination of each line, we can pass from

6 See Arnold (1978) or §1.D of GD for the definition of Poinaré invariants, Poisson brackets and canonical coordinates.

7 X-ray astronomers use this term in another, much looser, way.

the equation on the right of (47) to that on the left, and thus obtain a parametric representation of the orbital torus.

All these results extend to three-dimensional orbits. A three-dimensional orbit lies on a three-torus in six-dimensional phase space provided it admits three isolating integrals (H, I_2, I_3) such that $[I_2, I_3] = 0$.[8] There are three fundamentally different closed paths one can draw on a three-torus and the integrals $\oint \mathbf{v} \cdot \mathrm{d}\mathbf{x}$ around these paths define three actions which can then be complemented with three angle variables.

It is frequently convenient to imagine a potential's orbits as forming a three-dimensional continuum called **action space**. The three basic actions constitute Cartesian coordinates for this space. We shall see that in simple cases the key step in building a galaxy model is the correct distribution of the galaxy's stars through action space.

These results show that it is of fundamental importance for the phase-space structure of an orbit for the latter to have at least as many isolating integrals as it has spatial dimensions. What is the significance of an orbit's having *more* integrals? To answer this question consider the progression from Keplerian motion $\rightarrow$ general central motion $\rightarrow$ motion in a squashed potential. In every case the orbits are confined to tori. A Keplerian orbit admits five isolating integrals because all its three frequencies are equal—we have already seen that $\omega_r = \omega_\psi$, and to this we should add that the frequency ω_l of vertical oscillation of a star whose orbital plane is inclined to the plane $z = 0$ satisfies $\omega_l = \omega_\psi$. This degeneracy of the frequencies gives rise to two isolating integrals in addition to the actions $J_r = (2\pi)^{-1} \oint v_r \,\mathrm{d}r$, $J_a \equiv L_z$ and $J_l \equiv L - |L_z|$. These additional integrals may be taken to be $I_4 \equiv \theta_a - \theta_l$ and $I_5 \equiv \theta_r - \theta_a$.

We have seen that in a general central field of force $\omega_r \neq \omega_\psi$, though of course we still have $\omega_a = \omega_l = \omega_\psi$. Thus $I_4 \equiv \theta_a - \theta_l$ is an integral even in a non-Keplerian central field of force, but I_5 is not. E and the three components of $\mathbf{L}$ may be considered to be functions of these four integrals.

On squashing the potential parallel to the z-axis we break the degeneracy $\omega_a = \omega_l$ and thus lose I_4. The actions of nearly circular motion are now

$$J_r \simeq E_R/\kappa, \quad J_a = L_z \quad \text{and} \quad J_l = E_z/\nu. \tag{48}$$

In general any relationship $\omega_i/\omega_j = n_i/n_j$ for $i \neq j$ and integer n_i, n_j gives rise to an additional isolating integral $I = n_i\theta_j - n_j\theta_i$. When $\omega_i/\omega_j = a$ for irrational a we can only form the non-isolating integral $I' = a\theta_j - \theta_i$.

3.4.3 Irregular orbits

Life would be much simpler if all orbits were quasiperiodic. But alas, as Figure 13 shows, this is not the case. These orbits not have the beautiful regular structure in real space of the orbits of Figure 7, and a surface of section shows that this is because they have no isolating integral in addition to H. Yet the orbits of Figure 13 are in a potential that closely resembles that, (33a), which supports the quasiperiodic orbits of Figure 7. In fact, in polar coordinates (33a) can be written

$$\Phi_L(R, \phi) = \tfrac{1}{2} v_0^2 \ln \left[\tfrac{1}{2} R^2 (q^{-2} + 1) - \tfrac{1}{2} R^2 (q^{-2} - 1) \cos 2\phi + R_c^2 \right], \tag{49}$$

while the potential giving rise to the orbits of Figure 13 is

[8] This last property is expressed by saying that "I_2 and I_3 are in involution".

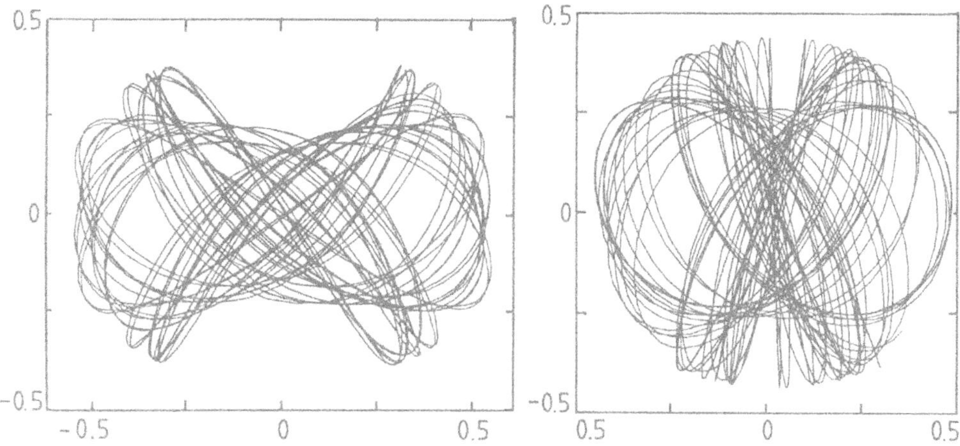

Fig. 13. Not all orbits are quasiperiodic. These orbits in (50) are slightly irregular.

$$\begin{aligned}\Phi_N(R,\phi) = \tfrac{1}{2}v_0^2 \ln \Big[&\tfrac{1}{2}R^2(q^{-2}+1) - \tfrac{1}{2}R^2(q^{-2}-1)\cos 2\phi \\ &- \left(\frac{R^3}{R_e}\right)\cos 2\phi + R_c^2\Big] \quad (R_e \text{ a constant}).\end{aligned} \tag{50}$$

The only difference between the potentials (49) and (50) is the addition to the logarithm of the latter of a term $\propto R^3$. How does this small addition so effectively lay waste the exquisite system of nested tori with which (49) fills phase space?

This is a very large question which I cannot answer adequately here for want of both space and knowledge. Figure 14, which is a surface of section for motion in (49) with a rather small value of the axis ratio q, hints at the conventional answer. Evidently for small q the phase space is not filled with the tori of box and loop orbits only, but with the nested sequences of tori belonging other orbit families. In Figure 14 the tori of each such family appear as curves surrounding a single dot, which is the consequent of a closed orbit. In the conventional picture this closed orbit is considered to be an orbit of one of the box or loop families on which a resonance condition $\omega_1/\omega_2 = n_1/n_2$ happens to be satisfied. Perturbation theory may then be used to show that this orbit traps neighbouring orbits into libration about itself and thus fills nearby phase space with a sequence of tori centred on itself. However, this process of resonant trapping rarely gives rise to the sort of orderly structure you see in Figure 14 since quarrels soon break out between closed orbits as to which pieces of phase space owe them allegiance; not only are there as many closed orbits as there are rational numbers on the real line, but fresh resonances arise as soon as a strong resonance has trapped a decent portion of phase space into forming a familly centred on itself since the frequencies associated with these tori will also often resonate. So just as in China the collapse of imperial authority in the early years of this century was followed by the war-lord period in which much of the country degenerated into anarchy as local strong-men struggled for supremacy, so resonant trapping soon gives rise to anarchy in phase space.

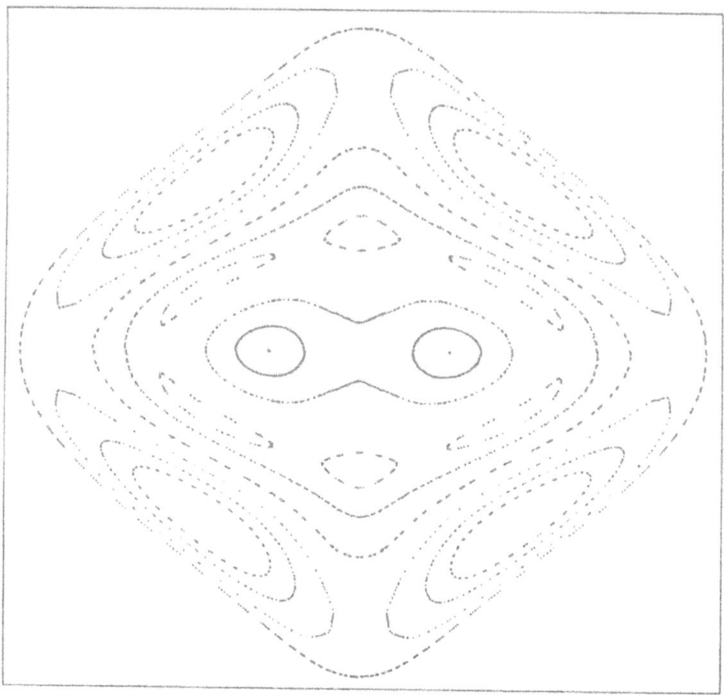

Fig. 14. An $(x, \dot{x})$ surface of section for motion in Φ_L when $q = 0.6$. The simple box/loop structure characteristic of Stäckel motion has been disrupted by the formation of several families of "resonant" orbits.

However, for galactic potentials the anarchy, or chaos, is never absolute. It tends to be localized around certain resonances and is rarely on a large scale unless the Hamiltonian has a saddle point, as (50) has and (49) has not. Since the Hamiltonian of motion in a rotating barred potential has saddle points near the "corotation" resonance, the latter is generally associated with much chaotic motion.

The study of the onset of chaos is currently a fashionable topic in so-called "non-linear physics".[9] An excellent introduction to the field will be found in Berry (1978). Anybody working in galactic dynamics should be aware of these developments, but one must beware that a great deal of nonsense is talked about chaos because (a) the numerical and mathematical sides of the field are too loosely coupled, and (b) the field relies heavily on polynomial perturbations of the harmonic oscillator, and this system is by no means generic.

---

9 Saying a process is non-linear says very little about it since essentially all physics is non-linear. Hence I doubt that much useful will emerge from generalized study of non-linear processes. Natural philosophy left philosophy in the dust precisely by narrowing its horizons to those relatively boring questions which *can* be answered.

4 Equilibrium Models

Having studied orbits in galactic potentials the next task is to populate these orbits so as to build up a realistic galaxy model. However, before we look at ways of doing this it is expedient to take a step backwards and derive the equation that must be satisfied by any collisionless stellar system, whether it is in equilibrium or not.

4.1 The Collisionless Boltzmann Equation

The great majority of stellar types endure for much longer than the characteristic dynamical time of a galaxy ($\approx 0.1\,\mathrm{Gyr}$). Hence to a fair approximation such stars are neither created nor destroyed as they flow through phase space under Φ. Consequently, thus the phase fluid formed by these stars must satify the six-dimensional analogue of the usual hydrodynamical continuity equation, $\dot{\rho} + \nabla \cdot (\rho \mathbf{u}) = 0$. We express this conclusion mathematically by defining $(\mathbf{x}, \mathbf{v}) \equiv \mathbf{w} \equiv (w_1, \ldots, w_6)$, and writing

$$\frac{\partial f}{\partial t} + \sum_{\alpha=1}^{6} \frac{\partial (f \dot{w}_\alpha)}{\partial w_\alpha} = 0. \tag{51}$$

But $\dot{\mathbf{w}} = (\dot{\mathbf{x}}, \dot{\mathbf{v}}) = (\mathbf{v}, -\nabla\Phi)$, so

$$\sum_{\alpha=1}^{6} \frac{\partial \dot{w}_\alpha}{\partial w_\alpha} = \sum_{i=1}^{3} \left(\frac{\partial v_i}{\partial x_i} + \frac{\partial \dot{v}_i}{\partial v_i} \right) = \sum_{i=1}^{3} -\frac{\partial}{\partial v_i} \left(\frac{\partial \Phi}{\partial x_i} \right) = 0. \tag{52}$$

Substituting (52) in (51) we obtain

$$\frac{\partial f}{\partial t} + \sum_{\alpha=1}^{6} \dot{w}_\alpha \frac{\partial f}{\partial w_\alpha} = 0. \tag{53}$$

This **collisionless Boltzmann equation** ("CBE"; also called the Vlasov equation) is usually written in one of three alternative forms:

$$\frac{\partial f}{\partial t} + \mathbf{v} \cdot \nabla f - \nabla\Phi \cdot \frac{\partial f}{\partial \mathbf{v}} = 0, \tag{54}$$

or

$$\frac{\mathrm{d}f}{\mathrm{d}t} = 0, \quad \text{or} \quad \frac{\partial f}{\partial t} + [f, H] = 0. \tag{55}$$

Equation (54) is just a straight rewrite of (53). The first of equations (55) follows from (53) when one notices that the combination $\frac{\partial}{\partial t} + \dot{\mathbf{w}} \cdot \frac{\partial}{\partial \mathbf{w}}$ is the total derivative along an orbit. The second of equations (55) reduces to (54) when uses (45) to evaluate the Poisson bracket with $H = \frac{1}{2}v^2 + \Phi$. We shall have occasion to use each form of the CBE.

I have derived the CBE by appealing to an intuitive notion of a stellar phase-space density. Actually no consistent interpretation of the quantity f that satisfies the CBE is possible in terms of an actual star density: A density of discrete objects can be defined only as an average through a sufficiently large volume and thus must necessarily be a reasonably smooth function of the phase-space coordinates $\mathbf{w}$. The solution f of the CBE, by contrast, becomes an ever more rapidly varying function of $\mathbf{w}$. This causes no difficulty if we interpret f as a ***probability*** density. Then f like the wave-function ψ of quantum mechanics becomes an unreal abstraction from which we extract real predictions by using it to calculate expectation values. But as soon as we "coarse-grain" an evolved f by averaging it through neat volumes around each point, it ceases to satisfy the CBE.

We shall need the CBE in polar coordinates. It is best not to mess with the chain rule but to proceed from the first of equations (55). For example, in cylindrical polars we have

$$\frac{\mathrm{d}f}{\mathrm{d}t} = \frac{\partial f}{\partial t} + \dot{R}\frac{\partial f}{\partial R} + \dot{\phi}\frac{\partial f}{\partial \phi} + \dot{z}\frac{\partial f}{\partial z} + \dot{v}_R\frac{\partial f}{\partial v_R} + \dot{v}_\phi\frac{\partial f}{\partial v_\phi} + \dot{v}_z\frac{\partial f}{\partial v_z} = 0$$

and

$$\dot{v}_R = -\frac{\partial \Phi}{\partial R} + \frac{v_\phi^2}{R} \quad ; \quad \dot{v}_\phi = -\frac{1}{R}\frac{\partial \Phi}{\partial \phi} - \frac{v_R v_\phi}{R} \quad ; \quad \dot{v}_z = -\frac{\partial \Phi}{\partial z},$$

so

$$\begin{aligned}\frac{\partial f}{\partial t} + v_R\frac{\partial f}{\partial R} + \frac{v_\phi}{R}\frac{\partial f}{\partial \phi} + v_z\frac{\partial f}{\partial z} + \left(\frac{v_\phi^2}{R} - \frac{\partial \Phi}{\partial R}\right)\frac{\partial f}{\partial v_R} \\ - \frac{1}{R}\left(v_R v_\phi + \frac{\partial \Phi}{\partial \phi}\right)\frac{\partial f}{\partial v_\phi} - \frac{\partial \Phi}{\partial z}\frac{\partial f}{\partial v_z} = 0.\end{aligned} \tag{56}$$

4.2 The Jeans Equations

Suppose we integrate the CBE over all velocities:

$$\int \frac{\partial f}{\partial t} d^3\mathbf{v} + \int v_i \frac{\partial f}{\partial x_i} d^3\mathbf{v} - \frac{\partial \Phi}{\partial x_i}\int \frac{\partial f}{\partial v_i} d^3\mathbf{v} = 0. \tag{57}$$

If we define

$$\nu \equiv \int f d^3\mathbf{v} \quad \text{and} \quad \overline{v}_i \equiv \frac{1}{\nu}\int f v_i d^3\mathbf{v}, \tag{58}$$

then (57) can be written

$$\frac{\partial \nu}{\partial t} + \frac{\partial(\nu \overline{v}_i)}{\partial x_i} = 0, \tag{59}$$

which is a real-space continuity equation. If we now multiply the CBE by v_j and integrate over all $\mathbf{v}$, we obtain

$$\frac{\partial}{\partial t}\int f v_j d^3\mathbf{v} + \int v_i v_j \frac{\partial f}{\partial x_i} d^3\mathbf{v} - \frac{\partial \Phi}{\partial x_i}\int v_j \frac{\partial f}{\partial v_i} d^3\mathbf{v} = 0. \tag{60}$$

The divergence theorem applied to velocity space enables us to transform last term

$$\int v_j \frac{\partial f}{\partial v_i} d^3\mathbf{v} = -\int \frac{\partial v_j}{\partial v_i} f d^3\mathbf{v} = -\int \delta_{ij} f d^3\mathbf{v} = -\delta_{ij}\nu.$$

Substituting this into (60) yields

$$\frac{\partial(\nu \overline{v}_j)}{\partial t} + \frac{\partial(\nu \overline{v_i v_j})}{\partial x_i} + \nu\frac{\partial \Phi}{\partial x_j} = 0, \tag{61a}$$

where

$$\overline{v_i v_j} \equiv \frac{1}{\nu}\int v_i v_j f d^3\mathbf{v}. \tag{61b}$$

This is an equation of momentum conservation. It can be converted to Euler's equation of fluid flow by subtracting $\overline{v}_j \times$ [the continuity eq. (59)].

For many purposes it is helpful to partition $\overline{v_i v_j}$ by defining

$$\sigma_{ij}^2 \equiv \overline{(v_i - \overline{v}_i)(v_j - \overline{v}_j)} = \overline{v_i v_j} - \overline{v}_i \overline{v}_j. \tag{62}$$

At each point $\mathbf{x}$ the symmetric tensor $\boldsymbol{\sigma}^2$ defines an ellipsoid whose principal axes run parallel to $\boldsymbol{\sigma}^2$'s eigenvectors and whose semi-axes are proportion to the square roots of $\boldsymbol{\sigma}^2$'s eigenvalues. This is the **velocity ellipsoid** at $\mathbf{x}$.

Starting from (56) we can derive in close analogy with the derivation of (61) the Jeans equations for steady-state axisymmetric system:

$$\begin{aligned}
\frac{\partial(\nu\overline{v_R^2})}{\partial R} + \frac{\partial(\nu\overline{v_R v_z})}{\partial z} + \nu\left(\frac{\overline{v_R^2} - \overline{v_\phi^2}}{R} + \frac{\partial\Phi}{\partial R}\right) &= 0, \\
\frac{\partial(\nu\overline{v_R v_\phi})}{\partial R} + \frac{\partial(\nu\overline{v_\phi v_z})}{\partial z} + \frac{2\nu}{R}\overline{v_\phi v_R} &= 0, \\
\frac{\partial(\nu\overline{v_R v_z})}{\partial R} + \frac{\partial(\nu\overline{v_z^2})}{\partial z} + \frac{\nu\overline{v_R v_z}}{R} + \nu\frac{\partial\Phi}{\partial z} &= 0.
\end{aligned} \tag{63}$$

Similarly, the single non-trivial Jeans equation of a steady-state spherical system reads

$$\frac{d(\nu\overline{v_r^2})}{dr} + \frac{\nu}{r}\left[2\overline{v_r^2} - \left(\overline{v_\theta^2} + \overline{v_\phi^2}\right)\right] = -\nu\frac{d\Phi}{dr}. \tag{64}$$

4.2.1 Application of the Jeans equation to spherical systems

As an example of the utility of the Jeans equations consider their application to observations of a spherical galaxy. We assume that there are no mean streaming motions in the system and that the two tangential velocity dispersions $\overline{v_\theta^2}$ and $\overline{v_\phi^2}$ are equal. Then writing

$$\beta \equiv 1 - \frac{\overline{v_\theta^2}}{\overline{v_r^2}}, \tag{65}$$

equation (64) becomes

$$\frac{1}{\nu}\frac{\mathrm{d}(\nu\overline{v_r^2})}{\mathrm{d}r} + 2\frac{\beta\overline{v_r^2}}{r} = -\frac{\mathrm{d}\Phi}{\mathrm{d}r}. \tag{66}$$

Suppose we were able to measure $\overline{v_r^2}$, β, and the luminosity density ν. Then we could derive the mass $M(r)$ interior to r from

$$M(r) = -\frac{r\overline{v_r^2}}{G}\left(\frac{\mathrm{d}\ln\nu}{\mathrm{d}\ln r} + \frac{\mathrm{d}\ln\overline{v_r^2}}{\mathrm{d}\ln r} + 2\beta\right). \tag{67}$$

Comparing this with the light $L(r)$ interior to r we might hope to obtain evidence for a massive black hole at the centre of galaxies such as M87 that have active galactic nuclei. Unfortunately we can measure only the surface brightness $I(R)$ and line-of-sight velocity

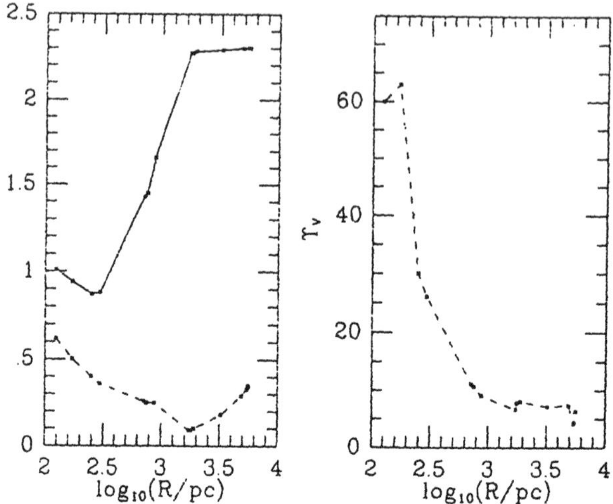

Fig. 15. If the velocity dispersion in Messier 87 were isotropic, the cumulative mass-to-light ratio $\Upsilon_V = M(r)/L_V(r)$ within radius r would rise steeply towards $r = 0$, suggesting that a massive dark object may lurk there.

dispersion $\sigma_p(R)$ at projected radius R. In the case $\beta = 0$ these are related to ν and $\overline{v_r^2}$ by

$$I(R) = 2\int_R^\infty \frac{\nu r\,\mathrm{d}r}{\sqrt{r^2 - R^2}} \quad \text{and} \quad I(R)\sigma_p^2(R) = 2\int_R^\infty \frac{\nu\overline{v_r^2} r\,\mathrm{d}r}{\sqrt{r^2 - R^2}}. \tag{68}$$

These Abel integral equations can be readily solved for $\nu(r)$ and $\overline{v_r^2}$. Sargent *et al.* (1978) derived by this route the cumulative mass $M(r)$ for M87 that is plotted in Figure 15. One sees that $M(r)/L(r)$ increases as $r \to 0$, suggesting the presence of a massive dark object at the galaxy centre.

In another approach to the analysis of the same data one assumes that the mass density $\rho(r) = \Upsilon\nu(r)$, where the mass-to-light ratio Υ is a constant. σ_p is now related to β, $\overline{v_r^2}$ and ν by

$$I(R)\sigma_p^2(R) = 2\int_R^\infty \left(1 - \beta\frac{R^2}{r^2}\right)\frac{\nu\overline{v_r^2} r\,\mathrm{d}r}{\sqrt{r^2 - R^2}}. \tag{69}$$

We may use this equation to eliminate β from the Jeans equation (66) to find

$$I\sigma_p^2 - G\Upsilon R^2\int_R^\infty \frac{\nu L(r)\,\mathrm{d}r}{r^2\sqrt{r^2 - R^2}} = R^2\int_R^\infty \left[2\nu\overline{v_r^2} + \frac{R^2}{r}\frac{\mathrm{d}(\nu\overline{v_r^2})}{\mathrm{d}r}\right]\frac{r\,\mathrm{d}r}{\sqrt{r^2 - R^2}}. \tag{70}$$

The left side of this equation contains the measured quantities I and σ_p, ν which we can obtain from the first of equations (68), and the undetermined constant Υ. Binney & Mamon (1982) showed that Υ emerges as an eigenvalue when one solves the right side for $\overline{v_r^2}$. β can then be obtained by substituting the recovered profile $\overline{v_r^2}$ into equation (67). If β satisfies the constraint $\beta \leq 1$ one has a viable model of the galaxy in which there is no dark object at the centre. The Sargent *et al.* observations of M87 pass this test.

This example illustrates a characteristic of the Jeans equations. Handy as they are for the interpretation of observations, they rarely yield a unique solution because they are an incomplete set of equations: the sets (59) and (61) contain four scalar equations linking ν, the three components of $\overline{\mathbf{v}}$ and the six independent components of $\boldsymbol{\sigma}^2$. To be sure, we could obtain further equations involving these quantities by multiplying the CBE by $v_j v_k$ and integrating over all $\mathbf{v}$. But the new equations would contain the third moments $v_i v_j v_k$ of $\mathbf{v}$ and we should still be short of equations. So we have to add something extra ourselves; say $\beta = 0$ or $\rho(r) = \Upsilon\nu(r)$, and our conclusions will be as plausible as our extra assumption.

4.2.2 Application of the Jeans equations to axisymmetric galaxies

From photometry of an axisymmetric galaxy and some assumed (constant) mass-to-light ratio Υ we can derive the galaxy's luminosity density $\nu(R, z)$ and potential $\Phi(R, z)$. Can we predict the velocity dispersion and mean-streaming velocity from these data? In the simplest possible model we assume

$$\sigma_{ij}^2 = \overline{(v_i - \overline{v}_i)(v_j - \overline{v}_j)} = \sigma^2 \delta_{ij}. \tag{71}$$

Then the first and last of equations (63) become

$$\frac{\partial(\nu\sigma^2)}{\partial R} - \nu\left(\frac{\overline{v}_\phi^2}{R} - \frac{\partial\Phi}{\partial R}\right) = 0 \quad \text{and} \quad \frac{\partial(\nu\sigma^2)}{\partial z} + \nu\frac{\partial\Phi}{\partial z} = 0. \tag{72}$$

Integrating the second equation we obtain the velocity dispersion as

$$\sigma^2(R, z) = \frac{1}{\nu}\int_z^\infty \nu\frac{\partial\Phi}{\partial z}\,\mathrm{d}z. \tag{73}$$

Substituting this result into the first of equations (72) we have

$$\overline{v}_\phi^2(R, z) = R\frac{\partial\Phi}{\partial R} + \frac{R}{\nu}\frac{\partial}{\partial R}\int_z^\infty \nu\frac{\partial\Phi}{\partial z}\,\mathrm{d}z. \tag{74}$$

Figure 16 shows the results of applying this procedure to NGC 4697 (Binney, Davies & Illingworth in preparation).

4.3 The Jeans Theorems

One form of the CBE simply states that f is constant along all orbits. Thus if f does not depend explicitly on time, it is an integral. Furthermore, since any function of integrals is itself an integral, any function of integrals solves the CBE. Thus we have

Jeans Theorem *Any steady-state solution of the collisionless Boltzmann equation depends on the phase-space coordinates only through integrals of motion in the galactic potential, and any function of the integrals yields a steady-state solution of the collisionless Boltzmann equation.*

As Jeans remarked in his (1915) statement of this theorem, this is a trivial but useful result. It does not tell us upon how many of a given set of integrals the distribution function (DF) of given galaxy is likely to depend. The following assures us that in ideal circumstances three independent isolating integrals would suffice:

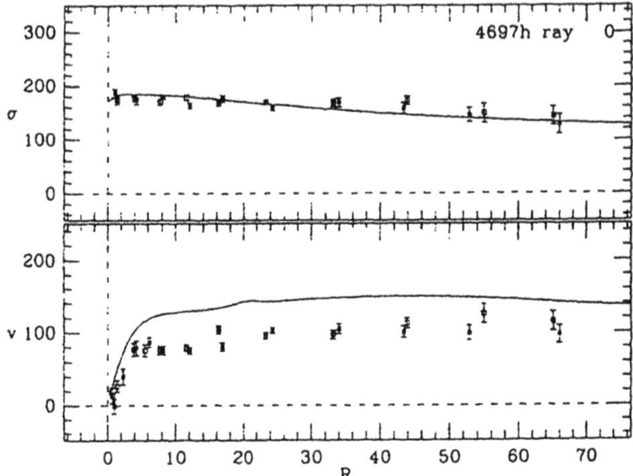

Fig. 16. The kinematics along the major axis of NGC 4697 predicted from CCD photometry and equations (73) and (74). The overall mass-to-light ratio has been chosen to optimize the fit between the predicted and observed minor-axis kinematics.

Strong Jeans Theorem *The DF of a steady-state galaxy in which almost all orbits are regular with incommensurable frequencies may be presumed to be a function only of three independent isolating integrals.*
The strong Jeans theorem is non-trivial. Its proof (§4.A of GD) exploits the fact that quasiperiodic orbits with irrational frequencies cover their tori uniformly. Notice the words "may be presumed". You can get correct results with a DF that is not of this special form, but you can get the same results with a simpler DF.

In applications of the Jeans theorems it is helpful to employ a potential and energy that are greater than zero for bound stars. So we define

$$\Psi \equiv -\Phi + \Phi_0 \quad \text{and} \quad \mathcal{E} \equiv -E + \Phi_0 = \Psi - \tfrac{1}{2}v^2. \tag{75}$$

Here Φ_0 is an arbitrary constant that we set to the value taken by Φ on the galaxy's boundary. Ψ satisfies Poisson's equation in the form

$$\nabla^2\Psi = -4\pi G\rho \quad \text{with} \quad \Psi \to \Phi_0 \quad \text{as} \quad |\mathbf{x}| \to \infty. \tag{76}$$

4.4 Distribution functions $f(\mathcal{E})$

The simplest systems have DFs that depend on energy only. These systems are very special because their velocity dispersion tensors $\boldsymbol{\sigma}^2$ are everywhere isotropic: if $f = f(\Psi - \frac{1}{2}v^2)$ then

$$\left.\begin{aligned} \overline{v_r^2} &= \frac{1}{\rho}\int dv_r dv_\theta dv_\phi v_r^2 f\left[\Psi - \tfrac{1}{2}\left(v_r^2 + v_\theta^2 + v_\phi^2\right)\right], \\ \overline{v_\theta^2} &= \frac{1}{\rho}\int dv_r dv_\theta dv_\phi v_\theta^2 f\left[\Psi - \tfrac{1}{2}\left(v_r^2 + v_\theta^2 + v_\phi^2\right)\right], \end{aligned}\right\} \qquad \text{so} \quad \overline{v_r^2} = \overline{v_\theta^2} = \overline{v_\phi^2}.$$

Given a DF $f(\mathcal{E})$ how do we discover what the corresponding $\rho(r)$ looks like? Well, Poisson's equation for our system can be written

$$\begin{aligned}\frac{1}{r^2}\frac{\mathrm{d}}{\mathrm{d}r}\left(r^2\frac{\mathrm{d}\Psi}{\mathrm{d}r}\right) &= -16\pi^2 G\int_0^{\sqrt{2\Psi}} f\left(\Psi - \tfrac{1}{2}v^2\right)v^2\mathrm{d}v \\ &= -16\pi^2 G\int_0^{\Psi} f(\mathcal{E})\sqrt{2(\Psi-\mathcal{E})}\mathrm{d}\mathcal{E}.\end{aligned} \tag{77}$$

The right side is a known function of Ψ so the equation is a non-linear differential equation for Ψ. One usually solves it from the inside out with initial conditions $\Psi = \Psi(0)$ and $\mathrm{d}\Psi/\mathrm{d}r = 0$.

4.4.1 Polytropes

The simplest dependence of f on $\mathcal{E}$ is as a power law:

$$f(\mathcal{E}) = \begin{cases} F\mathcal{E}^{n-3/2}, & \mathcal{E} > 0; \\ 0, & \mathcal{E} \le 0, \end{cases} \tag{78}$$

where F and n are constants. Then

$$\rho = 4\pi\int_0^{\infty} f(\Psi - \tfrac{1}{2}v^2)v^2\mathrm{d}v = 4\pi F\int_0^{\sqrt{2\Psi}} \left(\Psi - \tfrac{1}{2}v^2\right)^{n-3/2} v^2\mathrm{d}v.$$

With the substitution $v^2 = 2\Psi\cos^2\theta$ this becomes

$$\rho = c_n\Psi^n \quad (\Psi > 0), \quad \text{where} \quad c_n \equiv (2\pi)^{3/2}\frac{(n-\frac{3}{2})!F}{n!}. \tag{79}$$

c_n is finite for $n > \frac{1}{2}$. If we now eliminate ρ from $\nabla^2\Psi = -4\pi G\rho$, we obtain

$$\frac{1}{r^2}\frac{\mathrm{d}}{\mathrm{d}r}\left(r^2\frac{\mathrm{d}\Psi}{\mathrm{d}r}\right) + 4\pi G c_n\Psi^n = 0. \tag{80}$$

We can banish the constants from this equation by defining

$$s \equiv \frac{r}{b}, \quad \psi \equiv \frac{\Psi}{\Psi(0)}, \quad \text{where} \quad b \equiv \left(4\pi G\Psi(0)^{n-1}c_n\right)^{-1/2}. \tag{81}$$

In terms of these variables equation (80) becomes the Lane-Emden equation

$$\frac{1}{s^2}\frac{\mathrm{d}}{\mathrm{d}s}\left(s^2\frac{\mathrm{d}\psi}{\mathrm{d}s}\right) = \begin{cases} -\psi^n, & \psi > 0; \\ 0, & \psi \le 0. \end{cases} \tag{82}$$

that first arose in the theory of self-gravitating clouds of polytropic gas.

In stellar dynamics far and away the most important polytrope is that with $n = 5$, which is generally called *Plummer's Model* although it was actually first derived by Schuster (an Englishman). The solution to (82) when $n = 5$ is

$$\psi = \frac{1}{\sqrt{1+\frac{1}{3}s^2}} \quad \text{because then} \quad \frac{1}{s^2}\frac{\mathrm{d}}{\mathrm{d}s}\left(s^2\frac{\mathrm{d}\psi}{\mathrm{d}s}\right) = -\frac{1}{(1+\frac{1}{3}s^2)^{5/2}} = -\psi^5.$$

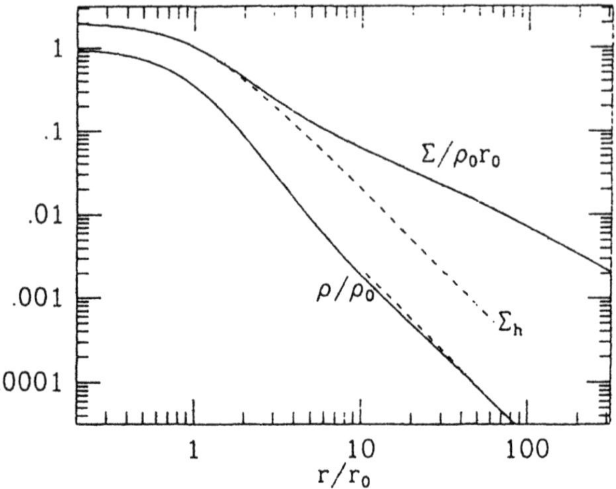

Fig. 17. Full curves show true and projected mass densities of the isothermal sphere. The dashed curves show (a) the singular isothermal sphere to which ρ asymptotes as $r \to \infty$ and (b) the projected modified Hubble profile.

Hence the density of Plummer's model is given by

$$\rho = c_5 \psi^5 = \frac{c_5 \Psi_0^5}{(1 + \frac{1}{3}s^2)^{5/2}}. \tag{83a}$$

The density extends to ∞ but the total mass is finite;

$$M_\infty = \frac{1}{G}\left(r^2 \frac{\mathrm{d}\Phi}{\mathrm{d}r}\right)_{r\to\infty} = \frac{\sqrt{3}\Psi_0 b}{G}. \tag{83b}$$

4.4.2 The isothermal sphere

In limit $n \to \infty$ the polytropes asymptote to the isothermal sphere, which has DF

$$f(\mathcal{E}) = \frac{\rho_1}{(2\pi\sigma^2)^{3/2}} e^{\mathcal{E}/\sigma^2} \quad (\rho_1, \sigma \text{ constants}). \tag{84}$$

The parameter σ sets the system's "temperature" T; for particles of mass m one has $\sigma^2 = k_B T/m$. With the DF (84) the mean-square speed of stars is everywhere the same:

$$\overline{v^2} = \frac{\displaystyle\int_0^\infty \exp\left(\frac{\Psi - \frac{1}{2}v^2}{\sigma^2}\right) v^4 \mathrm{d}v}{\displaystyle\int_0^\infty \exp\left(\frac{\Psi - \frac{1}{2}v^2}{\sigma^2}\right) v^2 \mathrm{d}v} = 2\sigma^2 \frac{\int_0^\infty e^{-x^2} x^4 \mathrm{d}x}{\int_0^\infty e^{-x^2} x^2 \mathrm{d}x} = 3\sigma^2. \tag{85}$$

Integrating (84) over velocities we find $\rho = \rho_1 e^{\Psi/\sigma^2}$, so Poisson's equation can be written as

$$\frac{\mathrm{d}}{\mathrm{d}r}\left(r^2\frac{\mathrm{d}\Psi}{\mathrm{d}r}\right) = -4\pi G\rho_1 r^2 e^{\Psi/\sigma^2}. \tag{86}$$

It is helpful to define the **King radius**

$$r_0 \equiv \sqrt{\frac{9\sigma^2}{4\pi G\rho_0}} \quad \text{and the dimensionless variables} \quad \widetilde{\rho} \equiv \rho/\rho_0 \text{ and } \widetilde{r} \equiv r/r_0. \tag{87}$$

Figure 17 shows the function $\widetilde{\rho}(\widetilde{r})$ that is obtained by numerically solving equation (86) in these variables. As $r \to \infty$, $\widetilde{\rho}$ asymptotes to

$$\widetilde{\rho} = \frac{2}{9\widetilde{r}^2}, \quad \text{that is,} \quad \rho(r) = \frac{\sigma^2}{2\pi G r^2}, \tag{88}$$

which is an exact solution of equation (86). The elementary solution (88) is called the **singular isothermal sphere.**

At $\widetilde{r} \lesssim 2$ a useful approximation to $\widetilde{\rho}$ is

$$\widetilde{\rho}(\widetilde{r}) \approx \widetilde{\rho}_h(\widetilde{r}) \equiv \frac{1}{(1+\widetilde{r}^2)^{3/2}} \quad \text{(modified Hubble profile)}, \tag{89a}$$

which projects to

$$\Sigma_h(\widetilde{R}) = \frac{2}{1+\widetilde{R}^2}. \tag{89b}$$

The dashed curve in Figure 17 shows the fit to the projected isothermal sphere that Σ_h provides.

The isothermal sphere has two main applications:

(i) Core fitting: one measures the central surface brightness $I(0)$, the King radius r_0 and the central line-of-sight velocity dispersion σ of a galaxy and then infers the central mass-to-light ratio Υ_0 from

$$\Upsilon = \rho_0/\nu_0, \text{ where } \rho_0 = \frac{9\sigma^2}{4\pi G r_0^2} \text{ and } \nu_0 = 0.495 I(0)/r_0.$$

(ii) Dark-Halo fitting: galaxies seem to have circular speeds that are remarkably radius independent. Except near their centres their potentials cannot differ greatly from that of the singular isothermal sphere.

4.4.3 King models

No real galaxy can extend to infinity. So it is natural to try to construct model galaxies that are isothermal-like close to their centres, but die away as $r \to \infty$ faster than $\rho \propto r^{-2}$. A DF for such systems is obtained by modifying the DF of the isothermal sphere so that it vanishes at a finite value of $\mathcal{E}$:

$$f_K(\mathcal{E}) = \begin{cases} \rho_1(2\pi\sigma^2)^{-3/2}\left(e^{\mathcal{E}/\sigma^2} - 1\right), & \mathcal{E} > 0, \\ 0, & \mathcal{E} \leq 0. \end{cases} \tag{90}$$

Systems with DFs of this type have come to be known as **King models**. One solves for the spatial structure of a King model by integrating (90) over all $\mathbf{v}$, substituting the result into equation (77) and integrating Ψ from some assumed central value Ψ_0 outwards. At some radius r_t the derived density vanishes and the integration of (77) ceases. r_t is

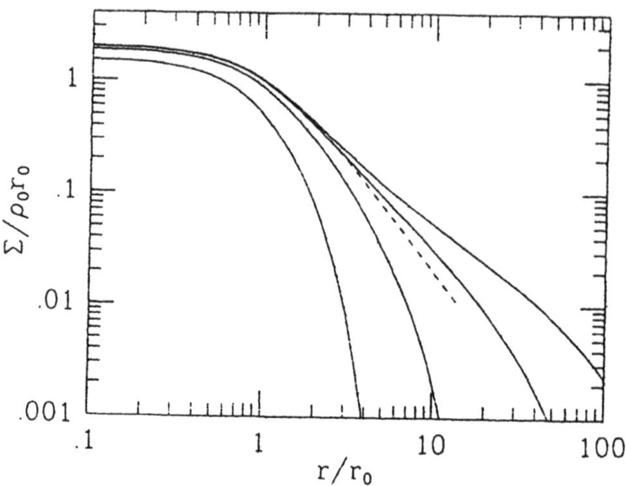

Fig. 18. Projected King models with $\Psi(0)/\sigma^2 = 12, 9, 6, 3$. The dashed curve shows the projected modified Hubble profile.

called the **tidal radius**. The **concentration** of the system $c \equiv \log_{10}[r_t/r_0]$ is a function of Ψ_0/σ^2—see Figure 18. The brightness profiles of elliptical galaxies are moderately well fitted by King models with $\Psi_0/\sigma^2 \simeq 10$ and $c \simeq 2.4$. Globular clusters can be fitted with King models with Ψ_0/σ^2 in the range $(3-7)$.

The King models constitute just one step in a potentially infinite hierarchy of lowered isothermals with DFs of the form;

$$f = \text{constant} \times \begin{cases} [e^{\mathcal{E}/\sigma^2} - 1 - (\mathcal{E}/\sigma^2) - \cdots] & \text{for } \mathcal{E} > 0, \\ 0 & \text{for } \mathcal{E} \leq 0. \end{cases} \tag{91}$$

The more terms in the series for e^x that we subtract from the exponential the more smoothly the DF goes to zero at $\mathcal{E} = 0$; if we simple truncate $e^{\mathcal{E}/\sigma^2}$ at $\mathcal{E} = 0$ we obtain the models of Woolley & Dickens (1961), while if we subtract from the exponential both 1 and x we obtain Wilson's (1975) DF. A stimulating discussion of this topic will be found in Hunter (1977).

4.4.4 Eddington inversion

Suppose we are given a density profile $\rho(r)$ and are asked to find a DF $f(\mathcal{E})$ that self consistently derives this profile. Can we oblige? Eddington (1916) showed that a given $\rho(r)$ is generated by a unique $f(\mathcal{E})$, although the latter is not necessarily non-negative. Thus, since Ψ is a monotone function of r, we may regard ρ to be a function $\rho(\Psi)$. Then equation (77) can be written

$$\frac{1}{\sqrt{8}\pi}\rho(\Psi) = 2\int_0^{\Psi} f(\mathcal{E})\sqrt{\Psi - \mathcal{E}}\,d\mathcal{E}. \tag{92}$$

On differentiation with respect to Ψ this becomes an Abel integral equation, with solution

$$
\begin{aligned}
f(\mathcal{E}) &= \frac{1}{\sqrt{8}\pi^2}\frac{\mathrm{d}}{\mathrm{d}\mathcal{E}}\int_0^{\mathcal{E}}\frac{\mathrm{d}\rho}{\mathrm{d}\Psi}\frac{\mathrm{d}\Psi}{\sqrt{\mathcal{E}-\Psi}} \\
&= \frac{1}{\sqrt{8}\pi^2}\left[\int_0^{\mathcal{E}}\frac{\mathrm{d}^2\rho}{\mathrm{d}\Psi^2}\frac{\mathrm{d}\Psi}{\sqrt{\mathcal{E}-\Psi}}+\frac{1}{\sqrt{\mathcal{E}}}\left(\frac{\mathrm{d}\rho}{\mathrm{d}\Psi}\right)_{\Psi=0}\right].
\end{aligned}
\tag{93}
$$

4.5 Distribution functions $f(\mathcal{E}, L)$

If we allow the magnitude L of the angular momentum vector to appear in f, we obtain models in which $\overline{v_\theta^2} = \overline{v_\phi^2} \neq \overline{v_r^2}$. The number of simple forms of f with which one can now play is legion. So rather than fool with any particular DFs, let me mention an elegant modification of Eddington's inversion that enables one to derive one-parameter families of DFs that all generate the same density profile $\rho(r)$; the systems generated by the members of such a family differ only in the degree of radial bias of their velocity dispersion tensors.

Suppose we assume that f depends on $\mathcal{E}$ and L only in the combination

$$
Q \equiv \mathcal{E} - \frac{L^2}{2r_a^2} \quad (r_a \text{ a constant}). \tag{94}
$$

Then we may analytically integrate out the direction of the velocity vector $\mathbf{v}$ to derive the relation

$$
\sqrt{32}\pi\int_0^{\Psi} f(Q)\sqrt{\Psi - Q}\,\mathrm{d}Q = \left(1+\frac{r^2}{r_a^2}\right)\rho(r) \equiv \rho_Q.
$$

But this equation states that the relation between $f(Q)$ and ρ_Q is identical with Eddington's relation (92) between $f(\mathcal{E})$ and ρ. Hence we may obtain $f(Q)$ from $\rho_Q(\Psi)$ by making the substitutions $\rho \to \rho_Q$ and $\mathcal{E} \to Q$ in (93).

Figure 19 shows the projected velocity dispersion profiles of three dynamical models of Jaffe's (1983) profile

$$
\rho = \frac{M}{4\pi r_J^3}\frac{r_J^4}{r^2(r+r_J)^2} \quad \Rightarrow \quad \Phi = \frac{GM}{r_J}\ln\left(\frac{r}{r+r_J}\right). \tag{95}
$$

Notice that the more radial the model's velocity dispersion tensor is, the more centrally peaked is its projected velocity dispersion profile.

4.6 Differential Energy Distribution

A quantity of considerable interest, particularly in relation to n-body models, is the differential energy distribution $\mathrm{d}M/\mathrm{d}\mathcal{E}$, which gives the mass in stars with binding energies near $\mathcal{E}$.

Suppose the DF is of the form $f(\mathcal{E}, L)$ and let v_t denote the magnitude of the component of $\mathbf{v}$ that is perpendicular to $\hat{\mathbf{e}}_r$. Then introducing polar coordinates (v, η) in the (v_r, v_t) plane by $v_r = v\cos\eta$, $v_t = v\sin\eta$, we may write the total mass as

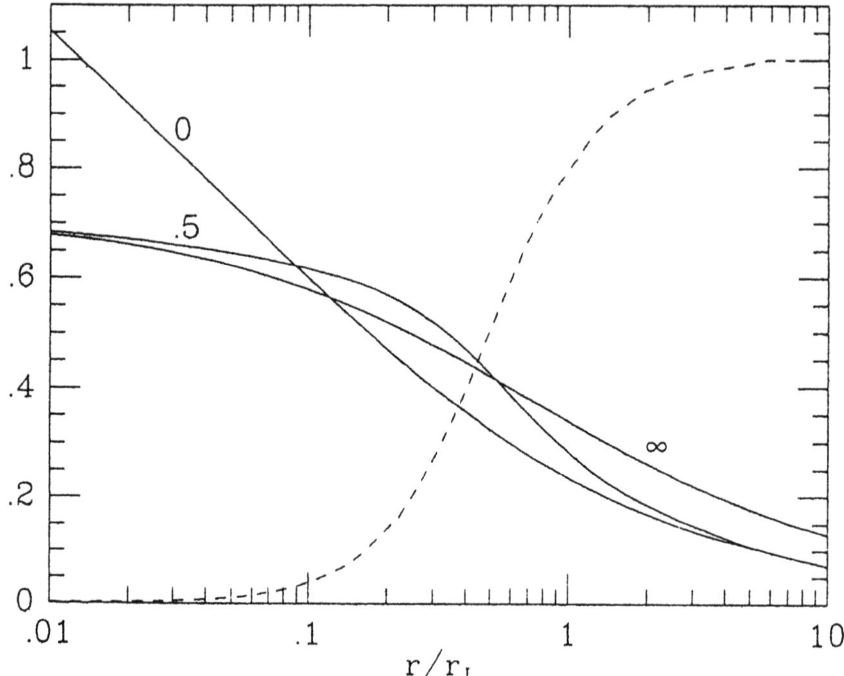

Fig. 19. The more radial a model's orbits, the steeper its projected velocity dispersion falls with radius. Full curves show the projected velocity dispersions of three dynamical versions of Jaffe's density profile; the curve marked "0" is for a model in which all orbits are precisely radial, while the curve marked "∞" corresponds to a model in which $\boldsymbol{\sigma}^2$ is everywhere isotropic. The dashed curve shows the form of the anisotropy parameter β in the model whose curve is marked ".5".

$$\begin{aligned} M &= 8\pi^2 \int_0^\infty r^2 \mathrm{d}r \int_0^{\Psi(r)} v \mathrm{d}\mathcal{E} \int_0^\pi f\left(\mathcal{E}, rv|\sin\eta|\right) \sin\eta \,\mathrm{d}\eta \\ &= 8\pi^2 \int_0^{\Psi(0)} \mathrm{d}\mathcal{E} \int_0^{r_m(\mathcal{E})} v r^2 \mathrm{d}r \int_0^\pi f\left(\mathcal{E}, rv|\sin\eta|\right) \sin\eta \,\mathrm{d}\eta, \end{aligned} \tag{96}$$

where $r_m(\mathcal{E})$ is defined by $\Psi(r_m) = \mathcal{E}$. From the second line of (96) it follows that

$$\frac{\mathrm{d}M}{\mathrm{d}\mathcal{E}} = 8\pi^2 \int_0^{r_m(\mathcal{E})} v r^2 dr \int_0^\pi f\left(\mathcal{E}, rv|\sin\eta|\right) \sin\eta \,\mathrm{d}\eta. \tag{97}$$

Figure 20 shows $\mathrm{d}M/\mathrm{d}\mathcal{E}$ for the two extreme Jaffe models of Figure 19. Evidently $\mathrm{d}M/\mathrm{d}\mathcal{E}$ depends strongly on the density profile $\rho(r)$ and only weakly on the system's velocity dispersion structure. Later I'll show how this fact can be used to derive approximate DFs for odd-shaped galaxies.

Notice that the curves in Figure 20 slope upwards towards $E = 0$, the energy of *unbound* stars. By contrast, the DFs of these models increase strongly towards $E = -\infty$, the energy of the most tightly bound stars. Thus even though the phase-space density is largest deep in the system's potential well, most stars are only loosely bound.

When f is a function $f(\mathcal{E})$ of $\mathcal{E}$ only, we can integrate immediately over η to find

$$\frac{\mathrm{d}M}{\mathrm{d}\mathcal{E}} = f(\mathcal{E})g(\mathcal{E}) \quad \text{where} \quad g(\mathcal{E}) \equiv 16\pi^2 \int_0^{r_m(\mathcal{E})} \sqrt{2(\Psi - \mathcal{E})} r^2 \mathrm{d}r. \tag{98}$$

The function g is the classical analogue of the density of states familiar from the statistical mechanics of quantum systems. It increases so strongly towards $\mathcal{E} = 0$ that its growth

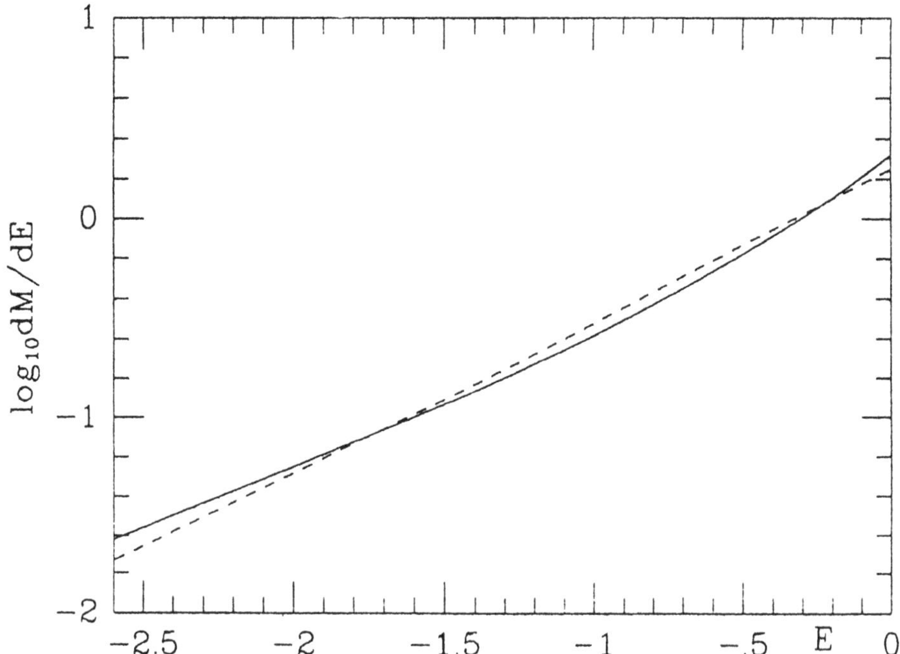

Fig. 20. The differential mass distribution depends only weakly on a system's dynamics. The full curve is for a Jaffe sphere with isotropic velocity dispersion while the dashed curve is for a Jaffe sphere with entirely radial orbits.

overwhelms the near-exponential decrease of f with decreasingas $\mathcal{E}$, with the result that $\mathrm{d}M/\mathrm{d}\mathcal{E}$ is largest near $\mathcal{E} = 0$.

4.7 Planar Systems

Motion in an axisymmetric planar system obviously admits two isolating integrals, E and L_z. The strong Jeans theorem assures us that we should be able to model most axisymmetric disks with functions $f(\mathcal{E}, L_z)$. Here are two examples.

4.7.1 Mestel's disk

It turns out that a disk in which the surface density Σ is inversely proportional to radius R generates exactly the same potential as a spherical body which has as much mass interior to a sphere of radius R as does the disk. Thus

$$\Sigma(R) = \frac{\Sigma_0 R_0}{R} \quad \Rightarrow \quad v_c^2 \equiv -R\frac{\partial \Psi}{\partial R} = 2\pi G \Sigma_0 R_0 \text{ is constant.} \tag{99a}$$

This object, which we call **Mestel's disk**, is self-consistently produced by DFs of the form

$$f(\mathcal{E}, L_z) = \begin{cases} F L_z{}^q e^{\mathcal{E}/\sigma^2} & L_z > 0 \\ 0 & L_z \le 0, \end{cases} \quad \text{where} \quad \begin{cases} q \equiv v_c^2/\sigma^2 - 1, \\ F \equiv \dfrac{\Sigma_0 R_0}{2^{\frac{q}{2}}\sqrt{\pi}\left(\frac{q-1}{2}\right)!\,\sigma^{(q+2)}}. \end{cases} \tag{99b}$$

From the form of f it follows that the free parameter σ is the velocity dispersion in the radial direction. The variable q is a measure of the disk's "coldness". In fact

$$\overline{v}_\phi = \frac{\sqrt{2}\,(\frac{1}{2}q)!}{(\frac{q-1}{2})!}\,\sigma. \tag{100}$$

Thus as $q \to \infty$, $\overline{v}_\phi \propto \sqrt{q}$ and the disk becomes entirely supported by centrifugal force, while $\overline{v}_\phi$ vanishes when $q = -1$.

4.7.2 Kalnajs' disk

If we project any homogeneous, axisymmetric ellipsoid of radius a parallel to its symmetry axis, we obtain surface density

$$\Sigma(R) = \Sigma_0\sqrt{1 - \frac{R^2}{a^2}}. \tag{101a}$$

The potential inside any homogeneous ellipsoid is always a quadratic function of the Cartesian coordinates. Thus when we squash our ellipsoid right down into its symmetry plane its potential at $z = 0$ and $R \leq a$ must be proportional to R^2. In detail

$$\begin{aligned}\Phi(R) &= \frac{\pi^2 G\Sigma_0}{4a}R^2 + \text{constant}\\ &\equiv \tfrac{1}{2}\Omega_0^2 R^2 + \text{constant}.\end{aligned} \tag{101b}$$

We call the potential-density pair (101) after Agris Kalnajs who investigated the stability of the models of this system which have DFs of the form

$$f(\mathcal{E}, L_z) = \begin{cases} F/\sqrt{(\Omega_0^2 - \Omega^2)a^2 + 2(\mathcal{E} + \Omega L_z)} & \text{if } \surd \text{ real},\\ 0 & \text{if } \surd \text{ imag.}\end{cases} \tag{102}$$

Here $F(\Omega)$ is a constant and Ω is a free parameter that turns out to be related to the mean streaming velocity simply by $\overline{v}_\phi = \Omega R$. Thus DFs of the form (102) encompass both hot disks and cold, centrifugally supported disks. In every case the radial and tangential velocity dispersions are equal: $\overline{v_R^2} = \overline{(v_\phi - \overline{v}_\phi)^2}$.

4.8 Three-Dimensional Aspherical Systems

According to the strong Jeans theorem we must in general use three isolating integrals in the DF of a three-dimensional system. Unfortunately, even in the axisymmetric case we can write down only two isolating integrals.

4.8.1 Systems with $F(E, L_z)$

The standard Jeans theorem permits us to construct some models with $f(E, L_z)$. Now it is easy to show that throughout any three-dimensional system with a DF of the form $f(E, L_z)$ we have $\overline{v_R^2} = \overline{v_z^2}$. This is bad news since near the Sun we know that for most populations $\overline{v_R^2} \simeq 2\overline{v_z^2}$. Despite this restriction, Jarvis & Freeman (1985) and Rowley (1988) have obtained remarkably accurate fits to observations of external galaxies using DFs of the form $f(E, L_z)$. Perhaps these galaxies are simpler than ours. Or perhaps similar observations of our Galaxy by a distant observer could also be fitted by assuming $f(E, L_z)$.

Lynden-Bell (1962), Hunter (1975) and Dejonghe (1986) have shown that from $\rho(R,z)$ and $\overline{v}_\phi(R,z)$ one can uniquely determine a corresponding DF $f(E,L_z)$. However, there is no guarantee that $f \geq 1$. Also Dejonghe shows that the analytic techniques employed in these demonstrations are in practice unstable. However, Newton (1986) has shown that Lucy's (1974) iterative method can be successfully applied to this inversion problem in practical cases.

4.8.2 Models that employ non-classical integrals

The most obvious way to model a stellar system with a general DF is to contstruct a suitable n-body model. However, n-body models suffer from several drawbacks: (i) they are plagued by small-number statistics; (ii) they are expensive; (iii) they are hard to characterize exactly; (iv) it is hard to tailor an n-body model to a particular set of observations. Consequently it is worth investigating other methods of generating general galaxy models. The starting point for such a model can be either a DF or a density distribution $\rho(\mathbf{x})$.

1. From $f \rightarrow \rho$: Since action-angle coordinates $(\mathbf{J},\boldsymbol{\theta})$ are canonical, the Jacobian $\partial(\mathbf{x},\mathbf{v})/\partial(\mathbf{J},\boldsymbol{\theta}) = 1$. Thus the phase-space volume occupied by a group of orbits with actions in $\mathrm{d}^3\mathbf{J}$ is

$$\tau = \int_{\text{orbits}} \mathrm{d}^3\mathbf{x}\mathrm{d}^3\mathbf{v} = (2\pi)^3\mathrm{d}^3\mathbf{J}, \tag{103}$$

where the second equality follows because each angle variable covers the range $(0,2\pi)$ around any orbit. Thus the mass in stars on orbits in $\mathrm{d}^3\mathbf{J}$ is $\mathrm{d}M = (2\pi)^3 f(\mathbf{J})\mathrm{d}^3\mathbf{J}$ and we see that $f(\mathbf{J})$ is, to within the constant factor $(2\pi)^3$, the density of stars in three-dimensional action space as well as in six-dimensional phase space. Hence if a galaxy's potential, like a Stäckel potential, admits global action-angle variables, then choosing a DF for the galaxy is equivalent to choosing a stellar density distribution for the system's action space. It is not hard to gain a feel for how the structure of a galaxy depends on where in action space we place its stars. From the spherically-averaged density profile $\rho(r)$ one can estimate $\mathrm{d}M/\mathrm{d}E$, that is how many stars should be placed on each surface of constant energy (in action space these are approximately planar triangles). Then one obtains the desired shape and velocity-ellipsoid structure by shuffling the stars around each $E =$ constant surface. In particular, we encourage radial bias in $\boldsymbol{\sigma}$ and ultimately prolate geometry of the entire galaxy by pushing the stars towards the J_r-axis. For examples of the use of this approach see Binney & Petit (1989), Binney (1987) or Ostriker, Binney & Saha (1989).

2. From $\rho \rightarrow f$: In a seminal paper Schwarzschild (1979) showed how a triaxial galaxy could be constructed with a given density profile $\rho(\mathbf{x})$. Since ρ is given one can immediately solve for the potential $\Phi(\mathbf{x})$ and calculate some orbits. Let the αth orbit spend a fraction $P_{\alpha\beta}$ of its time in a small cell of volume Δ_β centred on $\mathbf{x}_\beta$. Then if we assign mass m_α to this orbit, the density distribution to which this orbit gives rise is $\rho_\alpha(\mathbf{x}_\beta) = m_\alpha P_{\alpha\beta}/\Delta_\beta$. The condition that we generate the given density distribution at all spatial grid points by placing m_α stars on the αth of N orbits is therefore that the m_α satisfy

$$\rho(\mathbf{x}_\beta) = \sum_{\alpha=1}^{N} m_\alpha P_{\alpha\beta}/\Delta_\beta \quad (\beta = 1,\ldots,K). \tag{104}$$

It is tempting to set $K = N$ and solve these linear equations for the m_α. Unfortunately, if we did so we would generally find some of the m_α were negative, which is physically impossible. So Schwarzschild chose $N \simeq 2K$ and used the technique of linear programming so find (non-unique) sets of $m_\alpha \geq 0$ that satisfy equations (104). Statler (1987) has used this technique to explore in some detail the range of possible dynamical models of the perfect ellipsoids.

The major snag with Schwarzschild's technique is that the need repeatedly to invert $K \times K$ matrices during the linear programming phase severely restricts the resolution of the models that one can construct with it. Also in the final solutions adjacent orbits in action space are assigned weights m_α which are large in one case and zero in the other. Consequently, it is next to impossible to infer the structure of a smooth distribution function $f(\mathbf{J})$ for the system from the m_α (Newton 1986). Larger numbers of orbits can be employed and smoothness of the m_α guaranteed if one uses Lucy's (1974) iterative procedure to invert equations (104), but at the expense of settling for an approximate solution. Richstone & Tremaine (1988) have explored the use of entropy maximization as a device for obtaining smooth, non-negative m_α that satisfy equations (104).

5 Perturbation Theory

Let me now turn from the construction of equilibrium models to techniques for studying the evolution of such systems by perturbing them slightly. Consider first the perturbations that arise because galaxies are not the perfectly smooth things we have hitherto assumed, but contain lumps ranging from the size of an individual star to the size of a globular cluster or giant molecular cloud ($M \sim 5 \times 10^5\,\mathrm{M}_\odot$). The gravitational fields of these lumps deflect slightly the orbits of stars from those they would follow in a perfectly smooth potential. So at each time every star is on a well-defined orbital torus, but the label $\mathbf{J}$ of this torus gradually changes. In other words, stars gradually diffuse through action space, causing the action-space stellar density $(2\pi)^3 f(\mathbf{J})$ to obey a diffusion equation. We can use perturbation theory to determine this equation.

5.1 Orbit-Averaged Fokker-Planck Equation

Let $\delta P(\mathbf{J}, \mathbf{\Delta})$ be the probability that a star initially on the torus $\mathbf{J}$ moves to the torus $\mathbf{J} + \mathbf{\Delta}$ in time δt. Then

$$\int \delta P(\mathbf{J}, \mathbf{\Delta}) \mathrm{d}^3 \mathbf{\Delta}$$

is the total probability that a star on $\mathbf{J}$ is scattered from $\mathbf{J}$. Now we have seen that there are $(2\pi)^3 \mathrm{d}^3\mathbf{J} f(\mathbf{J}, t)$ stars on tori near $\mathbf{J}$. So the number of stars leaving these tori is

$$(2\pi)^3 \mathrm{d}^3\mathbf{J} f(\mathbf{J}, t) \int \delta P(\mathbf{J}, \mathbf{\Delta}) \mathrm{d}^3 \mathbf{\Delta}. \tag{105}$$

Similarly number of stars scattered onto $\mathbf{J}$ is

$$(2\pi)^3 \mathrm{d}^3\mathbf{J} \int f(\mathbf{J} - \mathbf{\Delta}, t) \delta P(\mathbf{J} - \mathbf{\Delta}, \mathbf{\Delta}) \mathrm{d}^3 \mathbf{\Delta}. \tag{106}$$

Equating the difference between these rates to the rate of accumulation of stars on $\mathbf{J}$ we have

$$\frac{\partial f}{\partial t} = \int f(\mathbf{J}-\mathbf{\Delta},t)\dot{P}(\mathbf{J}-\mathbf{\Delta},\mathbf{\Delta})\mathrm{d}^3\mathbf{\Delta} - f(\mathbf{J},t)\int \dot{P}(\mathbf{J},\mathbf{\Delta})\mathrm{d}^3\mathbf{\Delta}. \tag{107}$$

But since each individual scattering changes the actions by only a small amount, $\dot{P}(\mathbf{J},\mathbf{\Delta})$ is non-negligible only for small Δ. So we may expand $f(\mathbf{J},t)\dot{P}(\mathbf{J},\mathbf{\Delta})$ as a power series in $\mathbf{J}$:

$$f(\mathbf{J}-\mathbf{\Delta},t)\dot{P}(\mathbf{J}-\mathbf{\Delta},\mathbf{\Delta}) = f(\mathbf{J})\dot{P}(\mathbf{J},\mathbf{\Delta}) - \Delta_i\frac{\partial(f\dot{P})}{\partial J_i} + \tfrac{1}{2}\Delta_i\Delta_j\frac{\partial^2(f\dot{P})}{\partial J_i\partial J_j} + \cdots \tag{108}$$

Retaining the largest three terms we obtain the **orbit-averaged Fokker-Planck equation**

$$\frac{\partial f}{\partial t} \simeq -\frac{\partial(f\overline{\Delta}_i)}{\partial J_i} + \tfrac{1}{2}\frac{\partial^2(f\overline{\Delta^2}_{ij})}{\partial J_i\partial J_j}, \tag{109a}$$

where the **diffusion coefficients** are defined by

$$\overline{\Delta}_i(\mathbf{J}) \equiv \int \Delta_i\dot{P}(\mathbf{J},\mathbf{\Delta})\mathrm{d}^3\mathbf{\Delta}, \quad \overline{\Delta^2}_{ij}(\mathbf{J}) \equiv \int \Delta_i\Delta_j\dot{P}(\mathbf{J},\mathbf{\Delta})\mathrm{d}^3\mathbf{\Delta}. \tag{109b}$$

5.1.1 The diffusion coefficients

Before we can use (109a) to calculate the evolution of f we have to obtain expressions for the diffusion coefficients (109b) as functions of $\mathbf{J}$. We break Φ into a smooth, constant part and the small, fluctuating part that gives rise to the stellar diffusion:

$$\Phi(\mathbf{x},t) = \Phi_0(\mathbf{x}) + \delta\Phi(\mathbf{x},t) = \Phi_0 + \sum_{n_1,n_2,n_3}\delta\Phi_{\mathbf{n}}(\mathbf{J},t)\exp[i(\mathbf{n}\cdot\boldsymbol{\theta})], \tag{110a}$$

where $\mathbf{J}$ labels the tori of Φ_0. By the reality of Φ we have

$$\delta\Phi_{-\mathbf{n}} = \delta\Phi^*_{\mathbf{n}}. \tag{110b}$$

The Hamiltonian similarly decomposes

$$H(\boldsymbol{\theta},\mathbf{J}) = H_0(\mathbf{J}) + \delta\Phi(\boldsymbol{\theta},\mathbf{J}) \quad \text{with} \quad \omega_0(\mathbf{J}) \equiv \frac{\partial H_0}{\partial \mathbf{J}}. \tag{111}$$

We correspondingly decompose the coordinates of a star into unpeturbed and perturbed contributions

$$\begin{aligned}\mathbf{J}(t) &= \mathbf{J}_0 + \mathbf{\Delta}_1(t) + \mathbf{\Delta}_2(t) + \cdots\\ \boldsymbol{\theta}(t) &= \boldsymbol{\theta}_0 + \boldsymbol{\omega}_0 t + \boldsymbol{\theta}_1(t) + \cdots\end{aligned} \quad \text{where} \quad (\boldsymbol{\theta}_{\text{orbit0}} = \boldsymbol{\theta}_0 + \boldsymbol{\omega}_0 t,\ \mathbf{J}_{\text{orbit0}} = \mathbf{J}_0). \tag{112}$$

Here the subscript "orbit0" means along the unperturbed orbit. We can now find Δ_1 by integrating $\mathbf{J}$'s equation of motion along the unperturbed orbit:

$$\begin{aligned}\dot{\mathbf{J}} &= -\frac{\partial H}{\partial\boldsymbol{\theta}} = -\frac{\partial\delta\Phi}{\partial\boldsymbol{\theta}}\\ &= -i\sum_{\mathbf{n}}\mathbf{n}\delta\Phi_{\mathbf{n}}(\mathbf{J},t)e^{i\mathbf{n}\cdot\boldsymbol{\theta}}\end{aligned} \quad \text{so} \quad \mathbf{\Delta}_1(T) = -i\sum_{\mathbf{n}}\mathbf{n}\int_0^T \delta\Phi_{\mathbf{n}}(\mathbf{J}_0,t)e^{i\mathbf{n}\cdot(\boldsymbol{\theta}_0+\boldsymbol{\omega}_0 t)}\,\mathrm{d}t. \tag{113}$$

Squaring and averaging over the orbital phase $\boldsymbol{\theta}_0$, we have

$$\langle \Delta_{1i}\Delta_{1j}(T)\rangle_\theta = \sum_{\mathbf{n}} n_i n_j \int_0^T \mathrm{d}t \int_0^T \mathrm{d}t'\ \delta\Phi_{\mathbf{n}}(\mathbf{J}_0,t)\delta\Phi^*_{\mathbf{n}}(\mathbf{J}_0,t')e^{i\mathbf{n}\cdot\boldsymbol{\omega}_0(t-t')}. \tag{114}$$

We now ensemble-average equation (114) under the assumption that $\delta\Phi$ is a random variable with an autocorrelation of the form

$$c_{\mathbf{n}}(\mathbf{J},t-t') \equiv \overline{\delta\Phi_{\mathbf{n}}(\mathbf{J},t)\delta\Phi^*_{\mathbf{n}}(\mathbf{J},t')};\quad \text{[note that } c_{-\mathbf{n}}(\mathbf{J},v) = c^*_{\mathbf{n}}(\mathbf{J},v) = c_{\mathbf{n}}(\mathbf{J},-v)]. \tag{115}$$

On substituting this relation into the ensemble-average of equation (114), changing variables to $u_\pm \equiv t \pm t'$ and integrating over u_+, we find

$$\overline{\Delta_{1i}\Delta_{1j}} = \sum_{\mathbf{n}} n_i n_j \int_{-T}^{T} (T - |u_-|)c_{\mathbf{n}}(\mathbf{J}_0,u_-)e^{i(\mathbf{n}\cdot\boldsymbol{\omega}_0)u_-}\,\mathrm{d}u_-. \tag{116}$$

The diffusion coefficients $\overline{\Delta^2}_{ij}$ defined in equations (109b) are averages of the left side of (116) for intervals T sufficiently long that there is negligible correlation between the changes $\boldsymbol{\Delta}$ accomplished in one interval and in the next. In other words, we must evaluate the right side of this equation for some T greater than the characteristic autocorrelation time of $\delta\Phi$ and divide through by $2T$. For such large values of T the term on the right of (116) that is proportional to $|u_-|$ becomes negligible and we obtain finally

$$\overline{\Delta^2}_{ij} = \sum_{\mathbf{n}} n_i n_j \widetilde{c}_{\mathbf{n}}(\mathbf{J},\mathbf{n}\cdot\boldsymbol{\omega}) \quad \text{where} \quad \widetilde{c}_{\mathbf{n}}(\mathbf{J},\nu) \equiv \int_{-\infty}^{\infty} c_{\mathbf{n}}(\mathbf{J},t)e^{i\nu t}\,\mathrm{d}t. \tag{117}$$

Equation (117) enunciates a very important physical principal for it says that an *orbit absorbs power from an externally applied potential only at its resonant frequencies* $\mathbf{n}\cdot\boldsymbol{\omega}$. In particular, if the potential is periodic with a frequency that does not coincide with any of the $\mathbf{n}\cdot\boldsymbol{\omega}$, it does not affect the long-term structure of the orbit.

The physical origin of this principle is simple. An alternating force applied to a massive, free particle transfers no energy to the particle. If the perturbing potential were a fixed, periodic function of the spatial coordinates, the orbiting particle would perceive an alternating force and absorb no net energy. But if a spatially periodic potential oscillates, its pattern moves through space as a wave. If the speed of this wave is just right, an orbiting particle can ride on one of the slopes of this wave, being steadily ac- or de-celerated.

5.1.2 Relations between the diffusion coefficients

Prima facie one has to evaluate separately the first-order diffusion coefficients $\overline{\Delta}_i$ in addition to the second-order coefficients $\overline{\Delta^2}_{ij}$ that we have just evaluated in a special case. However, it turns out that there are often simple relations between these coefficients.

In the case we have just discussed of diffusion driven by an externally applied random potential, we have (Binney & Lacey 1988)

$$\overline{\Delta}_i = \tfrac{1}{2}\frac{\partial\overline{\Delta^2}_{ij}}{\partial J_j}. \tag{118)}$$

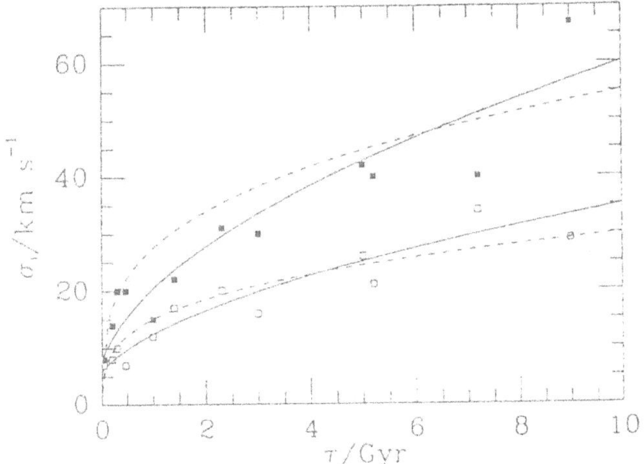

Fig. 21. The radial (filled points) and vertical (open points) velocity dispersions of populations observed near the Sun increase with age roughly as $\sqrt{t}$ (full curves). Dashed curves show fits of the form $\sigma \propto t^{0.3}$.

Equation (118) is not generally valid because in many interesting situations the perturbing potential $\delta\Phi$ is strongly correlated with the perturbed positions of typical stars in consequence of $\delta\Phi$'s being wholly or partly due to these perturbations. But we can write the Fokker-Planck equation (109a) in the form

$$\frac{\partial f}{\partial t} = -\frac{\partial S_i}{\partial J_i}, \tag{119a}$$

where the action-space flux $\mathbf{S}$ is defined by

$$S_i \equiv -f\overline{\Delta}_i + \tfrac{1}{2}\frac{\partial f\overline{\Delta^2}_{ij}}{\partial J_j}. \tag{119b}$$

If the scatterers are thermally distributed with temperature $T = (\beta k_B)^{-1}$, we know that f will attain equilibrium when it has reached the Gibbs distribution $f \propto \exp(-\beta H)$. Furthermore, by the principal of detailed balance the flux $\mathbf{S}$ must vanish in this limit. Setting $\mathbf{S} = 0$ in (119b) and using $\omega_j = \partial H/\partial J_j$, we conclude that the diffusion coefficients generated by any thermally distributed population of scatterers must satisfy

$$\overline{\Delta}_i = \tfrac{1}{2}\left(\frac{\partial\overline{\Delta^2}_{ij}}{\partial J_j} - \beta\omega_j\overline{\Delta^2}_{ij}\right). \tag{120}$$

A useful relation between the second-order diffusion coefficients generated by *any* scattering process is imposed by the requirement that the quadratic form $Q(\mathbf{x}) \equiv x_i\overline{\Delta^2}_{ij}x_j$ must be non-negative (Binney & Lacey 1988).

5.2 Schwarzschild's Distribution and the F-P Equation

Figure 21 shows that the velocity dispersions and ages of the stellar populations of the solar neighbourhood are approximately related by $\sigma \propto \sqrt{t}$. Presumably this correlation arises because each population is steadily diffusing away from the L_z axis of action space, and the older the population is the further it has diffused parallel to the J_r-axis and J_l-axis. Can we understand this result in terms of equation (109)?

In the first decade of this century K. Schwarzschild observed that at small peculiar speeds v the velocity distribution $f_0(\mathbf{v})$ of stars near the Sun can be fitted by

$$f_0(\mathbf{v}) \simeq \frac{n_0}{(2\pi)^{3/2}\sigma_R\sigma_\phi\sigma_z} \exp\left[-\tfrac{1}{2}\left(\frac{v_R^2}{\sigma_R^2}+\frac{v_\phi^2}{\sigma_\phi^2}+\frac{v_z^2}{\sigma_z^2}\right)\right]. \tag{121}$$

The epicycle approximation discussed in §3.3.2 should apply to these stars. Hence their orbits admit three approximate isolating integrals which we may take to be either L_z, E_R and E_z or the corresponding actions J_a, J_r and J_l [eq. (48)]. By the strong Jeans theorem it should be possible to model their DF by a function of these integrals, and it is easy to show[10] We need the result that (121) is just the restriction to $(R,z)=(R_0,0)$ of the DF

$$f_S(L_z,E_R,E_z) \equiv \frac{n_0}{(2\pi)^{3/2}\sigma_R\sigma_\phi\sigma_z} \exp\left[-\left(\frac{E_R}{\sigma_R^2}+\frac{E_z}{\sigma_z^2}\right)\right]. \tag{122}$$

The σ_i and n_0 are undetermined functions of L_z, but their dependence on L_z is sufficiently weak that the terms in the Fokker-Planck equation (109a) that involve L_z are smaller than the other terms in that equation by a factor $\lesssim (v/v_c)$. Dropping these terms the Fokker-Planck equation reduces to

$$2\frac{\partial f}{\partial t} = \frac{\partial}{\partial E_R}\left(D_{RR}\frac{\partial f}{\partial E_R}+D_{Rz}\frac{\partial f}{\partial E_z}\right)+\frac{\partial}{\partial E_z}\left(D_{Rz}\frac{\partial f}{\partial E_R}+D_{zz}\frac{\partial f}{\partial E_z}\right), \tag{123a}$$

where

$$D_{RR} \equiv \kappa^2\overline{\Delta^2}_{rr}\ ;\ D_{Rz} \equiv \kappa\nu\overline{\Delta^2}_{rl}\ ;\ D_{zz} \equiv \nu^2\overline{\Delta^2}_{ll}. \tag{123b}$$

The structure of the empirical distribution function (122) suggests that we seek solutions to (123) of the form

$$f(E_R,E_z,t) = \beta_R\beta_z e^{-(\beta_R E_R+\beta_z E_z)} \quad \text{with} \quad \beta(t), \tag{124}$$

where the β's preceeding the exponential are to ensure that the total number of stars $\int f\,\mathrm{d}^3\mathbf{J} = (\kappa\nu)^{-1}\int f\,\mathrm{d}E_R\mathrm{d}E_z\mathrm{d}L_z$ is constant. Plugging (124) into (123) we find

$$\begin{aligned}2\left(\frac{\dot\beta_R}{\beta_R}+\frac{\dot\beta_z}{\beta_z}\right)-2(\dot\beta_R E_R+\dot\beta_z E_z) &= -\beta_R\left(\frac{\partial D_{RR}}{\partial E_R}+\frac{\partial D_{Rz}}{\partial E_z}\right)\\ &\quad -\beta_z\left(\frac{\partial D_{zz}}{\partial E_z}+\frac{\partial D_{Rz}}{\partial E_R}\right)+\beta_R^2 D_{RR}+2\beta_R\beta_z D_{Rz}+\beta_z^2 D_{zz}.\end{aligned} \tag{125}$$

Equating coefficients of $\propto E_i^n$ on each side we conclude that

[10] This step involves the assumption that $\sigma_R/\sigma_\phi=\gamma$ as epicycle theory predicts.

$$D_{RR} = 2K_R E_R, \quad D_{Rz} = M, \quad D_{zz} = 2K_z E_z,$$
$$\sigma_R(t) = \frac{1}{\sqrt{\beta_R}} = \sqrt{K_R(t-t_0)} \quad (t_0 \text{ a constant}) \quad \text{and} \quad M = 0. \tag{126}$$

Thus the velocity distribution of stars near the Sun will retain Schwarzschild's form (121) as the dispersions σ_i grow providing the latter grow as $\sqrt{t}$. Fortunately, this is exactly the pattern of growth suggested by Figure 21. This evolution is driven by diffusion coefficients which are proportional to E_R and E_z, or equivalently to the corresponding actions. At present it is not clear exactly what physical process gives rise to such diffusion coefficients. Irregular spiral structure undoubtedly makes an important contribution, but generates diffusion coefficients that grow more slowly with the E_i than as a simple proportionality (e.g. Binney & Lacey 1988).

5.3 Stability of Spherical Systems

A few years ago it was discovered that spherical stellar systems that are defficient in tangential pressure spontaneously deform into bar-like configurations (e.g. Merritt & Aguilar 1985). As a second example of the uses of perturbation theory in stellar dynamics, I shall derive a criterion for determining when a spherical system is liable to thus spontaneously deform into a bar.

The physical idea underlying this criterion is the following. Suppose we remove all the mass from the particles of the spherical system under test and fix the removed mass in space so as to form exactly the original density profile $\rho(r)$. Then the system's now-massless stars will continue in their original orbits as if nothing had happened. Now we slowly distort the mass distribution into a slight bar. The response of the orbits to this change causes the overall stellar system to become barred also. Now imagine returning to each star its original mass. If this transfer of mass causes the system's potential to become still more barred, the original system was bar-unstable. In a notably elegant paper Goodman (1988) derives a quantitative criterion by which to judge whether the return to the stars of their mass causes the potential to become "still more barred".

We imagine a small exponentially growing perturbation $\delta\Phi$ to the original self-consistent potential; $\delta\Phi(\mathbf{x},t) = e^{st}\delta\overline{\Phi}(\mathbf{x})$ where s is in general complex. $\delta\Phi$ induces a change in the system's DF $\delta f(\mathbf{x},\mathbf{v},t) = e^{st}\delta\overline{f}(\mathbf{x},\mathbf{v})$, which in turn leads to a change in the density $\delta\rho = e^{st}\delta\overline{\rho}(\mathbf{x})$. We define a linear operator $\mathcal{R}$ by

$$\int \delta\overline{f}(\mathbf{x},\mathbf{v})\,\mathrm{d}^3\mathbf{v} = \delta\overline{\rho}(\mathbf{x}) \equiv \mathcal{R}(s)\delta\overline{\Phi}. \tag{127}$$

If s is the frequency of a normal mode, and thus $\delta\overline{\rho}$ is consistent with $\delta\overline{\Phi}$, then $\delta\overline{\rho}$ and $\delta\overline{\Phi}$ will also be related by

$$\delta\overline{\rho} = \frac{1}{4\pi G}\nabla^2\delta\overline{\Phi} \tag{128}$$

and the difference $\mathcal{M}(s)\delta\overline{\Phi}$ between (127) and (128) will vanish:

$$\mathcal{M}(s)\delta\overline{\Phi} \equiv \left(\mathcal{R}(s) - \frac{1}{4\pi G}\nabla^2\right)\delta\overline{\Phi} = 0 \quad \text{(for a normal mode).} \tag{129}$$

That is, s is a normal frequency of the system if $\mathcal{M}(s)$ has zero for an eigenvalue. Clearly if some real $s > 0$ satisfies this condition, the system must be unstable. More generally, Goodman shows that system unstable if one can find a functional form $\delta\overline{\Phi}(\mathbf{x})$ such that[11]

$$\langle \delta\overline{\Phi}, \mathcal{M}(s)\delta\overline{\Phi}\rangle < 0 \quad \text{for some real} \quad s > 0, \tag{130a}$$

where the inner product $\langle\cdot,\cdot\rangle$ is defined by

$$\langle \delta\overline{\Psi}, \delta\overline{\Phi}\rangle = \int \delta\overline{\Psi}^* \delta\overline{\Phi}\, \mathrm{d}^3\mathbf{x} \quad \text{for any} \quad \delta\overline{\Psi},\ \delta\overline{\Phi}. \tag{130b}$$

Proof: When we perturb the Poisson-bracket form (55) of the CBE for a Hamiltonian $H = H_0 + \delta\Phi$ and linearize, we obtain

$$\frac{\partial \delta f}{\partial t} + [\delta f, H_0] = [\delta\Phi, f_0]. \tag{131}$$

Now we break $\delta\overline{f}$ into pieces $\delta\overline{f}_\pm(\mathbf{x},\mathbf{v}) \equiv \frac{1}{2}[\delta\overline{f}(\mathbf{x},\mathbf{v}) \pm \delta\overline{f}(\mathbf{x},-\mathbf{v})]$ that are even and odd in $\mathbf{v}$ respectively. The Poisson bracket is odd in $\mathbf{v}$, so for even f_0 the perturbed CBE decomposes into

$$s\delta\overline{f}_+ + [\delta\overline{f}_-, H_0] = 0 \quad \text{and} \quad s\delta\overline{f}_- + [\delta\overline{f}_+, H_0] = [\delta\overline{\Phi}, f_0]. \tag{132}$$

On eliminating $\delta\overline{f}_-$ between equations (132) we have

$$s^2\delta\overline{f}_+ - [[\delta\overline{f}_+, H_0], H_0] = -[[\delta\overline{\Phi}, f_0], H_0]. \tag{133}$$

We solve this equation for $\delta\overline{f}_+$ by expanding $\delta\overline{f}_+$ and $\delta\overline{\Phi}$ in the action-angle coordinates of H_0:

$$\begin{aligned}\delta\overline{f}_+(\mathbf{J},\boldsymbol{\theta}) &= \sum_{\mathbf{n}} \delta\overline{f}_{\mathbf{n}}(\mathbf{J}) \exp(i\mathbf{n}\cdot\boldsymbol{\theta}) \quad \text{where} \quad \delta\overline{f}_{\mathbf{n}} \equiv (2\pi)^{-3}\int \mathrm{d}^3\boldsymbol{\theta}\, \exp(-i\mathbf{n}\cdot\boldsymbol{\theta})\delta\overline{f}(\mathbf{J},\boldsymbol{\theta}),\\ \delta\overline{\Phi}(\mathbf{J},\boldsymbol{\theta}) &= \sum_{\mathbf{n}} \delta\overline{\Phi}_{\mathbf{n}}(\mathbf{J}) \exp(i\mathbf{n}\cdot\boldsymbol{\theta}) \quad \text{where} \quad \delta\overline{\Phi}_{\mathbf{n}} \equiv (2\pi)^{-3}\int \mathrm{d}^3\boldsymbol{\theta}\, \exp(-i\mathbf{n}\cdot\boldsymbol{\theta})\delta\overline{\Phi}(\mathbf{J},\boldsymbol{\theta}).\end{aligned} \tag{134}$$

On substituting these expansions into (133) and equating coefficients of $e^{i\mathbf{n}\cdot\boldsymbol{\theta}}$ we find

$$\delta\overline{f}_{\mathbf{n}}(\mathbf{J}) = \frac{(\mathbf{n}\cdot\boldsymbol{\omega})(\mathbf{n}\cdot\partial f_0/\partial\mathbf{J})}{s^2 + (\mathbf{n}\cdot\boldsymbol{\omega})^2}\delta\overline{\Phi}_{\mathbf{n}}, \tag{135}$$

where use has been made of the result that for any $X(\mathbf{J},\boldsymbol{\theta})$ and $G(\mathbf{J})$ we have

$$[X, G] = \sum_{\mathbf{n}} i\left(\mathbf{n}\cdot\frac{\partial G}{\partial \mathbf{J}}\right) X_{\mathbf{n}}(\mathbf{J}) \exp(i\mathbf{n}\cdot\boldsymbol{\theta}). \tag{136}$$

To get $\delta\overline{\rho}$ from our expression (135) for the expansion coefficients of $\delta\overline{f}_+$, we need to integrate over all $\mathbf{v}$. A direct attack on this problem would involve integration over the Jacobian $\partial(\mathbf{v})/\partial(\mathbf{J})$, which is horrendous. To circumvent this difficulty we multiply

11 Actually, $\delta\overline{\Phi}$ has to satisfy certain additional constraints such as the mass-conservation constraint $\int \mathrm{d}S\cdot\boldsymbol{\nabla}\delta\overline{\Phi} = 0$.

(135) by an arbitrary function $\delta\overline{\Psi}(\mathbf{x})$ and integrate over all phase space. The change of variables $(\mathbf{x},\mathbf{v}) \to (\mathbf{J},\boldsymbol{\theta})$ is now trivial since $\partial(\mathbf{x},\mathbf{v})/\partial(\mathbf{J},\boldsymbol{\theta}) = 1$. There results

$$\begin{aligned}\langle \delta\overline{\Psi}, \mathcal{R}(s)\delta\overline{\Phi}\rangle &= \int \mathrm{d}^3\mathbf{x}\, \delta\overline{\Psi}^* \int \mathrm{d}^3\mathbf{v}\, \delta\overline{f}_+ = \int \mathrm{d}^3\boldsymbol{\theta}\mathrm{d}^3\mathbf{J}\delta\overline{\Psi}^*\delta\overline{f}_+ \\ &= (2\pi)^3 \sum_{\mathbf{n}} \int \mathrm{d}^3\mathbf{J} \frac{(\mathbf{n}\cdot\boldsymbol{\omega})(\mathbf{n}\cdot\partial f_0/\partial\mathbf{J})}{s^2 + (\mathbf{n}\cdot\boldsymbol{\omega})^2} \delta\overline{\Psi}^*_{\mathbf{n}}\delta\overline{\Phi}_{\mathbf{n}}.\end{aligned} \tag{137}$$

Since $\delta\overline{\Phi}$ and $\delta\overline{\Psi}$ are arbitrary functions, equation (137) is an expression for a general matrix element of the operator $\mathcal{R}$. For real s we clearly have $\langle\delta\overline{\Psi}, \mathcal{R}(s)\delta\overline{\Phi}\rangle = \langle\delta\overline{\Phi}, \mathcal{R}(s)\delta\overline{\Psi}\rangle^*$, that is, that $\mathcal{R}$ is Hermitian. But ∇^2 is also Hermitian.[12] So the operator $\mathcal{M}$ of equation (129), being the difference of two Hermitian operators, is itself Hermitian. Consequently $\mathcal{M}$ has a complete set of orthonormal eigenfunctions m_i with real eigenvalues λ_i. When we expand Goodman's criterion (130a) in this set it becomes

$$0 > \langle\delta\overline{\Phi}, \mathcal{M}(s)\delta\overline{\Phi}\rangle = \sum_i \lambda_i(s)\big|\langle m_i^*(s), \delta\overline{\Phi}\rangle\big|^2. \tag{138}$$

Thus if Goodman's criterion is satisfied, at least one of the $\lambda_i < 0$. In particular the smallest of $\mathcal{M}$'s eigenvalues $\lambda_{\min} < 0$. But as $s \to \infty$, $\mathcal{M}(s) \to -(4\pi G)^{-1}\nabla^2$, which is a positive operator.[13] Hence for sufficiently large real s, $\lambda_{\min} > 0$. Since $\lambda_{\min}$ must be a continuous function of s it follows that for some real positive s_0, $\lambda_{\min} = 0$ and s_0 is an eigenfrequency of the system. Hence when Goodman's criterion is satisfied the system is unstable.

5.4 Disk Stability and Spiral Structure

Self-gravitating systems are at once capitalist and conspiratorial. The monopoly capitalism in them manifests itself in the propensity of large galaxies to cannibalize their smaller and more fragile companions and of loose groupings of objects to merge together to form large clusters. The conspiratorial element shows up in various instabilities such as the Jeans instability that leads to star formation and the bar-forming instabilities of spherical and disk systems; whenever you make a system in which there is a high density of stars in some portion of phase space you are liable to find those stars conspiring together to subvert the structure you established. Hence if you wish to build a stable model you must bear in mind the maxim of the last pan-European government *divide et impera*. If the model is to be spherical, we must not assign too many stars to radial orbits, and if it is a disk, we must watch how many stars we assign to nearly circular orbits. To create harmony we must sow dissension. But what is the minimum level of dissension that is compatible with harmony in a disk?

Our best estimate of the answer to this question is provided by observations of disk galaxies for there is ample evidence that spiral galaxies are perpetually wracked by minor conspiracies, while lencticular galaxies are pretty harmonious. The conspiracies in spirals

12 By a double application of the divergence theorem $\int \psi^*\nabla^2\phi\,\mathrm{d}^3\mathbf{x} = \int \phi\nabla^2\psi^*\,\mathrm{d}^3\mathbf{x}$ for functions $\phi(\mathbf{x})$ and $\psi(\mathbf{x})$ that vanish sufficiently rapidly as $|\mathbf{x}| \to \infty$.

13 $-\int \phi^*\nabla^2\phi\,\mathrm{d}^3\mathbf{x} = \int \boldsymbol{\nabla}\phi^*\cdot\boldsymbol{\nabla}\phi\,\mathrm{d}^3\mathbf{x} > 0$.

consitute the spiral structure itself. As we saw in §5.2 they gradually sow dissension in the disk, thus, like yeast and men, poisoning the very medium that sustains them. In spirals fresh conspiracies continually grow nevertheless, since gas is constantly placing fresh stars on nearly circular orbits. Lenticulars lack gas and significant star formation, so in them conspiracies rarely prosper.

When people first experimented with n-body models of self-gravitating disks, they found their models were liable not so much to minor conspiracies as to violent revolutions no less momentous in their consequences than the French and Bolshevik revolutions. The crucial difference between these violently unstable n-body disks and real, marginally stable disks must lie in the spread in action space of the orbits bearing most of the mass; in real galaxies much of the mass must be on significantly eccentric orbits, either within the disk plane or in a thick component ("massive halo"). One important factor in the failure of early models was undoubtedly a tendency to assume that the radial velocity dispersions of disk stars are as small near the galactic centre as they are near the Sun.

In any event, it is clear that we need to be able to assess the stability of model disks. A vast amount of work has been done on this problem, yet the theory is still fragmentary because the calculations are very laborious. The problem is as follows.

We have an equilibrium model with DF $f_0(\mathbf{J})$ and Hamiltonian $H_0(\mathbf{J})$. For a rotating disk f_0 will not be even in the velocities since it will have a piece odd in L_z. So we can't use the trick Goodman employed to get from (131) to (135). The best we can do is to derive in the notation of §5.3

$$\delta\overline{f}_{\mathbf{n}} = \frac{\mathbf{n}\cdot\partial f_0/\mathbf{J}}{\mathbf{n}\cdot\boldsymbol{\omega}-\nu}\,\delta\overline{\Phi}_{\mathbf{n}}, \tag{139}$$

where I have made the substitution $s \to -i\nu$ so that $\delta f = \delta\overline{f}e^{-i\nu t}$ and $\delta\Phi = \delta\overline{\Phi}e^{-i\nu t}$. Thus the equation to be satisfied by the potential perturbation $\delta\overline{\Phi}$ of a normal mode is

$$(4\pi G)^{-1}\nabla^2\sum_{\mathbf{n}}\delta\overline{\Phi}_{\mathbf{n}}e^{i\mathbf{n}\cdot\boldsymbol{\theta}} = \delta\overline{\rho} = \sum_{\mathbf{n}}\int\frac{\mathbf{n}\cdot\partial f_0/\mathbf{J}}{\mathbf{n}\cdot\boldsymbol{\omega}-\nu}\,\delta\overline{\Phi}_{\mathbf{n}}e^{i\mathbf{n}\cdot\boldsymbol{\theta}}\,\mathrm{d}\mathbf{v}. \tag{140}$$

Given relations between $(\mathbf{x},\mathbf{v})$ and the action-angle variables $(\mathbf{J},\boldsymbol{\theta})$ of the unperturbed model, (140) constitutes a well defined eigenvalue problem for ν. If ν can have positive imaginary part, the system is unstable. In practice one neither has handy expressions for $\mathbf{J}(\mathbf{x},\mathbf{v})$ etc nor capacity to search for the infinite sequence of functions of three variables $\delta\overline{\Phi}_{\mathbf{n}}$. Consequently very little work has been done along these lines. What *has* been done is illustrated by the following bed-time story

5.4.1 Gröninger and the spectrum of waterstuff

Once upon a time far away on the planet Nirgends, which lies on the furthest edge of the Galaxy, there lived a race of beings, the Homoeoids, who had neglected both mathematics and engineering most shamefully. One day Gröninger, a brilliant young Homoeoid physicist then just at the end of his first year as a graduate student, became convinced that the frequencies of lines in the spectrum of waterstuff are differences between the frequencies ω for which it is possible to solve, subject to the boundary conditions $\psi \to 0$ as $r \to \infty$ and ψ finite at $r = 0$, an equation which we would write

$$\frac{1}{r^2}\frac{\mathrm{d}}{\mathrm{d}r}\left(r^2\frac{\mathrm{d}\psi}{\mathrm{d}r}\right) - \left(\frac{l(l+1)}{r^2} - \frac{q}{r}\right)\psi + \omega\psi = 0 \quad (0 < q, \text{ a constant, } l = 0, 1, \ldots). \tag{141}$$

Obviously, if Gröninger was right about this, he would win the Ignobel Prize while still young. So he worried away at this problem for several weeks before telling his supervisor, Personne, about it. But finally he had to admit himself defeated and went to explain his suspicions to Personne, who immediately saw the Ignobel implications. Unfortunately, as the Homoeoids had yet to produce equivalents of either Frobenius or Bardeen & Brattain, Personne didn't know of a simple route, be it analytic or numerical, to the magic frequencies. But as a graduate student at the Massimo Institute of Technology she had taken courses in continuum mechanics from Professors Thin and Loonry, and wasn't in the least perplexed by equation (141). "Gröninger," she said, "don't you see that your equation obviously derives from a wave equation? So your next step step should be to find the corresponding dispersion relation by assuming that $\psi \propto e^{ikr}$."

On hearing this Gröninger felt very small. A couple of days later he returned with the desired dispersion relation:

$$k^2 = \omega - \frac{l(l+1)}{r^2} + \frac{q}{r}. \tag{142}$$

"Good," said Personne, "it's exactly as I expected. When ω is negative k becomes imaginary and the waves evanescent at both small and large r. So all you have to do is to find the radii where $k = 0$ and then determine your magic frequencies by insisting that an integral number of half-wavelengths can be fitted between them." Gröninger called these special radii $r_\pm$ the inner- and outer-Sinbad resonances after his girl-friend. Then identifying k with the rate of change with r of the wave's phase, he found that the wave's phase change between Sinbad resonances is[14]

$$\Delta\phi = \int_{r_-}^{r_+} k(r)\,\mathrm{d}r = \int_{r_-}^{r_+} \sqrt{\omega - \frac{l(l+1)}{r^2} + \frac{q}{r}}\,\mathrm{d}r. \tag{143}$$

After many days of struggling in vain with this integral Gröninger was rescued by a friend who worked on the dynamics of the Molar system and spotted that the integral (143) is identical with that involved in the definition of the radial action of a planet of energy ω and angular momentum $\sqrt{l(l+1)}$ in orbit around a Mun of mass q/G. Thus he could figure that

$$\Delta\phi = \pi\left(\frac{q}{\sqrt{2|\omega|}} - \sqrt{l(l+1)}\right). \tag{144}$$

On setting $\Delta\phi = n\pi$ Gröninger concluded that the waterstuff's spectral lines should occur at frequencies

$$\Delta\omega = \tfrac{1}{2}q^2\left\{\frac{1}{[n+\sqrt{l(l+1)}]^2} - \frac{1}{[n'+\sqrt{l'(l'+1)}]^2}\right\} \quad (n, n', l, l' = 0, 1, \ldots). \tag{145}$$

In a way it was encouraging that this formula gave a reasonable account of the lowest-frequency lines in the spectrum, but those lines had not been accurately measured and Gröninger was very cast down to find that (145) did not model adequately the accurately measured, high-frequency lines of waterstuff. In his disappointment he called (145) the Bore quantum condition and relegated it to an appendix in his thesis. Soon he was working happily on more productive ideas and forgot all about the Bore condition and the Ignobel prize that never was.

---

14 We, who can solve the Airy equation to which (141) reduces near the Sinbad resonances (see Landau & Lifshitz 1977, §§47–49), know that Gröninger's value of $\Delta\phi$ is too small by $\pi/2$.

5.4.2 Conventional theory of spiral structure

The idea behind Personne's suggestion that Gröninger substitute $\psi \propto e^{ikr}$ into (141) is that ψ, being wave-like, oscillates with on a lengthscale $\lambda \ll r$. If this is so, occurrences of r (but not of $\mathrm{d}r$) in (141) can be approximately replaced by a constant mean value r_0. Then the equation is translationally invariant in r and thus must have eigensolutions of the form e^{ikr}. Once one knows the form of the solutions, determination of the frequencies is easy.

Similarly, we might return to (140) with the guess that some of the normal-mode potentials of a disk will be of the form $e^{i(kR+m\phi)}$. (The ϕ-dependence at least is guaranteed as the system is translationally invariant in ϕ.) This is generally called the "tight-winding approximation" (TWA) since it is implicitly assumed that the modes of interest have $2\pi m \ll kR$. Next we assume that the epicycle approximation is valid for all the disk's stars and use the resulting action-angle coordinates to solve the CBE. Taking $\delta\overline{\Phi}$ to be of form $\delta\overline{\Phi} = \epsilon e^{i(kR+m\phi)}$ and noting that the Cartesian and action-angle coordinates are related by [see eqs (25) and (30)]

$$R = R_g + a\cos\theta_r, \quad \phi = \theta_a + \frac{\gamma a}{R_g}\sin\theta_r, \quad \text{where} \quad a \equiv \sqrt{\frac{2J_r}{\kappa}}, \tag{146}$$

we find that $\delta\overline{\Phi}$'s action-angle representation is (see e.g. Binney & Lacey 1988)

$$\begin{aligned}\delta\overline{\Phi} &= \epsilon e^{ikR+m\phi} = \epsilon e^{ikR_g}\exp(ika\cos\theta_r)e^{im\theta_a}\exp\left(i\frac{m\gamma a}{R_g}\sin\theta_r\right)\\ &= \epsilon\sum_{l=-\infty}^{\infty}\exp\left[i(kR_g + l\alpha)\right]J_l(\mathcal{K}a)e^{i(l\theta_r + m\theta_a)},\end{aligned} \tag{147a}$$

where

$$\alpha(J_a) \equiv \arctan\left(\frac{m\gamma}{kR_g}\right) \qquad \text{and} \qquad \mathcal{K}(J_a) \equiv \sqrt{k^2 + \frac{m^2\gamma^2}{R_g^2}}. \tag{147b}$$

In other words

$$\delta\overline{\Phi}_{(l,m)} = \epsilon\exp\left[i(kR_g + l\alpha)\right]J_l(\mathcal{K}a), \tag{148}$$

so with (139) we have

$$\delta\overline{f}_{\mathbf{n}}(\mathbf{J}) = \epsilon\exp\left[i(kR_g + l\alpha)\right]J_l(\mathcal{K}a)\frac{\mathbf{n}\cdot\partial f_0/\partial\mathbf{J}}{\mathbf{n}\cdot\boldsymbol{\omega} - \nu} \quad [\mathbf{n} \equiv (l, m)]. \tag{149}$$

The perturbed surface density is

$$\begin{aligned}\delta\Sigma(R',\phi') &= \int \delta f\,\mathrm{d}^2\mathbf{v} = \frac{1}{R'}\int \delta f\,\delta(R - R')\delta(\phi - \phi')\,R\,\mathrm{d}R\,\mathrm{d}\phi\,\mathrm{d}^2\mathbf{v}\\ &= \frac{1}{R'}\int \delta f\,\delta(R - R')\delta(\phi - \phi')\,\mathrm{d}^2\mathbf{J}\mathrm{d}^2\boldsymbol{\theta}\\ &= \frac{1}{R'}\sum_{\mathbf{n}}\int \delta\overline{f}_{\mathbf{n}}e^{i(\mathbf{n}\cdot\boldsymbol{\theta} - \nu t)}\,\delta(R - R')\delta(\phi - \phi')\,\mathrm{d}^2\mathbf{J}\mathrm{d}^2\boldsymbol{\theta}.\end{aligned} \tag{150}$$

Inserting (146) and (149) into (150)

$$\delta\Sigma(R',\phi') = \frac{\epsilon}{R'} \sum_{l=-\infty}^{\infty} \int e^{i(kR_g + l\alpha)} J_l(\mathcal{K}a) \frac{\mathbf{n}\cdot\partial f_0/\partial\mathbf{J}}{\mathbf{n}\cdot\boldsymbol{\omega} - \nu} e^{i(\mathbf{n}\cdot\boldsymbol{\theta} - \nu t)} \\ \times \delta(R_g + a\cos\theta_r - R')\delta\Big(\theta_a + \frac{\gamma a}{R_g}\sin\theta_r - \phi'\Big)\,\mathrm{d}^2\mathbf{J}\mathrm{d}^2\boldsymbol{\theta}. \tag{151}$$

When we take advantage of the two δ-functions to integrate out J_a and θ_a, every occurence R_g (including those in the definitions of κ, α etc) should, strictly, be replaced by $R' - a\cos\theta_r$ and similarly for θ_a. However, the TWA allows us to neglect the small difference between R_g and R' except when multiplied by the large number k in an exponential. Then we obtain

$$\delta\Sigma(R',\phi') \simeq \frac{\epsilon}{R'} e^{i(kR' + m\phi' - \nu t)} \frac{\mathrm{d}J_a}{\mathrm{d}R_g}\bigg|_{R_g = R'} \sum_{l=-\infty}^{\infty} e^{il\alpha} \int \mathrm{d}J_r\, J_l(\mathcal{K}a) \frac{\mathbf{n}\cdot\partial f_0/\partial\mathbf{J}}{\mathbf{n}\cdot\boldsymbol{\omega} - \nu} \\ \times \int \mathrm{d}\theta_r \exp\Big[i\Big(l\theta_r - m\frac{\gamma a}{R'}\sin\theta_r - ka\cos\theta_r\Big)\Big]. \tag{152}$$

From equation (27) it is simple to show that $\mathrm{d}J_a/\mathrm{d}R_g = R_g\kappa/\gamma$. For f_0 it is natural to adopt Schwarzschild's DF in the form $f_0(\mathbf{J}) = (\gamma\Sigma_0/2\pi\sigma^2)e^{-\kappa J_r/\sigma^2}$ [cf. eq. (121)]. Finally, the last line of (152) involves exactly the same product of exponentials of $\cos\theta_r$ and $\sin\theta_r$ as occur in the first line of (147a). Reducing this product to an infinite sum of exponentials as in (147a) and doing the integral over θ_r yields

$$\begin{aligned} \delta\Sigma(R',\phi') &\simeq \frac{\epsilon\kappa^2\Sigma_0}{\sigma^4} e^{i(kR' + m\phi' - \nu t)} \sum_{l=-\infty}^{\infty} \frac{-l}{l\kappa + m\Omega - \nu} \int \big|J_l(\mathcal{K}a)\big|^2 e^{-\kappa J_r/\sigma^2}\,\mathrm{d}J_r \\ &= \frac{\epsilon\kappa\Sigma_0}{\sigma^2} e^{i(kR' + m\phi' - \nu t)} \sum_{l=-\infty}^{\infty} \frac{-lI_l(\chi)e^{-\chi}}{l\kappa + m\Omega - \nu} \\ &= \frac{\mathcal{K}^2\Sigma_0\delta\Phi}{\kappa^2(1-s^2)}\mathcal{F}(s,\chi), \end{aligned} \tag{153a}$$

where

$$s \equiv \frac{\nu - m\Omega}{\kappa}, \quad \chi \equiv \frac{\mathcal{K}^2\sigma^2}{\kappa^2}, \quad \mathcal{F}(s,\chi) \equiv 2(1-s^2)\frac{e^{-\chi}}{\chi}\sum_{l=1}^{\infty}\frac{I_l(\chi)}{1 - s^2/l^2}, \tag{153b}$$

and the integral has been evaluated with formula 6.615 of Gradshteyn & Ryzhik (1965).

From Poisson's equation it follows that our potential perturbation $\delta\Phi$ is generated by the perturbed surface density

$$\delta\Sigma = -\frac{|k|\delta\Phi}{2\pi G}. \tag{154}$$

Equating this to (153a) with $\mathcal{K}$ approximated by $|k|$ [see (147b)], we obtain the Lin-Shu-Kalnajs dispersion relation for tightly wound spiral waves:

$$\frac{|k|}{k_{\mathrm{crit}}}\mathcal{F}(s,\chi) = 1 - s^2 = 1 - \frac{(\nu - m\Omega)^2}{\kappa^2}, \quad \text{where} \quad k_{\mathrm{crit}} \equiv \frac{\kappa^2}{2\pi G\Sigma_0}. \tag{155}$$

This is the analogue of Gröninger's formula (142). Qualitatively it has something in common with (142): waves are trapped between radii at which $\kappa = \pm(\nu - m\Omega)$ and thus k vanishes. These are the radii of the inner and outer Lindblad resonances (ILR and OLR). But there are two important differences between (155) and (142) that arise because the function $k\mathcal{F}$ behaves qualitatively like $k - \epsilon k^2$, where $\epsilon > 0$:

(i) For each value of s for which one can solve (155) for k, there are *two* allowed values of k; one speaks of the long- and short-wave branches of the dispersion relation.

(ii) $k\mathcal{F}$ achieves a maximum for some value of k. Consequently it can happen that there is a range of radii around the **corotation resonance** (CR), at which $\nu = m\Omega$, for which the term proportional to $k\mathcal{F}$ cannot make up the difference between κ^2 and $(\nu - m\Omega)^2$. Indeed from (155) Toomre (1964) showed that if there is no such forbidden zone around the CR, the disk is unstable to some axisymmetric disturbance. Hence any real disk will have a forbidden zone for every value of ν.

The two branches of the dispersion relation merge at the edge of the forbidden zone around the CR. Consequently, waves are trapped not so much by two radii at which $k = 0$ as by one such radius and the radius at which long- and short-wavelength disturbances become identical. This leads to the following difficulty in carrying out for disks the analogue of Personne's programme for waterstuff. We would like to find the eigenfrequencies ν_i by equating to $2n\pi$ the total phase increment of a wavepacket that runs from one boundary to the other and back again. But as waves bounce off the edge of the forbidden zone around the CR, they move from one branch of the dispersion relation to the other. Consequently, the waves that return at the end of the day are qualitatively different from those which set out, and it is insufficient to impose a simple phase condition.

The following example will show that this problem is a real, physical difficulty rather than mathematical gibberish. Suppose we have a packet of short-wavelength leading ($k < 0$) waves of such a frequency ν that an ILR exists and follow the packet from just outside the ILR. It is not hard to show that the group velocity for such waves is positive and thus that the packet propagates outwards towards the CR. At the junction of the two branches of the dispersion relation the packet transfers to the long-wave branch and starts to propagate inwards back to the ILR as a long-leading wave. At the ILR it turns into a packet of outwardly propagating long-trailing waves. When it next reaches the CR it transfers to the short-wave branch of the dispersion relation and starts to move inward again as a short-trailing wave. Thus our packet, which started life moving outwards as a packet of tightly wound leading waves, is now moving inwards towards the ILR as a packet of tightly wound trailing waves. This progression is the mathematical expression of the natural tendency of structures in the disk to be wound up by differential rotation. Furthermore, one may show that as the packet of short-trailing waves approaches the ILR for the second time its wavelength gets shorter and shorter without limit and the energy of the wave is effectively thermalized. Thus the TWA predicts that our initial packet degenerates into heat rather than completing a perpetually self-renewing life cycle.

Does it follow from this analysis that disks have no spiral normal modes? Actually no for two reasons: (i) for sufficiently large ν the disk may have no ILR, and (ii) the TWA is generally badly violated on the long-wave branch of the dispersion relation. Using a different approximation scheme Goldreich & Lynden-Bell (1965) showed that self-gravity can significantly amplify wave packets as they pass from leading to trailing configurations—this effect is called **swing amplification**. Furthermore, it is likely that the breakdown of the TWA allows a portion of packets of short-trailing waves to pass right through the centre of a galaxy whose ILR is at $r \lesssim 2\pi/k$, to emerge as packets of short-leading waves which can evolve once more into short trailing waves and again pass through the centre. It is thought that these two effects combined make it possible to establish, for magic values of ν, closed loops of waves that run from the centre to the corotation barrier and back. The entire cycle involves both leading and trailing waves, but the latter dominate since they are the immediate products of the swing amplifier. Hence the observed predominance of trailing spirals on the sky.

5.4.3 Conclusion

Gröninger was dissatisfied with his approximate treatment of (141) since it gave poor estimates of the fundamental lines in the waterstuff spectrum. We have even less reason to be satisfied with the Lin-Shu-Kalnajs analysis of (140) which, unaided, is unable to give an adequate account of even the high-frequency normal modes; Personne's programme turns out to involve loosely-wound waves for which the LSK-dispersion relation is not really valid. Obviously more powerful techniques need to be developed for the solution of (140).

Meanwhile, is the TWA theory of spiral structure worth bothering with? Quantitatively the TWA theory is not a success. Yet it has played an important role in the study of galaxies by introducing a widely employed conceptual framework. Only after effective machinery for the evaluation of normal modes of disks is available will we know for certain whether this role has been beneficial. In any event one must not underestimate the con- and de-structive influence on progress in science of the conceptual frameworks that simple models introduce. Two examples will illustrate this point:

(i) High-energy physicists think always in terms of "particles" and "interactions" and yet these are really just elements (propagators and vertices) introduced during the iterative solution of a set of coupled non-linear integro-differential equations.

(ii) Isaac Newton spent vastly more time, thought and experimental effort on chemistry than on either physics or mathematics. Yet his incomparable mind, which both before and after his period as a chemist revolutionized mathematics and first demonstrated the possibility of exact science, achieved nothing of lasting value in Chemistry, whose foundations were to be laid by men of much smaller stature in the mid 18$^{\rm th}$c. Why did he fail so miserably? Because his conceptual framework was pre-Newtonian; brought up in the mystical, pre-Enlightenment mid 17$^{\rm th}$c. he thought in terms of the ancient alchemical concepts of corruption and redemption rather mechanistic causality.

Concepts such as Lindblad resonances and long- and short-waves introduced by the conventional theory of spiral structure may prove as useful as quarks or as obstructive as alchemy. Only the future and hard work will tell.

References

Arnold, V. I. 1978. Mathematical Methods of Classical Mechanics, (New York: Springer).
Berry, M. V. 1978. Topics in Nonlinear Mechanics, ed. S. Jorna (New York: American Institute of Physics), p. 16.
Binney, J. J. 1987. In The Galaxy, eds. G. Gilmore & R. Carswell (Dordrecht: Reidel) p. 399.
Binney, J. J. & Lacey, C. G. 1988. Mon. Not. Roy. Astron. Soc., 230, 597.
Binney, J. J. & Mamon, G.A. 1982. Mon. Not. Roy. Astron. Soc., 200, 361.

Binney, J. J. & Petit, J-M. 1988. In Dense Stellar Systems, ed. D. Merritt (Cambridge University Press), 1989.
Dejonghe, H. 1986. Physics Reports, 133, 217.
de Zeeuw, T. 1985. Mon. Not. Roy. Astron. Soc., 216, 273 & 599.
Eddington, A. S. 1916. Mon. Not. Roy. Astron. Soc., 76, 572.
Goldreich, P. & Lynden-Bell, D. 1965. Mon. Not. Roy. Astron. Soc., 130, 125.
Goodman, J. 1988. Astrophys. J., 329, 612.
Gradshteyn, I. S. & Ryzhik, I. M. 1965. Tables of Integrals, Series and Products, (New York: Academic).
Hunter, C. 1975. Astron. J., 80, 783.
Hunter, C. 1977. Astron. J., 82, 271.
Jaffe, W. 1983. Mon. Not. Roy. Astron. Soc., 202, 995.
Jarvis, B. J. & Freeman, K. C. 1985. Astrophys. J., 295, 314.
Jeans, J. H. 1915. Mon. Not. Roy. Astron. Soc., 76, 70.
Landau, L. D. & Lifshitz, E. M. 1977. Quantum Mechanics, (Oxford: Pergamon).
Lucy, L. B. 1974. Astron. J., 79, 745.
Lynden-Bell, D. 1962a. Mon. Not. Roy. Astron. Soc., 123, 447.
Merritt, D. & Aguilar, L. A. 1985. Mon. Not. Roy. Astron. Soc., 217, 787.
Newton, A. J. 1986. DPhil thesis, Oxford University.
Ostriker, J. P., Binney, J. J. & Saha, P. 1989. Mon. Not. Roy. Astron. Soc., 000, 00.
Rowley, G. 1988. Astrophys. J., 000, 00.
Richstone, D. & Tremaine, S. D. 1988. Astrophys. J., 327, 82.
Sargent, W. L. W., Young, P. J., Boksenberg, A., Shortridge, K., Lynds, C. R. & Hartwick, F. D. A. 1978. Astrophys. J., 221, 731.
Schwarzschild, M. 1979. Astrophys. J., 232, 236.
Spitzer, L. 1987. Dynamical Evolution of Globular Clusters, (Princeton: Princeton University Press).
Statler, T. 1987. Astrophys. J., 321, 113.
Toomre, A. 1964. Astrophys. J., 139, 1217.
Wilson, C. P. 1975. Astron. J., 80, 175.
Woolley, R. & Dickens, R. J. 1961. Roy. Greenwich Obs. Bulletin, No. 42.

This article was processed by the author using the TEX Macropackage from Springer-Verlag.

Stellar Populations in Galaxies: Structure and Evolution

James Lequeux
Groupe de Radioastronomie, Observatoire de Paris and Ecole Normale Supérieure - Paris, France

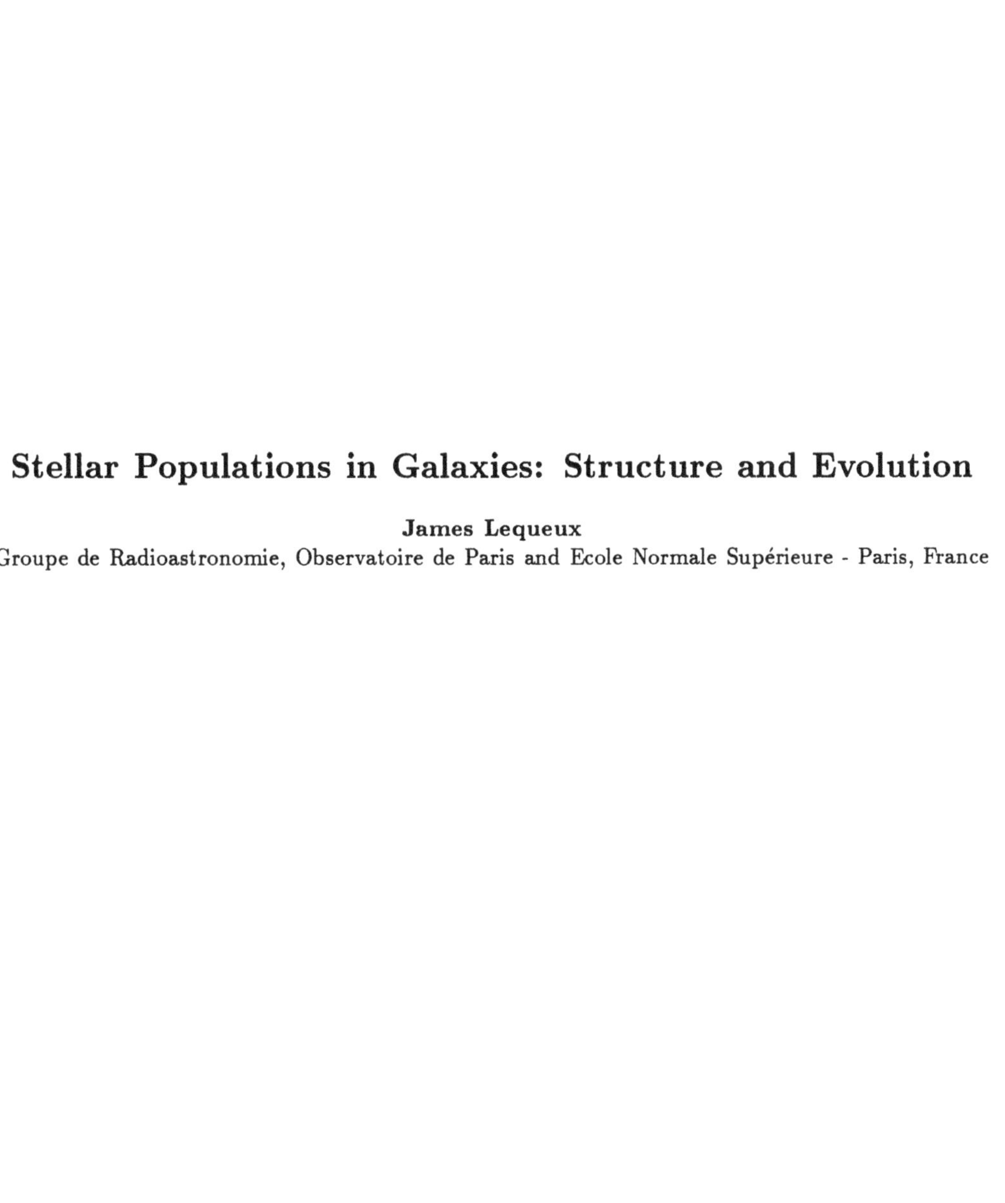

Foreword

This series of lectures concentrates on star formation and on the evolution of stellar populations in galaxies with a final chapter on chemical evolution. It assumes some previous knowledge of the general properties of stars and galaxies (classification, stellar populations, gaseous content) but otherwise is self-consistent. Emphasis is put on principles and methods, including difficulties and pitfalls, rather than on results; however some results will be presented mainly as examples and references will be given on a selection of the most important ones.

1 Stellar Evolution

Knowledge of stellar evolution is obviously a prerequisite to any study of stellar populations in galaxies. Here I will outline the evolution of stars of different masses with emphasis on the less well-understood aspects. Indeed there are still major uncertainties on some phases of stellar evolution that however contribute to a large amount to the integrated luminosity of a galaxy ; any astronomer should be aware of this problem. Also, some quantities like any color in the visible are insensitive to the temperature of the stars when they are hot, and stars almost similar in the visible may have vastly different properties in the far-UV; one should also be aware of this limitation and not try to draw too hastily conclusions e.g. on details of massive star formation in galaxies from colors alone.

In this lecture I will discuss the Hertzprung-Russell diagram –the basic tool of stellar astronomers– then discuss the main sequence, the post-main sequence evolution of massive stars and of intermediate-mass stars, and terminate by a few remarks on the role of binarity.

1.2 The Hertzprung-Russell (HR) diagram

Theoreticians use to present the results of their model calculations of the evolution of stars in a diagram where the total (bolometric) luminosity is plotted as a function of the effective temperature of the star. This is usually called the "theoretical HR diagram". The luminosity L is presented either in units of solar luminosity $L_\odot$,

$$1 L_\odot \simeq 4\,10^{33}\,\mathrm{erg}\;s^{-1} = 4\,10^{26}\,W,$$

or as the absolute bolometric luminosity,

$$M_{\mathrm{bol}} = -2.5 \log L + cst,$$

the constant being such that the absolute bolometric luminosity of the Sun is $M_{bol,\odot} = +4.69$.

The effective temperature T_{eff} is the temperature of a blackbody with the same radius as the star (assumed spherical) and the same luminosity :

$$L = 4\pi R^2 \sigma T_{\mathrm{eff}}^2$$

It should be remarked that the radius of a star is an ill-defined quantity in stars with extended thick envelopes, because radiation can come from different radii at different wavelengths. Theoreticians then chose an arbitrary radius (usually small) and their T_{eff} can be considerably larger than the radiation temperature at most or all wavelengths. A good example is that of Wolf-Rayet stars for which the theoretical T_{eff} (that corresponds to the radius of a dense core) is considerably higher than for example the temperature describing the far-UV (ionizing) properties of these objects ($10^5 K$ or more instead of $30000K$ or so).

Observers measure fluxes (or magnitudes) of stars at various wavelengths, as estimated above the terrestrial atmospheres, and quantities related to surface temperatures : colors or spectral types. They build HR diagrams in which they plot a magnitude (usually in the visible V band) as a function of a color (e.g. B-V) or of the spectral type. In order to relate the magnitude to the bolometric luminosity one must know the interstellar extinction and the distance of the star, usually expressed as the extinction-corrected distance modulus

$$(m - \mathrm{M})_0 = 5 \log \mathrm{d}(pc) - 5$$

Then the bolometric correction is required to relate the magnitude at a given wavelength to the bolometric magnitude, for example :

$$BC_{\mathrm{V}} = \mathrm{M_V} - \mathrm{M_{bol}}$$

When only the symbol BC is given it corresponds to the V magnitude. BC is almost zero for A, F and G stars that radiate most of their energy in the visible but can be as large as 3 magnitudes for cold M stars that radiate mostly in the near-IR, and even larger for the hottest O stars that radiate mostly in the UV.

Colors or spectral types are more directly related to T_{eff}. However the relation between color and T_{eff} is far from uniform. O and early B stars have almost the same continuum at wavelengths longer than say 1200 A, thus almost the same colors ($B - V \simeq -0.30$) : this is simply due to the fact that the Rayleigh-Jeans part of the blackbody spectrum at their very high temperatures covers this wavelength range, and these stars match not too badly blackbodies in this range (not beyond the Lyman discontinuity, however). Consequently it is almost impossible to classify those stars on the basis of colors alone : spectral classification is required, but is unfortunately difficult because of the weakness and paucity of the interesting lines. Moreover these stars are not terribly bright in the visible due to their large bolometric corrections. It is thus extremely hazardous to try to derive e.g. the ionizing properties of a stellar population from its colors in any observable wavelength range.

As a consequence of all this the observational and the theoretical HR diagrams of a stellar population look very different. This is most important for the massive, luminous stars. These stars evolve at roughly constant luminosity but their tracks in observational HR diagrams are highly curved due to bolometric correction, and compressed in the "hot" region so that the main-sequence part is faint and almost vertical ; the star becomes bright in the relatively unimportant $A - F$ evolutionary stage and is faint again in the region of M supergiants in spite of its evolutionary importance. Fig.1.1 from the excellent paper by Flower (1977), illustrates this.

1.3 The Main-Sequence (MS)

This is the part of the HR diagram where stars burn helium in their cores. L and T_{eff} do not vary much during this phase which lasts roughly for 90 per cent of the stellar lifetime. One can show that, roughly :

$$L_{\mathrm{MS}} \propto m^3,$$

where m is the mass of the star (this relation flattens at high masses). As at the time of central exhaustion of hydrogen, where the MS terminates, about 15 per cent of all the stellar hydrogen has been burnt, one can estimate easily the MS lifetime as a function of mass :

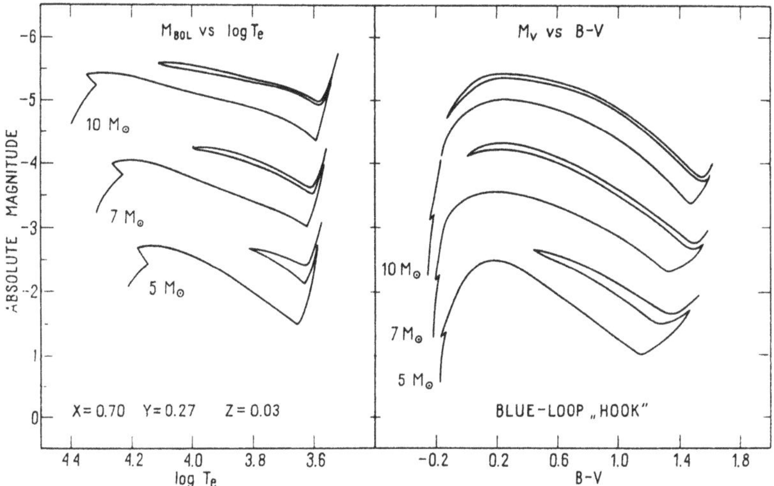

Fig. 1.1. Transformation of the theoretical HR diagram into an observational HR diagram M_V *vs.*$B-V$. The evolutionary tracks of stars of 5, 7 and $10m_\odot$ with the indicated composition and no mass-loss are indicated. From Flower (1977).

$$t_{\rm MS} \simeq 10^{10}(m/m_\odot)^{-2}yr$$

given the luminosity of the Sun, its mass $m_\odot = 2\,10^{33}\,\mathrm{g} = 2\,10^{30}\,\mathrm{kg}$ and the energy yielded by the combustion of hydrogen $6\,10^{14}\,\mathrm{J}$ per kg.

Consequently, any star with $m < 0.8m_\odot$ is still on the MS, given the age of galaxies $15\,10^9\,\mathrm{yr}$; only more massive stars have had time to evolve further.

During the recent years, it has become clear that the MS is not as narrow (in $T_{\rm eff}$ as well as in color or spectral type) as expected from conventional stellar evolution models. For example, the predicted MS for O–B stars should countain 80 to 90 per cent of all stars of the corresponding masses while observations show that it contains only 60 per cent, and the actual MS band should extend up to spectral type A0 ($T_{\rm eff} \simeq 10^4 K$). A much better agreement is produced when including stellar mass loss and a higher degree of mixing of the stellar interior e.g. via convective overshooting. Overshooting means that the inertia of packs of material moving up through convection is such that they are able to rise considerably further than the normal upper radius of convective zones, thus yielding increased mixing. See a discussion in the review article by Chiosi and Maeder (1986). Main-sequence widening is also present at lower stellar masses, presumably due only to enhanced mixing (by overshooting or other mechanisms) as mass-loss probably does not exist on MS at low masses (Mermilliod and Maeder, 1986). It should be noted that there are large uncertainties in the MS widening, both observationally and in the models that are adjusted empirically to match the observations : in particular, the influence of metallicity is not known.

1.4 Post-MS Evolution

Post-MS evolution is complicated and there are still major unsolved problems in this evolution. Good review papers are by Iben (1967, 1974), Iben and Renzini (1983) and Chiosi and Maeder (1986). Three kinds of stars should be separated, with different properties and evolution (fig.1.2).

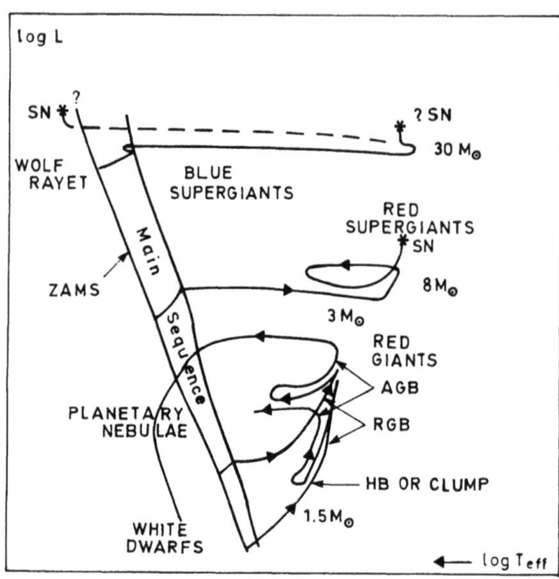

Fig. 1.2. Sketch of the evolution of stars of various masses in the theoretical HR diagram. ZAMS = Zero-Age Main Sequence; RGB = Red-Giant Branch (first ascent); HB = Horizontal Branch (reduced to a stationary position called the clump at high metallicities); AGB = Asymptotic Giant Branch (second ascent); SN = supernovae.

Massive stars with $m \gtrsim 8m_{\odot}$ burn H, then He, then C in non-degenerate cores and evolve rather smoothly at roughly constant luminosity until they end their life in a catastrophy : either they explode as a type II or Ib supernova, or (at least for the most massive ones) they collapse into a black hole. It is very hard to decide from observations between these two possibilities as their difference would not much affect the statistics of supernovae ; however it has large consequences on nucleosynthesis and chemical evolution of galaxies since the considerable production of heavy elements by massive supernovae might be strongly reduced in the second case.

Intermediate-mass stars $2.3m_{\odot} \lesssim m \lesssim 8m_{\odot}$ do not burn H and He in degenerate cores but experience carbon burning in degenerate matter for $m \gtrsim 4\,m_{\odot}$, resulting in a supernova exp losion. Stars with lower masses do not explode but end up as planetary nebulae nuclei then white dwarfs after substantial post-MS mass-loss. They first climb up the HR diagram as red giants, execute complicated loops and come back to an asymptotic giant branch (AGB) that extends the red giant branch (RGB) to higher luminosities. Then they turn to planetary nebulae, then white dwarfs.

Low-mass stars with $m \lesssim 2.3\,m_{\odot}$ ignite He in degenerate cores at the tip of the RGB (the helium flash). They then stay at a fixed luminosity ($M_{bol} \simeq 0$) for some time forming either the "clump" or a horizontal branch according to metallicity (respectively high or low); then after a short stay on the AGB they end as the previous ones.

For the two latter categories of stars the relative importance of the post-MS evolution, i.e. the ratio of energy produced in this place to that of energy produced during the MS, is larger than for the most massive stars and as these post-MS stages have relatively high luminosity they can dominate the energy produced by a galaxy at some stages. For example the AGB dominates the luminosity of a normal stellar population $\simeq 10^8$ years old and the RGB dominates after $\simeq 2\,10^9$ years (fig.1.3, from Renzini and Buzzoni, 1986).

The effects of mass loss and internal mixing are even more important during the post-MS evolution than on the MS. I discuss some aspects now.

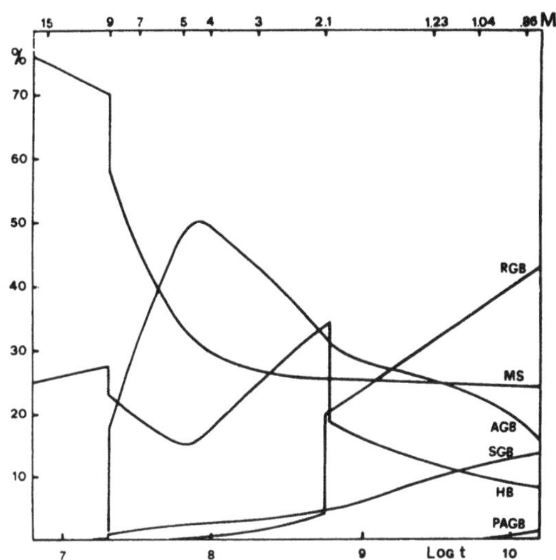

Fig. 1.3. Percentage contribution to the total (bolometric) luminosity of an ensemble of stars of various masses born together at time t = 0, as a function of time. The case corresponds to the composition X = 0.70, Y = 0.28, Z = 0.02 (see Chapter 5 for definitions) and the IMF has a uniform slope x = 1.35. MS = Main-Sequence, RGB = Red Giant Branch; HB = Horizontal Branch and all post-MS stages stars more massive than $9m_\odot$; AGB = Asymptotic Giant Branch; PAGB = Post-AGB stages for $m < 9m_\odot$; SGB = Sub-Giant Branch (stage between the MS and the location of the RGB, slow in low-mass stars). The mass of stars with total lifetime t is indicated on the top scale. This figure is schematic and some of its elements are uncertain, in particular the contribution of the AGB. From Renzini and Buzzoni, 1986.

1.5 Massive stars

If there were no mass loss and mixing, massive stars would experience core He-burning as red supergiants and also near the center of the HR diagram. However all massive stars are now known to experience strong mass loss at all stages of their lifes and may lose half of their masses or more in their lifetines. This does not affect strongly their central evolution (i.e. the regions where energy is generated) but affects very much the stellar envelopes (see Maeder, 1981a). The most massive stars may lose entirely their envelopes in the red supergiant phase and become then reduced to a small central core, very hot as the energy production is essentially unchanged. The external regions of this core are enriched in products of nucleosynthesis, essentially N in early stages (produced by the CNO cycle) and C in later stages (produced by He burning). These are Wolf-Rayet stars, respectively WN and WC. It is clear that in this case part or most of the core He burning occurs when the star is very hot thus at the extreme left of the HR diagram. Stars more massive than $\sim 60m_\odot$ may even never become red supergiants (however the upper limit for supergiants may result from instabilities : Maeder, 1983a).

This has of course important consequences on the appearance of the HR diagram. Also the supernovae (if any) that end the life of the stars are not of the same type : type II if the star exploded as a red supergiant, type Ib if it exploded as a Wolf-Rayet star (the difference being the presence or absence of an extended envelope). The relatively high frequency of type Ib supernovae shows indirectly the importance of mass loss. SN 1987A is an intermediate case since its progenitor was an evolved blue (B3I) supergiant of about $20\,m_\odot$ presumably on its way to the left of the HR diagram.

However there may be an intrinsic scatter in the importance of stellar mass loss, and mass loss seems to depend very much on metallicity. The best evidence for this is the highly variable ratio between Wolf-Rayet and red supergiants in galaxies (or part of galaxies) of various metallicities : this ratio is large in metal-rich galaxies and becomes quite small in metal-poor galaxies (Azzopardi et al. 1988a). This can be simply interpreted as due to an increase of mass loss with metallicity. This phenomenon should affect in a way qualitatively clear but difficult to predict well quantitatively the population of the HR diagrams. Mass loss also produces changes into the stellar nucleosynthesis, although not very important.

1.6 Intermediate and small mass stars

Post-MS evolution is not too badly understood for low-mass stars, although uncertain rates of mass loss on the RGB makes this important part of the evolution difficult to predict accurately. Also, some features of the evolution are not clear (e.g. possible important mixing at the He flash, or a possible reapparition of the blue HB at very high metallicities). The situation is much worse for intermediate-mass stars as the very important stage of AGB evolution is not quantitatively understood in spite of much theoretical effort. Mixing on the AGB produces a dredge-up of freshly produced carbon to the surface ; a carbon star is formed when C/O becomes larger than 1. Carbon stars are easily recognizable and are excellent probes of late stellar evolution. There are major discrepancies between model predictions and observations (e.g. in the Magellanic Cloud clusters) : there is a lack of bright AGB stars, and carbon stars are observed at much too faint magnitudes. Both phenomena may perhaps be understood as due to strong mass-loss or extra mixing on the AGB (Bertelli et al., 1985) but this is rather controversial. However observations of intermediate-age clusters in the Large Magellanic Cloud show that their light is very much contributed by AGB stars, thus this contribution cannot be much smaller than predicted in fig.1.3.

It is likely that mass loss depends on metallicity thus that metallicity affects evolution like for the more massive stars. Moreover, metallicity strongly affects colors on both the RGB and AGB, due to its effects on opacity and blanketing. This can be used with profit to derive the metallicity of a given population (see e.g. Azzopardi et al., 1988 for the galactic bulge and Mould and Kristian, 1986 for the old populations of M31 and M33).

1.7 Role of stellar multiplicity

50 per cent of stars independently of mass are members of physical (bounded) multiple systems, mostly binary. Moreover, massive star tend to occur in (unbounded) compact groups. Both phenomena have strong effects on the HR diagram. Obviously the unresolved brightest groups of stars will look brighter and less numerous than the same population if the stars were well separated. Moreover evolution of close binaries is rather different from that of single stars, due to mass transfer from one star to the other. In particular, an evolving massive star in a close binary system, will "try" to become a supergiant but will instead pour out its expanding envelope onto its companion ; thus one effect of binarity will be to decrease the number of red supergiants. Another one resulting from mass transfer on a white dwarf is the formation of type I supernovae. There is no room here to discuss in details the principles and consequences of close binary evolution (there is an extensive literature on the subject). What should be reminded at this stage is that it introduces further difficulties in the interpretation of HR diagrams that one generally tends to ignore as a first approximation –something that one should probably not do !

2. Tracers of Star Formation in Galaxies

2.1 Introduction and Definitions

In this chapter, I will concentrate on the various ways of determining the Present Day Star Formation Rate (PDSFR) in galaxies. The history of star formation will be dealt with in chapter IV and V. It should be stated right from the start that we have essentially no information on the PDSFR from low-mass stars (say with $m \lesssim 2m_{\odot}$), even in our Galaxy. Thus all I will say concerns higher-mass stars.

Let me define two quantities describing star formation. The Star Formation Rate (SFR) is the amount of mass converted into stars per unit time $\mathrm{SFR}(t) = \mathrm{d}M(t)/\mathrm{d}t$. This quantity can refer to the unit volume, area, mass of gas or total mass according to the author. The Inital Mass Function (IMF) describes the relative proportions of stars of different masses. It is generally defined as :

$$\psi(m) = \mathrm{d}n(m)/\mathrm{dln}(m) = m\mathrm{d}n(m)/\mathrm{d}m$$

where n is the number of stars of mass m. The IMF is normalized such that

$$\int_{m_l}^{m_u} \psi(m)\mathrm{d}m = 1$$

m_l and m_u being the lower and the upper mass of formed stars respectively. If the IMF does not depend on time, the mass of stars formed between masses m and $m + dm$ is obviously $\psi(m)SFR(t)\mathrm{d}m$.

Determining the IMF is a very difficult task except perhaps locally in our Galaxy : see the big review by Scalo (1986). A power-law parametrization $\psi(m) \propto m^{-x}$ is sufficient for high-mass stars (the IMF must flatten at low masses), and the various authors agree on $x \simeq 1.5$ to 1.7 (not inconsistent with the early determination $x = 1.35$ of Salpeter). It appears almost impossible to determine separately m_u and x : star counts are affected by small-number effects near m_u, and moreover an increase of m_u or a decrease of x produce essentially the same secondary effects. Finally the assumption of a constant IMF, although practical, is not really justified : one expects in particular metallicity to affect m_u (or x), an effect that seems observed at least in the star clusters that ionize extragalactic HII regions.

Let us discuss now the various methods to determine the PDSFR. The results will be discussed in the next chapter.

2.2 PDSFR from Star Counts

This is obviously the only direct method ; unfortunately it is applicable only to those galaxies nearby enough for the brightest stars to be resolved (essentially those of the Local Group of galaxies).

The method is the following :

1) Build a HR diagram (e.g. V vs $B - V$) by observations of a galaxy or part of a galaxy (CCDs are ideal for that).

2) Assess completeness (to a given magnitude and possibly color) and correct for incompleteness when possible (this is generally not possible !)

3) Eliminate those parts of the HR diagram that are contaminated by galactic foreground stars ; the blue part is much less contaminated than the red part, thus a cut-off at say $B - V \simeq 0.3$ is indicated, leaving essentially the (widened) main sequence.

4) Count stars in identical parts of the HR diagram of different galaxies, yielding relative values of the PDSFR : in effect, the accessible parts concern in general massive stars and they yield the PDSFR as the lifetime of these stars is quite short.

This assumes of course that the distances are known, as the HR diagrams must be put in an absolute magnitude scale. This is not as trivial as one would imagine, as the distance of the studied galaxies is often derived from the apparent magnitude of their brightness stars, hence a kind of circular argument. One has to assume that the IMF and the stellar evolution are the same in the different systems studied. This is far from granted : indeed the IMF may well depend on metallicity and perhaps on other factors (environment) while we know for sure that massive-star evolution depends on metallicity (see the previous chapter). Still the HR diagrams of galaxies as different as our Galaxy and the two Magellanic Clouds that have quite lower metallicities are remarkably similar (see Humphreys and Mc Elroy, 1984 ; note however that the Magellanic Clouds may have more massive stars than indicated due to star crowding). Also the luminosity functions (proportion of stars with different absolute magnitudes e.g. M_V or M_B or M_{pg}) are rather similar for most galaxies where they could be determined (see e.g. Lequeux, 1986 ; Freedman, 1985 ; Pierre and Azzopardi, 1988 ; Aparicio et al., 1988). This is rather surprising, especially when one realizes that the observed stars are mostly late MS or post MS stars. It might be that nature conspires in mutually cancelling the effects of variations in the IMF and of variations in the evolutionary properties ?

In any case, many other difficulties make this kind of observation quite uncertain. Image crowding and stellar clustering are major difficulties, particularly troublesome if one compares galaxies at different distances and observed with different seeings. In spite of advanced methods of image processing, the limit of completeness due to image crowding is barely better than magnitudes $\simeq 21$ although CCDs allow to reach much fainter magnitudes. The image quality of the Hubble Space Telescope should yield a big improvement of the situation.

Nevertheless, this method seems to produce results good to a factor $\simeq 2$, a result which is not worse than that of the other methods I will describe now.

2.3 PDSFR from Lyman Continuum Flux

Only O stars and early B stars are hot enough to generate photons at wavelengths shorter than the Lyman limit at 912A (actually the contribution of B stars is rather small). These photons ionize the hydrogen of the interstellar gas. Recombination produces cascades and all hydrogen lines are emitted, in particular the Balmer lines (these are called for this reason recombination lines). As ionization and recombination equilibrate each other, the flux in any recombination line is proportional to the Lyman continuum flux to a good accuracy. One has for example for the Balmer β line (H_β) :

$$\frac{N_C'}{1ph\ s^{-1}} = 2.47\,10^{56}\,\frac{I(H_\beta)}{1erg\ cm^{-2}s^{-1}}\left(\frac{T_e}{10^4 K}\right)^{-0.09}\left(\frac{D}{1kpc}\right)^2$$

where N_C' is the number of Lyman continuum photons absorbed by the gas each second, $I(H_\beta)$ the flux of H_β photons observed from the Earth, T_e the electron temperature of the gas and D the distance to the source. The flux in the other lines can be estimated using the tables of Brockelhurst (1971) for the Balmer and Paschen lines and Giles (1977) for the Brackett lines in the IR (the list of wavelengths is in Lang, 1980, p.119).

In order to derive the PDSFR from the observation of the flux of one recombination line, one has to :

1) Correct this flux for interstellar extinction : this is somewhat problematic (see below).

2) Calculate N_C'

3) Estimate the number of Lyman photons N_C emitted by the stars. N_C is larger than N_C' as dust mixed with the gas absorbs part of the photons. Usually one takes $N_C = 2N_C'$ from a study by Smith et al (1978).

4) Estimate the number of stars : one has to chose an IMF, decide is one is dealing with a burst (all stars born at the same time) or with a continuous star formation (or "extended burst" : stars formed at a constant rate starting at a given time) or with a more complicated situation (not recommended as this introduces too many parameters !). Then one has to estimate the flux of Lyman continuum photons using model atmosphere (Kurucz or Mihalas) or empirical data (e.g. Panagia, 1973). Fortunately these determinations do not depend much on this choice (this would not be the case if one was dealing with helium-ionizing photons). For an example, see Lequeux et al. (1981).

There are two major problems in this derivation. One is that measurements of Balmer-line fluxes are difficult, especially for not very active galaxies where the line/continuum contrast is small ; errors by a good factor 2 are possible when filter photometry on and off the line is used (one can do better with sweeping Fabry-Perot interferometers but there are not many measurements yet). The other problem is the extinction correction to be applied. Extinction can be derived from the observation of the ratio of fluxes of two Balmer lines, making use of the fact that extinction varies with wavelength. One has for example

$$\log\left[F\left(H_\alpha\right)/F\left(H_\alpha\right)\right] = \log\left[I\left(H_\alpha\right)/I\left(H_\beta\right)\right] + 0.335C(H_\beta)$$

where F are the observed fluxes, I the theoretical fluxes $[I\left(H_\alpha\right)/I\left(H_\beta\right) \simeq 2.86$ in general]; $C\left(H_\beta\right) = \log\left[I\left(H_\beta\right)/F\left(H_\beta\right)\right]$ is the logarithmic extinction coefficient. The relation between visual extinction and C is : $A_V = 2.2C\left(H_\beta\right)$. Another way (possible for HII regions) is to compare recombination line fluxes with radio continuum fluxes, as free-free radio emission from an ionized gas is also proportional to the ionizing photon flux. Numerically, one has

$$\frac{F_\nu}{1Jy} = 3.28\,10^9 \left(\frac{\nu}{1GHz}\right)^{-0.1} \left(\frac{T_e}{10^4 K}\right)^{0.34} \frac{I\left(H_\beta\right)}{1erg\,cm^{-2}s^{-1}}$$

F_ν being the radio continuum flux at frequency ν . $I\left(H_\beta\right)$ calculated from this formula can then be compared to the observed $F\left(H_\beta\right)$ to obtain $C\left(H_\beta\right)$.Unfortunately, the two determinations of $C\left(H_\beta\right)$ rarely coincide, even in single HII regions, and it is hard to make a choice (see e.g. Lequeux et al. 1981). The problem is particularly severe for galaxies as a whole ; Kennicutt et Hodge (1986) tend to adopt a large correction in the case of the Magellanic Clouds (1 magnitude of extinction at H_α) but this seems too much.

It would be better to use radio continuum fluxes directly to determine N_C', thus avoiding the problem with extinction. Unfortunately this is only possible for relatively bright HII regions. The radio continuum of galaxies is a mixture of thermal (free-free) and non-thermal (synchrotron) radiation which are very hard to separate from each other.

2.4 PDSFR from far-UV measurements

Only hot stars radiate in the far-UV. These can be either young, massive O and B stars, or evolved objects like horizontal-branch stars or nuclei of planetary nebulae. The latter class of objects gives only a minor contribution except where there is no young population at all (globular clusters and may be bulges of spiral galaxies and elliptical galaxies). Thus in general far-UV radiation gives a good measure of the PDSFR for massive stars. While the Lyman continuum photons are dominated by O stars, the far-UV photons are dominated by early B main-sequence stars and thus relate to a somewhat fainter part of the IMF.

The method is straightforward : measure flux at some wavelength, then correct for extinction and estimate the numbers of stars by using model calculations. These model calculations are relatively easy as the far-UV fluxes of stars of different types are well known (see e.g. Nandy et al., 1976). I find useful to note that the far-UV fluxes can be fitted by empirical relations like :

$$M_{2030\mathrm{A}} - M_{bol} = -1.00 + 11.0\,(\log T_{\mathrm{eff}} - 4.28)^2$$

that allow after some transformations to express the emitted energy as :

$$\log I_{2030\mathrm{A}}\left(\mathrm{erg}\, s^{-1} A^{-1}\right) = -4.40\,(\log\, T_{eff} - 4.28)^2 + \log\, L/L_{\odot} + 30.19$$

giving a sufficient approximation. Note also that contrary to optical astronomers far-UV observers have the good idea to define magnitudes from received fluxes !

$$\log\, f_\lambda\left(\mathrm{erg\, s^{-1} A^{-1} cm^{-2}}\right) = -0.4\, m_\lambda - 8.44$$

The problem with extinction is much worse than with the recombination lines, as extinction in the far-UV is much larger than in the visible : thus the results will be rather uncertain. On the other hand, the measurements are probably more accurate. A strange fact is that the UV radiation from stars ionizing large extragalactic HII regions do not seem to be much reddened (in the far-UV) : the observed spectra are not far from a Rayleigh-Jeans spectrum as expected without extinction, while interstellar reddening should flatten it considerably and produce a 2200 A absorption feature. Presumably this is due to a selection effect related to strong inhomogeneities in the extinction : far-UV radiation is visible only from stars sitting in extinction "holes" while radiation from stars located behind obscured regions is completely hidden. This makes the problem even worse. An attempt at a statistical correction in galaxies has been made by Donas et al. (1987).

2.5 PDSFR from far-IR radiation ?

Far-IR emission comes from dust heated by stellar radiation. As dust absorbs more the far-UV photons than the visible ones, one may expect dust heating to be dominated by UV photons, and far-IR radiation to trace the PDSFR for massive stars. Can we assess this quantitatively ?

A first problem is to quantify the far-IR radiation, which has most often be measured only in the four bands of the IRAS satellite (12, 25, 60 and 100 μm). An empirical extrapolation often used is that of Helou et al. (1985) :

$$\frac{F_{\mathrm{FIR}}}{1Wm^{-2}} = 1.26\; 10^{-14}\left(2.58\,\frac{S_\nu(60\mu m)}{1Jy} + \frac{S_\nu(100\mu m)}{1Jy}\right)$$

the S_ν being the flux densities as given in the IRAS catalogues. Based on the FIR spectrum of the center of our Galaxy, which is known over the whole range of FIR wavelengths, Boulanger and Perault (1988) rather use :

$$F_{\substack{\mathrm{FIR} \\ Wm^{-2}}} = 10^{-26} \sum_{\mathrm{4bands}} \nu S_\nu$$

Table 2.1. Local distribution of FIR luminosity

Component	FIR/M_{gas}	Surface FIR luminosity
"CIRRUS" = diffuse gas	$2\ L_{\odot}/M_{\odot}$	$10\ L_{\odot}/pc^2$
+ some molecular clouds		
Cold Molecular Clouds	$1\ L_{\odot}/M_{\odot}$	$1.5\ L_{\odot}/pc^2$
Star-forming regions	$\gg 1\ L_{\odot}/M_{\odot}$	$3.0\ L_{\odot}/pc^2$
	(Variable)	———
	TOTAL	$14.5\ L_{\odot}/pc^2$

Table 2.2. Origin of FIR galactic radiation (This seems to apply both to the solar vicinity and the active SF region ~ 5 kpc from the galactic center)

Heating stars	Fraction of FIR radiation	
O	0.17	
B0 - B0.5	0.21	0.53 (O to late B)
late B	0.15	
A	0.10	
FGKM	0.16	
KM giants	0.21	
	1.00	

ν being the frequency corresponding to the nominal center of each band. I tend to prefer the second formulation as more appropriate for galaxies, but the differences are not very large.

In order to appreciate how much FIR flux comes from young massive stars, I quote some results from Boulanger and Perault et al. (1988).

The second table shows that about half of the galactic FIR radiation actually comes from young massive stars. Thus the FIR luminosity is a good measure of the PDSFR for galaxies that have substantially more star formation than our Galaxy, i.e. $L_{FIR} \gg L_{visible}$ ... unless there are other processes for heating the dust, e.g. non-thermal UV radiation from an active nucleus. In the case of a low SFR (like M31) the use of L_{IR} to estimate the PDSFR is not recommended.

When L_{IR} can be used as a tracer, a useful formula for the steady-state case (continuous SF older than about 10^8 years) is given by Scoville and Young (1983) for a "normal" IMF

$$dM(\mathrm{O,B,A\ stars})/dt = 7.710^{-11} L_{IR}/L_{\odot}(M_{\odot} yr^{-1})$$

We have also to apply a correction for the fraction of stellar radiation that does not heat the dust. In the case of our Galaxy this fraction is about 0.5, but it may vary from galaxy to galaxy. One may try to use the same receipe as Donas et al. (1987) used for the far-UV radiation.

2.6 Other possible tracers of PDSFR

These have been discussed by Lequeux (1979). Supernovae and their remnants may be useful, although type I Supernovae do not have massive progenitors ; however the statistics are poor and incomplete. Cepheids cannot be used. The non-thermal radio radiation shows a beautiful correlation with the FIR luminosity at le ast for galaxies actively forming stars (see the next chapter) and it is tempting to use it as a PDSFR tracer ; the idea is that the relativistic electrons that are responsible for this radiation originate in young objects (supernova remnants). However we do not really understand why this correlation is so tight.

It is not recommended to use any luminosity in the visible as a PDSFR tracer, except in extreme cases of starburst galaxies where we can be reasonably certain that most of the light comes from the starburst (but be careful about the extinction correction that may be very large in such cases, and also to the possible non-thermal contribution of an active nucleus).

2.7 Time evolution of some of the PDSFR tracers

It is useful to give some information on the time evolution of at least the two main-tracers : the Lyman continuum and the far UV fluxes -Relevant tables can be found in Lequeux et al. (1981) and Lequeux (1988). I assume a burst of star formation at time $t = 0$ (fig.2.1).

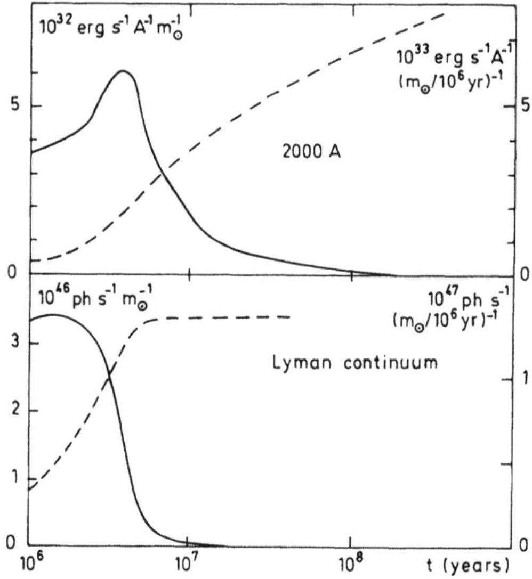

Fig. 2.1. Lyman continuum flux and far-UV flux at 2000 A for an ensemble of stars all born together at t = 0 (full lines) and for stars forming at a constant rate starting at t = 0 (dotted lines). The initial mass function is defined by $x = 1.5$ for $m > 1.8m_\odot$, $x = 0.6$ for $m < 1.8m_\odot$, $m_u = 110m_\odot$, $m_l = 0.007m_\odot$. The atmosphere models used for estimating the Lyman continuum flux are those of Mihalas.

The Lyman continuum flux stays about constant then drops precipitously after about $4\,10^6$ years due to the disappearance of O stars from the main sequence. The far-UV flux first increases slightly, due to the evolution of hot stars inside the main sequence, then drops much more slowly as the contribution of B stars is much more important. The time evolution in the case of continuous star formation is easily derived from integration (fig.2.1). A steady-state is reached soon for the Lyman continuum photons.

3. The Present Rate of Star Formation in Galaxies

This chapter is an application of the principles developed in the previous one, and is intended as an illustration of what can (or cannot) be done. I will discuss first the irregular galaxies, then the "normal" spiral galaxies and finally those galaxies that have a particularly high PDSFR : the "starburst" galaxies.

3.1 Irregular Galaxies

The best known of the irregular galaxies are our two closest neighbors : thet Large Magellanic Cloud (LMC) at about 50 kpc distance and the Small Magellanic Cloud (SMC) at 70 kpc. They offer the opportunity of studying massive star formation in conditions different from those in our Galaxy : in particular the abundances of heavy elements are about 3 and 10 times lower respectively, and much more of their mass is under the form of gas.

The following table compares the PDSFR per unit mass of gas (a reasonable normalization as stars are made from gas) to that in our Galaxy near the Sun. Molecular hydrogen is included for the Galaxy and not included (for lack of data) for the LMC and SMC ; these galaxies are however believed to contain relatively less H_2 molecules. The data are from Lequeux (1984) with some minor changes : the total mass of the SMC is a rough estimate from the velocity dispersion of old objects (planetary nebulae), and the data for the Lyman continuum flux of the LMC are from the H_α observations of Kennicutt and Hodge (1986), respectively uncorrected for exctinction (lower number) and corrected for an extinction $A_V = 1$ magnitude (higher number)

Table 3.1. Comparison of PDSFR per unit mass of gas

	GALAXY, 1 kpc^2	LMC	SMC
Total mass	$9\ 10^7$ $M_\odot$	$6.1\ 10^9$ $M_\odot$	10^9 $M_\odot$
Gas mass M_{gas}	$6\ 10^6$ $M_\odot$	$7\ 10^8$ $M_\odot$	$6.5\ 10^8$ $M_\odot$
$N_{bright\ stars}/M_{gas}$	1	1.2 - 1.6	0.15 - 0.27
N'_C/M_{gas}	1	1.5 - 3.0	0.28
L_{1690A}/M_{gas}	1	1.3	0.30
N_{SNR}/M_{gas}	(1)	(1.8)	(0.4 - 0.6)

In this case we have 3 good tracers of PDSFR : bright star counts in similar portions of the HR diagram, Lyman continuum photons and far-UV luminosities, and some information on supernova remnants. It is seen that all determinations are consistent with each other. This suggests that the IMF/evolution combinations are the same within the large errors in all three cases, yielding similar stellar luminosity functions in spite of very different metallicities. This rather surprising, but perhaps not really significant result has already been discussed in the previous chapter. In any case, it appears that the LMC is producing 1.5 times more massive stars per unit mass of gas than the solar neighborhood, and the SMC 3 times less. The physical reasons for this difference are not really known. Note also that there is much gas left in the SMC but less in the LMC.

Stars counts are also possible in other irregular galaxies in the Local Group with the result that the PDSFRs per unit mass of gas are similar within a factor 3 (this encompasses both Magellanic Clouds) : see Lequeux (1986), but this study would be worth re-doing using more modern data. As for the Magellanic Clouds, there is no sign of significant differences in the luminosity functions.

Statistics on PDSFRs on larger samples have been obtained using far-UV observations (Donas et al. 1987) : they also indicate a PDSFR/M_{gas} ratio uniform within a factor 3, and somewhat smaller than for spiral galaxies, by a factor 2–3. Note that for working with any such correlation it is better to work with apparent quantities (as seen from the Earth) rather than with absolute quantities in order to avoid biases and also to be less impressed by the natural correlations that exist between any pair of parameters in galaxies.

There is less H_α photometry but the agreement is good with the results from the far-UV. IRAS data are scarce as the FIR emission of irregular galaxies is weak, obviously due to a lack of dust linked to their low metallicities. It is remarkable and not explained that their dust is somewhat hotter than that of spirals ($60\mu m/100\mu m$ ratio higher).

Finally I would like to mention that massive star formation looks qualitatively different in irregular and spiral galaxies. Molecular clouds are very small, there is no equivalent to the giant molecular clouds in our Galaxy ; on the other hand, there is no equivalent in our Galaxy to the hypergiant HII regions 30Dor in the LMC or CM 39 in NGC44 49. Finally there are young globular clusters in the LMC, SMC (and also in the late-type spiral M33) with no equivalent in our Galaxy.

3.2 "Normal" Spiral Galaxies

As expected, our Galaxy is the best documented of all. The PDSFR has been deduced from the Lyman continuum flux, itself derived from the thermal radio continuum emission of HII regions by Smith et al. (1978). It is interesting to follow what they have done. They estimated a total ionizing flux for what they call the giant HII regions of $N_C' = 2 10^{52}$photon s^{-1} for the whole Galaxy. They correct this by a factor 2 to take into account dust absorption inside the HII region, yielding $N_C = 4.1\ 10^{52}$ph s^{-1}.

They estimate a further contribution from smaller HII regions of $0.7\ 10^{52}$ph s^{-1} and another one from the galactic center HII regions of $0.7\ 10^{52}$ph s^{-1}. This gives a total $N_C = 5.5\ 10^{52}$ph s^{-1}. For a "normal" IMF their model calculations give a production rate per unit mass of star formed of $2.2 10^{46}$ph s^{-1}M$_\odot^{-1}$. However the lifetime of an HII regions is not longer than about $5\ 10^5$yr after which it is dispersed by the combined effect of increased pressure and stellar winds : thus we see the effect of O stars only during this time. Consequently the SFR is, with their adopted IMF :

$$\mathrm{d}M_*/\mathrm{d}t = \frac{N_C}{2.2\ 10^{46} \times 5\ 10^5} = 5\,\mathrm{M}_\odot\,\mathrm{yr}^{-1}$$

for the whole Galaxy. Actually O stars continue to radiate and ionize the gas when the HII region has dispersed so that the total Lyman continuum flux is about $45\ 10^{52}$ph s^{-1}, mostly outside HII regions, yielding a general ionization in the diffuse interstellar medium and a weak diffuse H_α emission that has indeed been observed.

It is also possible to obtain from the distribution of HII regions the radial distributions of the PDSFR : it is found to peak in a ring about 5 kpc in radius (fig.3.1). The distribution of atomic hydrogen is much more uniform and the SFR is clearly not correlated with it. Conversely, the distribution of molecular gas as derived from that of the 2.6 mm line of the CO molecule or from γ-ray observations with the COS B satellite is similar to that of the PDSFR : a not unexpected result as we know that stars form in molecular clouds. The

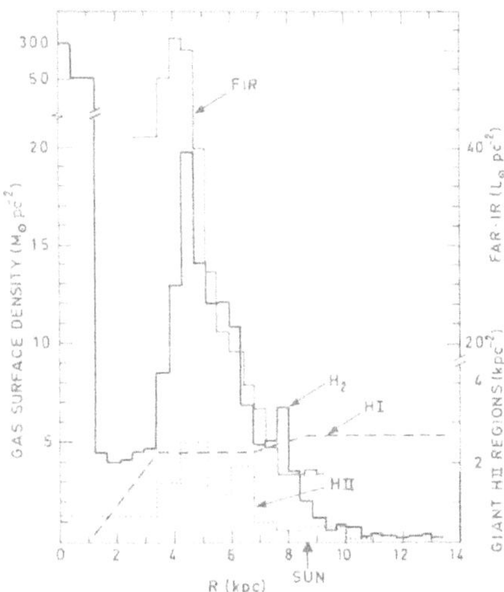

Fig. 3.1. Comparison of the H_2, HI, giant HII regions and far-IR emission surface densities in the Galaxy. Values for H_2 and HI include a 1.36 correction for the mass of helium. The HI distribution is from Burton and Gordon (1978), that for H_2 is from Clemens, Sanders and Scoville (1986), that for the giant HII regions from Smith et al. (1978) and that for the far-IR emission from Perault et al. (1988).

large-scale distribution of the far-IR radiation is also similar, showing that it is not such a bad tracer of PDSFR after all.

The Andromeda Nebula M31 is another well-studied case (Walterbos and Schwering, 1987). The PDSFR is rather low in this galaxy as L_{IR}/L_B is 1/10 of galactic. Star formation occurs mainly in a ring about 10 kpc from the center, as can be seen in the images of various PDSFR tracers. The dust is colder than in the Galaxy (smaller 60/100 μm flux ratio), may be due to a smaller contribution of hot star-forming regions. There are unfortunately no quantitative measurements of the PDSFR yet.

M51, M83, NGC6946: For these rather extended Sc galaxies, the PDSFR could be mapped using various tracers, usually H_α, and compared to the distributions of atomic and molecular hydrogen as described from 21-cm and CO observations respectively (Scoville and Young, 1983; Combes et al. 1978 and Young, 1988). There are also FIR and radio continuum distributions available. In every case, the PDSFR as well as the latter distributions follows the distribution of molecular hydrogen, not that of atomic hydrogen, confirming that massive stars at least form in molecular clouds as observed in our Galaxy.

M33, a nearby Sc galaxy, also has a lot of data, although CO observations are very incomplete. H_α, far-UV radiation, as well as FIR and thermal radio continuum radiation (available in this case), all have exponential distributions and follow each other quite well but not at all the atomic hydrogen. There are however interesting small but significant differences in the scale lengths of these components that have several possible explanations (Rice et al., 1988).

More detailed studies of the distributions of all these components are possible. The best case is probably M51 for which detailed CO observations have been recently obtained by Garcia-Burillo and Guélin with the 30-m IRAM millimeter telescope, yielding in particular for the first time a good figure for the arm/interarm contrast in CO.

Statistics on PDSFRs in spiral galaxies have been made on various samples. A correlation within a factor 3 exists between the PDSFR as derived from far-UV of H_α photometry (Donas et al, 1987 and Buat and Deharveng, 1988, table 2) and the atomic hydrogen content. We have seen previously that (for spirals, not for irregulars !) the PDSFR is better correlated with H_2 (or CO) than with atomic hydrogen in galactic dicsc, one may expect a tight correlation between Hα (or far-UV flux) and CO. Good quality data are still too scarce for checking this, unfortunately. The correlations between the FIR radiation and H_α (Persson and Helou, 1987) and between the FIR and far-UV fluxes (Buat and Deharveng 1988, fig.1a) are not very good, showing again that the FIR radiation is not a very good tracer of PDSFR for "normal" galaxies. Similarly, FIR and CO fluxes are well, but not very well correlated (Solomon and Sage, 1988). It is remarkable that the FIR/CO relation is a factor 2 higher than the same relation for galactic molecular clouds, an indication of the contribution of the "cirrus" (diffuse) component of the interstellar medium to the FIR emission of galaxies. Similarly the correlation between FIR and the total mass of interstellar gas $M(HI) + M(H_2)$ seems better than with any of these two parameters alone. A very recent study (Buat, Thesis) shows that this is also the case for the far-UV radiation, thus the PDSFR. This appears to be in contradiction with the spatial correlation PDSFR/CO noted in several galaxies.

3.3 Starburst Galaxies

These are galaxies with a very high PDSFR. For a description of what can happen in a starburst see Larson (1987) and Rieke et al. (1988).

They are detected in various ways :

1) Strong $H\alpha$ or UV emission (far or even near). This is a powerful method which has led to the discovery of many starbust galaxies that are called blue compact galaxies or extragalactic (isolated) HII regions (more or less synonymous), Haro, Markarian or KISO galaxies, etc. This has drawbacks however as galaxies with a high amount of internal extinction cannot be found in this way ; also galaxies with active galactic nuclei (AGN) often exhibit an UV excess and/or strong emission lines and are included in the above samples although they may not experience a strong burst of SF.

Far-IR emission is a more clear-cut way of discovering starburst galaxies. Many have been found as IRAS sources and are often called IRAS galaxies. Every galaxy with $L_{FIR} \gg L_{visible}$ is likely to be a starburst galaxy or to be an AGN galaxy or both. In a pure starburst, FIR radiation is almost entirely due to reradiation by dust of the light of the many massive stars formed in the burst. Such galaxies also show a strong continuum radio emission that is well correlated with FIR emission (see Helou et al., 1985 or de Jong et al., 1985) : thus strong radio emission is also a way to discover starburst galaxies. However some AGN galaxies share the same FIR/radio correlation while others have an excess in radio emission. Starburst galaxies are also strong X-ray emitters, but this is also the case for AGNs.

It is interesting to elaborate on the FIR properties of starburst galaxies. Their FIR spectrum was defined by the fluxes in the four IRAS bands (12, 25, 60 and 100 μm) is well distinct in pure starbust cases from that of "normal" (quiet) galaxies, Seyfert (AGN) galaxies and quasars (Rowan-Robinson, 1987) : see fig.3.2.

However there are intermediates between normal galaxies, starburst galaxies, AGNs and quasars. This shows in particular that starburst galaxies may contain an AGN, and also that catalogued AGN galaxies may also be starburst. It may well be that the "typical" Seyfert FIR spectrum showed fig.3.2 results from the superimposition of a quasar and a starburst ; however the 60/100 μm flux ratio for Seyferts is larger than for both quasars and pure starbursts, indicating a higher dust temperature, may be due to an increased radiation field or smaller dust grains, with respect to starbursts -the "pure" quasars have no grain emission. Most interesting also is the relation between quiet and starburst galaxies. Starburst galaxies

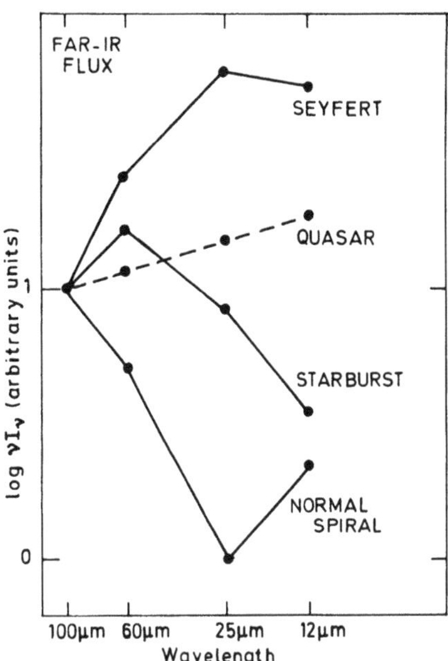

Fig. 3.2. Typical far-IR spectra of normal spiral galaxies, starburst galaxies, Seyfert galaxies and quasars, normalized at 100 μm. Adapted from B. Rowan-Robinson (1987).

have a larger 60/100 μm ratio indicating hotter dust, while their 12/25 μm flux ratio is considerably smaller. The increased 60/100 μm flux ratio corresponds to hotter "normal" dust due to an increased interstellar radiation field itself due to the high PDSFR (or/and AGN ?). The drop of the 12 μm flux corresponds to the disappearance of very small particles heated by single UV photons, the Polycyclic Aromatic Hydrocarbons or PAHs (see the lectures by H. Habing) : these are presumably destroyed in the intense UV radiation of the young, hot stars or/and by shocks. These properties are found at small scale in some regions of the Galaxy (Helou, 1988).

I would like to mention that the blue compact galaxies, although clearly starburst galaxies, emit little FIR radiation : this is obviously due to their lack of dust, itself due to their lack of heavy elements. This is the same explanation as for the weak FIR emission of quiet irregular galaxies (most blue compact galaxies are just irregular galaxies experiencing a starburst).

Correlations between the FIR flux and various quantities can be attempted for starburst galaxies. There are not enough far-UV data to do that, but Moorwood et al (1986) present a poor correlation between FIR and $H\alpha$ fluxes of IRAS galaxies (corrected for extinction as well as they could). Why this correlation is poor may be due either to incorrect corrections for extinction or to the presence of AGNs. The correlation of FIR and HI (21 cm) fluxes is very poor, a result not unexpected (some starburst galaxies show no detectable HI at all !) : Dickey et al., 1987. Starburst galaxies exhibit up to an order of magnitude more FIR radiation per unit CO luminosity, thus per unit mass of molecular hydrogen, than quiet galaxies (Solomon and Sage, 1988), a result not unexpected either. Finally there is an excellent and not so well explained correlation between the FIR and the radio continuum fluxes, as I already mentioned. One may assume that the relativistic electrons responsible for the non-thermal part of the radio emission are produced by supernova explosions thus proportional to the

PDSFR (after some delay following the beginning of the burst). The thermal part is less problematic.

The mass of stars formed in starbursts is enormous. The FIR luminosity can be as high as a few $10^{12} L_\odot$. The most "economical" way to radiate such a luminosity is to have a short burst observed during the first $4\,10^6$ years or so. With a standard IMF ($x = 1.5, m_u = 110\,m_\odot$), the bolometric luminosity per unit mass formed is about $750 L_\odot / m_\odot$, thus the mass of stars formed may be as high as a few $10^9 m_\odot$. This is the gas content of an average spiral galaxy ! There are two ways to reduce the mass of stars formed for a given luminosity : increase the upper mass limit m_u or decrease the slope x of the IMF), or increase the lower mass limit m_L. The first possibility is excluded, and indeed observation of rather low gas excitation in some starbursts goes against m_u being higher than $30 - 40\,M_\odot$ (or against the IMF being flatter) : see Scalo, 1986. If $m_u = 30\,M_\odot$ the bolometric luminosity is reduced by a factor 3 and this makes the problem correspondingly 3 times worse. A better way is to increase m_l; there are physical arguments (Silk, 1987) and observational support (Scalo, 1987) for that. If $m_l = 20\,m_\odot$ one gains a factor 10 in the mass but this is clearly an extreme possibility. It seems quite hard to go any further unless there is a hidden AGN after all (the energy conversion in an AGN is presumably much more efficient as it seems that the gravitational energy can be used). In any case, it is clear that strong starbursts cannot last for very long and also must be rather rare phenomena. For an application of these notions, see e.g. Augarde and Lequeux (1985).

I do not want to discuss here the origin of starbursts. There is an abundant literature on this subject (see e.g. T.X. Thuan et al., eds., 1987) ; the general opinion is that they are triggered by galaxy encounters, the most powerful starbursts being triggered by direct collisions. Note that AGNs also seem to be triggered by encounters. The relation between starbursts and AGNs is a fascinating subject for future studies.

4. Photometric and Spectral Evolution of Galaxies

4.1 Principles of Color and Spectral Synthesis

Up to now we have dealt only with present-day star formation in galaxies. Of course, we wish to know the past history of star formation. The tools for this are the color and spectral synthesis, and to some extent the chemical evolution that will be dealt with in the last chapter of this course.

The idea of synthesis is to reproduce the colors and spectra of galaxies by a superimposition of stars or star clusters of various ages and metallicities. Inverting colors into basic components is not tractable, but ther e have been attempts at inverting spectra. The problem with synthesis is that there are many free parameters and one should impose reasonable guidelines in order to limit their number. The simplifying assumptions that are usually made are :

1) The IMF is constant in space and time. This is a very critical assumption which is probably not true. Metallicity may affect it, in particular : Campbell (1988), Viallefond and Stasinska (1988). There are a few attempts to escape this assumption, in particular the bimodal star formation models developed by Güsten and Mezger (1983) and Larson (1986).

2) The star formation rate SFR(t) has a smooth variation, with may be very few bursts superimposed. This is also somewhat arbitrary.

3) A chemical evolution model is sometimes coupled with the history of the galaxy in order to yield a variation of metallicity with time, allowing to take these variations into

account. This is better than nothing, but also dangerous as the chemical models are uncertain. As we will see later, infall of gas on the galaxy or galactic winds can change drastically this evolution. Many people do not include metallicity variations in their models.

The ingredients are :

1) An IMF (or two in the case of bimodal SF models)

2) Evolutionary tracks of stars at various metallicities. There are problems here that have been discussed in Chapter I. Stellar mass-loss and mixing have very important effects and are not too well known, and are moreover metallicity-dependent. Post-MS evolution is still poorly understood, especially on the AGB. These problems are especially troublesome for interpreting the IR radiation of galaxies.

3) A stellar spectral library. There have been much recent improvements in this matter : see e.g. the UV library of Fanelli et al. (1987) or the visible library of Pickles (1985a). The main problem is the lack of stellar spectra with metallicities larger than solar.

An alternative recently developed by Alloin and Bica (see e.g. Bica, 1988) is to use a library of spectra of clusters. In the LMC there are clusters with a large range of ages and metallicities which can be supplemented by globular clusters of our Galaxy. This bypasses the problems with IMF and evolutionary tracks at least to some extent. Unfortunately there is also a lack of information for high metallicities.

4) Finally estimates and corrections for interstellar reddening are required in the studied galaxies (this can be difficult) and even for the construction of the library : for example we can find little-reddened O stars only in the Magellanic Clouds.

There are two approaches to the problem :

1) Evolutionary synthesis consists in fitting spectra of galaxies with a "reasonable" model of evolving populations. This is the only possibility when only colors are available (color synthesis). This method yields insight on the evolution but is not always foolproof.

2) Population synthesis or "optimized synthesis" is an attempt at an inversion of the observed spectrum using a library of components with various ages and metallicities. As there may be many solutions, one has to be very careful in interpreting the results and conservative in selecting as little free parameters as possible.

4.2 Color Evolution of Galaxies

Color evolution of galaxies is treated by evolutionary synthesis. A galaxy can be considered as the sum of "star clusters" of different importances and ages, a star cluster being defined as an ensemble of stars formed simultaneously according to an IMF. These "clusters" are left to evolve using theoretical tracks and their contributions to each considered wavelength band are added. Usually the only free parameter is the time history of the formation of these clusters. One assumes the IMF to be constant and neglects metallicity effects as a first approximation.

In order to consider what happens, it is useful to work in a color-color diagram (e.g. U-B vs. B-V) and to consider how the representative point of a star cluster evolves : very young clusters have very blue colors ($U - B \simeq -1.1$, $B - V \simeq -0.2$) and they become redder and redder in both colors to reach $U - B \simeq 0.4$, $B - V \simeq 0.9$ at about 10^{10} years, the colors of globular clusters (fig.4.1).

On the other hand, if stars are formed at a uniform rate from time $T = 0$ (this can be envisioned as a superimposition of clusters regularly distributed in age), the colors start from the same point at $t = 0$ but evolve slowly : after 10^{10} years one still has $U - B \simeq -0.25$, $B - V \simeq 0.4$ (fig.7). The vast majority of galaxies have colors intermediate between those of old single clusters and old continuous star formation with a uniform rate. This

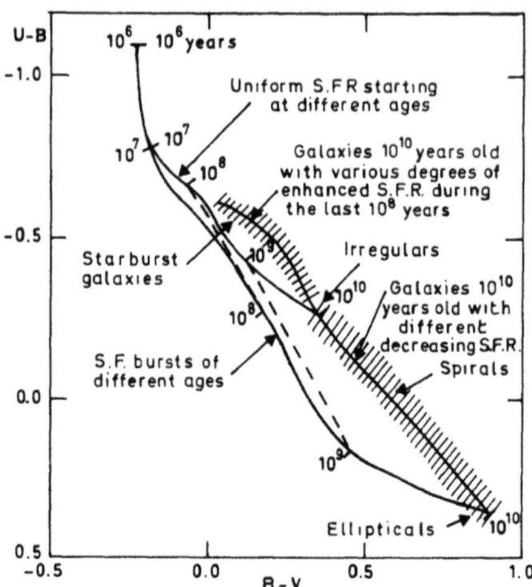

Fig. 4.1. Evolution of galaxies in a color-color diagram. The tick marks indicate ages in years. The hatched areas represent the locations of observed colors of galaxies, from ellipticals (bottom right) to spirals, then irregulars then galaxies with present starbursts. See text.

suggest that SFR(t) is a uniform or decreasing function of time, assuming that galaxies are very old. The bluest galaxies (irregulars) have had a roughly uniform SFR while the reddest ones (ellipticals) have had a strongly decreasing SFR(t) with much star formation at the beginning and no or little afterwards. Spirals galaxies are intermediate. These notions, as well as most of the substance of this section, have been introduced in the fundamental papers of Searle et al. (1973) and Larson and Tinsley (1978). I strongly recommend reading these papers.

Larson and Tinsley (1978) demonstrate the following important property related to $U-B$, $B-V$ colors, which has been extended through numerical simulations to <u>any</u> color-combination by Rocca-Volmerange et al. (1981) : in any two-color diagram the position of a galaxy that experienced a relatively smooth SFR(t) depends only on the ratio R of integrated SF to PDSFR :

$$R = \frac{\int_0^{now} \mathrm{SFR}(t)\mathrm{d}t}{\mathrm{PDSFR}} (\mathrm{Gyr})$$

the PDSFR being an average over the last 10^8 years or so. This of course assumes a constant IMF and ignores metallicity effects.

This idea has been applied for example to the whole available set of colors of the Magellanic Clouds by Rocca-Volmerange et al. (1981) : the following values of R give a good fit of all colors :

$$\mathrm{LMC} : R = 7 - 10\,\mathrm{Gyr}$$

$$\mathrm{SMC} : R = 9 - 20\,\mathrm{Gyr}$$

I want to insist on the meaning of these numbers. R tells us <u>nothing</u> about age and remote SFR history : $R = 10\,\mathrm{Gyr}$ may mean for example either an age of 10 Gyr and a uniform SFR, <u>or</u> an age of 5 *Gyr* and a decreasing SFR, etc. The SFR can have been smooth or in several isolated bursts with little SF in between. What imports is that there has not been a recent SF burst : the interpretation would have to be different in such a case. Any other information should come from independent data like spectral synthesis, distribution of the age of clusters, turn-off of the luminosity function, abundance of AGB stars, in particular carbon stars. In the case of the LMC the last two criteria suggest a major episode of SF $3-5\,10^9$ years ago (Stryker, 1984 ; Frogel and Blanco, 1983).

If the considered galaxy is presently experiencing a starburst, it is clear that its color will be bluer as the contribution of the burst will add to the light of the "basic" galaxy. The locus of such galaxies in a color-color plot is a band extending that of "quiet" galaxies towards the upper left of the color-color diagram (fig.4.1). This is however a somewhat oversimplified view of the actual situation. If the burst was relatively short and is seen sometimes after it occured, its colors have turned redder. This introduced a dispersion in the color-color diagram for starburst galaxies, which has been modeled in detail by Larson and Tinsley (1978) : fig.4.2.

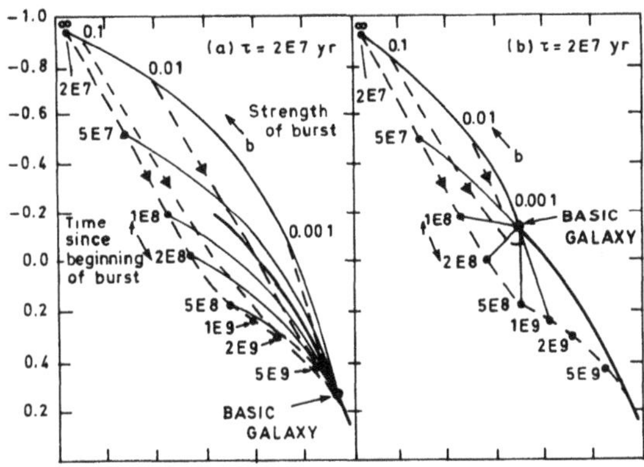

Fig. 4.2. Evolution of galaxies experiencing a starburst in a color-color diagram. The duration of the "burst" (in fact a short continuous SF phase) is $2\,10^7\,yr$; its strength is expressed by the parameter b, the mass of stars formed divided by the mass of stars ever formed in the galaxy (the IMF is assumed constant). t is the time elapsed since the beginning of the burst. (a) is for a red underlying galaxy, (b) for a blue underlying galaxy. From Larson and Tinsley, 1978.

Larson and Tinsley (1978) have indeed remarked that the colors of interacting galaxies from the Arp's atlas are much more scattered in the color-color diagram than the colors of quiet galaxies. This is historically the first evidence that interaction between galaxies triggers starbursts, an evidence quite well documented since.

4.3 Spectral Evolution and Synthesis

I will now discuss some aspects of spectral evolution and synthesis of galaxies, that is better discussed on examples than in an abstract way.

The first example is that of "young" galaxies with blue colors like blue compact galaxies. These have UV spectra very much like those of giant extragalactic HII regions e.g. in M33, clearly dominated by hot, massive stars. Can we say something about the very recent history of SF ? It is clear that not much can be extracted from colors even in the far-UV as stars from O3 to B4 have almost the same continuum spectrum in the region which can be observed with IUE ($\lambda > 1200 A$). Only lines can be used. Fanelli et al. (1988) have recently applied population synthesis techniques to several blue compact galaxies. The best fit they find is invariably with a discontinuous set of hot stars (e.g. : O stars, not early B stars, but late B stars again etc.). This points to a discontinuous SFR consisting in recurrent bursts. Earlier star formation cannot be studied in the far-UV and requires visible and IR observations. The problem now is to know whether these systems are young (i.e. no star formation previous to the recent burst, or series of bursts) or old. Some blue compact dwarfs have rather red colors and are certainly old but a few don't and deserve detailed observations.

The second example I will discuss is the far-UV excess of some elliptical galaxies. This excess varies from galaxy to galaxy and can be due either to young hot stars or to evolved hot populations (e.g. post-AGB stars or accreting white dwarfs in binary systems). However the signal-to-noise ratio of the IUE observations of such systems is too small to allow to see the spectral lines. There is no surprise in these conditions that attempts to match the continuum spectrum by evolutionary synthesis have not produced clear-cut results (Nesci and Perola, 1985).

Spectral synthesis of elliptical galaxies in the visible produces more interesting results because lines can be seen. Some show Balmer lines in their spectra, indicating the presence of relatively young stars (A stars mostly). These stars may have formed from gas captured from another galaxy or from accretion of intergalactic gas (accretion flows are seen in particular around the brightest elliptical galaxies in clusters). Evolutionary synthesis studies shows that this is far from exceptional and that young and intermediate populations do exist in a large fraction of ellipticals (see e.g. Pickles, 1985b ; Bica, 1988).

Another kind of astrophysical objects for which these techniques have been applied is the central regions of spiral galaxies. They turn out to be a mixture of ages with a strong, old metal-rich component (Pritchett, 1977 ; Bica, 1988).

4.4 Galaxies at Large Redshifts

Although reasonable spectroscopy is now possible with large telescopes for the brightest elliptical galaxies up to redshifts of 0.5 (see an example in Hamilton, 1985) such data are still very scarce because of the long exposure times required and one is still limited in general to low-resolution spectrophotometry. Clearly this will be a major field of interest for observations with the future large telescopes like the ESO VLT. Another limitation is the quality of the photometric evolutionary models. Those by Bruzual (e.g. 1983) are very popular although they suffer from various known problems (lack of AGB in particular). More recent models like those of Guiderdoni and Rocca-Volmerange (1987), see also Rocca-Volmerange and Guiderdoni (1988), differ very appreciably from Bruzual's models. An embarassment is that at high redshifts the UV excess of elliptical galaxies (which is known to be variable from galaxy to galaxy) comes into the visible and modelization becomes extremely uncertain. Recent studies do not reach agreement, although it is clear that one sees the effects of evolution and that there must have been star formation after the initial burst even for elliptical galaxies. Some interesting papers are Lilly and Longair (1984), Eisenhardt and Lebofsky (1987), Guiderdoni and Rocca-Volmerange (1987), Rocca-Volmerange and Guiderdoni (1987), etc.

5. Chemical Evolution of Galaxies

All the elements we see in nature have been synthetized in stars from hydrogen and helium with the important exceptions of most of the main isotope of helium itself, ^{4}He, ^{3}He, deuterium D = ^{2}H and a significant fraction of ^{7}Li which were produced in the Big Bang. Thus most elements as seen in stars in the interstellar matter and in the solar system tell us about the past history of nucleosynthesis, thus about past SF. Studies of chemical evolution aim primarily at accounting the distribution and abundances of these elements in galaxies. An interesting although somewhat outdated review of the field has been given by Audouze and Tinsley (1976). Before describing how models of chemical evolution are built and giving some results, I will briefly describe the ingredients : measurements of abundances of the elements and nucleosynthesis.

5.1 Which abundances are measured, and where ?

An excellent review of the patchwork of observational data on abundance measurements is that by Pagel and Edmunds (1981). The most abundant element, that serves of reference, is hydrogen ; hydrogen makes about 90 per cent by number and 70 per cent by mass of the atoms in the Universe. Then comes helium, the other elements being by far less abundant. The latter are often called "metals" and "metallicity" is synonymous to abundance of heavy elements. Amongst those, the triad C, N and O dominates (O being generally more abundant than the two others), then Si and Fe.

Abundances can be given, by number of atoms, with reference to hydrogen. By convention one usually takes $n(\mathrm{H}) = 10^{12}$ so that abundance of element A is then expressed in logarithmic form as $12 + \log[n(\mathrm{A})/n(\mathrm{H})]$. For example, if the abundance of oxygen is $5\,10^{-4}$ by number it can also be given as $12 + \log(5\,10^{-4}) = 8.70$.

Abundances can also be given by mass. In this case X designates the mass fraction of hydrogen (of the order of 0.70), Y that of helium (of the order of 0.28) and Z that of all heavy elements (about 0.02 near the Sun). It is useful to remember than $Z(\mathrm{oxygen}) \simeq 0.45\,Z$ (all metals) in not too extreme situations in galaxies.

A frequently used notation is the logarithmic abundance per number with reference to the "cosmic" (or better solar) abundances :

$$[\mathrm{A/H}] = \log[n(\mathrm{A})/n(\mathrm{H})] - \log[n(\mathrm{A})/n(\mathrm{H})]_\odot$$

One finds also the simplified notation [A], meaning the same thing.

HII regions and planetary nebulae are good places to determine abundances from emission lines intensities in their spectra. The abundances of H^+ and He^+ are derived from their recombination lines (the Balmer series for hydrogen). A small correction may be necessary for He^0 that is not observed and for He^{++} that gives faint recombination lines (especially $\lambda 4686\,A$), requir ing some knowledge of the physical conditions in the nebula. This is the only safe determination of the abundance of He: do not believe too hard in determinations in stars e.g. from the appearance of HR diagrams of clusters, this indirect method being far from foolproof. Fortunately the determination of the abundances of He in nebulae can be accurate to a few per cent. The abundances of heavier elements come from forbidden or semi-forbidden lines. The abundance of oxygen is relatively easy to obtain as O^0 is scarce and lines of both O^+ and O^{++} are seen. However it is not so well determined when it is large for reasons connected with the physics of the gas. N is more difficult, as well as C (abundance usually derived from the semi-forbidden line CIII $\lambda 1909$). The abundances of Ar and Ne can be determined, while that of S is somewhat problematic for various reasons. It should be remembered that the abundances of He, C, N and O in planetary nebulae may no represent those of the material from which the corresponding star was born, due to stellar nucleosynthesis and mixing. N and C in particular can be much affected.

Within exceptions (evolved hot massive stars going back to the left part of the HR diagram, red giants and AGB stars) the abundances at the surface of stars are those of the interstellar material from which they were formed. Abundances of most elements can be obtained in the Sun and represent those in the local interstellar medium at the birth of the solar system $4.6\,10^9$ years ago (some abundances come from ratios of heavy elements in meteorites).

This set of abundances is refered to as the "solar", or "cosmic", or "standard" abundances. In other stars than the Sun, the situation is different. The abundance of He is usually impossible to measure except in the hottest stars. That of C, N and O (already difficult in the Sun) can be obtained at great efforts in hot stars and more easily in cold stars here molecular bands are seen ; however in the latter objects they may be affected by nucleosynthesis in the star itself. What is less problematic is the abundance of Fe and its group that can be derived from the many lines of these elements in intermediate temperature stars. These lines are so numerous that they affect strongly the continuum of the stars though blanketing such that the abundance of iron can be derived to some extent from spectrophotometry and even photometry ; this requires however absolute calibration using high-resolution spectra os some stars of the same type. It should be remarked finally that abundance determinations are more uncertain in stars with extended atmospheres like supergiants, due to the difficulties in modelling line formation in such atmospheres.

Abundances of some elements in very hot gas can be derived from X-ray spectroscopy. It has been found in this way that Fe is overabundant in some young supernova remnants. The most interesting application for evolution of the galaxies is the unexpected discovery of iron lines in the very hot gas of rich clusters of galaxies (Holt and Mc Cray, 1982). This yields [Fe/H] $\simeq -0.3$, half the standard abundance. As the total mass of this hot gas is roughly equal to that of all galaxies in the cluster, it is clear that this gas must play a very important role in galaxy evolution : presumably it was ejected from galaxies at early times, and it is presently accreted by the most massive cluster galaxies forming new stars there.

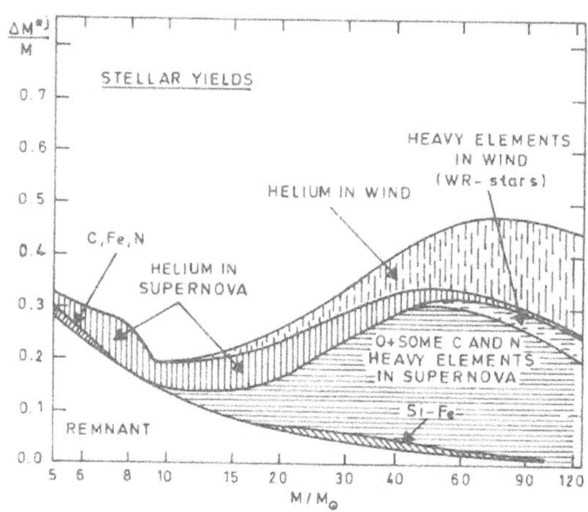

Fig. 5.1. Production of helium and elements by stars of various initial masses. The contributions to C and N are uncertain. Adapted from Chiosi and Maeder (1986).

Abundances in spiral and irregular galaxies can be relatively easily derived from spectrophotometry of their HII regions (and also for the nearest galaxies, planetary nebulae). This concerns essentially He, C, N, O and a few other elements. Abundances in the central regions of spirals do not differ by large factors from the standard abundances but there are in many cases gradients, the abundances being smaller in the external parts (this is the case for our Galaxy). Irregular galaxies have usually low abundances, with marked differences in the C/O and N/O ratios. Two blue compact galaxies and one dwarf irregular galaxy have [O/H] as low as -1.5, but no galaxy has been found to-date with lower abundances. Some abundance determinations have been made for individual stars in the Magellanic Clouds, with in some cases discrepancies with the HII-regions abundances ; they give in any case the only information we have on the Fe abundance in the Clouds. Essentially all that we know on abundances in elliptical galaxies comes from spectrophotometry of their integrated light and concerns Fe and to some extent Mg (giant ellipticals have [Fe/H] > 0) ; we have no knowledge in particular of the abundance of oxygen. One of the main problems in discussing abundances in galaxies is : does the abundance which is measured in one constituent (say the HII regions) represent abundances in general ? It is clear that this is not the case for our Galaxy where stars exhibit a variety of abundances which may differ considerably from those in HII regions, something that can be easily accounted for by differences in ages. A more difficult problem is that of abundances in the Magellanic Clouds. While abundances in HII regions are remarkably uniform in each Cloud, abundances in some young stars which reflect those of the interstellar matter from which they were born recently seems to differ appreciably : this is the case in particular for two stars in the SMC (Spite et al., 1986 ; Russell et al., 1988). This may indicate that the SMC at least is not so well mixed as was previously thought, and is a warning as well as an incentive for further studies.

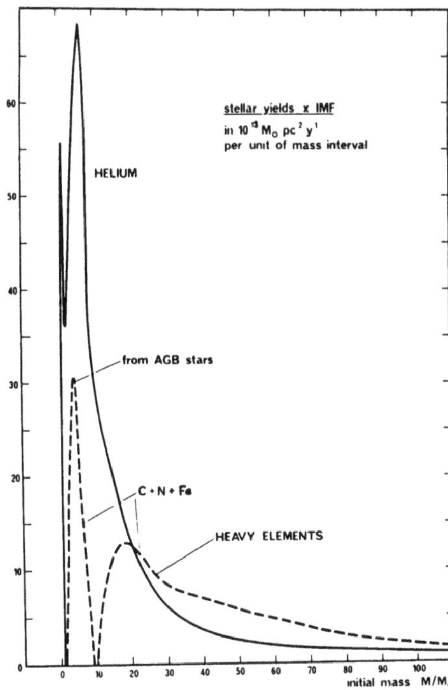

Fig. 5.2. Contribution of stars of different masses to the yield in heavy elements. This is obtained by multiplying the yields by the initial mass function. The numerical values have to be taken with caution due to uncertainties in the nucleosynthesis. Adapted from Maeder (1983b).

5.2 Where are the Elements and their Isotopes Formed ?

There is an enormous amount of work on this very complex subject and I will give here only a few simple ideas that will be useful for what follows. Review papers include Audouze and

Tinsley (1976) and more recently Truran (1984) ; see also the textbook by Audouze and Vauclair (1980) and the nice small review by Truran and Thielemann (1987).

Initially there was only hydrogen. Big-bang nucleosynthesis produced most ^{4}He and ^{3}He, probably all deuterium and a significant part of ^{7}Li. Their abundances give strong constraints upon cosmology and particle physics.

The observed abundances of the rare elements ^{6}Li, ^{9}Be, ^{10}Be and ^{11}B are fully consistent with production through the interaction of cosmic rays with the interstellar medium.

All the other elements are produced in stars (stars also destroy some of the light elements mentioned above, particularly D that does not survive through passage in any star). As stars derive most of their energy from combustion of H into He, it is clear that they should release He. However their integrated contribution adds not more than some 20 percent to the big-bang He in most evolved galaxies. O, Mg, Si, S and many other elements are synthetized mainly in massive stars ($m \geq 10\, m_{\odot}$: fig.5.1) and are released at least in part during the supernova explosion that ends the life of at least the less massive of these stars. Some Fe is also produced into the explosion. If the most massive stars end as black holes instead of supernovae, this will significantly decrease the heavy element production. Production by stars of various masses has of course to be integrated over the IMF, weighting heavily towards stars of lower masses (fig.5.2).

C and Fe appear to be mostly produced in lower-mass stars, the former during the AGB phase and the latter by type I supernovae. The situation with N is complicated (see later).

Elements that are produced directly from H (and He) in stars are called primary elements. They can thus be synthetized in first-generation stars as well as in later generations. Secondary elements are formed from pre-existing primary elements and are thus expected to appear later during the evolution of galaxies. One should have :

$$\frac{n(\text{primary})}{n(\text{secondary})} \propto \frac{n(\text{primary})}{n(\text{H})}$$

However there are few real secondary elements. ^{13}C, N and the s-process elements like Ba formed by capture of neutrons were considered some years ago to be pure secondary elements. We now think that N can be partly primary, i.e. formed by the CNO cycle from fresh ^{12}C generated in the same (massive) star. This may be also the case for ^{13}C. On the other hand, elements like C or Fe that come mostly from low-mass stars, although they are primaries, occur late in the evolution of galaxies due to the long time of these stars and behave roughly as secondaries !

5.3 Principles of Chemical Evolution

The evolution of a galaxy is schematized in fig.5.3. The galaxy can be considered as a box initially made of gas (H and He) from which stars are formed. Stars form heavy elements, and return enriched matter into the interstellar medium especially at the end of their lifes : remember that the stellar lifetimes increase drastically with decreasing stellar mass, roughly as m^{-2}. There is usually a remnant left after the death of the star : these remnants trap mass and do not participate to the further evolution of the galaxy. Modelling what happens in this box when assuming that it does not exchange matter with the external world is building a closed-box model of evolution. Such models are useful for a basic understanding but may well be wrong even as zero-order approximations of the reality. Indeed, matter can be ejected from galaxies (especially from systems or at early stages) : the abundant hot gas with half-solar metallicity that is seen in clusters of galaxies has probably be ejected by the galaxies during

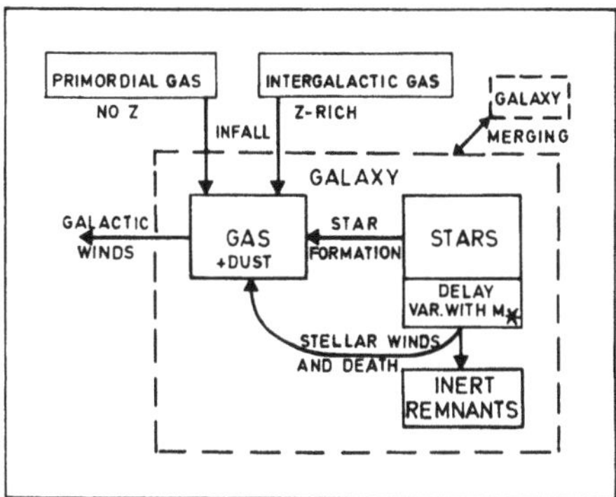

Fig. 5.3. Scheme of the chemical evolution of a galaxy. See text.

their early evolution. On the other hand, galaxies are known to accrete external gas : this is clear for the biggest galaxies in clusters that accrete the hot gas just mentioned with rather high rates (several tens or even hundreds of solar masses per year). Galaxies can also accrete primordial matter (not yet chemically processed). There may also be exchanges of matter between different parts of a galaxy (e.g. between the halo and the disk or via radial flows in spiral galaxies). Finally galaxies can merge together during encounters, in part or totally. All these complications make evolution of galaxies, and in particular chemical evolution, a very complex subject. If one realizes that in an external galaxy there are only a few parameters that enter as inputs in chemical evolution models –mass of gas, total mass (plagued with the problem of the dark matter) and metallicity in the gas and possibly in the bulk of the stars– one sees that it is hopeless to try to derive the galaxy evolution (past history of SF, for example) from chemical evolution models. They can only provide useful constraints. This lowers considerably the optimism that prevailed in the early days of chemical evolution.

It seems worth however to discuss the basic equations of chemical evolution. I will not describe for the sake of simplicity the bimodal star formation models in which the low-mass stars and high mass stars are formed in different contexts and at different rates, although these models are physically sound and have many attractive properties properties : the basic papers (Güsten and Mezger, 1983 ; Larson, 1986, see also Scalo, 1987) will be easily understood from the principles that will be described now. I will thus assume here a constant IMF.

The first basic equation concerns the evolution of the mass of gas + dust (here called simply "gas") and reads :

$$\mathrm{d}M_g/\mathrm{d}t = -\mathrm{SFR}(t)\int_{m_l}^{m_u}\psi(m)dm + \int_{m(t)}^{m_u} R(m)\psi(m)\mathrm{SFR}\left[t - T(m)\right]dm + A(t)$$

Time t is taken as zero at the formation of the first stars.

The first term is the loss of gas through star formation. The integral it contains can be reduced to 1 if the IMF $\psi(m)$ is normalized as usual. The second term expresses the return of mass to the interstellar medium. $R(m)$ is the quantity of mass returned by a star of mass m. One assumes for simplicity that a star of mass m expells all this gas at the end of its life which lasts $T(m)$: this neglects stellar mass loss but is not a bad assumption as mass loss occurs either in short-lived massive stars or in the late phases of the evolution of low-mass stars. At time t, only the stars formed $t - T(m)$ ago return mass, and only stars with masses large enough to have had time to evolve and return mass : hence the lower limit of integration masses $m(t)$ which is the mass of a star with lifetime t. The last term represents the exchange of gas with the outside world (positive sign for accretion).

Similar equations can be written for the mass of an element in the interstellar matter. If Z is the mass fraction of such an element, one has :

$$\mathrm{d}(ZM_g)/\mathrm{d}t = -Z\mathrm{SFR}(t)\int_{m_u}^{m_u}\psi(m)dm + \int_{m(t)}^{m_u}R_Z(m)\psi(m)\mathrm{SFR}\left[t - T(m)\right]dm + A_Z(t)$$

$R_Z(m)$ and $A_Z(t)$ expressing respectively the mass of elements Z returned by a star of mass m and exchanged from the external world.

This set of equations can of course be solved numerically, but it is useful to give a simple analytical solution valid in the <u>instant recycling approximation</u>. This approximation neglects the lifetime of the stars, thus sets $T(m) = 0$. This is not as bad as one would think if one is willing to limit the study to elements produced by high-mass stars only (like oxygen) as those stars have very short lifetimes. Numerical experiments show that the results are reasonable for not too evolved galaxies ($M_{gas}/M_{stars} > 0.1$) : see e.g. Alloin et al., 1979. Consider first the closed-box model with $A(t)$ and $A_Z(t) = 0$. One has :

$$\mathrm{d}M_g/\mathrm{d}t = -\mathrm{SFR}(t)\int_{m_l}^{m_u}\psi(m)\left[1 - R(m)\right]\mathrm{d}m.$$

If we set $\mu = M_g/M_{tot}$ (note that M_{tot} ignores dark matter and is the total mass of <u>stars</u> + gas alone) and

$$\alpha = \int_{m_l}^{m_u}\psi(m)R(m)\mathrm{d}t(\simeq 0.2 \text{ for a "normal" IMF})$$

this equation reduces to

$$\mathrm{d}\mu(t)/\mathrm{d}t = -\mathrm{SFR}(t)\left[1 - \alpha\right]$$

Similarly the equation for an heavy element reads

$$\mathrm{d}(\mu Z)/\mathrm{d}t = -\mathrm{SFR}(t)(Z - a),$$

setting $a = \int_{m_l}^{m_u} R_Z(m)\psi(m)\mathrm{d}m(\simeq 0.01$ for all heavy elements).

Developing the first member and using the equation for μ one has

$$\mu dZ/\mathrm{d}t = -\mathrm{SFR}(t)(Z - a)$$

Multiplying by $\mathrm{d}t/\mathrm{d}\mu = -\left[\mathrm{SFR}(t)(1-\alpha)\right]^{-1}$ one eliminates the time and SFR and get:

$$\mu\mathrm{d}Z/\mathrm{d}\mu = -Z\alpha/(1-\alpha) + a/(1-\alpha)$$

The quantity

$$p = a/(1-\alpha)$$

is called the <u>yield</u>. This is the mass of heavy elements returned per stellar generation (mass normalized to 1) per unit <u>net</u> mass turned into stars. p is of the order of 0.01 for the sum Z of all heavy elements (half of which is oxygen, roughly) and similar or slightly higher for helium Y.

We note that $Z\alpha \ll a$ as $Z \simeq 0.01$, $\alpha \simeq 0.2$ and $a \simeq 0.01$. Then the second term of the equation above can be neglected and we have the simple solution :

$$Z = p\ln(1/\mu)$$

taking $Z = 0$ at the start.

This equation shows that for relatively evolved systems the abundances will not differ enormously from p. This is what is observed in galaxies.

One can also calculate the fraction of all stars formed that have an abundance $< Z$: this reads :

$$\frac{S}{S(\text{now})} = \frac{1 - \mu_{\text{now}}^{Z/Z_{\text{now}}}}{1 - \mu_{\text{now}}}$$

An analytic solution also exists when there is accretion of metal-free gas such that it exactly compensates for star formation :

$$A(t) = \text{SFR}(t) \text{ and } A_Z(t) = 0$$

This gives :

$$Z = p\left[1 - exp - (\mu^{-1} - 1)\right]$$

In this case, a steady state is soon reached where $Z = p$. It is intuitively clear that accretion of metal-free gas will continuously dilute the enriched gas and keep Z from growing very much. The fraction of stars abundances smaller than Z reads :

$$S/S_{\text{now}} = -\left(\mu^{-1} - 1\right)^{-1} \left\{1 - (Z/Z_{\text{now}})\left[1 - exp - \left(\mu^{-1} - 1\right) \ \right]\right\}$$

Similarly, accretion of metal-rich gas accelerates the initial chemical evolution but tends to slower it down later. Conversely, galactic winds (mass loss from the galaxy) tend to accelerate chemical evolution via a decrease of the gas content.

5.4 Confrontation to Observations

In general, observations converge in showing that metallicity increases when the gas fraction decreases, in agreement with the predictions of the simple model. However there is a large spread in the Z/μ relation, showing that things are more complicated. This spread may be due to poor evaluations of the total masses in stars : dynamical studies of galaxies yield gravitational masses that contain an unknown and probably variable amount of dark matter, while we are only interested in the total mass of stars + gas. For the purpose of evolutionary calculations it may be better to estimate the mass of stars from their luminosity assuming a mass/luminosity ratio for the galaxy. But dark matter cannot be the whole story, as we will see now.

Let us look first at the galactic disk. There is an abundant literature on this subject (see Audouze and Tinsley, 1976, and for recent work Tosi, 1988 and references herein). It is clear in this case that closed-box models do not work. The metallicity distribution of low-mass stars shows a strong lack of low-metallicity objects that cannot be reproduced by such models. Infall models with no metals and a more or less constant rate do not work either. A better fit can be obtained assuming e.g. that the initial abundances were non-zero due to strong pre-galactic stellar evolution, or with an infall model and a large spread of abundances at any time. Bimodal star-formation can also give a good fit. The time variation of metallicity is also known in the disk and is so fast at early stages that it excludes closed-box models again. The situation is more satisfying with infall or bimodal star formation models. Finally the radial abundance gradient seen in the galactic disk, although expected in local closed box models (the abundances decrease and the $M_{\text{gas}}/M_{\text{tot}}$ ratio increases toward the external parts of the Galaxy), is not fitted quantitatively with such models. There are again ways out : infall, radial flows, bimodal star formation, etc. According to Tosi (1988) the only thing which can really be excluded apart from closed-box evolution of the galactic disk is infall with high metallicity ($Z \geq 0.4\, Z_{\odot}$). Note that dark matter is not a dominant constituent of the disk and cannot be involved to alleviate the problems.

Irreguliar galaxies are relatively unevolved objects that can be thought of to be more easy places to test chemical evolution models. However there is a large dispersion in their Z/μ correlation. Moreover the application of the closed-box model formula gives values for the yield p quite variable (corresponding to this correlation) but generally much smaller than the predictions of nucleosynthesis with normal IMFs : p of the order of 0.003 instead of 0.01. There are several possible explanations, one being the existence of a large amount of dark matter contributing to the mass. This should be a big effect amounting to a factor 50 in some cases. However this is sometimes excluded by the rather large values of M_{gas}/M_{tot} observed in some cases. After all, the solution may have rather to be sought in accretion (Matteucci and Chiosi, 1983) or galactic winds : we shall see later a possible example of the role of winds.

Let us discuss now a few more specific problems concerning individual elements.

The helium problem. Standard stellar evolution theory with mass loss predicts a stellar production of helium about equal to the metal production : i.e. to the existing helium (mostly primordial) a mass ΔY is added, approximately equal to the mass ΔZ of heavy elements (Maeder, 1981b). Observations of irregular or blue compact galaxies together with observations of more metal-rich HII regions should allow a test of this prediction, as the abundance of oxygen then Z can be determined accurately, and also that of helium ; however since the added helium is only a small part of the primordial helium, the latter observations have to be very accurate. Note that the problem is very important for cosmology, as an extrapolation of the Y/Z relation to $Y = 0$ would yield the mass fraction of helium created in the Big Bang. The difficulties of the measurements is such that the prefered values of $\Delta Y/\Delta Z$ have evolved during the last decade between 3-4 and 0] The present prefered value is 3.2 ± 0.5 (1 σ ; Peimbert, 1987). If real, the discrepancy with the models must be explained. This may involve more mixing (convective overshooting ?) in those intermediate mass stars that synthetize the bulk of the new He, hence an enhanced production of this element. Another solution is to admit that the most massive stars end as black holes instead of exploding, decreasing the production of oxygen without affecting much that of helium (see also at the end of this chapter another possibility).

The Carbon abundance. Observations show that C/O is low at low abundances, increases and then saturates at high Z. This could not be explained in old models where carbon was mainly produced in massive stars like oxygen. But the cross section in the $^{12}C(\alpha,\gamma)^{16}O$ reaction that determines the ratio of C to O in those stars has been increased substantially and we now believe that most C is produced in intermediate-mass stars, thus appears later in the evolution. This is in better agreement with observations (Peimbert, 1987).

The Nitrogen abundance. Observations show that N/O is low, but rather constant at low abundances (see e.g. Campbell et al., 1986) while this ratio increases at higher abundances (see e.g. Pagel and Edmunds, 1981). The first observations suggest a primary element and the second a secondary element ! Actually it seems that N is both primary and secondary : high-mass stars appear to produce a small amount of primary nitrogen and low and intermediate-mass stars a larger amount of secondary one. Low-abundance galaxies usually have had a more or less constant SFR, and may even be rather young, thus N production should be dominated by high-mass stars because of their shorter lifetimes, while lower-mass stars dominate the production of N in more evolved galaxies.

The O/Fe problem. It has been known from some time that [O/Fe] > 0 (oxygen less deficient than iron) in old galactic stars with very low metallicities (Spite and Spite, 1985). The acknowledged explanation is that a first stellar population consisting mainly of massive stars (the so-called Population III) has injected more oxygen than iron in the gas from which the observed stars have formed. A recent observation of Fe abundance in relatively young stars of the Magellanic Clouds (Russell et al., 1988; Spite et al., 1988) indicates that $[O/Fe] < -0.3$ in the Clouds (the oxygen abundance is derived from observations of the HII regions). This discrepancy is unexpected and raises a big problem : given the relatively young stellar population in these galaxies one rather expects $[O/Fe] > 0$. A tentative explanation proposed by Russell et al. (1988) is that a fraction of the oxygen-rich material expelled by type II supernovae actually leaves the galaxies instead of being mixed to the interstellar medium, through holes dug out in this medium by collective star explosions. This might also explain the small yield in heavy elements (essentially oxygen !) seen in irregular galaxies, and may be the high $\Delta Y/\Delta Z$ ratio since helium, like Fe, is mainly produced by relatively low-mass stars. Clearly chemical evolution is a much involved topic...

References

1 ALLOIN, D., COLIN-SOUFFRIN, S., JOLY, M., VIGROUX, L., 1979, Astron. Astrophys., 78, 200
2 AUDOUZE, J., TINSLEY, B.M.: 1976, Ann. Rev. Astron. Astrophys., 14, 43
3 AUDOUZE, J., VAUCLAIR, S.: 1980, An Introduction to Nuclear Astrophysics, Reidel.
4 AUGARDE, R., LEQUEUX, J.: 1985, Astron. Astrophys., 147, 273
5 APARICIO, A., GARCIA-PELAYO, J.M., MOLES, M.: 1988, Astron. Astrophys.Suppl., 74, 375
6 AZZOPARDI, M., LEQUEUX, J., MAEDER, A.: 1988a, Astron. Astrophys., 189, 34
7 AZZOPARDI, M., LEQUEUX, J., REBEIROT, E.: 1988b, Astron. Astrophys., 202, L27
8 BERTELLI, G., BRESSAN, A.G., CHIOSI, C.: 1985, Astron. Astrophys., 150, 33
9 BICA, E.: 1988, Astron. Astrophys., 195, 76
10 BOULANGER, F., PERAULT, M.: 1988, Astrophys.J., 330, 964
11 BROCKELHURST, M.: 1971, Monthly Not. Roy. Astron. Soc., 153, 471
12 BRUZUAL, G.: 1983, Astrophys.J., 273, 105
13 BUAT, V., DEHARVENG, J.M.: 1988, Astron. Astrophys., 195, 60
14 CAMPBELL, A.: 1988, Astrophys.J., in press
15 CAMPBELL, A., TERLEVICH, R., MELNICK, J.: 1986, Monthly Not. Roy. Astron. Soc., 223, 811
16 CHIUSI, C., MAEDER, A.: 1986, Ann. Rev. Astron. Astrophys., 24, 329
17 COMBES, F., ENCRENAZ, P.J., LUCAS, R., WELIACHEW, L.: 1978, Astron.Astrophys., 67, L13
18 DE JONG, T., KLEIN, U., WIELEBINSKI, R., WUNDERLICH, B.: 1985, Astron. Astrophys., 147, L6
19 DICKEY, J.M., GARWOOD, R.W., HELOU, G.: 1987, in Star Formation in Galaxies, ed. C. Lonsdale, NASA CP-2466, p.575
20 DONAS, J., DEHARVENG, J.M., LAGET, M., MILLIARD, B., HUGUENIN, D.: 1987, Astron. Astrophys., 180, 12
21 EISENHARDT, P.R.M., LEBOFSKY, M.J.: 1987, Astrophys.J., 316, 70
22 FANELLI, M.N., O'CONNELL, R.W., THUAN, T.X.: 1987, Astrophys.J., 321, 768
23 FLOWER, P.J. : 1977, Astron. Astrophys., 54, 31
24 FREEDMAN, W.L.: 1985, Astrophys.J., 299, 74
25 FROGEL, J.A., BLANCO, V.M.: 1983, Astrophys.J.Letters, 274, L57
26 GILES, K.: 1977, Monthly Not. Roy. Astron. Soc., 180, 57P
27 GUIDERDONI, B., ROCCA-VOLMERANGE, B.: 1987, Astron. Astrophys., 186, 1
28 GUSTEN, R., MEZGER, P.G.: 1983, Vistas in Astron., 26, 159
29 HAMILTON, D.: 1985, Astrophys.J., 297, 371
30 HELOU, G.: 1988, in Interstellar dust, ed. L. Allamandola and X. Tielens, Kluwer, in press
31 HELOU, G., SOIFER, B.T., ROWAN-ROBINSON, M.: 1985, Astrophys.J.Lett., 298, L7
32 HOLT, S.S., McCRAY, R.: 1982, Ann. Rev. Astron. Astrophys., 20, 323
33 HUMPHREYS, R.M., MC ELROY, D.B.: 1984, Astrophys.J., 284, 565
34 IBEN, I., 1967, Ann. Rev. Astron. Astrophys., 5, 571
35 IBEN, I., 1974, Ann. Rev. Astron. Astrophys., 12, 215
36 IBEN, I., RENZINI, A.: 1983, Ann. Rev. Astron. Astrophys., 21, 271
37 KENNICUTT, R.C.Jr., HODGE, P.W.: 1986, Astrophys.J., 306, 130
38 LANG, K.R.: 1980, Astrophysical Formulae, Springer-Verlag
39 LARSON, R.B.: 1986, Monthly Not. Roy. Astron. Soc., 218, 409
40 LARSON, R.B.: 1987, in Starbursts and Galaxy Evolution, ed. T.X.Thuan et al., Editions Frontières, p.467
41 LARSON, R.B., TINSLEY, B.M.: 1978, Astrophys.J., 219, 46
42 LEQUEUX, J.: 1984, in Structure and Evolution of the Magellanic Clouds, ed. S. van den Bergh and K. de Boer, Reidel, p.67

43 LEQUEUX, J.: 1986, in Spectral Evolution of Galaxies, ed. C. Chiosi and A. Renzini, Reidel, p.57
44 LEQUEUX, J.: 1988, in Millimetre and Submillimetre Astronomy, ed. R.D. Wolstencroft and B. Burton, Kluwer, p.249
45 LEQUEUX, J., MAUCHERAT-JOUBERT, M., DEHARVENG, J.M., KUNTH, D.: 1981, Astron. Astrophys., 103, 305
46 LILLY, S.J., LONGAIR, M.S.: 1984, Monthly Not. Roy. Astron. Soc., 211, 833
47 MAEDER, A., 1981a, Astron. Astrophys., 102, 401
48 MAEDER, A.: 1981b, Astron. Astrophys., 101, 385
49 MAEDER, A., 1983a, Astron. Astrophys., 120, 113
50 MAEDER, A.: 1983b, in Primordial Helium, ESO Workshop, Garching, p.89
51 MATTEUCCI, F., CHIOSI, C.: 1988, Astron. Astrophys., 123, 121
52 MERMILLIOD, J.C., MAEDER, A.: 1986, Astron. Astrophys., 158, 45
53 MOULD, J., KRISTIAN, J.: 1986, Astrophys.J., 305, 591
54 MOORWOOD, A.F.M., VERON-CETTY, M.P., GLASS, I.S.: 1986, Astron. Astrophys., 160, 39
55 NANDY, K., THOMPSON, G.I., JAMAR, C., MONFILS, A., WILSON, R.: 1976, Astron. Astrophys., 51, 63
56 NESCI, R., PEROLA, G.C.: 1985, Astron. Astrophys., 145, 296
57 PAGEL, B.E.J., EDMUNDS, M.G.: 1981, Ann. Rev. Astron. Astrophys., 19, 77
58 PANAGIA, N.: 1973, Astron.J., 78, 929
59 PERAULT, M., BOULANGER, F., PUGET, J.L., FALGARONE, E.: 1988, submitted to Astrophys.J.
60 PEIMBERT, M.: 1987, in Star-Forming Dwarf Galaxies, ed. D.Kunth et al, Editions Frontières, p.403
61 PICKLES, A.J.: 1985a, Astrophys.J.Suppl., 59, 33
62 PICKLES, A.J.: 1985b, Astrophys.J., 296, 340
63 PIERRE, M., AZZOPARDI, M.: 1988, Astron. Astrophys., 189, 27
64 PRITCHETT, C.: 1977, Astrophys.J.Suppl., 35, 397
65 RENZINI, A., BUZZONI, A.: 1986, in Spectral Evolution of Galaxies, ed. C. Chiosi, A. Renzini, Reidel, Dordrecht, p.195,
66 RICE, W.L., BOULANGER, F., VIALLEFOND, F., SOIFER, B.T., FREEDMAN, W.L.: 1988, in preparation
67 RIEKE, G.H., LEBOFSKY, M.J., WALKER, C.E.: 1988, Astrophys.J., 325, 679
68 ROCCA-VOLMERANGE, B., LEQUEUX, J., MAUCHERAT-JOUBERT, M.: 1981, Astron. Astrophys., 104, 177
69 ROCCA-VOLMERANGE, B., GUIDERDONI, B.: 1987, in Starbursts and Galaxy Evolution, ed. T.X. Thuan et al., Ed. Frontières, p.501
70 ROCCA-VOLMERANGE, B., GUIDERDONI, B.: 1988, Astron. Astrophys. Suppl., 75, 93
71 ROWAN-ROBINSON, M.: 1987, in Starbursts and Galaxy Evolution, ed. T.X. Thuan et al., Editions Frontières, p.235
72 RUSSEL, S.C., BESSELL, M.S., DOPITA, M.A. : 1988, in Galactic and Extragalactic Star Formation, ed. R.E. Pudritz and M. Fich, Kluwer, p.601
73 SCALO, J.M.: 1986, Fundamentals of Cosmic Physics, 11, 1
74 SCALO, J.M.: 1987, in Starbursts and Galactic Evolution, ed. T.X.Thuan et al., Editions Frontières, p.445
75 SCOVILLE, N.J., YOUNG, J.S.: 1983, Astrophys.J., 265, 148
76 SEARLE, L., SARGENT, W.L.W., BAGNUOLO, W.G.: 1973, Astrophys.J., 179, 427
77 SILK, J.: 1987, in Star-Forming Regions, ed. M. Peimbert, J. Jugaku, Reidel, p.663
78 SMITH, L.F., BIERMANN, P., MEZGER, P.G.: 1978, Astron. Astrophys., 66, 65
79 SOLOMON, P.M., SAGE, L.J.: 1988, Astrophys.J., in press
80 SPITE, M., SPITE, F.: 1985, Ann. Rev. Astron. Astrophys., 23, 225
81 SPITE, M., CAYREL, R., FRANCOIS, P., RICHTER, T. and SPITE, M.: 1986, Astron. Astrophys., 86, 168
82 SPITE, F. SPITE, M., FRANCOIS, P., 1989 : Astron. Astrophys., in press
83 STRYKER, L.: 1984, in Structure and Evolution of the Magellanic Clouds, ed. S. Van den Bergh and K. de Boer, Reidel, p.79
84 THUAN, T.X., MONTMERLE, T., TRAN TRANH VAN, T. eds.: 1987, Starbusts and Galaxy Evolution, ed. Frontières, Paris
85 TOSI, M.: 1988, Astron. Astrophys., 197, 33
86 TRURAN, J.W.: 1984, Ann. Rev. Nucl. Part. Sci., 34, 53
87 TRURAN, J.W., THIELEMANN, F.K.: 1987, in Stellar Populations, ed. C.A. Norman et al., Cambridge University Press, p.149
88 VIALLEFOND, F., STASINSKA, G.: 1988, in preparation
89 WALTERBOS, R.A.M., SCHWERING, P.B.W.: 1987, Astron. Astrophys., 180, 27
90 YOUNG, J.S.: 1988, in Galactic and Extragalactic Star Formation, ed. R.E. Pudritz and M. Kluwer, p.579

This article was processed by the author using the TEX Macropackage from Springer-Verlag.

The Interstellar Medium

H. J. Habing
Sterrewacht, Leiden, the Netherlands

1. INTRODUCTION THROUGH SOME PSYCHOLOGY AND SOME HISTORY

Astronomy as a science is thousands of years old, the study of the interstellar medium only a few centuries. The reason for this discrepancy is obvious: go outside on a clear night and look up. You see thousands of stars but no indication of interstellar matter, although... . Definitely here in the Alps you see the Milky Way, huge, splendid and magnificent. What is it? A large agglomerate of stars or diffusely lighted matter, illuminated interstellar fog? The answer to this question is by no means trivial; see further down. What concerns me here is that the question never was asked until around 1930 the existence of interstellar dust had been discovered. Why not? The thought could have occurred -but did not- as soon as it had become clear what the stars were: distant suns. I think the reason is psychological and my argument stems from many years of experience in explaining interstellar matter to laymen: people refuse, at first, to believe that it exists, they donot want to accept that the sky is polluted by matter between the stars; the sky is a splendid experience- an escape from the deficiencies of our earth. Accepting the existence of interstellar matter demands to give up a well liked myth. As long as I keep my story limited to beautifully shining nebulae (observe the Orion Nebula through even an small telescope and you know what I talk about) the emotions and preconditioned opinions of my relations are not offended, but interstellar dust is too much! Recently I have horrified my friends with the information that very likely the interstellar gas contains large molecules, PAH's, that also occur in the exhaust of automobiles. They think I should leave astronomy.

Problems with friends aside: What is the light of the Milky Way made of? The answer depends a bit on where you look and at what wavelength, but if you take a Palomar Sky plate then most of the flux in the plate, after subtraction of contributions by the earth atmosphere (night sky) and by dust in the solar system (zodiacal light), is in stellar point sources, although a significant fraction (up to 40%, Mattila and Scheffler, 1978) of diffuse emission may be found. In the infrared diffuse emission dominates over stellar fluxes, e.g. in 12μ maps made by IRAS; it will also be much stronger in the ultraviolet, e.g. near the 220nm resonance of interstellar dust- but that has to be confirmed by observations.

The Milky Way is thus a doubtful sign of the presence of interstellar matter and for an observer without a telescope there is very little other evidence. The history of interstellar matter begins with the discovery of diffuse nebulae, soon after the demonstration by Galilei of the usefulness of the telescope for astronomical studies- 1610 is the oldest reference to the Orion nebula and the first, still remaining drawing has been published by Huygens in 1656 - see O. Gingerich (1982). The true nature of the diffuse nebulae (gas excited by blue stars) became certain only much, much later, when in 1864 spectroscopy by the British astronomer Huggins turned up several emission lines, most of unknown origin but a few definitely identified with hydrogen. The next discovery of fundamental significance concerned the rarified material in front of all distant stars. The material exists not only where we see it in emission, but also in most other directions where the material reveals its presence only through faint, small absorption lines in the spectra of background stars. The first to state this fact clearly was Hartmann in 1904, but it took until the mid twenties before his conclusion was generally accepted. Even more evasive was the presence of interstellar dust. The existence of a some general interstellar absorption at all (visual) wavelengths was feared for a few tens of years around the turn of this century, but the phenomenon could not be proven. In the same period Barnard in the USA and Wolf in Germany argued from photographic studies that in some directions dense clouds absorb all background stars. This gave urgency to the question whether a more rarified form of this absorbing matter also occurs in the more general interstellar space? It were Trumpler's observations on distant galactic open clusters that gave the first, decisive demonstration of interstellar extinction in 1930. Soon it was realised that the large wavelength range over which extinction occurs implies the presence of solid state particles significantly smaller than the wavelength of observation. They occur everywhere, and there are enough to block our line of sight to the galactic center. Going back over history, it is clear that the presence of matter between the stars was not something dreamed of by astronomers and then happily discovered; the subject came into existence only because observers accepted facts unwillingly observed.

Here I will not go deeper into the history of interstellar matter except for one historical event: the discovery of radio astronomy by Jansky and Reber in the thirties, followed by radio spectroscopy starting with the 21cm line first measured in 1951 after its prediction by van de Hulst in 1944. The success of radio astronomy was a factor of significance in the development of UV, X-ray and infrared astronomy, and without those the subject of today would lose a lot of its meaning.

The subject of interstellar matter has expanded vastly over the last twenty years. It has become impossible for a teacher to cover all its subtopics to a desirable depth in an average course. In writing these lecture notes I have therefore adopted the philosophy that a distinction should be made between the subject of "interstellar matter in general" and specialized subjects, connected to, but recognizable as a separate unity: star formation, circumstellar shells, planetary nebulae and supernovae. These specialized subjects are almost completely left out. I have also limited myself to our Galaxy. Interstellar matter in other galaxies is a hot topic at this moment, of which the development is too rapid to review here.

I finish this introduction by summarizing the basic observations of the interstellar medium: The interstellar matter may appear as extended emission- either as continuum emission (radio, X-ray) or in a line ($H\alpha$, 21cm HI). Molecular clouds appear, for example, in radio lines. The radiation may be polarized- magnetic fields are involved. In a spectral line the polarization may vary over the line profile- Zeeman splitting appears in that way. Alternatively interstellar matter appears through absorption- again via a continuum or via a line. Small, faint absorption lines in optical spectra of stars are the oldest example, but also the 21cm line of HI may be seen in absorption against a strong radio source. X-ray sources in the galactic plane suffer strong continuum absorption by interstellar hydrogen, helium, carbon etc. In fact, X-ray spectra can be used to measure very accurately the total amount of interstellar matter between the source and us. In all absorption studies there is always the question: is the effect that we see truly interstellar, or is it somehow related to peculiarities in the star or in the source? The time-proven argument for an interstellar origin is the demonstration that the effect occurs (statistically) in all stars and becomes stronger when the star is more distant or, rather, is more deeply embedded in the Galaxy.

In the following I first discuss (Section 2) the physics of the interstellar gas- those physical processes that one might invent, as it where, a priori. In Section 3 I discuss the distribution of the interstellar matter in our Galaxy. In section 4 there is special attention for the diffuse, neutral component of the interstellar gas, and in Section 5 various theoretical explanations or interpretations of the interstellar medium are discussed. Finally in Section 6 I sum up some topics that had, unfortunately, to be left out of my lectures.

This chapter is introductory only; it aims at giving several recent view points and results, and it tries to put those in some historical perspective. Yet the reader who really wants to study the subject in detail is referred to other, more extensive literature:

First, the theory of the physics of the interstellar medium is very well dealt with in the classic book, that is close to being perfect, except that it is difficult to read:

L.Spitzer, 1978, "Physical Processes in the Interstellar Medium", Wiley and Sons (New York).

Much more accessible, but also more limited in scope is

D.E. Osterbrock, 1974, "Astrophysics of Gaseous nebulae", Freeman and Cy (San Francisco).

Osterbrock has some, and Spitzer has practically no observational information. Luckily that information is now around in three recent books each consisting of many chapters written by separate combinations of authors (almost all chapters are worth reading; most of them I find excellent):

D.J. Hollenbach and H.A. Thronson, 1987, "Interstellar Processes", Reidel (Dordrecht);

G.E. Morfill and M. Scholer, 1987, "Physical Processes in Interstellar Clouds", Reidel (Dordrecht);

G. Verschuur and K.I. Kellermann, 1988, "Galactic and Extragalactic Radio Astronomy", Springer Verlag (Heidelberg)

PHYSICS OF THE INTERSTELLAR MEDIUM

2.1. The ingredients

Interstellar space is pervaded by atoms and molecules and by small solid state particles, called interstellar dust. In addition there is electromagnetic radiation (direct and scattered star light), there are magnetic fields, and there are relativistic moving atomic nuclei, the cosmic rays. I will come back somewhat later to these additional forms of energy, and I concentrate first on the interstellar matter. In this subsection I consider its composition and some of the immediate implications thereof. In the next subsections it will be considered in how far the interstellar matter can be treated as a gas.

2.1.1. Atoms

Most of the matter is in the form of atoms and molecules; the dust particles constitute roughly 1% of the mass and since each dust particle is made out of thousands of atoms, there are very few dust particles per free atom. Analysis of stellar spectra led already in the 1920's to the conclusion that in stars the most abundant element is hydrogen: by number 90% of all atoms. Helium provides about 9% of all atoms and the other elements together provide about 1%. Because stars, such as O and B stars were recently made out of interstellar material this matter must be of the same composition. Direct evidence confirming this conclusion comes from absorption lines of interstellar matter seen against bright stars: a large number of elements have been detected. The most important lines are the so called "resonance lines" -those where the transition starts in the atomic ground state. Most resonance lines lie in the far UV, below 300 nm, and it was the opening of this window to a spectrograph of high resolution in 1975 by the satellite Copernicus that gave an enormous impetus to the study of the elemental composition of the interstellar matter. Hydrogen is seen either through the Lyman lines, e.g. Ly α at 121.6 nm, or through electronic transitions of H_2. The strength of the Ly α line when compared to those of other elements is so overwhelming that this proves immediately that in interstellar space a lopsided elemental distribution exists similar to that in stars, but with some very interesting differences.

Abundances for the interstellar medium have been reviewed by Cowie and Songaila (1986) -see their table 3 for a compilation. For every million interstellar Hydrogen atoms, they give 250 Carbon, 80 Nitrogen, and 500 Oxygen atoms, and within the errors this agrees with what is found in the Sun. These three elements are the most abundant (except for hydrogen and helium). Other elements, however, show interstellar abundances considerably lower than the solar values; this is expressed in a so called "depletion factor", which is the ratio between the interstellar and the solar (or "cosmic") abundance. For Mg the depletion factor is only 0.5, which value is quite well determined. For Al the factor is between 0.07 and 0.01. For Na the factor is variable from direction to direction, with a maximum of 1.0 (no depletion) and a minimum of 0.03; something similar holds for K. Extreme is Ca, with depletion factors again varying from place to place, but between 0.04 and 0.0001! Another well studied element with very small depletion factors is Ti: between 0.4 and 0.001. How to interprete this depletion? The generaly accepted hypothesis is that the atoms are locked up in interstellar grains, and thus taken out of the gas phase. When a star forms the dust particles are all destroyed and normal gas abundances are restored: in stellar atmospheres of newly born stars we see always the normal or "cosmic abundances" (unless the stars are very old, in which case the elemental composition reflects that of the interstellar gas of the past). Direct confirmation is not possible: there is at present no way to analyse the composition of the grains except qualitatively for the presence of C and Si. Indirect evidence supports the hypothesis: interstellar lines with high velocities have much larger depletion factors, i.e. more "normal" abundances. Since the gas seen there must have passed through a shock it is likely that a fraction of the grains has been destroyed and the elements returned to the gas phase. Similarly lines supposed on other grounds to correspond to a much warmer gas (the so called intercloud medium) yield higher abundances, suggesting that also there some of the grains have been destroyed. It is clear that interstellar depletion will be a powerful way to study what ingredients go into grains. Unfortunately the subject is difficult to study and the most important observations, high resolution spectroscopy in the (rocket) UV can not be done at present.

The dominance of hydrogen allows us to characterize regions of the interstellar medium according to the condition of hydrogen: ionized, atomic, and molecular. The first two forms, ionized and atomic hydrogen are mutually exclusive- where the one occurs the other does not; the reason lies in the source of the ionization (energetic photons) as will be explained in section 2.3. We thus distinguish between so called "HI" and "HII" regions, meaning that hydrogen is either completely neutral (HI) or completely ionized (HII). A large fraction of interstellar space (>80 % ?) is filled with low density atomic and ionized hydrogen (see sections 3.3 and 4). A somewhat different, and more complex situation occurs with respect to molecular hydrogen (H_2). It occurs predominantly in "molecular clouds", that fill only a relatively small fraction of interstellar space. Inside the clouds molecular H_2 seems to occur in pockets of high density that are embedded in a sea of less dense atomic hydrogen. Associated with molecular clouds are dense HII regions bordering on or even inside molecular clouds; in fact the dense HII regions are parts of molecular clouds, ionized by the hot stars that recently formed there. The Orion Nebula is a prime example; emission nebulae well known from the early (e.g. the NGC) surveys are likewise associated with molecular clouds.

Here I inject a brief comment on the measurement of interstellar hydrogen. Atomic hydrogen is detected directly through the 21cm line; this allows us to study hydrogen in its most common state. Molecular hydrogen can be observed directly, but only rarely in its most common state, the ground vibrational and ground rotational state. As a consequence ground state molecular hydrogen is observed via an easily detected tracer molecule: CO. Ground state rotational transitions of H_2 are strongly forbidden because the homonuclear molecule has a very high degree of symmetry and thus no permanent electrical dipole, unlike so many other molecules consisting of two different atoms (CO!). For the same reason other homonuclear molecules cannot be detected in emission : N_2, O_2 and C_2. Some molecular hydrogen is observed in absorption against background stars. Because this involves electronic transitions the lines are in the far UV and one needs bright hot stars as background objects: hence our information is restricted to nearby lines of sights through relatively transparent regions of space! Molecular hydrogen is also detected through vibrationally excited lines in the near infrared; these lines need fairly high excitation energies and hence come from rather excited regions. Ionized hydrogen has no spectral lines, only continuum radiation, but luckily the gas can be seen in recombination lines, lines of highly excited atomic hydrogen. Such excited hydrogen occurs only in regions where hydrogen is predominantly ionized, and thus are these lines of neutral hydrogen excellent tracers of ionized hydrogen! Examples are the Balmer lines at visual wavelengths, Bracket and Pfund lines in the near infrared and the lines due to transitions between very highly excited states: the radio recombination lines (H109α).

2.1.2. Dust particles

The sizes and the composition of the interstellar dust particles remain uncertain since little direct information is available. The piece of information that is the oldest and probably still the most convincing about the nature of the "absorbing" material is the so called "extinction curve": see figure 2.1.

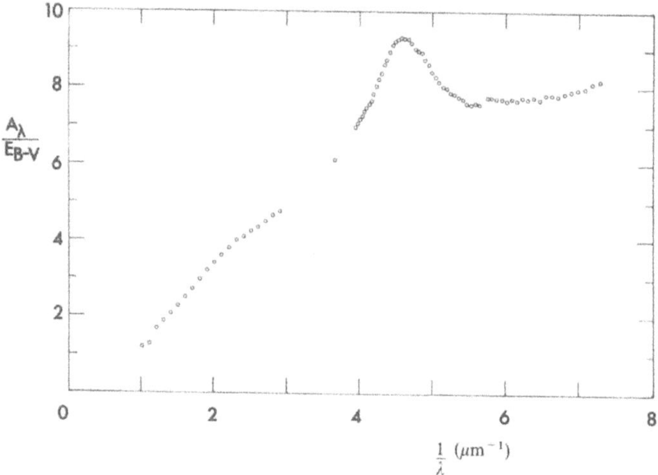

Figure 2.1. *The extinction curve: the function that describes how extinction varies with wavelength. Vertically is given the number of magnitudes that a star looks dimmer because of interstellar extinction, A_λ; it has been normalized to a standard amount of material by dividing it by A_v-A_B. The horizontal axis is measured in inverse micron. The curve presented here is based on observations (Nandy et al., 1975).*

It shows that the "absorbing" material is effective over a large wavelength range, and that the "absorbing" properties vary smoothly. It was realised soon after the discovery of interstellar "absorption" that a small population of solid state particles with a distribution of sizes, but all smaller than the wavelength of light, gives the desired effect. The particles "block" the light by scattering it in other directions or by absorption; when the particles are spherical the effect can be described exactly by classical electromagnetics (Mie theory); when the particles are non-spherical but still small compared to the wavelength, approximate theories are available- see van de Hulst (1981) or Spitzer (1978). Theory shows that at visual wavelengths most of the electromagnetic radiation is removed by scattering, and that only a small fraction of the radiation is absorbed, that is: only a small fraction of the visual photons are added to the internal energy of the particle; in the ultraviolet absorption is much more important. The dimming of distant stars, and their reddening, by interstellar dust grains is thus only partially an absorption effect. From now on the term "interstellar extinction" will be used to describe the weakening of stellar light by interstellar particles; extinction is the sum of scattering and absorption. Interstellar extinction of the stellar light also leads to a small amount of polarization. The explanation of this effect requires (1) that the particles are non-spherical, so that they scatter differently in different directions, and (2) that a magnetic field exists to align the particles (see also in section 2.1.3).

The major questions concerning the dust particles are: what are they made of, how many are there and what are their shapes and sizes? These questions prove difficult to answer from the information available: without genuine samples in the laboratory the objects remain somewhat elusive. Until recently the best (understood) information was the curve of fig. 2.1 -a worthwhile analysis of this curve is that by Matthis et al. (1977), which led to the now well known MNR model. Taking particles of definite shapes (spheres and cylinders) of well defined materials (graphite and various silicates and magnetites) they calculate the extinction per particle of a given size and then reconstruct the extinction curve by adding samples of different particles. They find that they always need at least two types of particles -and always carbon in combination with one other material; carbon in particular is held responsible for the "hump" in the extinction curve at 220 nm. The size distribution is given by a power law and cut-offs at small and at large sizes. The cut-offs at both ends are poorly constrained by the analysis of MNR -one could add smaller particles and larger particles without changing the extinction curve of figure 2.1 very much.

New developments since the MNR paper concern the existence of very small particles – diameter below 5 nm. How these affect the MNR results is well described in a paper by Draine and Anderson (1985). There are two new pieces of evidence -not listed in chronological order. First is the discovery of significant dust emission at rather short infrared wavelengths -especially between 10 and 30 μm, where IRAS gave ubiquitous information. Dust particles that radiate effectively at these wavelengths must be at a temperature of 100 k or more. This is much more than the equilibrium temperature of grains heated by the interstellar radiation field. Here an old idea by Greenberg (1968) becomes interesting: very small particles (smaller than $\sim$ 10 nm) have so little heat capacity that the absorption of a single UV photon will increase their temperature momentarily to rather high values. They will distribute the energy over internal degrees of freedom and radiate it then away in smaller energy packages. In a short time they will cool. Although individual particles thus lack an equilibrium temperature, a large collection of particles will have an equilibrium *distribution* of temperatures, and a small fraction will always be warm. The hypothesis is thus that the dust emission below $\sim$ 30 μm is due to this warm fraction of the smallest particles.

The second new piece of evidence concerns the detection of interstellar and circumstellar emission bands in the range from 2 to 20 μm. Their broadness suggest an origin in small solid state particles, or in very large molecules. Leger and Puget (1986) and Allamandola et al. (1986) have suggested that large molecules of hydrogenated cyclic carbon molecules (Polycyclic Aromatic Hydrocarbons, or PAH's) are responsible. Allamandola et al. have remarked the similarity between these particles and those in the exhaust of automibiles ("autoexhaust along the Milky Way" is the title of their paper). Although this hypothesis is attractive, it is not yet proven beyond doubts. One expects that ISO, the next European infrared satellite, will be of great help in studying the possible presence of PAH in interstellar space.

In their analysis Mathis et al. made rather definite choices of the materials from which the grains are made. Is there direct support for these choices? Yes, there is some direct evidence. Against some infrared sources a strong absorption band is seen centered at 9.7μm; it is assigned to absorption by silicates in an amorphous solid state. There is also an absorption line at 3.0 μm due to H_2O-ice: it is seen in a few very thick and cool clouds. Yet the conclusion that interstellar particles consist of silicates and of ice is not the one generally accepted. There is more, although less direct information: large numbers of very cool giant stars of high luminosity are found that are surrounded by ejected gases in which dust particles have formed. Such stars are very likely the source of the interstellar dust particles. Most such stars have the silicate 9.7μm band in their infrared spectrum in emission; however a small, but significant fraction of the stars do not show the silicate band, but have instead an emission band at 11.3μm attributed to SiC. It is assumed that the dust particles in those cases are rich in carbon; graphite particles are a possibility but the carbon might also be amorphous. Therefore it appears likely that there are two kinds of grains: "graphite" and "silicate" and that they are produced in dying red giant stars. In very dense clouds other atoms will collide with the particles and stick on to them, thus forming a coating. This might then explain the ice band at 3.1μm. The coating may become photolysed by ultraviolet photons from distant stars when the grain wanders outside of the cloud (inside the cloud it is protected against this radiation). The question of coating leads to a messy situation, and to many interesting questions. Convincing and decisive answers have not yet been given (see Tielens and Allamandola, 1987).

2.1.3. Other forms of interstellar energy

In interstellar space there are other forms of energy in addition to the matter just described: magnetic fields, photon fields and cosmic ray particles. In section 3.5 the little that is known about their galactic distribution is presented. Here I mention only some local observations. I start with magnetic fields. The most direct evidence for the existence of a magnetic field is the Zeeman splitting observed in interstellar emission lines of HI at 21cm, and in lines of OH. The effect is difficult to measure because the Zeeman splitting is smaller than that due to turbulent velocities, and –when it is seen– is probably uncharacteristically strong. Other pieces of evidence for the existence of magnetic fields are: (i) The direction of linear polarization of the radio emission from a distant object, e.g. a radiogalaxy, varies with the wavelength as λ^2. The explanation is that the electromagnetic waves travel through an electron gas that contains a weak magnetic field. Consider first what happens when the magnetic field is zero. The electric field of the e.m. waves forces the electrons to oscillate with the wave. Because the plasma provides a natural oscillation frequency for the electron, called the electron frequency ω_e, and because the frequency of the e.m. wave is different (higher), the motion of the wave through the plasma is retarded and the result can be described by assigning a refractive index to the plasma. Because e.m. waves of different frequency are affected differently, we see the diffraction e.g. in the different arrival times of the same pulse from a pulsar when we observe it at different frequencies. (From the relative delay between the pulse arrival time at two different frequencies one can derive the total number of electrons along the line of sight; this is usually called the *Dispersion Measure* $DM = \int n_e \, dl$.) This diffraction effect becomes more complex when a magnetic field is present, because the electron oscillations are then subjected also to a Lorentz force. The net result is that two waves of the same frequency, but of opposite circular rotation, will experience a small difference in refractive index. If one now considers two circularly polarized waves, of opposite rotation, moving through this medium, then one wave will experience a lower than average refractive index, the other a higher value, and at the end of their path through the medium the first wave will be ahead of the other. A linear polarized wave travelling through the gas can be thought of as consisting of two opposite circular waves with exactly the same phase, and when the linear wave leaves the gas, one of the circular components has been retarded with respect to the other, and the direction of the linear polarization vector has rotated. This rotation is called the Faraday effect. The angle of rotation is proportional to λ^2 RM where the *Rotation Measure* is given by $RM = \int n_e \, B_{//} \, dl$; here n_e is the electron density and $B_{//}$ the component of the magnetic field, parallel to the line of sight. For a thorough explanation of the concepts of "DM" and "RM" see the book by Rybicki and Lightman (1979), their chapter 8.

Faraday rotation is a well observed phenomenon. Its diagnostic value for the interstellar magnetic field is of great importance, but only for lack of better diagnostics: the measurement gives RM, which is just the sum over a very long pathlength. $B_{//}$ may have positive and negative values along the line of sight ("field reversals"), often n_e is unknown, and part of the effect may not occur in the interstellar medium but inside the source. For these reasons the most useful observations are those of the Faraday rotation seen in pulsars. First, pulsars seem not to have internal Faraday rotation and second, for pulsars one can also measure DM , which is the integral of the electron density. When both RM and DM are known, there is more confidence in the average value of $B_{//}$. Unfortunately the pulsar measurements are very difficult and therefore only a few RM and DM values have already been measured.

(ii) Radio radiation at long wavelengths from inside the Galaxy is produced by cosmic ray electrons that are accelerated in the magnetic field of the Galaxy -for a detailed description of this synchrotron emission the reader is referred again to Rybicki and Lightman (1979), chapter 6. The emissivity of a certain volume depends on the strength of the magnetic field, and on the energy density of the relativistic electrons. The intensity of the radiation in a given direction (and this is the quantity that one measures) is again an integral along the line of sight, but instead of $B_{//}$ it involves $B_\perp$, the component of the magnetic field perpendicular to the line of sight. Because one has hardly any information on the cosmic ray electron density, it is impossible to estimate the value of $B_\perp$ from the observed intensity -but one may obtain useful information on the direction of $B_\perp$.

(iii) The earliest demonstration of the existence of the interstellar magnetic field comes from the linear polarization of the light of distant stars. The effect is always small: at most a few percent; this indicates already that it is difficult to measure, and that measurements are limited to relatively bright stars. Discovered unintentionally in 1949 by Hall and by Hiltner it took some time before the explanation was generally accepted: interstellar grains are non-spherical and tend to be lined up by the magnetic field -each particle scatters differently in different directions, and because the particles are systematically aligned, a large collection of particles will have a collective effect. As it turns out the interstellar extinction is smaller when the E vector is parallel to the smaller axis of the grain. This smaller axis tends to line up with the magnetic field -and thus the polarization (the difference between the two extinctions) is proportional to $B_\perp$. Optical polarization measurements indicate clearly the direction of $B_\perp$, but they are little informative about the strength of $B_\perp$. An average value can sometimes be estimated from considerations on how the alignment is maintained. The direction of $B_\perp$ derived from many observations indicate that in the Solar neighbourhood the magnetic field contains a systematic component in the galactic plane, that is directed roughly into the local spiral arm, $l \approx 90°$.

Heiles (1987) reviewing all evidence concludes that there is a random component of the magnetic field in addition to the systematic component and that on average the two may well be equal. A representative range of possible magnetic field strengths is 1.5 to 3μG.

A second form of interstellar energy, is the "radiation field", the intensity of the electromagnetic radiation; it is a quantity of great interest, because of its influence on various molecular and atomic processes. Direct measurements are only possible in the solar neighbourhood. The radiation density has been measured directly by space crafts in orbit and it is also calculated from the known distribution of stars. The topic has been treated by a large number of authors; my favourite paper remains that by Mathis et al. (1983). A characteristic of the energy density is that it deviates strongly from that of a black body, because it is the sum of the radiation from many different stars- from a few very luminous, hot ones and from many cool, but not so luminous stars, plus (1) the 2.7k background, and (2) the soft X-ray background derived from extragalactic sources.

Figure 2.2 derived from Black (1987) gives an overview. The flux integrated over all wavelengths is 0.08 erg cm^{-3}, and this corresponds to an energy density of 0.83eV cm^{-3}. The reader may answer for herself the following question: what is the temperature of a black body field with the same energy density? And for that temperature what would be the black body energy density at $\lambda = 100$, 500 and 1000 nm? This should give you some feeling how large the local radiation field deviates from a blackbody. Finally, it is of some importance to realize that large fluctuations in the energy density may be present. The local value may differ from the value at some other point; for example strong variations occur inside dark clouds, where the field will be much weaker, and near an OB association, where the field will be much stronger -see section 3.5.

I come now to the third remaining form of energy: the cosmic-ray particles. The peculiar name reminds one of the original discovery: photographic emulsions never exposed to light were nevertheless found to be slightly developed. "Natural rays" like γ-rays were suspected, but in the 1920's experiments in balloons by V. Hess showed conclusively that

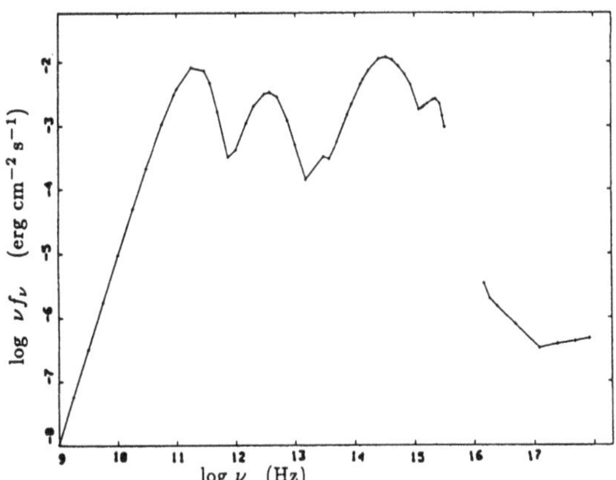

Figure 2.2. *The interstellar radiation field. At a point in interstellar space within the solar neighbourhood (but far enough away that the Sun is an unimportant contributor) the radiation flux f_ν, averaged over all directions, is given here as a function of frequency (in fact νf_ν is shown in the figure). The figure is due to Black (1987).*

the "rays" were "cosmic". It are atomic nuclei and electrons of very high energies -from a few MeV up to more than 10^{20}eV! Most of the energy (about 1 eV/cm^3) is contained in protons with energies between 1 and 10 GeV. The atomic composition of the rays is remarkably similar to that of the Sun, if one normalizes at the carbon abundance. The conclusion is that cosmic rays are probably injected into the interstellar medium from the atmospheres of stars with relatively normal abundances -see Bloemen (1987) for more details.

Cosmic rays are part of the interstellar medium and should be studied as such and not only for their own sake. Here I will restrict myself to their various effects on cosmic rays on the other ingredients: being electrically charged particles they influence and they are influenced by magnetic fields. Thus they exert some pressure on the magnetic field; this "cosmic ray pressure" equals 1/3 of the energy density. Cosmic ray pressure may "blow up" certain magnetic configurations. Secondly the lower energy protons from the cosmic rays (E<100MeV) will ionize the interstellar gas. The effect is very small compared to ionization by stellar photons, but deep inside molecular clouds where photons cannot penetrate, the cosmic ray ionization still occurs and provides just those few ions necessary to start some important chains of molecular reactions. Thirdly, as noticed before the electrons from the cosmic rays emit synchrotron radiation when they interact with magnetic fields. The interpretation of the galactic synchrotron radiation is difficult, however, because its intensity depends on both that of the magnetic field and on the density of the cosmic ray electrons. Without further information on both ingredients, one cannot find the energy density of either. Luckily, in recent years, a new analysis tool has come available to help out on this point: γ-radiation. This radiation is produced primarily in the interstellar medium, in collisions between cosmic rays protons and low-energy atoms of the interstellar gas (mainly H atoms). Because the density of the interstellar gas is reasonably well known (via 21cm H° and mm CO observations) one thus obtains some information on the distribution of the cosmic-ray protons throughout the Galaxy -see further in section 3.5.

2.2. The statistical properties of atoms and molecules

Given that there are atoms, molecules and small particles- what are the statistical laws that rule their velocity distributions, their fractional ionisations, their recombination and dissociation rates? We are acquainted with four statistical distribution laws, that are valid at least under laboratory conditions. One often assumes their validity in stellar interiors. These "Big Four" are:

1. The Maxwell-Boltzmann distribution of random velocities.
2. The Planck law for the isotropic distribution of photon energies.
3. The Boltzmann distribution over excited energy states.
4. The Saha equation for the distribution over various states of ionization.

In astronomy one likes to use the adjective "thermal" to indicate that these four statistical laws apply: one speaks of thermal conditions, a thermal gas and of thermalisation. Well, interstellar space is not "thermal": In interstellar space only distribution law ♯1 is almost always valid: The velocities of individual atoms and electrons are distributed according to Maxwell-Boltzmann. This implies that the concept of a "kinetic temperature" almost always makes sense; it is the value of the parameter used to describe the Maxwell distribution, just like in ordinary gas kinetics. The reason that Maxwell's law is valid is that dynamic relaxation occurs on a short time scale: the collision frequency between the atoms (molecules, ions, electrons etc.) is sufficiently high that a non-equilibrium distribution of the atomic kinetic energies is washed out quickly before other processes might perturb it. As a simple argument use the following: take the Bohr radius (0.1 nm) to calculate the collisional cross section of an atom and give the atom a random velocity of 1 kms^{-1}; with a density of 1 cm^{-3} the collision frequency is 1 per 1000 yr which is very short compared to most interstellar processes. Only inside shock waves (see section 2.5) the events are happening so fast that on occasion one has to take into account non-equilibrium velocity distributions.

That the distribution of photon energies is not according to Planck's law, has been discussed already in Section 2.1.3. The fact has far reaching effects and is probably the main cause why interstellar material is so distinctly "non-thermal". The photon distribution in interstellar space consists of two components: one are the photons of the 2.7K background radiation, remnants of the big bang; this component is to a high degree Planckian but it dominates only in the submm-wavelength range. Second is the contribution of the light from all the stars seen from that point in interstellar space: it is the sum of a large number of black-body distributions of different temperatures and at very different distances. This sum is very much non-Planckian. At visual and at UV wavelenghts this second component dominates the photon distribution, and thus processes induced by these photons are "non-thermal". On the other hand, processes induced by collisions with other particles are usually "thermal". Since the population of excited states is the result of non-thermal photon and thermal, collision induced processes, these populations have rarely their thermal values. This explains why the distribution of the excited states is not given by the Boltzmann distribution, nor the distribution over ionisation states by the Saha equation. In fact, because of the low temperatures and the rarity of collisional and radiative excitation events, atoms (ions) are often in the ground state of the ion. Excited levels can be populated only when they are very close to the ground state and this population is only significant when the life time of that level against spontaneous decay is sufficiently large, that is when $1/A$ is large, where A is the Einstein coefficient for spontaneous transition. A similar argument can be used for an estimate of the state of ionization- see section 2.3; but there the important point is that recombination is always very improbable compared to other transitional possibilities. It is interesting to notice that spontaneous emission introduces an absolute time scale $(1/A)$ against which one can measure the rapidity of other atomic processes.

As a consequence of all this, the term "temperature" has only a very limited meaning in interstellar space; usually one means the " kinetic temperature", as defined above. There is at least one other temperature that has a good physical meaning: the temperature of a dust particle, here often symbolized as T_d. The particles absorb interstellar photons, and add the energy to their internal energy; ultimately the energy is emitted again, but this happens only after the absorbed energy has been distributed over a large number of internal degrees of freedom (lattice vibrations). Thus the emission process can be described as a thermal proces at a definite temperature, which then is called the dust temperature. It is quite normal that this dust temperature differs significantly from the kinetic temperature of the surrounding gas atoms. This has sometimes interesting consequences- see section 2.4.

Sometimes a formal "temperature" is introduced to describe the outcome of a non-thermal process- for example to describe the population ratio between two atomic states: if the populations of the levels are n_i and n_j respectively, one defines an excitation temperature via $kT_{ex,ij} = \ln(n_1 g_2/n_2 g_1)$, just as in Boltzmann's distribution function. Of course this temperature has no physical meaning and the concept is only useful when it can be compared to a really physical temperature, for example when in some limit T_{ex} will approach T_g. The new comer in Astrophysics Wonder Land is to be warned that the word "Temperature" is often used in a non-physical sense, and like for all other slang, its unbridled usage may lead to sloppy thinking.

Because the excitation and ionisation conditions are not given by some temperature, how do we find them? The answer is in the hypothesis of statistical equilibrium: write down the rate equations for the population of e.g. all excitation levels

$$\frac{dn_1}{dt} = (A_{21} + I_\nu B_{21} + C_{21})n_2 - (I_\nu B_{12} + C_{12})n_1$$

and then assume that all time derivatives are zero; one is then left with a set of equations that together describe the population levels n_i $(i = 1...)$. The equations are not linear, because the collisional rate coefficients C_{ij} depend on the particle density. The absorption and stimulated emission coefficients depend on the radiation field at the position under consideration. Solving the set of equations one obtains the population distribution n_i at this given position.

Let me give you here two examples of results obtained by solving such rate equations: C^+ (158μm) and H° (21cm).The ground state of hydrogen is split into two "hyperfine" levels, separated by 1420 MHz ($\lambda = 21$cm; $E = 6 \times 10^{-6}$eV). The spontaneous transition probability is very small: $A = 2.85 \times 10^{-15}$ s^{-1} (the life time for spontaneous decay is thus 12 Myr). The cross section for inelastic collisions with other hydrogen atoms is rather large ($\sim$ 8 Bohr areas at $T_k \sim 100$ k; Allison and Dalgarno, 1969), and thus the atoms are excited and deexcited by collisions. Deexcitation by emission of a 21cm line photon is exceptional! As a consequence, the distribution N_1, N_2 of the atoms over the two energy states is thermal: $N_2 = (g_2/g_1)\, N_1 \exp(-h\nu/kT_k)$, where g_2 and g_1 are statistical weights, and where T_k is the kinetic temperature of the hydrogen atoms.

Next C^+: Its ground state is split into two fine structure levels separated by 1.90 THz ($\lambda = 157.7\mu$m; E = 0.008 eV). The spontaneous transition probability is much larger: $A = 2.3 \times 10^{-6}$ s^{-1}. Excitation and deexcitation by collisions are rapid processes (Hayes and Nussbaumer, 1984), but deexcitation by spontaneous emission is still faster than that by collisions (as long as the density of the colliding agent free electrons, is smaller than about 1 cm^{-3}; this is often the case in interstellar regions with neutral hydrogen). Because excitation occurs at a thermal rate, but deeexcitation does not, the equilibrium result cannot be described as thermal -and there is thus no physically realistic temperature that describes the excitation!

I like to warn the reader: thermal conditions make calculations often rather easy; yet never take for granted that interstellar conditions are thermal. Always check whether this is true by solving the statistical equilibrium equations - often a tedious, sometimes a difficult, and always a necessary task.

2.3. Ionization and recombination

Ionization and recombination are only introduced here. For a fuller explanation the reader is referred to Spitzer's (1978) textbook or to the small, but very well written book on ionized nebulae by Osterbrock (1973).

I will start, somewhat unconventionally, with the process of recombination. When an ion and an electron collide, at least three processes may follow: (i) there may be an "elastic" collision, in which there is no loss of kinetic energy (actually, a collision is never completely elastic because the pair forms an electric dipole and always some energy is radiated away as bremsstrahlung or free-free emission but the amount of energy lost is very small); (ii) the ion may become excited and produce some line photons, and (iii) there may be capture of the electron, sometimes in the lowest orbit of the atom or ion, but often in an excited orbit. This third incident is then followed by radiative transitions through a kind of cascade downwards into lower states, a process of capture called "radiative recombination"; of the three processes it has often the lowest probability. Recombination has been well analyzed and calculated very accurately for hydrogen and helium; the recombination rates of other ions can often easily be derived from those for hydrogen (see Osterbrock's book). The recombination probability is usually expressed in a coefficient α, such that $\alpha n_e n_i$ is the number of recombinations per cm^{-3} per sec when n_e and n_i are the density of electrons and ions, respectively α has dimensions cm^3 s^{-1} and is a function of temperature because the velocity of the colliding electron determines (partially) the probability of capture. Recombination is a process with a low probability; the cross section for capture of an electron by a proton is of the order of $10^{-20} cm^2$ -much smaller than the cross section for other electron- hydrogen collisions.

Often one finds reference to "Menzel's case B" or to "Case B" in short -sometimes "Case A" is mentioned. The following is implied: The total recombination coefficient is the sum over all levels of separate probabilities of recombination into each level. This sum is called "Case A". Under interstellar conditions, the (very few) neutral hydrogen atoms in the ionized gas are all in their ground state. Consequently, when an ion captures an electron and jumps directly into the ground state, a photon is produced that can ionize any neighbouring neutral atom. That process is very likely and therefore this specific capture is followed immediately by photoionization of another atom, so that recombination directly into the ground state does not contribute to the total rate. Therefore when in an HII region one sums to obtain the full recombination coefficient it is best to ignore capture directly to the ground state. The sum without groundstate capture is called "Case B".

In addition to radiative recombination there is another recombination process that involves not only one free electron but also one of the bound electrons of the ion; this process is of importance at high temperatures and is called dielectric recombination; see Jacobs (1985) and references therein.

Ionization occurs by photons, by free electrons and by cosmic rays. Ionization requires much energy- hydrogen for example requires 13.6 eV which is equivalent to the average kinetic energy of atoms in a gas of 160,000k; the ionization of O^{4+} requires ten times as much energy- nevertheless O^{5+} exists in the interstellar medium. Photoionization leads usually only to relatively low states of ionization: ionization potentials below, say, 50 eV, are typical for O stars: sources of still harder photons are very rare or not powerful, or both. Higher stages of ionization are found in the interstellar medium (the O^{5+} ions testify to this); they are produced by collisions with free electrons in gas of typical 10^6k.

Photoionization of hydrogen by typically O and (early) B star shows a remarkable property first proven by Stromgren in 1939: if one moves away from an ionizing star the degree of ionization is at first very high, almost, but not quite completely, 100%. At some distance, r_s, from the star the ionization drops down to 0% over a very short distance: there is a sharp boundary between ionized (HII) and non-ionized (HI) gas. In the ideal case of pure hydrogen, distributed uniformily, the distance r_s is determined by the total number of ionizing photons emitted by the star N(Lyc) and the density of the gas, n, and may be written, approximately, as

$$N(Ly_c) = \frac{4}{3}\pi r_s^3 \alpha n^2$$

where α is the recombination coefficient for case B. (Because α depends weakely on the kinetic temperature, T_k, of the gas, r_s depends also weakly on T_k -it is usually sufficiently accurate to take $T_k = 10^4$, a good average value). If one knows the detailed spectral class of the star and thus its temperature and its luminosity, one knows N(Lyc), and it is easy to calculate r_s for a given value of n. If $n = 1$ cm^{-3} is taken, the ensuing radius r_s is called the Strömgren radius,s. In the following table three examples are given.

Table

Spectral Type	log N (Ly$_c$)	$r_s(n{=}1)$
05 V	49.7	108 pc
09 V	48.2	44 pc
B0.5 V	46.8	12 pc

Stars of later spectral type have very small Strömgren radii and are not able to ionize a significant amount of interstellar medium.

The essence of Strömgren's proof is easy to understand, and is important to know: Consider an ionizing star in hydrogen gas of a uniform density $n = 1$ cm^{-3}. The photoionization cross section, σ, of a neutral hydrogen atom reaches its maximum at the threshold, $\lambda = 91.2$nm: $\sigma = 6.3 \times 10^{-18}$cm^2, and decreases like ν^{-3} for higher frequencies. Near the star the hydrogen will be almost completely ionized: if the degree of ionization is x then $\epsilon = 1 - x$ is a very small number (e.g., $\sim 10^{-5}$). The mean free path of a photon at 91.2nm is then $1/(\sigma n \epsilon) = 0.04\epsilon^{-1}$: it is large (i.e. $>$ 1pc) as long as $\epsilon < 0.04$. At some distance away from the star the ionizing flux decreases, and hence ϵ will increase; as soon as ϵ exceeds this critical value of say 0.04 the ionizing photons cannot penetrate much further, and suddenly over a very short distance, typically of the order of $1/(\sigma n) \approx 0.04$ pc, there is a transition from $\epsilon \ll 1$ to $\epsilon \approx 1$ (from $x \approx 1$ to $x \ll 1$). Stromgren proved the theorem using the equation of radiative transfer read section 2.3 in Osterbrock to see the proof). The crux is in the small free path of ionizing photons in neutral hydrogen. The argument would not hold if the density is rather low (see section 3.3.2.) or if the ionization were done by much harder photons, say 10 times more energetic. In the latter case ν is 10 times higher, and the photoionization cross section and the mean free path are each 1000 times larger: such photons penetrate neutral gas more easily, and the transition layer is very thick. However, the hottest main sequence star is cooler than 50,000k, and not able to produce many hard photons. There are a few stars hot enough to provide very hard photons, e.g., the nuclei of some planetary nebulae and some other white dwarfs, but they are not luminous enough to be of importance. Therefore the distinction between HII and HI regions remains of fundamental importance. (However, see the reference to the work by Mathis in section 3).

Ionization requires high energies and thus, when it is done by collisions with thermal particles, the gas has to have a high temperature, say well over 20,000k. Degrees of ionization have been found that are higher than what stellar photons will give -the solar system is inside such a hot bubble (see section 3.5) and collisional ionization in a 10^6 k gas appears the most likely explanation. When collisions are the cause of ionization the degree of ionization is in first order independent of the density of the gas and only a function of temperature. The reason is that the ionization/recombination equilibrium can be written as

$$n_e \cdot n(X^0) \cdot C(T_k) = n_e \cdot n(X^+) \cdot \alpha(T_k) \text{ or } \frac{n(X^+)}{n(X^0)} = \frac{\alpha(T_k)}{C(T_k)}$$

Here n_e is the electron density, $n(X^0)$ and $n(X^+)$ are the densities of the ions X^0 and X^+, and α and C are the recombination and ionization rate coefficients. Both rate coefficients depend on given atomic parameters and on the kinetic temperature of the gas. It thus follows that the ionization ratio is independent of the density. In second order small deviations occur.

The last ionization process of some significance is ionization by cosmic rays. Cosmic rays of low energy (below, say, 10 MeV) have a significant cross section for collision with gas atoms. In the collision a secondary electron of high energy is produced that can interact again with the gas. A full description is therefore necessarily complicated. Usually the ionization process is sufficiently well described as an ionization rate that is proportional to the density of atomic (neutral) hydrogen, n_H, multiplied by a proportionality constant ς, dimension s^{-1}. Cosmic ray ionization is too infrequent to be effective as a source of energy into the gas; however, they are very important in molecular clouds, because they ionize (a few) atoms deep in the interior of the cloud where critical molecular reactions can now start that require charged particles.

2.4. Heating and Cooling

In section 2.2 it has been argued that the interstellar gas has a well defined kinetic temperature. How high is it and what determines its value? Well, the value of the temperature is the outcome of a balance between heating and cooling.

Consider first the heating. There is heating macroscopically and microscopically. Under the first term one reckons adiabatic heating by compression (which may happen following the passage of a shock wave), dissipation of magnetic energy by damping of Alfven waves, and dissipation of turbulence (the last concept is particularly vague- see section 2.6). The magnitude of these heating processes depends on the actual geometric and kinematic conditions and I will not discuss them here further, but I stress their occasional importance. Consider next the microscopic heating, which is important under most interstellar conditions. In the first example stellar background light is converted into (thermal) kinetic energy of electrons via photoionization: when an atom or ion absorbs a photon with more energy than the ionization potential, I_{pot}, the difference $\Delta E = h\nu - I_{pot}$ is given to the (now free) electron. The difference is easily 1 eV or more and that exceeds the average kinetic energy of the thermalized electrons. Calculations show that the excess energy of the free electron is distributed among all thermal electrons by elastic collisions. Thus stellar radiation energy is converted into thermal kinetic energy of the gas atoms. Stationarity requires that later on the just ionized atom will recombine and radiate away the excess energy; thus in a full cycle some energy is lost again by radiation. Fortunately recombination will take place with an electron from the thermal pool, and the emitted photon will thus have an energy equal to kT_k, where T_k is the gas kinetic temperature. Therefore the net gain in this ionization-recombination cycle is $\Delta E - kT_k$. In this example it is not essential that the ionized particle is an atom, ion or molecule: the photon may also ionize a dust particle. Dust particles are rare, but they have the advantage that their ionization potential is quite small. Especially the smallest grains, large in number, may well the most important heaters of the cold gas (de Jong, 1980; Puget et al., 1985; Lepp and Dalgarno, 1988) (see section 3.2. for the definition of cold gas).

Conversion of radiative energy into kinetic energy is also the case in the second example of microscopic heating. As argued before, interstellar dust particles reach an internal equilibrium temperature through absorption and emission of stellar light. This temperature, T_d, is likely to differ from the kinetic temperature of the surrounding gas, T_g, because usually the two are determined by independent processes. Under certain circumstances, however, there may be a coupling between the two. For example in dense shells surrounding cool giant stars the dust particles attain often temperatures of 100 to 300 k by absorption of infrared photons, but the same photons cannot heat the molecules in the gas. When a molecule hits a dust particle there is a finite chance that it will stick to the grain. After some time it is released again, but it will depart with a kinetic energy kT_d; because this is much higher than kT_g, an energy $k(T_d - T_g)$ has been transfered to the gas.

The third example of microscopic heating is the conversion of cosmic ray energy to gas kinetic energy. It is similar to the first example: a cosmic ray particle ionizes an atom, and a free electron is produced with a kinetic energy larger than kT_g. The free electron then collides with other gas particles and looses its excess energy. A complicating factor in the cosmic ray example is that the excess energy of the free electron is sufficient for further ("secondary") ionization- there is thus a cascade of processes following the primary event. For a few years, around 1970, cosmic ray heating was fashionable; the fashion was supported because there was very little information about the energy density of the low energy end of the cosmic ray spectrum- and that is the important part of the energy spectrum. Subsequent studies of the chemistry of certain diatomic molecules have made it clear that cosmic rays are insufficient by a factor of ten to provide enough energy to the interstellar gas; photoionization of grains (see before) is at present more attractive- but again, perhaps only until we will know better.

Consider now cooling. Macroscopic cooling may occur through adiabatic expansion. Here I will not discuss it further (see above under macroscopic heating). Microscopic cooling takes place by conversion of gas kinetic energy into radiative energy, under the supposition that the so created photon then leaves the gas (low optical depth in the line). The basic physics are given by an example: start out with a C^+ ion in its ground state -the most likely state. Frequently the ion collides inelastically with a free, thermal electron and then transits to an excited state 0.008eV above the ground level. As discussed at the end of section 2.2, in the next step the ion makes a spontaneous transition back to the ground state under emission of a 157.7μ photon. The net outcome of this two step process is that a photon has been created at the expense of kinetic electron energy. If this photon then can leave the area freely, cooling of the electron gas is the consequence. Whether the photon can leave depends on the opacity, and thus upon local circumstances. A useful description of various heating and cooling processes is to be found in Black (1987); further reference is to Dalgarno and McCray (1972) and to Spitzer (1978).

2.5. Dynamics

In subsection 2.2 it has been argued that the atoms, ions and electrons of the interstellar gas have a Maxwellian velocity distribution; hence it makes sense to talk about the pressure of the gas. Then the laws of hydrodynamics also apply -only in the coupling between matter and radiation the very specific properties of the interstellar gas come into consideration. Hydrodynamics is often not taught at all during a university education in physics, or it is restricted to hydrodynamics of incompressible fluids, like water, and that is of no use to interstellar gas: there are fundamental differences between gas and water, compressible and incompressible fluids, of which the most obvious is the occurrence of shock waves in gas dynamics. Another important difference is the influence of magnetic fields on the behaviour of the gas.

There are three fundamental laws based on the conservation of matter, of momentum and of energy which can be expressed in the form of partial differential equations. I will not reproduce them here. A very clear introduction has been given by F.D.Kahn in the les Houches summerschool of 1975 on interstellar matter -see the references. If you are not satisfied with Kahn's presentation, and you want to know more: Landau and Lifshitz have a very good textbook on hydrodynamics. I feel the need, though, to discuss briefly the significance of sound waves and the phenomenon of shock waves. Even when you will never write down the partial differential equations just mentioned, you will always encounter shock waves, as soon as you study interstellar matter -e.g. by looking over a few Sky Survey prints.

Far away from external forces the fundamental hydrodynamical flow equations just mentioned have a solution in which all parameters are constant: the density, ρ, the velocity, V, the temperature, T, the pressure, P, are all constant in space and time. This is of course a trivial solution! Less trivial is the study of what happens when the density is perturbed by such a small amount that deviations from equilibrium are small compared to the equilibrium value ($d\rho/\rho$ is small). How will these density variations change in time? It turns out that a deviation will spread out through the whole region, that is, there is a wavelike solution to the equations: the small disturbance propagates like a wave; the wave velocity is given by $c^2 = \frac{dP}{d\rho}$; here c is the famous velocity of sound. Clearly it is the velocity by which a small density perturbation will propagate through the gas. It is a kind of resonance velocity for the gas and it plays a very fundamental role: one distinguishes between subsonic and supersonic motions. Often this distinction is expressed via the Mach number, which is the ratio between the velocity and the local velocity of sound; M=1 indicates the sound velocity, subsonic velocities have M<1, supersonic M>1. Subsonic motions sometimes compare well with those in incompressible hydrodynamics, but supersonic motions are totally different; if they occur, shock waves usually occur.

Shock waves can be introduced as follows: imagine the trivial solution just described in which density, pressure and temperature are constant in time and space. Now consider the question: can we have two such solutions different from each other, each in its own volume, separated by a sharp boundary? The answer is yes: The fundamental hydrodynamic equations allow one to have two very different gasses living next to each other, only separated by a thin layer. Two forms of coexistence are possible.

The first and easiest to imagine is one in which matter does not move through the layer: there is no net transport. From the point of view of the gas there is only one requirement: pressure equality on the two sides of the layer, the density and the temperature may be quite different on the two sides; there is no connection. The layer is called a *contact discontinuity* and may be thought of as a thin membrane, or more politically, as an Iron Curtain. Contact discontinuities are frequently encountered in astrophysics.

The second form of coexistence is when there is net transport of particles (atoms, ions) through the separating layer; then one speaks of a *shockwave*. The same three laws (conservation of matter, momentum and energy) are valid, but they take a slightly different form since they bridge a true discontinuity. Imagine two planes, one called S_1, immediately upstream of the discontinuity, the other, S_2, immediately downstream from discontinuity. Call V the velocity of the gas, P its pressure, ρ its density; then we require that all the matter that passes through S_1 per second will also pass through S_2: conservation of matter, or

$$\rho_1 V_1 = \rho_2 V_2$$

Similarly require that the momentum transported by the gas through S_1 equals that through S_2: conservation of momentum, or

$$P_1 + \rho_1 V_2^2 = P_2 + \rho_2 V_2^2$$

Finally the amount of energy transported through S_1 equals that through S_2 -conservation of energy, or:

$$\rho_1 V_1 (\frac{1}{2} V_1^2 + \frac{3}{2}\frac{P_1}{\rho_1}) + P_1 V_1 = \rho_2 V_2 (\frac{1}{2} V_2^2 + \frac{3}{2}\frac{P_2}{\rho_2}) + P_2 V_2$$

(If you want to have this exposed in more detail: see Kahn's 1975 lecture). Accept the postulate that the density on the downstream side of a shock is always larger than on the upstream side, then it can be proven that $M_1 > 1$, where M_1 is the upstream Mach number (see above), and that $M_2 < 1$. Hence the gas moves supersonically into a shock, and subsonically out of it. It can be proven that $\rho_2 = \frac{4}{1+3/M_1^2}\rho_1$, and thus $\rho_2 \leq 4\rho_1$; only when $M \to \infty$ will ρ_2 approach $4\rho_1$. The jump in density is thus at most a factor 4, but the jump in pressure can be very large: $P_2 = (5M_1^2 - 1)P_1/4$, and because $M_1 > 1$ we always find $P_2 > P_1$.

A shock is thus a layer of negligible thickness in which the parameters ρ, P, V, T of the gas abruptly change. From a microscopic point of view the gas atoms were happily moving at the same velocity, with only a small random component corresponding to thermal motions. After the shock a large fraction of the systemic velocity has been converted in random thermal motions: the temperature has gone up dramatically. In a shock the entropy of a gas is suddenly, and irreversibly increased.

In astrophysical circumstances this model is too simple: because of the shock several things happen downstream and also upstream -and these things happen so close to the shock that the observer cannot separate the various layers. In this sense an observationally determined shock front is thicker, and contains more phenomena. But the motor of all the events is the shock front as just described. Briefly, these are the extra things that happen: downstream the gas that has just been made very hot will start to radiate because of collisions between the gas atoms. This will cool the gas, and while it moves downstream and cools, it will become compressed -thus ρ_2 will quickly increase, while T_2 will decrease - their product remains approximately constant. Upstream the gas flowing toward the shock may be subjected to photons radiated downstream, and this may lead to pre-ionization and predissociation.

A further complication will be the presence of magnetic fields. Suppose that the magnetic field lines are parallel to the shock front; just after the shock the gas has been compressed and because of the coupling of gas and field, the field has also been compressed. If the field is weak, which it often is, then the increase in fieldstrength has not yet any influence on the gas: $P_2 > B_2^2/8\pi$. But farther downstream B increases, because of further compression of the gas, and ultimately $B^2/8\pi > P$: the magnetic field acts as a cushion and prevents a further increase of the density.

This is the point to leave the subject. For a good and much more detailed description of shocks under interstellar conditions, I refer the reader to the review chapter by Shull and Draine (1987).

2.6 Turbulence

The reader who has read section 2.5 may have obtained the impression that regular, luminar flows are the underlying basis of galactic fluid dynamics. This impression is justified: gasdynamics often makes this assumption tacitly. Yet we know from observations how irregular flows may be -inhomogeneities are frequent, and unpredictable. This leads then to the problem of turbulence, one of the large, unsolved problems in physics. I can add nothing to solve the problem, but I refer the reader to the very worthwhile review, although difficult to read, by Scalo (1987).

3. THE GLOBAL DISTRIBUTION OF INTERSTELLAR MATTER

3.1. Overview

In this section I want to give observational facts concerning the distribution of the interstellar matter in our Galaxy. Four major components are distinguished: (1) the molecular gas, with hydrogen predominantly in the form of H_2 and at temperatures between 20 and 50k; (2) the atomic gas (H°) with a temperature below 10^4k; (3) the photo ionized gas (H^+) with a kinetic temperature of, say, around 10^4k, and (4) the hot gas, with H^+ at temperatures probably exceeding 10^5k. The first two components contain most (>90%?) of the mass of interstellar matter ($\sim 3.5 \times 10^9\ M_\odot$) with about an even distribution (50/50) between the two. Components (2) and (4) take up almost all of the interstellar space, but how this space is distributed between them is a matter of debate. Only a very small amount of mass will be contained in component (4); in some of its aspects it is fundamentally different from the rest, as will be demonstrated farther below. The molecular gas is probably the easiest to describe and this will be done in subsection 3.2. In subsection 3.3 I describe component (3) and in subsection 3.4 component (2). Component (4) is discussed in subsection 3.5; we know of this component only because the Sun happens to be inside a blob of it, and thus this discussion is at the same time a description of the local surrounding. The local surrounding is therefore a place of general interest!

Large scale surveys have been made of various components of the interstellar gas. Distinguish between continuum surveys and spectral line surveys, and begin with the latter. Of the various line surveys I select only the three most extensive: the CO (1-0) line at 2.7 mm, that traces H_2; the 21cm line of H°, and the H166α line that traces H^+ of component (3) (but not the H^+ of component (4)). The basic observation is that of a spectral line and thus the measurement consists of an intensity in a given direction as a function of frequency. Frequency can be translated into velocity via the Doppler relation and the basic piece of information is thus intensity, I, versus velocity, V, in a certain direction given by galactic longitude, l, and latitude, b. Interstellar gas takes part in galactic differential rotation, and in the ideal case that the rotation is in perfect circles around the galactic centre the rotational velocity is given by the the centripetal force of the matter inside the circle. In reality the rotation is not perfect, but discussing things only in zeroth order, we may ignore the deviations. (Readers that become suspicious already at this point are referred to the discussion by Burton, 1988). Differential rotation will give gas at a certain distance a well defined radial velocity at which it will appear in the line profile. In *principle* this allows one to derive the distance of the gas; in *practice* there are complications. The most severe is duplicity: in the direction of the inner Galaxy, thus in directions at $l < 90^\circ$, differential rotation predicts a maximum radial velocity along the line of sight which coincides with that of the point closest to the galactic centre (the "subcentral point"). Velocities below the maximum then correspond to one of two possible, locations -each well defined : to a point closer than the subcentral point and to a point farther away. The situation is graphically explained in figure 3.1.

For some features the ambivalence is solved by considerations about the extent in latitude: if the feature is quite broad in latitude it is probably on the near side, if it appears

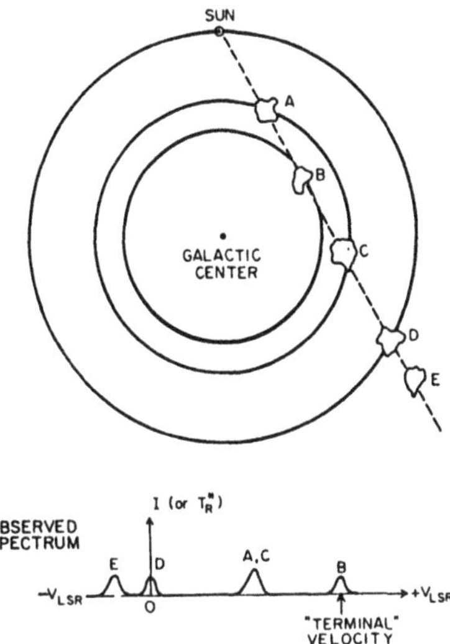

Figure 3.1. *(from Scoville and Sanders, 1987) Suppose that clouds of interstellar matter are located along a line of sight at the positions A through E. Due to galactic rotation each has a definite radial velocity with respect to the Sun. If a cloud emits a spectral line the Doppler effect will shift the velocity of each line; the net result will be the "observed spectrum" shown in the lower part of the figure. Cloud B is closest to the galactic center at the so called tangential point; it has the highest radial velocity. A and C are at the same distance from the center and have the same radial velocity; hence from the observation one cannot derive their distance in a unique way.*

narrow, it is probably on the far side. Another, better argument is occasionally available: one can make an association with an object (an early type star or an HII region) for which some distance indication can be obtained. A brief word here about presentations. Because we measure the intensity, I, as a function of three variables: l, b, and V, and because we can show in a graphical diagram the dependence of I of at most two variables we need more than one diagram to fully represent the data. Thus three diagrams are in use, in each of which I is given as a function of two other variables: (l, V), or (l, b), or (b, V) diagrams. Teaching practice has convinced me that the (l, V) and (b, V) diagrams need some time and exercise to get used to; once you understand them, you will find them indispensable. As an exercise in understanding (l, V) diagrams ask yourself what the locus is of the radial velocities (as seen from the Sun) of a ring of material at some given distance, R, from the galactic center, when that material moves in perfect circles.

Figure 3.2 shows a set of CO (1-0) line profiles, all in the galactic plane. Notice the variation in appearance as one moves from one longitude to the other. An even more elaborate demonstration of the systematics of galactic rotation is seen in fig. 3.3.

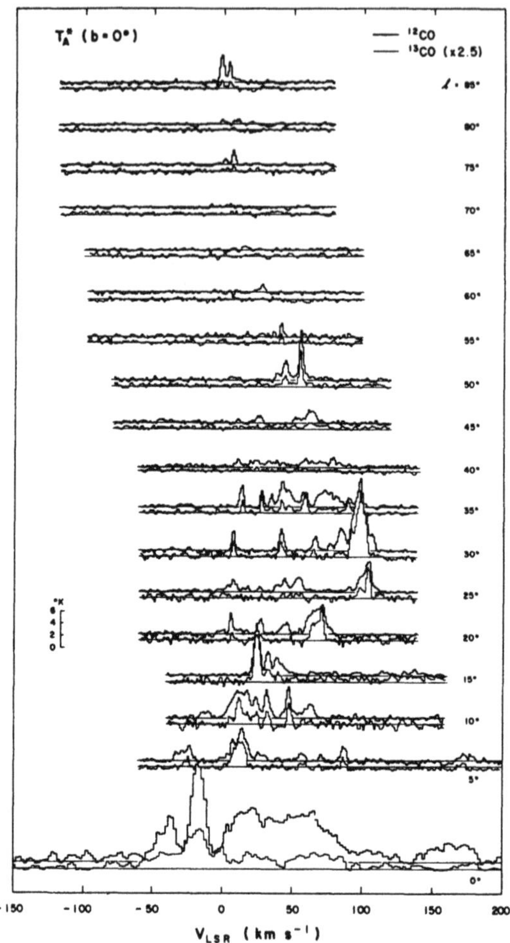

Figure 3.2. *(from Scoville and Sanders, 1987) Spectral line measurements of* ^{12}CO *and* ^{13}CO *(J=1-0) line in the galactic plane at various, consecutive longitudes. Only positive radial velocities occur; this means that all gas is within a circle around the galactic center with radius* $R_\odot$*, the distance of the Sun from the center; most of the gas appears to be between* l *= 20° to 50°, and this implies that most gas is confined to a ring between 3 and 6 kpc from the galactic center.*

It gives you a splendid conformation of galactic differential rotation: intense radiation by matter at zero velocity (or close to it) all around the sky: local material. Distant material is contained in the sinusoidally confined region from $l = 90°$, $V = 0$ km s^{-1} to $V = +120$ km s^{-1} at $l = 30°$, to $V =$ -120 km s^{-1} near $l = 330°$, and up to zero velocity again at $l = 270°$. Finally, notice the huge velocity extent near the galactic centre, showing that this is a very special place indeed.

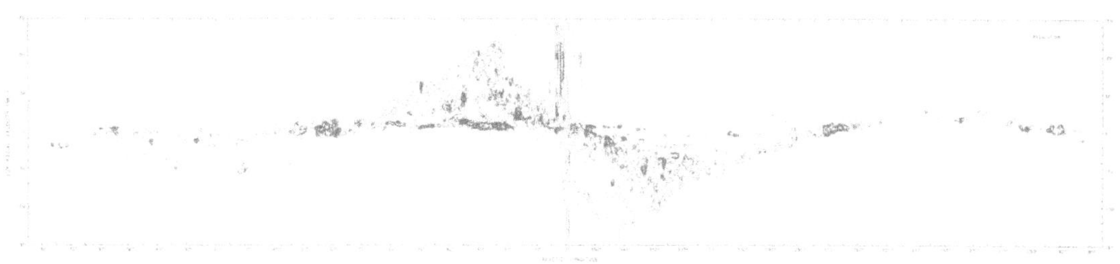

Figure 3.3. *(from Dame et al., 1987). A compilation of* ^{12}CO *(J=1-0) spectral line observations showing contours of intensity versus longitude and velocity- a so called* $l-V$ *diagram. All observations have been made in the galactic plane. Compare with figure 3.2 to better understand the present figure.*

In addition to line radiation there is continuum emission -radio continuum surveys, infrared surveys and γ-ray surveys are examples. To derive the spatial distribution for this case is one dimension more difficult than for line emission, because the velocity information is missing completely. Thus the basic information consists of an intensity I as a function of l and b. More or less successful attempts have been made to derive the distribution of the emisivity, ϵ, as a function of R and z (cylindrical coordinates in a system with the galactic centre as origin and the z-axis in the direction of the galactic rotation axis): as a first guess cylindrical symmetry is assumed, but iterations can often be made -those interested should turn to Beuermann et al. (1985 -radiocontinuum surveys) or to Deul (1988 -IRAS infrared continuum survey).

I close this introduction with a summary of what is coming: see figures 3.4 and 3.5.

Figure 3.4 (from Burton, 1988) gives the distribution of densities in the galactic plane as a function of R -the distance to the center of our Galaxy. I will comment on this figure in the subsequent subsections.

Figure 3.5 (from Dame et al., 1987) gives a synoptic view of four different sky surveys. Keep in mind that the top diagram is based on *optical* observations and thus does not penetrate deeply into the Galaxy; the other three diagrams contain information from far beyond the galactic center. Notice that in all four diagrams the radiation is strongly limited to the galactic plane ($b = 0°$), is centered on $l = 0°$ and more or less symmetric around this longitude. Notice also that most of the emission is concentrated between $l = \pm 90°$, indicating that the Sun is close to the outer boundary. If the 21cm map would have been added to figure 3.5, you would find that it is much more extended in longitude -the H° layer extends farther out -as you will find confirmed in figure 3.4.

3.2. The galactic distribution of molecular hydrogen.

Roughly one half of the interstellar matter in our Galaxy is in the form of discrete clouds of molecular hydrogen, H_2. In this subsection the reader is given an idea of the large scale,

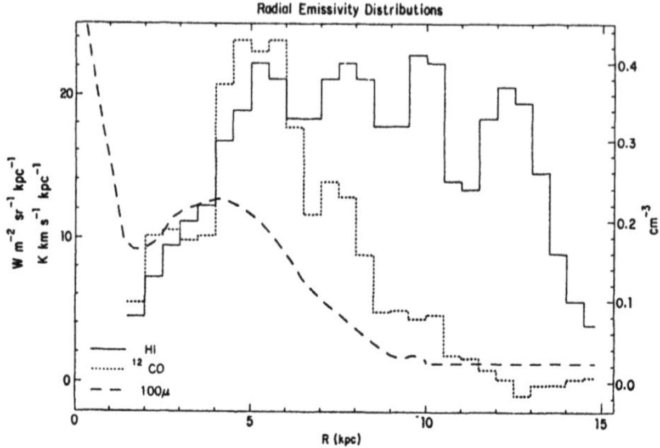

Figure 3.4. *(from Burton, 1988). The emissivity of various components of the interstellar hydrogen gas shown as a function of galactic radius. Below 5 kpc the emissivity of both molecular and atomic gas decreases, but (this is not shown) the emissivity reaches a maximum again at the galactic center itself. The molecular gas reaches a maximum emissivity around 6 kpc and then drops rapidly; there is very little gas seen outside of 10 kpc. The atomic gas has a constant emissivity from 5 to about 13 kpc and then begins to decrease slowly.*

overall, galactic distribution of the H_2. The discussion will be based only on observations of two lines of CO, at 2.6 and at 1.3mm and it is assumed that there is a fixed, one-to-one correspondence between CO and H_2.

Analyses of the galactic distribution of the CO emitting gas are best done from complete data sets: full coverage of the galactic plane (out to several degrees in latitude), and full coverage in velocity. The first of these requirements has been proven to be very difficult to meet: up to now radio antennas are one pixel detectors (in contrast to the eye or to photographic plates) and in practice this pixel is small: a large pixel may be a few arcmin in diameter (i.e. a few mm on a Sky Survey Plate!). To collect all the data and to reduce them is time consuming. Therefore surveys have become "complete" only gradually over many years. Nevertheless several key features of the derived distribution have not changed for several years and may be considered well established.

To begin with, take a closer look at figure 3.2. One notices that each line appears to consist of discrete features; only at the galactic center they melt together. The discreteness of these features leads to the conclusion that the CO gas (and thus the H_2 gas) is contained in discrete units, called "clouds", or sometimes, "cloud complexes". As we will see, this is much less the case for the H° distribution. It is now quite clear that the CO-clouds seen in figure 3.2 are observed in our Solar Neighbourhood as dark clouds -the ones that Barnard and Wolf studied photographically early in this century. I refer to the very readible paper by Dame et al. (1987) in which they identify all CO clouds of low velocity $< 20\mathrm{kms}^{-1}$ with well known dark clouds, or complexes of dark clouds. Read that paper and satisfy yourself with the conclusion that what you see on photographs agrees with what one measures in a CO survey. This paper gives a truly beautiful panorama of our Galaxy.

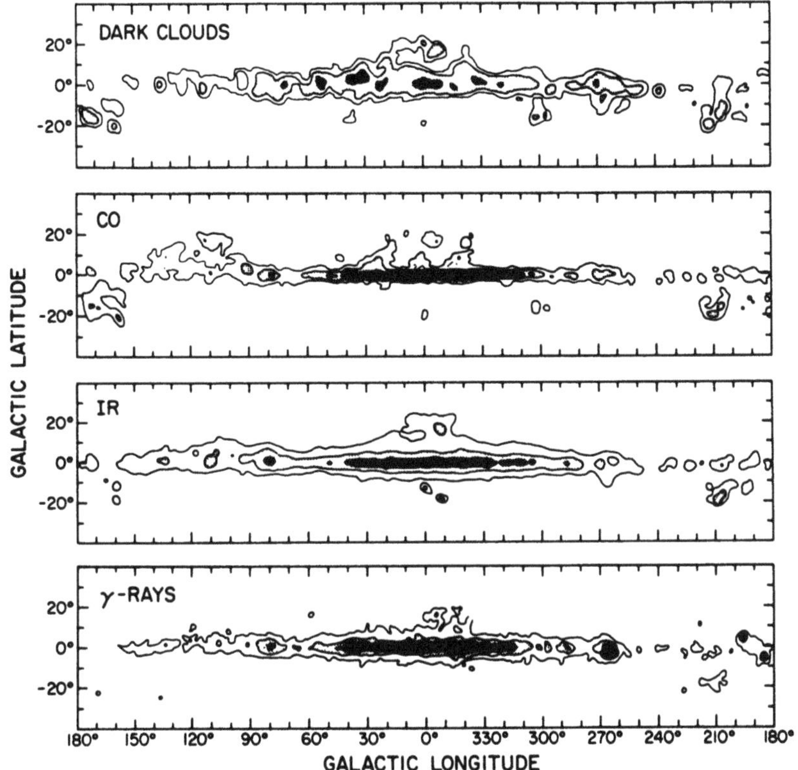

Figure 3.5. *(from Dame et al., 1987). A compilation of four contour diagrams of intensity versus longitude and latitude ($l-b$ diagrams) describing different tracers of the interstellar gas. Dark clouds are noticed mainly from dark areas on photographic plates. CO means: line fluxes (line profiles integrated over velocity); it traces the molecular clouds. IR means: emission at 100 micron; it traces emission by dust. γ rays trace a product of interstellar density and cosmic ray density.*

The large scale properties of the CO-cloud ensemble can be described as follows (Dame et al., 1987; Scoville and Sanders, 1987): the clouds form a flat disk with an scale height (full-width at half maximum) varying from $\sim$ 60 parsec in the inner Galaxy to $\sim$ 150 parsec at the position of the Sun and farther out. Most CO is contained between 4 and 8 kpc, a region often called "the molecular ring". There is some, but little CO outside the solar radius, that is at $R \geq 8.5$ kpc (see figure 3.4). An important question concerning the large scale distribution is whether the clouds are confined to spiral arms? Scoville and Sanders (1987) show that in the inner Galaxy cool clouds, i.e. clouds with a kinematic temperature below 15 k, scatter very much in the (l, V) diagram, but that hotter clouds in the diagram are arranged in longish structures, indicating that their motions are ordered and that the clouds are arranged in spiral arms. Dame et al. (1987) discuss also the nearby CO clouds and conclude that even the clouds within 1 kpc are arranged in spiral structure -see figure 3.6.

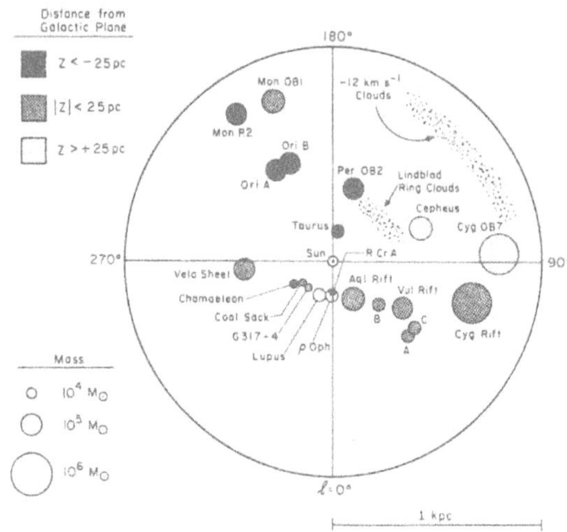

Figure 3.6. *(from Dame et al., 1987). A map of a circular area in the Galactic plane, within 1 kpc from the Sun. Shown are several large molecular clouds or cloud complexes. Spiral structure can be recognized. Several objects have names indicating that they are not only seen in molecular lines at radio wavelengths but also optically ("Coal Sack", "Cygnus Rift").*

Catalogues of individual clouds are given by Scoville et al. (1986), by Grabelsky et al. (1988) for the Carina region, and by Dame et al. (1987) for the local region, i.e. within 1 kpc. Comparing the lists of clouds by Grabelsky et al. and by Dame et al. one is struck by the difference in mass between the two sets (the cloud masses of the two lists have been derived in similar ways, and ought to be comparable); whereas Grabelsky has 42 clouds with masses ranging from $0.7 \times 10^5 M_\odot$ to $109 \times 10^5 M_\odot$, with a median value of $7.0 \times 10^5 M_\odot$, Dame et al. find a range 0.03×10^5 to $8.7 \times 10^5 M_\odot$, with the median at $1.0 \times 10^5 M_\odot$. These differences will be partially due to selection: in the distant Vela arm studied by Grabelsky et al. small clouds will remain unnoticed; they might be present but escaped detection. However, the absence of heavy clouds in the local surroundings is probably a real effect: if they existed, they would have been seen. One conclusion is thus that further inward in the Galaxy very massive clouds ($> 10^6 M_\odot$) occur more frequently than at our distance from the center. The same conclusion holds, mutatis mutandis, for HII regions, as we will see in subsection 3.2.

There are some other, quite interesting conclusions about our Solar Neighbourhood from the work by the Columbia group (Dame et al., 1987). The first is that the local CO gas follows Gould's Belt -a belt of OB stars, dust and gas extending over the full sky, resembling a disk tilted over some 30° with respect to the galactic equator. The origin of this Belt is unknown, but the reality of the feature is not in doubt. Second, locally (that is, within 1 kpc) there appears to be four times more gas in the northern than in the southern hemisphere. This is partially explained by the Sun's position with respect to two nearby pieces of spiral arms – divide figure 3.6 in a northern part and a southern part by lines from the galactic center in the direction $l = 220°$ and $l = 20°$, and you see the difference.

Larson (1981) has shown that the internal densities, n, and the velocity dispersions (linewidths, ΔV) of clouds scale with the size l of a cloud, as $n \propto l^{p}$, $\Delta V \propto l^{r}$. He also shows that the mass of clouds scales with ΔV. The last relation has an interesting physical explanation: consider the cloud as a spherical object with turbulent parts, and use the virial theorem to work out the theoretical relation between $\Delta V, l$ and $M(\Delta V =$ (G $M/l)^{1/2}$) where G is the gravitational constant). Then it turns out that this is also, approximately, the relation that is observed: The conclusion is that the molecular clouds are bound together by their large mass and that there is no need for an external pressure to keep the cloud together. Larson's relation has now been tested on a large number of clouds (see e.g. Grabelsky et al., 1988) and found to be approximately correct. The significance of the relations found by Larson may be very deep -see Scalo (1987).

3.3. The galactic distribution of ionized gas (H^+)

In this section I discuss the gas that is ionized by photons with energies of at most a few times the Rydberg energy; the gas has a temperature between, say 7000 and 12,000k. Much hotter ionized gas also exists: see section 3.5. The 10^4k ionized gas manifests itself by radiation coming from encounters between electrons and ions: (1) free-free radiation or bremsstrahlung, a continuum emitted by electrons and photons (or other ions) passing each other; (2) recombination line radiation, when the passing electron is caught by an ion and an highly exited atom is the product; (3) collisionally excited lines emitted by various non hydrogenic ions from levels very close to the ground level. Examples of (2) are the Lyman, Balmer, Paschen and Brackett lines, but also the recombination lines from very highly excited levels of hydrogen and with wavelengths in the cm range. Examples of (3) are forbidden lines from S^+, O^+, O^{++}, Ne^+, etc. – for a (still) useful compilation see Petrosian (1970).

Here I discuss the global distribution of ionized hydrogen. Since visual observations are so severely limited too small (galactic) distances by interstellar extinction, one has to turn primarily to radio observations; hopefully in the near future, when ISO flies, infrared spectroscopy will also be possible. Radio continuum studies can in principle be used, but because the continuum is always the sum of bremsstrahlung and synchrotron radiation it is very difficult to obtain a convincing picture of the distribution of either. Of more use are radio recombination lines, for example photons produced in the transition $n = 158$ to $n = 157$ (called H 157α) with a wavelength of 18 cm according to Rydberg's equation. These radio recombination lines are very weak and more difficult to measure than the continuum emission, but they are much more valuable because they contain information on the radial velocity of the gas. A few surveys in radio recombination lines have been made; the sampling has been much less complete than for the CO mm line and for the 21cm line.

Basically one distinguishes between discrete sources ("HII-regions") and extended emission. The discrete sources are (pieces of) molecular clouds ionized by young, massive stars just formed. One of the best studied nearby HII regions is the Oron nebula at a distance of 500 pc -see for a recent review Genzel and Stutzki (1989). In this case it is clear that not all of the hard radiation emitted by the stars is used up for photoionization in the nebula. A significant fraction of the hard photons escape the local surroundings. Such photons are thus available for photoionization elsewhere. The case is not only true for Orion, but is much more general: Leisawitz and Hauser (1988) analyze the heat input into molecular clouds by young stars associated with the cloud using IRAS infrared data. They estimate that only 1/3 to 1/5 of the stellar output is dumped with the molecular cloud; most of the output escapes into the interstellar medium. Thus it is likely that in the Galaxy there exist large areas of low density hydrogen, ionized by these stray photons.

I will now discuss first the discrete regions and second the extended emission. Two very useful review papers are that by (Gordon 1988) and those by Kulkarni and Heiles (1988a, 1988b).

3.3.1. Discrete HII regions

The first radio surveys were at meter wavelengths and showed exclusively non-thermal, synchrotron radiation. In 1958 Westerhout made the first radio continuum survey at cm wavelengths. He discovered a large number of sources with a "thermal spectrum", sources to be identified with ionized nebulae. Many of the "strongest" HII regions in the Galaxy carry the name from the Westerhout catalogue (W3, W33, W49, W51). This survey demonstrated for the first time that radio sources could reveal the HII regions deep inside the Galaxy that are hidden to the optical telescope -a complete census of HII regions is possible. In subsequent years more radio surveys were made and, as already mentioned, the recombination line surveys proved to be very worthwhile (Wilson et al., 1970; Reifenstein et al., 1970).

Figure 3.7 shows the distribution of discrete HII regions in the plane of the Galaxy, based on optical observations for the nearer objects and radio emission for the objects further into the Galaxy (Georgelin and Georgelin, 1976). Figure 3.7 shows clearly that the stronger sources (bigger symbols) are confined to spiral arms, whereas the weaker sources (smaller symbols) donot show this tendency. We have seen the same before in section 3.2: molecular clouds inside spiral arms are more massive than outside; similarly HII regions are "stronger".

A "strong" HII region means that the radioemission (corrected for distance) is strong; physically this implies roughly the following: bremsstrahlung photons or recombination line photons are emitted in proportion to the rate of electron/ion encounters, thus they are both proportional to the product of n_e n_i V, where V is the volume of the nebula and where n_e and n_i are the density of electrons and ions, respectively. The total flux of bremsstrahlung from a discrete HII region is thus proportional to the total number of recombinations inside the nebula. Because the rate of recombination equals that of photoionization (large scale equilibrium must be supposed), we conclude that the total amount of radio radiation received is proportional to the total number of ionizing photons absorbed. An HII region is a counter of ionizing photons! Larger symbols in fig. 3.7.

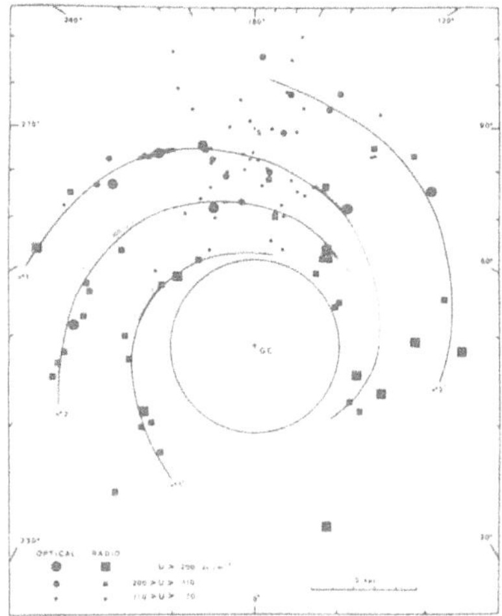

Figure 3.7. *(from Georgelin and Georgelin, 1976) A map of the galactic plane with the positions of discrete nebulae of ionized hydrogen (H^+ or HII regions). The regions have been seen optically in $H\alpha$ and/or in a radio recombination line. The quantity U is a measure for the consumption (and thus for the production) of Lyman continuum photons inside the nebulae.*

mean larger productions of hard (UV) photons. In this way it is possible to derive the distribution of newly born ionizing stars in the inner Galaxy.

Individual HII regions have been studied in various ways -optically, radioastronomically and in the infrared. There is a plethora of information, and yet fundamental questions can be answered only partially: in every observation with sufficient accuracy the object appears to be inhomogeneous: often the ionized gas is created inside a cave, carved out of a molecular cloud by the newly formed stars -the Orion nebula is a prime example (see e.g. Gordon, fig. 2.28). The gas flow is inhomogeneous and consists of clumps and low density regions. No wonder that the models proposed are never fully satisfactory; the reality here is too complex to be summarized.

Models may never be fully satisfactory, they could still be adequate for some purpose. Are they? Gordon gives a very useful example -two different models to explain the radio recombination line radiation from Orion. The two models are of similar complexity and both explain the observations sufficiently accurately and yet: they differ significantly. This tells me that I want to be careful about conclusions drawn from HII regions -especially if HII regions are compared for which only a limited amount of observations are available. This remark bears upon the following point: how strong radio recombination lines are in relation to the radio continuum is a function of the electron temperature: a low electron temperature (say 8000k) gives (relatively) strong lines, a high electron temperature gives weak lines. Thus the ratio is a measure of the electron temperature, T_e. Now it has been found that the temperature so derived is higher when the HII is farther from the galactic center, and lower when it is closer, see fig. 3.8.

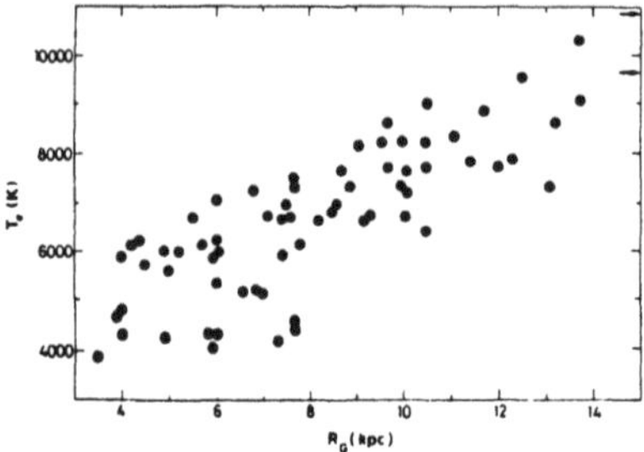

Figure 3.8. *(from Shaver et al., 1983). The kinetic temperature, T_k, of the electrons in HII or H^+ regions increases when one moves outward in the Galaxy: this figure shows how T_k varies with R, the distance to the Galactic center.*

If the effect is real, than this has important consequences. But is it real? The authors in this case (Shaver et al., 1983) have been careful in their methods and think the effect to be real but I remain cautious. In any case the interpretation of the effect is less debatable: it probably means that ionized gas farther out in the Galaxy (at larger R) cools less efficiently, because the ions to do this (see the discussion in section 2.4) are less abundant: an increase in R implies an increase in T_e and thus a decrease in abundance. The "metallicity" in our Galaxy decreases with R. Figure 3.8 seems to be the strongest piece of evidence in favour of this expected result. I am willing to look at it as interesting, but not yet fully convincing evidence. I hope that ISO will be able to measure the variation of metallicity much more directly through infrared lines from excited ions.

3.3.2. The diffuse photoionized gas

Are there large volumes of interstellar space filled by low density (say: $< 1cm^{-3}$) ionized gas outside of the dense HII regions? This question is probably as old as the discovery of the interstellar lines by Hartmann in 1904. Photons to ionize low density hydrogen gas are available: they escape from HII regions- see the introduction to section 3.3. Definitive evidence that such volumes exist was harder to get. Probably the best evidence that we have at presence are observations of weak, and broad optical and radio recombination lines: The radio lines were detected simultaneously by Gottesmann and Gordon (1970) and by Jackson and Kerr(1971). Especially the broadness of the lines indicates that the distribution of the ionized gas is very extended; because the lines are so weak little gas appears to be involved.

Figure 3.9 gives an overview of the values measured for the H166α line by Lockmann (1976). As Gordon (1988) makes clear, one measures only the product $n_e^2\ T_e^{-3/2}\ d$, where d is a distance along the line of sight, n_e the electron density and T_e the electron temperature. Because one does not know the strength of the continuum emission one cannot separate temperature and density effects and the basic parameters of the medium can only be derived by indirect arguments. Therefore there is great interest in recent measurements

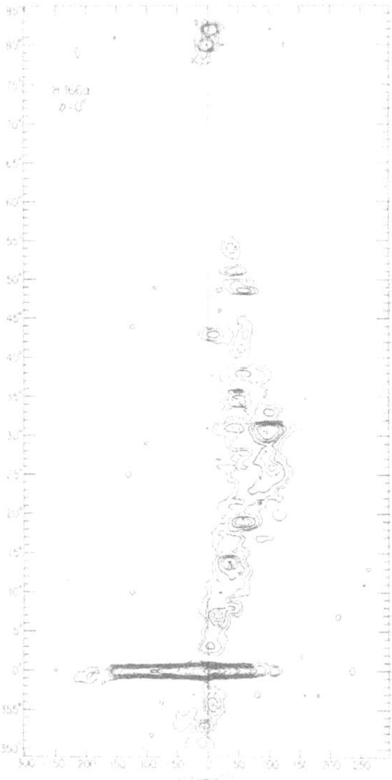

Figure 3.9. *(from Lockman, 1980) The $l - V$ diagram (see the legend to figure 3.3) of the intensity of the H166α recombination line. It has the same features as figure 3.3 but much less detail, largely because the signal is so weak.*

of optical lines of the diffuse ionized gas by Reynolds and coworkers (see e.g. Reynolds, 1983, 1984, 1985). The measurements refer only to the solar neighbourhood, but it seems fairly safe to assume that these lines originate in a similar medium as the diffuse radio recombination lines although there might be a difference in density in the sense that the radio lines refer to regions of somewhat higher density and degree of ionization than the optical lines. Mathis (1986) has pointed out that if this medium (which he calls "DIG", or Diffuse Ionized Gas) is ionized by distant O stars (and he estimates that there are enough such stars), then the ionization conditions are quite different from those in normal, i.e. dense HII regions where the ratio between the density of ionizing photons and the mean electron density is much higher. This has interesting consequences for the line strengths of ionic lines relative to the Balmer lines in the DIG, and indeed, Reynolds' measurements then confirm the conclusion that the DIG is ionized by stray photons from distant O stars. Another conclusion from Mathis' work is that the fractional ionization of the gas is significantly lower than in the dense HII regions, where it is >99.9%. The low density

DLG gas is rather similar, but perhaps of even lower density than what Mezger (1978) has called the ELD regions ("Extended Low Density HII").

It has already been remarked that the recombination lines by themselves give multi interpretable information. Therefore another piece of evidence is of great importance: from measurements of the delay of pulsar signals as a function of the frequency of the observation one can derive the total number of free electrons between the pulsar and us (see section 2.1.3). An analysis of this number as a function of pulsar distance and galactic coordinates then gives the distribution of the free electrons averaged over volume elements of some size. The result obtained by Lyne et al.(1985) is given here:

$$n_e = [0.025 + 0.015\exp(-|z|/70\mathrm{pc})][\frac{2}{1 + R/R_\circ}]\mathrm{cm}^{-3}$$

Especially the first term is interesting because of its large scale height in the z-direction, i.e. perpendicular to the galactic plane. The scale height is even infinite in this equation but that is only appearance: this component of the free electron distribution extends beyond the pulsar distribution (say beyond 400 pc). The second term in the equation above is thought to be associated with HII regions. The pulsar evidence for the free electrons is strong. But it is not proven (although I think that it is likely) that the electrons seen against the pulsars are the same as those measured by the recombination lines.

The properties of the diffuse ionized gas are elusive. Tentatively the following picture emerges: there is a diffuse medium, with electron densities as derived from the pulsar measurements. Optical recombination lines confirm that the gas is probably ionized by stray photons coming from distant O stars. The radio recombination lines suggest that it is wide spread throughout the Galaxy. The pulsar measurements give evidence that the ionized gas has a very wide distribution in the direction perpendicular to the galactic plane. The relation with the neutral gas is obscure at present. It is obvious that the subject needs more study. In this respect it is significant that in the very good book "Galactic and Extragalactic Radio Astronomy" two subsequent excellent chapters (one by Gordon, the other by Kulkarni and Heiles) each discuss the diffuse ionized medium, but do not refer to each other, and even contain non-overlapping lists of references. Perhaps the problem of the distribution of diffuse ionized gas can be solved best by observing other galaxies –e.g. M31 is a prime candidate for optical and radio observations (R. Braun, private communication).

3.4. The galactic distribution of the atomic component (H°)

The neutral atomic gas in the Galaxy has been studied extensively, because the 21cm line is easily observed in emission and does not suffer from interstellar extinction. Gas at the other side of the Galaxy is often detected. The basic measurement exists, as before, of an intensity as a function of velocity. This can be translated into atomic hydrogen density, n_H, as a function of distance along the line of sight. Many such analyses have been made; for a review see Burton, 1988. A fundamental difference between 21 cm H° line profiles and 2.7 mm CO line profiles is that the 21 cm line profiles are so much more smooth:

the CO profiles appear ragged, indicative of the presence of discrete clouds. The presence of discrete clouds in H° is well proven -the evidence emerges when 21 cm absorption line profiles are discussed; see section 4. However, a large fraction of the H° has a smooth, continuous distribution. The question "cloud" or "continuum" has been discussed from the moment of discovery of the 21 cm line (in 1951) and the discussion is not yet closed. In any case, the observed smoothness of the line profiles is not in debate, but it causes a large stumbling block for analyses. The ambivalence about the distance derived from radio velocity (see the introduction to this section), and the possible effect of deviations from pure rotational velocities, make the detailed derivation of n_H cumbersome (see e.g. Burton, 1988). In addition one has to take into account that optically thick, cold parcels of gas may block the view to deeper layers, without the possibility of noticing such parcels directly. All in all there are quite a few problems attached to the derivation of the interstellar atomic hydrogen density and for example the existence of a spiral pattern in the H° distribution is an unsolved problem. Yet the situation is not so bleak that a few good numbers cannot be derived.

In the inner Galaxy, that is, within the solar circle, the HI gas is confined to a thin (full width half maximum: 370 pc), flat (midpoint within a few parsec from the geometric plane) layer. The constant thickness is remarkable: the thickness is the consequence of an equilibrium between the gravitational force trying to bring the gas in the plane, and turbulent gas motions or magnetic and cosmic ray pressures acting to blow it up. The gravitational force increases strongly when one moves radially inward, but the turbulence remains the same. The constant thickness thus indicates that the increase in gravitational force is off set exactly by an increase in the cosmic ray pressure or in magnetic field strength, a very happy coincidence! (In the side line I remark that also the disk of population I stars has a constant thickness, probably because an increase in random motions of the stars perpendicular to the galactic plane off sets the increase in gravitational force). Inside a circle of about 3.5 kpc from the centre there is hardly any (atomic, ionic or molecular) gas, except when one approaches the centre within 500 pc. (The gas very close to the centre has some different properties and will be discussed separately in section 3.6.) An intriguing property of the gas appears when the highest velocity in the emission line profile is considered; this velocity corresponds to the point along the line of sight closest to the galactic center (the so called sub-central point): although the gas is very distant it is found at relatively high latitudes, and the conclusion is that there is neutral gas at large distances from the plane: about 13% of the atomic gas is at z >500pc, and thus in the lower parts of the galactic halo. The thickness of the layer of H° cannot be described by a single gaussian, but in addition an exponential function is needed with a scale height of approximately 500 pc (Lockman, 1984). Of course one is reminded of the similar large extent of the free electron gas, as derived from pulsar measurements (see section 3.3.2). There is additional, though less convincing evidence that the same large extension still exists at the solar circle, or at least above our heads, in the directions of the galactic poles. Outside the solar circle the gas remains, at first, in the same layer in which now some extended segments of spiral arms are seen. The thickness of the layer rises slowly with R, until approximately R=15 kpc where the layer bends away from the geometrical plane and increases dramatically in thickness ("the galactic warp"). In the northern hemisphere the warp is towards the north galactic pole, in the south it is directed to the southern poles (see Burton, 1988, section 7.5).

In the 1950's Muench (see Muench and Zirin, 1961) discovered interstellar absorption lines in early type stars far from the galactic plane and argued that the corresponding interstellar clouds are probably at large (kiloparsec?) distances from the galactic plane. (This discovery led Spitzer in 1956 to propose the existence of a gaseous halo around our Galaxy- see the discussion in section 5.4.) In the early 1960's a search began in the Netherlands for 21cm line emission from such halo clouds. Large numbers have been found of so-called "High Velocity Clouds", objects at high galactic latitudes with velocities up to several hundred km s^{-1}. It is still not established beyond doubt that these clouds are connected with our Galaxy although that hypothesis is generally adopted- see the discussion in Burton (1988, his section 7.5.3). If they belong to our Galaxy they fall into it, and carry with them a significant amount of interstellar matter, about one solar mass per year, and in the Hubble time these clouds add more than the total amount of interstellar matter presently in our Galaxy.

From 4 to 13 kpc the column density, n_H integrated over the thickness of the layer, is quite constant and estimated at $4.5x10^{20} cm^{-2}$. The total amount of hydrogen in our Galaxy is estimated at $3.6x10^9$ $M_\odot$, or only a few percent of the stellar population. In deriving these values the optical depth effects have been taken into account, as best as possible.

3.5. The hot component and the local neighbourhood

About the immediate surroundings of the Sun, say within a few hundred parsec, there is information available in much more detail than about regions further away. It therefore is worthwhile to try to paste these data together. Such a formulation shows already that the result will have quite some uncertainty but see for yourself the results in a few recent reviews: Cowie and Songaila (1986), Cox and Reynolds (1987) and Savage (1987). The information that we have are 21cm emission line observations, CO radio line observations of a few nearby molecular clouds, measurements of interstellar absorption lines in a large number of stellar spectra, measurements of interstellar extinction and polarization and measurements of soft X ray emission of a local nature. The analysis of the emission (CO, 21cm, soft X-rays) is hampered by the unknown distance of the radiating gas: because the gas is so nearby *systematic* effects on the velocity are smaller than random effects; since distance is always derived from a systematic effect the derivation fails for small distances. This problem is less serious in the absorption measurements: there the distance of the star is an upper limit. Searching then for interstellar lines in the spectra of nearby stars the surprise is that there are no lines or at best very weak lines. Nevertheless it has taken some time, and probably some courage, to reach the conclusion that there is no, or very little interstellar mattter witin say 100 parsec from the Sun: the Sun is inside a hole in the interstellar medium. Polarization measurements (Tinbergen, 1982) of nearby stars confirm that there is very little interstellar material within about 50 parsec; atomic densities of less than $0.3cm^{-3}$ are indicated. A third piece of the puzzle, overlooked at first, is that 21cm observations at high galactic latitudes had shown that in certain areas all the gas is at velocities of 30-50 km s^{-1} and that it looks as if the local low-velocity gas has been removed by some agent and put at intermediate velocities (Wesselius and Fejes, 1973).

In spite of these indications it came as a surprise when highly ionized, interstellar atoms were discovered in spectra taken with the Copernicus satellite in the far UV: at about 103nm two lines were detected of O^{5+}. The lines are interstellar, because they were found in many nearby stars and the line strength increased with stellar distance. The ionization potential of the ion is 114 eV and thus the presence of very hot (10^5k) gas is indicated (Jenkins, 1978a, 1978b). At the time of this discovery there came a matching one: soft X ray emission is seen over a large fraction of the sky and is almost certain of local origin. These two different pieces of evidence fitted the jig saw puzzle and lead to the hypothesis that the Sun is inside a bubble (the "local bubble") of very hot, highly ionized gas ($T > 10^5$k), see figure 3.10.

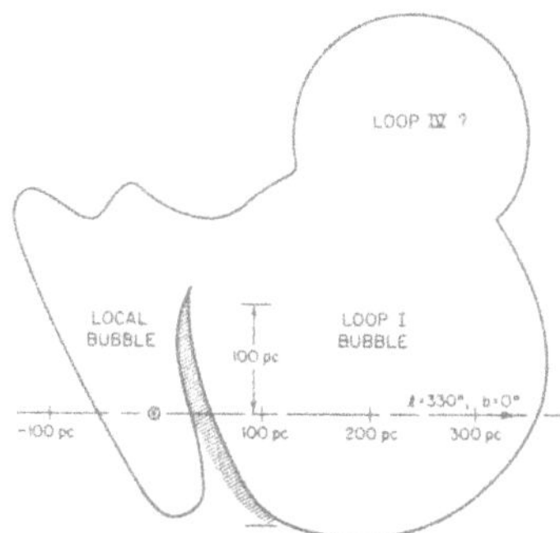

Figure 3.10. *(from Cox and Reynolds, 1987) Schematic presentation of two hot bubbles of supernova heated gas,the Sun being inside one of the two (the Local Bubble). The presentation is only a sketch; several aspects are still being debated.*

Since these discoveries it has become clear that even closer to the Sun (say within a few parsecs) there is some neutral, atomic hydrogen, but its density is very low; it is sometimes called the local "fluff".

Another intriguing component that has not yet been fitted into the picture is the discovery by the IRAS satellite of interstellar cirrus, thin threads of cold material, ubiquitous and local: distance estimates indicate distances below 100 pc. Part of the cirrus material is even molecular. Is there cirrus inside the local bubble? Other questions also appear: How did the bubble come about? And is the Sun located in unique surroundings? To both questions the answer is unknown, but some hypotheses have been made: the bubble could have been produced by a supernova explosion several tens of million years ago and has been rejuvenated since then by another supernova that went of inside the dying bubble, as we will see in section 5. This situation may not be unique: interstellar space might be full of hot bubbles; it has been proposed that the interstellar medium contains a network of bubbles connected by tunnels.

3.6. The Galactic Center

In sections 3.2 and 3.3 it has been discussed that the interstellar matter distribution in our Galaxy resembles a disk with a hole: for R <3.5 kpc there is a significant decrease in gas density. Only when one approaches the center within ≈1500pc does the gas density increase again. A good explanation for the hole does not exist. The mass of the Galaxy, i.e. its stars, do not show the hole at all. The edge of the hole seems to coincide with the beginning of the galactic bulge, i.e. the region where the gravitational potential becomes more spherical.

The gas inside the hole, and especially the dense, intensively radiating gas close to the galactic center has been the subject of many studies. The results are complex – a statement that does not imply that the results are not understood. In fact it is remarkable how many of the complicated measurements have found a solid interpretation. Because of this complexity and the fact that the galactic center is only one of the topics to be discussed in a short course, I will skip most of the detailed discussions that the reader can find elsewhere. There is an excellent review paper by Genzel and Townes (1987) and a beautifully complementing paper by Liszt (1988). Here is a summary:

Between 1500 and 200 pc from the center there is much atomic hydrogen gas which appears to form a thin disk tilted over 30° with respect to the galactic plane. The disk is not in pure rotation, there are non-circular motions; one possibility is that the matter follows elliptical orbits around some barlike mass distribution. In total an amount of about 4 x $10^7 M_\odot$ of atomic gas is involved. There is much more molecular gas (Liszt gives an estimate of 4 x $10^9 M_\odot$) but its velocity/ longitude/ latitude diagram has not yet been interpreted- it is very complex.

Within 1.5° (or 260 pc) again the picture changes. Liszt (1988) gives a beautiful mosaic of the various centers of activity (SgrA, B, C, D, E) all stretched out like pearls on a string along the galactic plane between l = +1.5° and l =-1.5 °. Sgr B is a well studied complex of molecular clouds and HII regions, several of which are very small, and probably recently formed around newborn O stars; the molecular cloud Sgr B has long been a goldmine for futher discovery of molecules. Sgr A is where the center of our Galaxy is. Its most intriguing component is a time variable radio source of very small dimensions known as "Sgr A West". It is generally expected to be radiation from matter in the immediate surroundings of a black hole, which then probably is the center of our Galaxy. Very precise position measurements obtained with VLBI techniques may ultimately help to prove that it is also the dynamical center and not only "the thing in the middle". In addition to Sgr A West there are several other components to be distinguished in Sgr A, for example the source Sgr A East which is thought to be the remnant of a supernova. Surrounding the source is a rotating disk of molecular material of about 5 parsec diameter or 2 arcmin. Streams of ionized gas have been seen that may spiral inward from this disk to the center.

There is a temptation to go on, but I stop: the reader is strongly advised to read the two review papers mentioned above and all the other papers he will then be refered to.

3.7 The galactic distribution of the magnetic field, the radiation field and of the cosmic rays

We know very little about the interstellar magnetic field elsewhere in the Galaxy. In principle three of the four methods described in section 2.1.3 namely the three radioastronomical tools (21cm Zeeman splitting, Faraday rotation, and synchrotron emission) could be used to derive information. In practice this has turned out to be very difficult, and only very general statements can be made. Analysis of the synchrotron emission by two groups (Phillips et al., 1981, and Beuermann et al., 1985) although differing in many aspects in their outcome, agree in broad terms (Heiles, 1987): in the Galactic plane the synchrotron emissivity decreases outward with R extending to $R \sim 20$ kpc, and decreases away from the galactic plane. If one then takes into account the little that is known about the cosmic rays, one concludes that the decrease in synchrotron emissivity is almost totally the result of the decrease in relativistic electrons; thus it follows that the magnetic field stays constant over the disk of our Galaxy. In the z direction we have no information about variations in the electron density and conclusions about the magnetic field cannot be reached. There is one more piece of evidence about the large scale structure of the magnetic field: radio galaxies seen at low galatic latitudes in directions $l < 90°$ donot show very large values of RM, values that would be expected if the magnetic fields were always directed along galactic circles and in the direction of galactic rotation: field reversals apparently occur (Heiles, 1987).

The photon field, or the interstellar radiation density, cannot be measured directly in the inner part of the Galaxy. Fortunately it can be calculated with some confidence from the distribution of stars and of interstellar matter -a good paper is by Bloemen (1985), but see also Mezger et al. (1983). The field grows in strength when one moves inward, roughly by a factor of 3 when one moves toward 1/2 $R_\odot$ ($R_\odot$ is the distance of the Sun to the galactic center). An important point not made by Bloemen or by Mezger et al. is that such calculations give only the average field, and that large statistical fluctuations are to be expected in the ultra violet, because these hard, and important photons originate in O and early B stars, which are rare and clustered. The somewhat softer photons of visual and red wavelengths are probably distributed more smoothly, because they come from more abundant stars of later spectral types. For a quantitative statement on the fluctuations see Habing (1988). The importance of OB associations as strong sources of UV photons also appears in the analysis of the infrared radiation by dust -see a paper by Boulanger and Pérault (1988).

The galactic distribution of the cosmic rays can be crudely derived from the observed diffuse γ ray emission, when one knows the galactic distribution of the interstellar gas -see Bloemen (1987) for a clear description of the complicated procedures. The results give a slow decrease of the cosmic rays in the plane of the Galaxy; if $\rho(R)$ is the cosmic ray density then $\rho \propto \exp(-R/L)$ where $L > 18$ kpc for cosmic ray protons and $L \sim 5$ to 11 kpc for the cosmic ray electrons. Both scale lengths are larger than those of other galactic components (stars, molecular clouds, supernovae?). A puzzling result is the low γ-ray emissivity at the galactic center, but that may have nothing to do with a lower cosmic ray density at the center.

4. CIRRUS AND THE INTERCLOUD MEDIUM

From the foregoing discussions it appeared that interstellar space contains two extreme components; I mean on the one hand the cold, dense, massive molecular clouds (particle densities between 10^3 and $10^5 cm^{-3}$ and kinetic temperatures between 10 and 50 k) and on the other hand the hot (10^6 k) bubbles of 0.001 cm^{-3}. In between these two extremes there is the neutral, atomic hydrogen gas and the photoionized H^+ gas, together containing about half of all the interstellar matter – a quantity not to be overlooked. In section 3.3 I have already discussed that the photoionized H^+ gas has two components: strong, localized HII regions that are associated with molecular clouds and an extended diffuse component which is seen in faint radio and optical recombination lines. Here I will not say more about this diffuse component and will concentrate on the atomic hydrogen: is it cloud like, is it extended? Does the intercloud medium consist of atomic gas, or is it ionized? In section 4.1 I discuss first first what we know about the small scale distribution of H° from the analysis of the 21cm line and in section 4.2 what we know about the clouds seen in optical interstellar absorption lines – sometimes refered to as translucent clouds to indicate that their optical depth is quite low.

4.1. Small scale structure of the atomic gas

Evidence about structure in the gas in between molecular clouds and hot bubbles comes for an important part from 21cm line observations. The H° gas emits because of a transition between two levels in the ground state of hydrogen, separated by about 10^{-6} eV. The transfer of population between these two levels and the resulting equilibrium distribution is governed by collisions between atoms (see section 2.2); because the kinetic energy of the atoms is much larger than the level separation the number of atoms in the upper level is always three times that in the lower level (three is the ratio between the statistical weights of upper and lower level). The emission coefficient is then proportional to the density of the gas, and independent of the kinetic gas temperature. This is a difference with respect to the absorption line coefficient; that coefficient is determined by (g_u n_u - g_l n_l), where the g's are statistical weights, the n's indicate the level occupation and u and l refer to "upper" and "lower". This quantity between parentheses is proportional to n_u and to $1/T_k$: the lower the temperature the stronger the absorption. Therefore one tends to see cold gas more easily in absorption in the 21cm line than in emission; in absorption one does not see warm gas at all. If in a given direction one knows the absorption line profile and one can estimate (by interpolation) how the emission line profile would have looked like, then one can derive the kinetic temperature of the H° atoms by comparing emission and absorption temperatures- see e.g. Kulkarni and Heiles (1988).

Those who have studied 21cm emission line profiles, have always been struck by how smooth the profiles are and how smoothly they appear to vary from position to position. But the conclusion is probably only valid when one considers the distribution over large parts of the Galaxy. When one restricts oneself to the distribution of gas in our neighboorhood - by looking only at the higher galactic latitudes or at low velocities- then the sky is not regular at all, but shows a wild picture of threads, filaments and very complex structures- see e.g. Kulkarni and Heiles, their figure 3.4. Historically it has always been tempting to compare the interstellar medium with the weather and in that way to think of clouds. Such a word often introduces in the conscious an image, that is round and soft. That, then is a clearly wrong as one realizes when one looks over Heiles' beautiful displays. A better association is with cirrus clouds: thin threads often connected in complicated ways. Very likely it is significant that the word cirrus reappears in connection with the infrared interstellar emission (see section 4.2).

Very important independent information about the H° gas is obtained when one measures 21cm absorption line profiles and compares those with corresponding emission line profiles. Absorption line profiles (absorption against discrete sources of radio emission: galaxies, supernova remnants etc.) are quite different in appearance from emission lines: whereas individual emission line profiles appear smooth, in absorption there are, along a line of sight, at most a few, very discrete components: see figure 4.1.

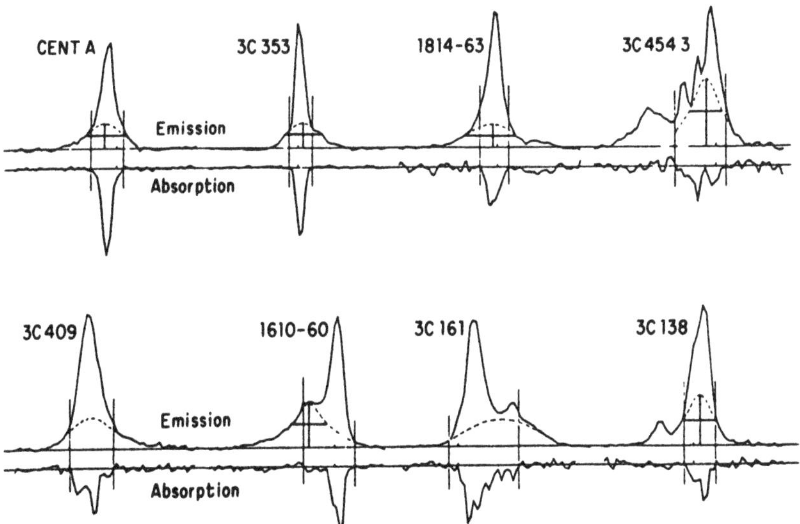

Figure 4.1. *from Radhakrishnan et al., 1972). Eight 21cm absorption line profiles with emission line profiles in the same direction obtained by interpolation between profiles in surrounding directions. The absorption profiles always show components also seen in emission, but the absorption profiles are simpler.*

This difference between emission and absorption line profiles, coupled with the theoretical insight in the difference between absorption and emission coefficients that I just have described, led to the idea that the H° gas was divided into two different components: cold clouds, seen mainly in absorption, and a warm, extended medium seen only in emission and called the intercloud medium (Clark, 1965; Mebold et al. 1974). Clark called this the "raisins' pudding" concept, a name that for many summarized the idea beautifully. But more detailed studies softened the contrast between the raisins and the pudding. There clearly is some H° gas so warm and with such a small absorption coefficient that it is never seen in absorption and only appears in emission. But the clouds, the discrete absorption components, show a range in kinetic temperature, which correlates well with the optical depth of the cloud (Lazareff, 1975; Payne et al., 1983) (figure 4.2) and it may well be that there is a continuous distribution in temperature and all the gas is in clouds without a significant amount of neutral intercloud medium - the warmest clouds cannot be seen in absorption, only in emission.

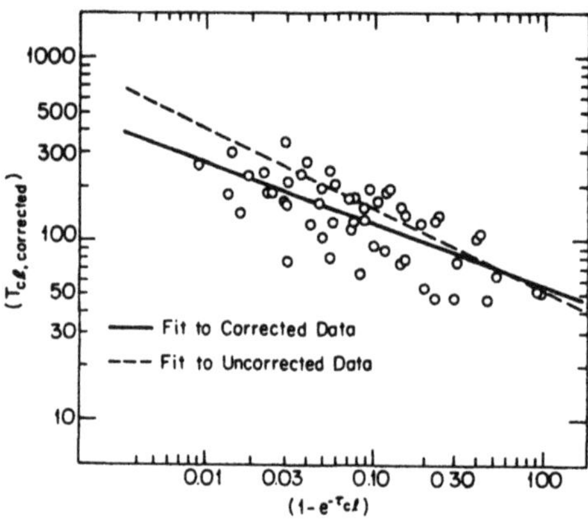

Figure 4.2. *(from Payne et al., 1983). Kinetic temperatures of H^0 gas as derived from a comparison of emission and absorption spectra and shown as a function of the optical depth of the absorption feature. A correlation is present: colder clouds have higher optical depth.*

The concept of a bimodal distribution, cool clouds and a warm intercloud medium, is certainly robbed of its simplistic charm; the two are extremes in a continuous distribution. Payne et al. (1983) suggest the possibility that much of the warmer gas is actually in warm envelopes surrounding the cores of cooler clouds, but they also keep open the possibility that some of their "independent, non- absorbing" gas is some kind of intercloud medium.

The work by Payne et al. is meticulous and the conclusions are reached carefully. Yet the interpolation method they use to calculate the emission profile at the position of the absorbing source introduces systematic errors. How serious these are appeared in a paper by Kalberla et al. (1985) who derived the emission profile in a different and better way, namely via high angular resolution measurements. For that purpose they combined observations with the Westerbork Synthesis Radio Telescope and with the 100m Effelsberg telescope. The paper concerns the analysis of the brightness distribution of the 21cm emission line over an 0.5° field centered on a continuum point source, 3C147. There is some extended emission with little structure, but most of the emission is concentrated in clumps of gas (see figure 4.3).

Comparing the properties of these clumps with the absorption peaks there is a close correspondence for each clump between the gaussian components of the emission and the absorption profile. Deriving the kinetic temperature of the clumps only low values are found (34 to 74 k)- a range much lower than that found by Payne et al.. But there is more,

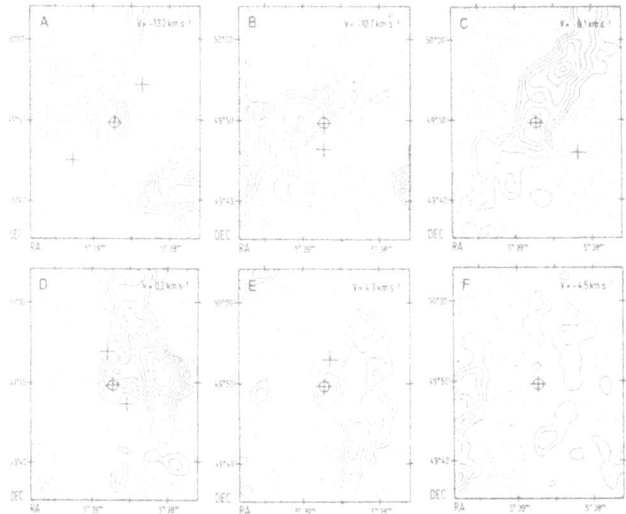

Figure 4.3. *(from Kalberla et al.,1985). Six maps of the 21cm line emission, in one small field of about (15 arcmin)^2 centered at the background radio source 3C147 (position indicated by a circled cross). Assuming that the gas is within 500 parsec, the field size is 2.2 parsec. This determines also the sizes of the "clouds" of 21cm emission. Each map is made at a radial velocity selected because at that velocity there is strong absorption seen against 3C147.*

and this may explain the difference: Kalberla et al. conclude that the emission line profile contains much gas associated with clumps, gas that does not show up in absorption. They suggest that as much as 80% of the emission associated with a clump, may not be seen in absorption, because its temperature is too high; the cold clumps are only the cores of much more and warm associated material. It is the high temperature of the associated material that introduces systematic errors in the temperature determinations of Payne et al. Yet, Kalberla et al. and Payne appear to agree on the point that much of the 21cm emission is associated with cold cores of atomic hydrogen gas. There is one obvious objection to a generalisation of the conclusions reached by Kalberla et al.: they refer to only one field in the sky, and is that field representative?

4.2. Infrared cirrus and optical interstellar lines

To the first observers who saw the IRAS results the 100μm measurements were a great surprise: point sources were seen, but there was also structure on much larger scales, from several arcminutes to several degrees. At first there was a fear that the result was instrumental, but the measurements turned out to repeat themselves exactly: the large scale emission had to come truly from the sky. Displaying the results in a sky frame gave the impression of threads and filaments, in short "cirrus" (Low et al., 1984). Most of the sky is covered by this emission although there are cirrus free areas. Already Low et al realised that there is a close correspondence between the infared cirrus and the threads and other features that Heiles had seen in the 21cm line. And even where at first sight this connection appeared not to exist, closer inspection turned up a corresponding 21cm line feature (Deul, 1988). Another discovery is that the infrared cirrus can be traced on photographic survey plates, especially on the more recent ESO and SERC plates. Cirrus here appears as faint extended "emission"- see figure 4.4 (de Vries and le Poole, 1985).

These authors show conclusively that what one sees is galactic light scattered by dust and not true emission (the phenomenon of cirrus on photographic plates had been noticed already by Sandage (1976) but the significance did not stick in the litterature). A surprise was that denser part of several cirrus clouds coincides with molecular clouds (Magnani et al., 1985). A catalogue of such clouds has recently been published by Désert et al. (1988). Because the clouds are optically thin (A_v is usually below 1.5) one expects the interstellar UV photons to penetrate the cloud easily and to destroy all molecules, but for some not yet understood reason this doesnot happen. Recently de Vries and van Dishoeck (1989) reported the detection of optical absorption lines from a cirrus cloud seen against an accidentally present background star; CH is strongly present and definitely, although weakly, one finds CH^+. The stellar distance can be estimated and so there is now a good upper limit to the distance of the cloud; it agrees with an earlier determination based on star counts. Cirrus clouds can apparently be studied with a large variety of methods and the work is only beginning; I expect that it will be an important topic for the next few years.

Interstellar absorption lines in stellar spectra were discovered by Hartmann in 1904. It took quite some while before the interstellar nature of these faint, narrow lines was generally accepted. And it has remained very difficult to tie the information drawn from interstellar lines to that obtained in other ways, e.g. to combine optical with 21cm line observations. Two line profiles (one radio, one optical) taken in the same direction often agree sufficiently to be sure that one samples the same part of the interstellar medium, and yet the differences are too large to combine results reliably: does one really see the same gas in emission in the 21cm line as one sees optically in absorption against that star? I have the optimistic feeling that the infrared/optical cirrus may well be the required step in between. The cirrus, whose outlines can be traced so well on photographic plates and on IRAS 100μm maps, is probably material identical to that which shows up in the optical absorption lines, and in the 21cm line (cf. section 4.1, the study by Kalberla et al.).

A good demonstration of how optical interstellar lines are analysed and what information such an analysis may yield are the studies on the absorption spectrum of the interstellar gas in front of the bright O-star ς Ophiuchi. In modern times the story begins with the analysis by Herbig in 1968, the first to note the very high densities required to explain the strength of the Na and K lines. Newer analyses include the many lines in the ultraviolet detected by the Copernicus satellite and a few new optical lines in the far red (van Dishoeck and Black, 1986; Viala et al., 1988). One starts with the measurements of many lines: the Lyman lines of H°, lines from various bands of H_2, lines of HD, CO, CH, CH^+, C, C^+, C_2, O, N, CN, Cl, Cl^+, Na. The observed line strengths are then compared to predictions based on a model that consists of a symmetric plan parallel layer of gas, radiated upon by the interstellar radiation field. By varying the central density and the

Figure 4.4. *(from C.P. de Vries, private communication). Two prints of the same part of a SERC Sky Survey plate. In the second print the low intensities are enhanced after digitization of the first print. Cirrus clouds show up very clearly. There is a very good correspondence with the 100μm IRAS measurements.*

temperature in the gas van Dishoeck and Black (1986) and Viala et al. (1988) find that for certain values of the density and the temperature the model predicts the observed line strengths rather well. There remain, however, at least two points to worry about: 1) although both sets of authors are successful, the two models are quite different from each other; 2) neither model explains all the line strengths: most notorious is the case of CH^+ which is observed at a much greater strength than theory can predict. A critical discussion of the differences between the two sets of authors is given by van Dishoeck and Black (1988). There is agreement that the central density is quite high (n_H =225 cm^{-3}), which means that the depth along the line of sight is less than 1 parsec, certainly much smaller than the other dimensions of the corresponding gas layer: the gas must form a thin sheet or filament (actually, this had already been concluded, but on less firm ground by Herbig in 1968). Van Dishoeck and Black stress the point that the detailed models they make, interesting already by themselves, are also of great value for understanding the formation and destruction of such complex contraptions as large molecules. Especially in such "diffuse" clouds as the one in front of ζ Oph one can expect to be able to check models that explain these molecules.

4.3. Now what about the intercloud medium?

I started this chapter asking what the intercloud medium is like, the gas not inside the hot bubbles and not in the molecular clouds- the ordinary HI gas. This has led to a discussion of cirrus clouds, and how these may contain much of the hydrogen that one sees in emission. At the same time the cirrus may also be the place where the optical absorption lines originate. Now a cirrus like distribution seems to be one that fills only a small fraction of space, that is, its filling factor is probably rather small. Thus I come back to the question: what fills the space between the bubbles, the molecular clouds and the cirrus? There are several different answers to this question and there is not a single one convincing: (1) bubbles may be so frequent that there is no further space to put anything in; (2) warm atomic hydrogen, with temperatures of several thousands of degrees may fill the open space; Kalberla et al. and Payne et al. have seen some of it (see section 4.1); (3) there is also the diffuse ionized medium seen in recombination lines of hydrogen; the most rarefied parts are seen in Hα and the denser parts in H166α and in other radio recombination lines (see section 3.3.2). And let us not forget that answers (2) and (3) might conceivably be combined into a medium partially ionized by stray radiation from O type stars. Clearly we donot know how most of the interstellar space is filled. The importance of this will become apparent in the discussions of section 5.3.2.

5. THEORY OF THE INTERSTELLAR MATTER.

5.1. What kind of theory can one expect?

Look outside on a day with clouds and some sunshine, and observe the complexity – of a single cloud or of the system of clouds. How would you describe this? How much can one predict further developments, the weather? No less is the difficulty of finding an adequate theory for the interstellar medium. The problem emerged already from the early observations of interstellar lines in stellar spectra: rather than forming a single, broad line – indication of a smoothly distributed medium – the lines soon were resolved in several narrow components, with each component apparently referring to a discrete parcel of gas somewhere along the line of sight. Analysis of a large number of interstellar lines (and the addition of some information obtained from the statistics of interstellar extinction) led Spitzer in the late fourties to formulate a "standard cloud model": interstellar matter was described as a collection of spherical clouds, embedded in and in pressure equilibrium with a hotter ionized medium of lower density. This model is now rejected by probably all workers in the field, and it is easy to criticise. Yet, I think that the debate about it had a healthy influence on the evolution of our insight. This suggests that in interstellar medium research no progress is made unless there is a theoretical concept in the back of ones mind –and it does not matter whether one is for this concept or against. Spitzer's rudimentary theory has now been replaced by a few others that are more sophisticated. However the definitive theory has yet to emerge.

What kind of theory can one expect? Like the Earth atmosphere the interstellar medium is a thin gas subject to all kinds of forces. Interstellar gas is definitely a non –linear system, where small, and at first negligible causes can dominate later events. It is thus impossible to find a fully deterministic theory, that for a given input would describe at all times how the interstellar medium looks like. The best we may hope for, so I think, is a more or less statistical, global description that in a general way will tell us what happens; to return to the weather comparison: it is feasible to have a theory that explains why England and the Netherlands have even more rain and south westerly winds during some part of the year than during another; but it is excluded to have a theory that gives detailed predictions from day to day about each cloud and its shape and speed. Once we have a global theory for the interstellar medium we will be in a much better position to understand the role of the interstellar medium in the history of our Galaxy and in the shaping of other galaxies.

Viewed on a galactic scale, the interstellar matter has a disk like distribution, that is sytematically deformed at the outer edge. The disk is in differential rotation. Near the center there is a hole and a then a concentration of gas at the center itself. We donot know what makes (or made) the hole and how old it is. There is some gas in the halo above the disk, and exchange of matter and energy between halo and disk takes place. Is that exchange only a marginal effect or one of great importance? The gas in the disk is constantly subjected to the radiation of stars, and especially hard photons that ionize hydrogen can change the state of the gas (pressure and temperature) by large factors. We distinguish between cold clouds of predominantly molecular hydrogen (with associated HII regions), of cold clouds of atomic hydrogen, of diffuse gas, part of which is atomic, and part is ionized (the boundaries between these two are unknown, but may be sharp) and hot bubbles of highly ionized gas. Apart from photon input there is also much mechanical input: supernovae go off and blow bubbles, but also winds from massive stars will create bubbles and push the interstellar gas around. Spiral arms do form and perturb the flow. Magnetic fields are present and even when they do not dominate the events, they certainly have some regulating effect; and finally cosmic rays affect the magnetic fields. Interstellar matter disappears in newly formed stars and a fraction of the matter reemerges with a changed atomic composition in supernova explosions and through the winds of red giants. All these aspects shall be dealt with in the ultimate theory of the interstellar medium, the theory that is yet to be invented.

5.2. Sources and sinks of energy and matter

5.2.1 Photons.

In section 3.7 I have mentioned already the interstellar radiation field as a source of energy for the interstellar gas. Let us now look at the overall situation. How much energy is available and where is it put in? Most important are the photons that can ionize hydrogen, i.e. with $\lambda < 91.2$ nm. Estimates of how many such photons are being produced inside our Galaxy have been derived from radio continuum measurements by Mezger and Smith (see e.g. Guesten and Mezger (1982). For example Mezger(1978) estimates that in the Galaxy (excluding the galactic center) $3\text{x}10^{53}$ Lyman continuum photons are produced by O stars, ionizing about $2\text{x}10^{8}\,M_{\odot}$ mass of hydrogen. Only 1/6 of the photons is absorbed in dense gas in HII regions, the rest (5/6) escapes. Leisawitz and Hauser (1988) discuss the fraction of escaping photons and estimate that it is between 2/3 to 4/5. There are other hot stars in the Galaxy: central stars of planetary nebulae and their successors, the white dwarfs. These hot objects are small fry next to the O stars, but occur much more frequently. A simple comparison shows, however that planetary nebulae produce less UV photons than main sequence, O-type stars and white dwarfs still less. From the comprehensive set of basic data on Lyman photons producing stars in Guesten and Mezger (1982) I estimate that there are about 3000 stars of spectral type O9 and earlier in the Galaxy (excepting the galactic center region). Each has a luminosity exceeding $1.7\ 10^{5}\,L_{\odot}$. In contrast there are with some uncertainty about 20,000 planetary nebulae in the Galaxy (Maciel, 1989); assume that on average these have a luminosity of 5000 $L_{\odot}$, and it is clear that the total luminosity of the massive stars exceed that by planetary nebulae by at least a factor 5. Clearly in the galactic plane the planetary nebulae produce less UV light than OB stars. White dwarfs have such low luminosities ($\sim 0.01 L_{\odot}$, see Liebert et al., 1988), that although they live very long ($\approx 10^{9}$.yr), they are again a factor 5 less influential than the planetary nebulae. There is the possibility that far away from the galactic plane, where OB stars occur only incidentally, white dwarfs and planetary nebulae may provide a few very hot photons; it may be, however, that also there the white dwarfs and planetary nebulae are surpassed, this time by quasars.

Where does all the energy go, or rather, what photons emerge from the gas and escape from the Galaxy? Calculations about the neutral and warm interstellar medium point toward the C^+ 158 μm line as one of the major escape channels. Direct observation of this line in our Galaxy, but also in other galaxies is expected to be of great importance. Through its measurement we will learn much better where the energy emerges, and our insight in the total energy balance will undoubtedly improve. Technically this infrared line is difficult to observe but we are now close to a breakthrough and for example ISO will do a lot of important research. First measurements of the line (Stacey et al., 1983, 1985; Crawford et al., 1985) support this conclusion. Lines of O° at 63 and 146μm may be of equal importance to the C^+ 158μm line.

An important problem with only an uncertain resolution is the cooling of the hot bubbles (Shapiro and Field, 1976). Very hot gas is a poor cooler, and the question is: if much of the supernova energy in the form of kinetic energy is transformed into thermal energy of the hot gas, how is a balance obtained? Can the hot gas get rid of the newly injected energy before the next supernova goes off? Radiation cannot do this and the solution that seems to be preferred is to have the energy transfered by conduction into cool gas or to have the bubbles float into the halo and there to cool off by adiabatic expansion.

5.2.2 Input of energy through motions.

In this subsection I want to mention briefly various dynamical processes that dump (kinematic or thermal) energy into the interstellar gas. First of all there are supernova explosions. As far as the interstellar gas is concerned a supernova is the sudden injection of a large amount of kinetic energy in a negligible small volume. At first this consists of a very fast, spherical expansion of the outer layers of the exploding star. Interstellar matter is swept up; when the mass of swept up matter equals that of the exploding star a new phase starts, usually called the Sedov phase in which the radius, R, of the shock front that advances in the interstellar medium varies as $R \propto t^{2/5}$. At the beginning of this Sedov phase the shocked interstellar gas is too hot to lose a significant amount of its thermal energy by radiation, but radiation losses gradually increase and when they become dominant the Sedov phase ends and another regime starts. At that moment what has been created is a large bubble of very hot gas surrounded by a layer of shocked gas. Most of the gas inside the bubble is of interstellar origin! Typically one assumes that a supernova injects of the order of 10^{51} erg into the interstellar medium, and the rate at which this happens in our Galaxy is rather uncertain but is usually adopted to be one event in 20 to 50 years. This means that supernovae are important energy suppliers for the interstellar medium. However, supernovae inject their energy only locally whereas ordinary stars inject their (photon) energy much more homogeneously . Another aspect is that there are at least two types of supernovae and those of Type II are exploding stars of high solar mass, they will thus be O stars when still on the main sequence. Such stars are predominantly found in groups and thus one expects Type II supernovae to go off in each others neighbourhood and at relatively short time intervals. Type I supernovae have low mass stars as their origin; they will be more evenly spread throughout the Galaxy and their explosions are uncorrelated events.

Remnants of supernovae can be very well studied via their synchrotron radio emission, via observations of optical filaments, and via their X ray emission. These give a good indication of the havoc made in the interstellar medium. A good review of supernova remnants is given by Reynolds (1988), another is by McCray (1987). In theoretical studies of the supernova influence on the interstellar medium this wealth of information is reduced to a few very schematic statements, and I often fear that the simplification is too severe. As an example consider the following argument (originally due to McKee and Ostriker, 1977): If the probability of a supernova going off within a distance R is given by a rate $Q(R)$ and if R^* is the maximum size of a supernova remnant then f =1-exp[-$Q(R^*)$] is the fraction of space filled by supernova remnants. It is easy to argue that R^* is so large that f is almost 1: most of interstellar space will then be filled by supernova remnants (hot bubbles). But much depends on the initial situation: did the supernova go off in a very thin, or in a rather dense interstellar medium?

Another important source of high velocity motions are the fast winds produced by hot stars, especially the O stars – see the review paper by McCray and Snow (1979). These stars lose a significant amount of mass at very high velocities – see the review by Cassinelli (1979); if one estimates its mass loss rate at 7x10^{-6} $M_\odot$/yr, its outflow velocity at 2000 km s^{-1} then each star dumps about 8x10^{36} erg s^{-1} in the interstellar medium, which is, of course, only a small fraction (less than one 1%) of its energy output in photons. Assuming that there are about 10^4 such stars inside the Galaxy (see the introduction to section 5; I have increased the number given there to include also the B0 stars). I estimate that the OB stars put about 1x10^{41} erg s^{-1} into the interstellar medium via their stellar winds. Compare this with 10^{51} erg per supernova explosion, once per 20 years, which translates into 16x10^{41} erg s^{-1}. Although the supernovae have a higher input rate, it is likely that the more gentle input by the OB stars lead to lower radiation losses of the shocked gas and hence to a much more efficient transfer of kinetic energy by stellar winds to the interstellar gas.

Two other dynamical forces that shape the interstellar medium will be mentioned, without further discussion: magnetic fields, and spiral arm pertubations of the gravitational field. They should both be taken into account, but it is not at all clear how. Finally I mention the subject of the high velocity clouds: clouds detected in the 21 cm line outside of the plane of the Galaxy with high velocities (up to 200 km s^{-1}, see section 3.4.). They may plunge into the Galaxy and produce a large local perturbation.

5.2.3. Interstellar matter into stars; stars into interstellar matter

Interstellar gas disappears when a star is born. Some of the matter will be returned, via stellar winds or when the stars die as supernovae (the more massive stars do this) or when the lighter stars turn into red giants that turn into planetary nebulae under ejection of a large fraction of their mass. The disappearance and later reappearance of interstellar matter is a cycle of fundamental importance in understanding the evolution of galaxies. In this summerschool it is treated in detail by Lequeux in his notes on stellar populations. I will therefore skip the subject here, and extract one point that connects with the overall estimates made in the previous section: in our Galaxy there yearly disappears about 5 $M_\odot$ of interstellar matter into newly formed stars.

There is one other aspect of the cycle that concerns us here: the formation of dust particles. Their formation requires relatively warm (1000k), relatively dense (10^6 cm^{-3}) regions. Those conditions occur in the stellar winds of cool giants. There is now a large amount of well understood evidence that shows that red giants loose much mass indeed, before they turn into planetary nebulae and then into white dwarfs. The winds flowing from these stars are slow (and thus dense) and cool: dust particles easily condense out. Observations show that the outflowing gas can be characterized according to the question whether there is more oxygen than carbon or less. In an atmosphere not enriched by nuclear synthesis oxygen dominates, but when convection in deep layers has added products of nucleosynthesis then carbon dominates. This simple change in the balance of the trace elements has dramatic consequences for the appearance of the star: the chemistry in the outer atmospheric layers is quite different between the two abundance states. This is reflected in the kind of dust particles that forms: "in oxygen rich envelopes" silicate type dust grains form, in "carbon rich" envelopes graphite type particles are born. One of the remarkable features of galactic evolution is that in our solar neighboorhood there are relatively few carbon stars; in the galactic center region the carbon stars are even more rare. But in the Magellanic Clouds carbon rich stars occur more frequently than oxygen rich stars. This suggests that metal abundance has some influence on the formation of Carbon stars.

We are left with the question whether enough dust grains form from condensations in the winds of red giants. When we assume that yearly 2 planetary nebulae are formed inside our Galaxy, that their average main sequence mass was 2 $M_\odot$ and that ultimately 0.6 $M_\odot$ is retained forever in the white dwarf that remains behind, then red giants return 2.8 $M_\odot$ to the interstellar medium. This is significantly less than what disappears into newborn stars (5 $M_\odot$, see above) and the amount of interstellar matter in our Galaxy constantly decreases. As there is about 10^{10} $M_\odot$ of interstellar matter in our Galaxy, the supply is enough for 5×10^9 yr, when the present day rate would continue. This time is less than the Hubble time, and thus we have a problem. The problem occurs in identical form if we ask ourselves where interstellar grains are being made and whether enough of them are being made. The answer is no: there is a net disappearance of interstellar gas and thus also of grains, unless the matter returned by red giants is a factor of two richer in dust grains than the gas from which the star formed.

5.3. Local Theories of the interstellar medium

The term "local theory" is used here to indicate that the interstellar medium is treated as a closed system of limited extent. The theory considers only the galactic gas layer, and does not consider large scale galactic exchanges, e.g. between halo and disk; neither is the role of the galactic centre discussed. "Local" is not used in any depreciative sense: it is quite possible that global theories can be built from local theory. Two main theories will be discussed.

5.3.1. The two-phase model

The model is based on heating of the interstellar matter by low energy cosmic rays and was first proposed by Field et al. (1969; "FGH") see also Field (1976). The model has been changed and reformulated several times using different heating mechanisms, but the essential features have always been retained and for educational reasons it appears justified to discuss the old "FGH" paper. The starting point is the partial ionization and heating of neutral atomic hydrogen gas by low energy cosmic rays, and the cooling of the gas by trace elements. As stated in section 2.3 the heating rate of the gas, symbolized by Γ erg.cm^{-3}.s^{-1}, is proportional to the density of the gas, $n_{\rm H}$, and to the ionization rate, ς. The cooling rate of the gas, symbolized by Λ erg.cm^{-3}.s^{-1}, and determined by inelastic collisions between the free electrons and trace elements, such as C^+, is proportional to $n_{\rm H}^2$ squared and further depends on T. Thermal equilibrium requires $\Lambda = \Gamma$ and this leads to a relation between n_H and T: to each density there belongs a certain equilibrium temperature and hence a certain gas pressure, $n_{\rm H}\ kT$. Thus one may have a stationary interstellar medium in which various "phases" coexist in pressure equilibrium- see figure 5.1 where these phases have been indicated by roman numerals.

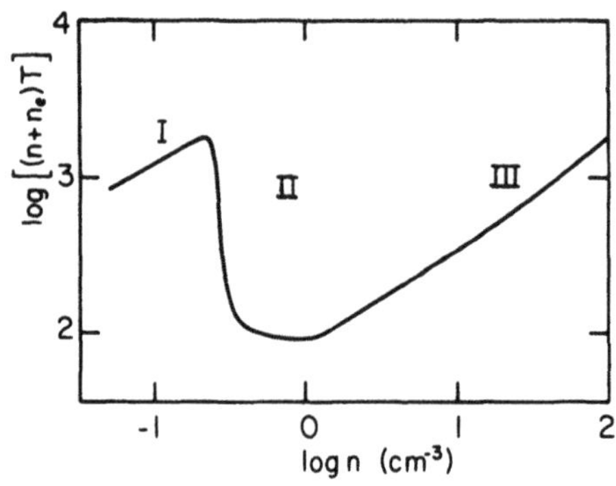

Figure 5.1. *(from Field et al., 1969). The "pressure" of the interstellar medium (actually the product of density and temperature) as a function of the density of the gas, when the gas is heated by low energy cosmic rays. The Roman numerals I, II, III denote three phases that can co-exist in pressure equilibrium. Phase II is however "thermally unstable" and any gas in this phase will go into I or into III.*

Closer study of each phase shows that some are not stable: in a small compression around point II the radiative losses increase by a larger factor than the energy gains, so that the pressure continues to decrease and the condensation continues to grow. The equilibrium at point II is called "thermally unstable" – see Field (1965). Points I and III are stable and hence can represent real phases of the interstellar gas. Part of the success of the FGH theory was that by selecting the right pressure it explained most interstellar observations then known. A weak point was the ad-hoc assumption of a rather large cosmic ray flux ($\varsigma = 10^{-15}$s^{-1}). It is now clear, e.g. from studies of chemical equilibria in molecular clouds, that ς is smaller by at least a factor of 10. In addition a totally new phase of the interstellar matter has been discovered, that does not occur in the FGH proposal: the hot interstellar medium (see section 3.5). This phase and its possible origin in supernova remnants indicate the need for a dynamic model rather than a more or less static model, like FGH. Nevertheless thermal instabilities may play a significant role in seperating phases, and to have this outlined may be one of the major gains obtained from the FGH model. The important question remains about the source of energy that heats the cool gas and by what mechanism the energy is transfered- if, as it appears, the comic rays are insufficient. The question is unanswered at the moment, but there is a good candidate: photoionization of interstellar grains (de Jong, 1980), especially when the very small grains are as efficient as has been proposed (Puget, Leger, Boulanger, 1985; Lepp and Dalgarno, 1988). The mechanism has the same dependence on density as cosmic ray heating and the general conclusions about the two phase model remain valid (Shull, 1987).

5.3.2. The three-phase model

Cox and Smith (1974) noticed that the spheres created by supernova shells may last so long that there is a finite probability that a second supernova will go off inside. This second explosion will find itself in surroundings quite different from those of the first explosion; it will reenergize the remnants of that explosion and enlarge the volume. A third supernova may follow and so on: Cox and Smith suggested that a tunnel system of supernova remnants could grow, "much like the holes in a Swiss cheese". This suggestion has subsequently been taken up by McKee and Ostriker (1978) who constructed a supernova regulated, dynamic model for the interstellar medium. They propose that most of interstellar space (80%?) is taken up by the hot interstellar medium. Neutral clouds are continuously being born out of radiatively cooling shells of supernovae, and they disappear by evaporation after they have entered the hot medium. Each cloud consists of a core of neutral hydrogen, surrounded by a layer of warm neutral gas, which in turn is surrounded by warm ionized gas - see figure 5.2.

The basic dynamic equilibrium is between formation and destruction of the clouds, whose origin is in new supernova remnants. The basic parameters in this model are the energy input by supernovae (their frequency times the average energy per supernova), the density of the "intercloud" medium into which the supernova expands, and the radiation field, especially in the far UV, since it has significant effects on the neutral material. A critical question is whether supernova events are uncorrelated or whether they occur in bursts, e.g. in young associations of massive stars. The supernova rate is uncertain. An overall value for the whole Galaxy may be estimated, but the distribution of the probability over the face of the Galaxy is totally unknown; also correlation between supernova events is likely (at least for supernovae of type II). The density of the intercloud medium may be something of a chicken and egg problem: if the density is low, the supernova remnants expand sufficiently fast and the density will remain low. If it was high to begin with the firecracker will stop before it took off. There is thus too much uncertain about the basic parameters of the model to feel comfortable.

The three-phase model leads to a number of quantitative predictions that are in general agreement with observed values. There are, however, some observational objections (see Shull,1987), of which I mention especially those concerning the distribution of the neu-

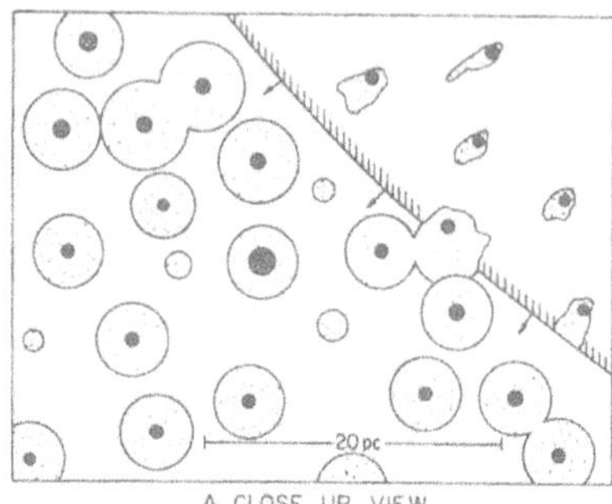

Figure 5.2. *Figure 5.2 (from McKee and Ostriker, 1977) The small scale structure of the interstellar medium: a region of 30pc x 40pc is shown. A supernova produced shockfront moving from the upper right hand corner towards left below overruns several small clouds of neutral hydrogen. Each cloud consists of a core of cold neutral hydrogen surrounded by a warm halo. Once the clouds enter the shockheated region they are compressed and begin to evaporate.*

tral atomic material: in the model the cold and the warm material are both supposedly contained in small clouds contrary to the evidence from the 21cm line data; the observations (see section 4.2) yield much larger quantities of warm neutral hydrogen and cold hydrogen associated with a cloud (see section 4.2) than is predicted by the theory. An interesting improvement of the model has been proposed by Cowie (1987). Cowie also discusses the dissipation of the energy input by the supernovae; it is too slow. He suggests that the bubbles rise into the halo and burst, leading to a galactic fountain, much along earlier ideas expressed by Shapiro and Field (1976).

5.4. Haloes and fountains

The discovery in the fifties of interstellar clouds at large (kpc) distances from the galactic plane (see section 3.4) raised the problem how these clouds could survive if they existed in vacuum: if they were not confined they would expand in a relatively short time and disappear from sight. In 1956 Spitzer proposed that the clouds are confined by a thin, high temperature gas, the gaseous halo. In brief he proposed that the gas should be in hydrostatic presence equilibrium, with a temperature of 10^6k and a density at the base of 0.0005 cm^{-3}. Later Field (1965) showed that such a halo is thermally unstable and might ultimately shrink or expand indefinitely. A next major step came when Shapiro and Field (1976) realised (i) that hot bubbles blown by supernovae have a thermal pressure much larger than the rest of the interstellar medium and thus cannot be confined by that medium and (ii) that the cooling of the bubble gas would take a very long time, longer probably than it would take the next supernova to go off inside the bubble and thus add new thermal energy. A steady dynamical equilibrium could then not be reached. Shapiro and Field therefore proposed that the bubbles would float upward, outside of the disk of the Galaxy, a phenomenon they termed "galactic fountain".

An elaborate model of a galactic fountain was then presented by Bregman (1980) and by Habe and Ikeuchi (1980) – see figure 5.3.

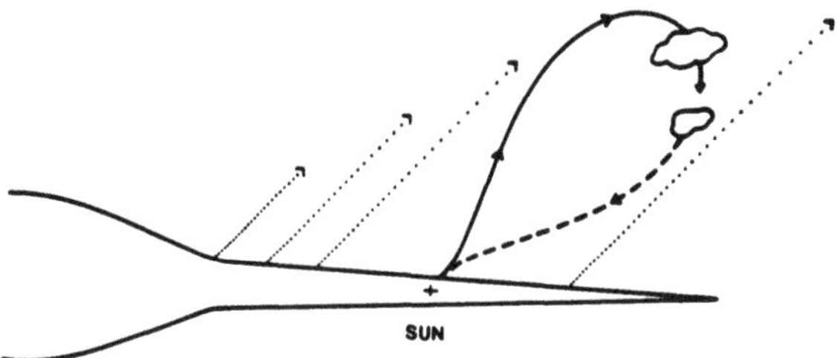

Figure 5.3. *(from Bregman, 1980) Hot gas rises from the disk of our Galaxy, cools by adiabatic expansion, condenses into "high velocity" clouds and then falls back.*

Bregman's major concern seems to have been to explain the high velocity clouds. A problem not solved by him (the observations were barely known) is the presence of ions such as C^{3+}, Si^{3+}, and N^{4+} seen via interstellar absorption lines against stars in the Magellanic clouds with IUE; these ions would be quickly ionized in a fountain. Bregman suggests that the ions are boundary layers between halo gas and clouds that have begun to condense out. In recent years Chevalier and Fransson (1984; see also Fransson and Chevalier, 1985) have proposed that cosmic rays support the halo; the rays diffuse slowly outward from the galactic plane and above a distance of some 400 parsec they would support a gas of rather uniform density (estimated at 0.0016 cm-3). This very thin gas is rather cool (12,000 k above 1 kpc) and is photoionized by quasars and active galaxies! Such a gas seems to offer a better explanation for the presence of the ions seen by IUE.

6. SUMMING UP WHAT HAS NOT BEEN DISCUSSED

In a limited course one can teach only a limited amount of matter. There are several items that I didnot discuss but that would have been included had there been more time available:
(1) The origin and future of the galactic disk of interstellar matter: was it born as thin and as flat as it is now, and did it take a long time to settle? Why is it warped at the outside? Why is there a 3kpc hole in the center?
(2) The origin and maintenance of spiral arms. The gas disk of our Galaxy has a tendency to form spiral arms. How are spiral arms shaped? What is the difference between the interstellar gas inside and outside spiral arms?
(3) Gas in external galaxies. What are the similarities and what are the differences between various (types of) galaxies and is our Galaxy different from the rest? Much progress in the field is to be expected from extragalactic studies. Global properties are better determined in other galaxies than in ours- for example the content and maintenance of spiral arms is much better studied in other Galaxies. Recently much detailed information has come available about other galaxies in the local group, notably the Magellanic Clouds, M33 and M31.
(4) Molecular clouds are a very important component of the interstellar medium. Yet I have not discussed them in detail, but only mentioned their overall galactic distribution. The topic is of great importance but also tends to develop into a subject of its own. Directly connected with the molecular clouds are the HII regions and the subject of the formation of stars. This last subject is of fundamental importance for all our ideas on the history of galaxies; there is considerable progress and it develops very strongly into a subject by itself.
(5) Supernova remnants have always been beautiful subjects of interstellar matter research. They are now being observed in X- rays and at radio and optical wavelengths. They also tend to form a subject of their own, uncoupled from the rest of interstellar matter. Somewhat connected with this topic is the study of bubbles blown by the fast winds from hot main sequence stars.
(6) My own favourite is the study of cool circumstellar envelopes. They are seen around late type giants and can be observed in a large number of ways. Their importance lies in the production of dust and in the amount of matter they return to the interstellar medium. Planetary nebulae are related objects.

As you see, dear reader, there is much to be discovered in the litterature beyond what I have taught. I am convinced that there is even more to be discovered outside the litterature and in nature. I wish you good luck!

REFERENCES

Allamandola, L.J., Tielens, A.G.G.M., Barker, J.R. 1985, *Astrophys. J. (Letters)* **290**, L25

Allison, A.C., Dalgarno, A. 1969 , *Astrophys. J.* **158**, 423

Beuermann, K., Kanbach, G., Berkhuysen, E.M. 1985 , *Astron. Astrophys.* **153**, 17

Black, J.H. 1987, in *"Interstellar Processes"*, eds. D.J.Hollenbach and H.A.Thronson, p.731.

Bloemen, H. 1985 , *Astron. Astrophys.* **145**, 391

Bloemen, H. 1987, in *"Interstellar Processes"*, eds. D.J. Hollenbach and H.A. Thronson, (Reidel), p. 143

Boulanger, F., Pérault, M. 1988 , *Astrophys. J.* **330**, 964

Bregman, J. 1980 , *Astrophys. J.* **236**, 577

Burton, W.B., 1988, in *"Extragalactic and Galactic Radio Observations"* eds. G.L. Verschuur and K.I. Kellermann (Springer), p. 295

Cassinelli, J.P., 1979 , *Ann. Rev. Astron. Astrophys.* **17**, 275.

Chevalier, R.A., Fransson, C. 1984, *Astrophys. J. (Letters)* **279**, L43

Clark, B. 1965 , *Astrophys. J.* **142**, 1398.

Cowie, L.L., Songaila, A. 1986 , *Ann. Rev. Astron. Astrophys.* **24**, 499

Cowie, L.L. 1987, in *"Interstellar Processes"*, eds. D.J. Hollenbach and H.A. Thronson, p.245

Cox, D.P. 1981 , *Astrophys. J.* **245**, 534

Cox, D.P., Smith,B.W. 1974 , *Astrophys. J. (Letters)* **189**, L105

Cox, D.P., Reynolds, R.J. 1987 , *Ann. Rev. Astron. Astrophys.* **25**, 303

Crawford, M.K., Genzel, R., Townes, C.H., Watson, D.M., 1985 , *Astrophys. J.* **291**, 755

Dalgarno, A., McCray, R.A. 1972 , *Ann. Rev. Astron. Astrophys.* **10**, 375

Dame, T.M., Ungerechts, H., Cohen, R.S., de Geus, E.J., Grenier, I.A., May, J., Murphy, D.C., Nyman, L.A., Thaddeus, P. 1987 , *Astrophys. J.* **322**, 706

Désert, F.X., Bazell, D., Boulanger, F. 1988 , *Astrophys. J.* **334**, 815

Deul, E., 1988, *" Interstellar Dust and Gas in the Mily Way and M33"*, thesis, Leiden University

van Dishoeck, E., Black, J.H. 1986, *Ap.J. Suppl. Ser.* **62**, 109

van Dishoeck, E., Black, J.H. 1988, in *"Rate Coefficients in Astrochemistry"*, eds, T.J. Millar and D.A. Williams (in press)

Draine, B.T., Anderson, N. 1985 , *Astrophys. J.* **292**, 494

Field, G.B. 1965 , *Astrophys. J.* **142**, 531

Field, G.B. 1976, in *"Atomic and Molecular Physics and the Interstellar Matter"*, eds. R. Balian, P. Encrenaz, J. Lequeux, North Holland Publ. (Amsterdam), 469

Field, G.B., Goldsmith, D., Habing, H.J. 1969 *Astrophys. J. (Letters)* **155**, L49

Fransson, C., Chevalier, R.A. 1985 , *Astrophys. J.* **296**, 35

Genzel, R., Townes, C.H. 1987 , *Ann. Rev. Astron. Astrophys.* **25**, 377

Genzel, R., Stutzki, J. 1989 , *Ann. Rev. Astron. Astrophys.* **29**, (in press)

Georgelin, Y.M., Georgelin, Y.P. 1976 , *Astron. Astrophys.* **49**, 57

Gingerich, O. 1982, in *"Symposium on the Orion Nebula to honor Henry Draper"* eds. Glassgold, A.,E. Huggings, P.J., Schucking, E.L., p. 308

Gordon, M.A. 1988, in *"Galactic and Extragalactic Radio Astronomy"*, eds. G.L. Verschuur and K.I. Kellermann (Springer), p. 37

Gottesmann, S.T., Gordon, M.A. 1970, *Astrophys.J. Letters* **168**, 299, L93

Guesten, R., Mezger, P.G. 1982, *Vistas in Astronomy,* **26**, 159

Habe, A., Ikeuchi, S. 1980, *Progr.Theor. Phys.* **64**, 1995.

Habing, H.J. 1988, in *"Mm and Submm astronomy"*, eds. R. Wolfstencroft and B. Burton (Reidel), p. 207

Hayes, M.A., Nussbaumer 1984, , *Astron. Astrophys.* **134**,, 193

Heiles, C. 1987, in *"Interstellar Processes"*, eds. D. Hollenbach and H.A. Thronson (Reidel), p. 171

Herbig, G.H., *Zeitschrift fuer Astrophysik*, **68**, 243,

van de Hulst, H.C., 1981, *"Light Scattering by Small Particles"*, Dover editions

Jackson, P.D., Kerr, F.J. 1971 , *Astrophys. J.* **168**, 29

Jacobs, V.L. 1985 , *Astrophys. J.* **296**, 121

Jenkins, E.B. 1978a , *Astrophys. J.* **219**, 845

Jenkins, E.B. 1978b , *Astrophys. J.* **220**, 107

de Jong, T. 1980, *Highlights in Astronomy* **5**, 301.

Kahn, F. 1975, in *"Atomic and Molecular Physics and the Interstellar Matter"*, eds. R. Balian, P. Encrenaz, J. Lequeux, North Holland Publ. Amsterdam, p. 535

Kalberla, P.M.W., Schwarz, U.J., Goss, W.M. 1985 , *Astron. Astrophys.* **144**, 27

Kulkarni, S.R., Heiles, C. 1988a, in *"Galactic and Extragalactic Radio Astronomy"*, eds. G. Verschuur and K. Kellermann (Springer), p. 95

Kulkarni, S.R., Heiles, C. 1988b, in "Interstellar Processes", eds. D.J. Hollenbach and H.A. Thronson (Reidel), p. 87

Lazareff, B. 1975 , *Astron. Astrophys.* **42**, 225

Leger, A., Puget, J.L. 1984, *Astron. Astrophys. (Letters)* **137**, L5

Leisawitz, D, Hauser, M.G. 1988 , *Astrophys. J.* **332**, 954

Lepp, S., Dalgarno, A. 1988 , *Astrophys. J.* **335**, 769

Liebert, J., Dahn, C.C., Monet, D.G. 1988 , *Astrophys. J.* **332**, 891

Liszt, H.S. 1983 , *Astrophys. J.* **275**, 163

Liszt, H., 1988 in *"Galactic and Extragalactic Radio Astronomy"*, eds. G.L. Verschuur and K.I. Kellermann (Springer), p.359

Lockman, F.J. 1976 , *Astrophys. J.* **209**, 429

Lockman, F.J. 1980, in *"Radio Recombination Lines"*, ed. P.A. Shaver, (Reidel)

Low, F.J., D.A. Beintema and 15 others 1984, *Astrophys. J. (Letters)* **278**, L19

Lyne, A.G., Manchester, R.N., Taylor, J.H. 1985, *Monthly Notices Roy. Astron. Soc.* **213**, 613

Maciel, W., 1989, in *"Planetary Nebulae"*, IAU Symposium 131, ed. S. Torres-Peimbert, Kluwer Academic Publ., 73

Magnami, L., Blitz, L., Mundy, L. 1985 , *Astrophys. J.* **295**, 402

Mathis, J.S., Mezger, P.G., Panagia, N. 1983 , *Astron. Astrophys.* **128**, 212

Mathis, J. 1986 , *Astrophys. J.* **301**, 423

Mattila, K., Scheffler, H. 1978 , *Astron. Astrophys.* **66**, 211

McCray, R., Snow, T.P., 1979 , *Ann. Rev. Astron. Astrophys.* **17**, 213

McCray, R., 1987, in *"Physical Processes in Interstellar Clouds"*, eds. G.E. Morfill and M. Scholer (Reidel), p. 95

McKee,C.F., Ostriker, J.P. 1977 , *Astrophys. J.* **218**, 148

Mebold, U., Hachenberg, O., Laury-Micoulat, C.A., 1974 , *Astron. Astrophys.* **30**, 329

Mezger, P.G., Mathis, J., Panagia, N. 1983 , *Astron. Astrophys.* **128**, 212

Mezger, P.G. 1978 , *Astron. Astrophys.* **70**, 565

Muench, G., Zirin, H. 1961 , *Astrophys. J.* **133**, 11

Nandy, K., Thompson, G.I., Jamar, C., Monfils, A., Wilson, R. 1975 , *Astron. Astrophys.* **44**, 195

Norman, C.A., Ikeuchi, S. 1988, in *"Kerr Symposium on the Outer Galaxy"*, eds. L. Blitz and F.J. Lockmann (in press)

Osterbrock, D.E., 1974, *"Astrophysics of Gaseous Nebulae"*, Freeman and Cy.

Payne, H.E., Salpeter, E.E., Terzian, Y. 1983 , *Astrophys. J.* **272**, 540

Radhakrishnan, V., Murray, J.D., Lockhart, P., Whittle, R.P.J. 1972, *Astrophys. J. Suppl. Ser.* **24**, 15

Phillips, S. Kearsey, S., Osborne, J.L., Haslam, C.G.T., Stoffel, H. 1981 , *Astron. Astrophys.* **103**, 405

Puget, J.L., Leger, A., Boulanger, F. 1985, *Astron. Astrophys. (Letters)* **142** L19

Reifenstein, E.C., Wilson, T.L., Burke, B.F., Mezger, P.G., Altenhoff, W.F. 1970 , *Astron. Astrophys.* **4**, 357

Reynolds, R.J. 1983 , *Astrophys. J.* **268**, 698

Reynolds, R.J. 1984 , *Astrophys. J.* **282**, 191

Reynolds, R.J. 1985 , *Astrophys. J.* **294**, 256

Reynolds, S.P., 1988, in *"Extragalactic and Galactic Radio Astronomy"* eds. G.L. Verschuur and K.I.Kellerman (Springer), p. 439

Rybicki, G.B., Lightman, A.P. 1979 *"Radiative Processes in Astrophysics"* (Wiley)

Sandage, A., 1976 , *Astron. J.* **81**, 954

Savage, B.D., Mathis, J.S., 1979 , *Ann. Rev. Astron. Astrophys.* **17**, 73

Savage, B.D., 1987 in *"Interstellar Processes"*, eds. D.J.Hollenbach and H.A.Thronson (Reidel), p. 123

Scalo, J.M. 1987, in *"Interstellar Processes"*, eds. D.J.Hollenbach and H.A. Thronson (Reidel), p. 349

Scoville, N.Z., Sanders, D.B. 1987, in *"Interstellar Processes"*, eds. D.J. Hollenbach and H.A. Thronson (Reidel), p. 21

Shapiro, P.R., Field, G.B. 1976 , *Astrophys. J.* **205**, 762

Shaver, P.A., McGee, R.X., Newton, L.M., Danks, A.C., Pottasch, S.R. 1983, *Monthly Notices Roy. Astron. Soc.* **204**, 53

Shull, J.M., Draine, B.T. 1987, in *"Interstellar Processes"*, eds. D.J. Hollenbach and H.A.Thronson (Reidel), p. 283

Shull,J.M. 1987, in *"Interstellar Processes"*, eds. D.J. Hollenbach and H.A. Thronson (Reidel), p.225

Spitzer, L. 1956 , *Astrophys. J.* **124**, 20

Spitzer, L., 1978, *"Physical Processes in the Interstellar Medium"* (John Wiley and Sons)

Stacey, G.J., Smyers, S.D., Kurtz, N.T., Harwit, M. 1983, *Astrophys. J. (Letters)* **268**, L99

Stacey, G.J., Viscuso, P.J., Fuller, C.E., Kurtz, N.T. 1985 , *Astrophys. J.* **289**, 803

Stromgren, B., 1939 , *Astron. J.* **89**, 256

Tielens, A.G.G.M., Allamandola, L.J. 1987 in *"Interstellar Processes"*, eds. D.J. Hollenbach and H.A. Thronson (Reidel), p. 397

Tinbergen, J. 1982 , *Astron. Astrophys.* **105**, 53

Viala, Y.P., Roueff, E., Abgrall, H. 1988 , *Astron. Astrophys.* **190**, 215

de Vries, C.P., le Poole, R.S., 1985, *Astron Astrophys. (Letters)* **145, L7**

de Vries, C., van Dishoeck, E. 1988, *Astron. Astrophys. (Letters)* **203**, L23

Wesselius, P., Fejes, I. 1973 , *Astron. Astrophys.* **24**, 154

Wilson, T.L., Mezger, P.G., Gordner, F.F., Milne, D.K., 1970 , *Astron. Astrophys.* **6**, 364

Zel'dovich, Ya. B., Raizer, Yu.P., 1968, *"Elements of Gasdynamics and the Classical Theory of Shock Waves"* (Acad. Press.)

Zweibel, E.G. 1987, in *"Interstellar Processes"*, eds. D.J. Hollenbach and H.A. Thronson, p. 195.

Images in Astronomy: an Overview

Pierre Léna
Université Paris 7 et Observatoire de Paris

1. What is an Image ?

The astronomical sky is a two-dimensional distribution of intensity of electromagnetic radiation.

We define the specific intensity I ($\boldsymbol{\theta}$,parameters), where $\boldsymbol{\theta}$ is the angular direction. Units are $W\,m^{-2}\,sr^{-1}\,Hz^{-1}$ or photons $s^{-1}\,m^{-2}\,sr^{-1}\,Hz^{-1}$.

The parameters may be : wavelength λ (or radial velocity), polarization, time...etc.

In practice, the signal is received from the source at $\boldsymbol{\theta}$ in a finite angle $\Delta\boldsymbol{\theta}$. One measures the spectral illumination

$$\iint_{\Delta\boldsymbol{\theta}} I(\boldsymbol{\theta}, \text{param.}) d\boldsymbol{\theta}.$$

The real image is degraded with respect to the source $O(\boldsymbol{\theta})$:

• the finite size of optical instruments creates DIFFRACTION. This is an ultimate physical limitation due to the nature of light.

• the finite quantity of energy reaching the detector creates NOISE. This is an ultimate physical limitation due to the signal itself.

• There are distorsions of the signal between the source and the instrument, due to various causes : atmospheric turbulence, scintillation due to interplanetary medium, gravitational lensing...

The image process starts from the object $O(\boldsymbol{\theta})$ and produces the image $I(\boldsymbol{\theta})$. Conversely, the restoration process recovers an estimate of the object $O(\boldsymbol{\theta})$, from the image $I(\boldsymbol{\theta})$.

An image may be studied in the Fourier domain. Let define Fourier transform of $F(x)$ by

$$\tilde{F}(s) = \int_{-\infty}^{+\infty} F(x) \exp(-2i\pi sx) dx$$

and extend to two-dimensions

$$\tilde{F}(\boldsymbol{f}) = \iint F(\boldsymbol{\theta}) \exp(-2\pi i \boldsymbol{\theta}.\boldsymbol{f}) d\boldsymbol{\theta}$$

We define $\tilde{I}(\boldsymbol{f})$ as the complex image spectrum.

Property : any image given by a physical system has a spectrum going to zero beyond some cut-off frequency.

$$\tilde{I}(\boldsymbol{f}) = 0 \text{ for } |\boldsymbol{f}| \geq f_c$$

As a consequence, by application of the sampling theorem (Shannon), $I(\boldsymbol{\theta})$ is fully determined by discrete samples distant of $1/2f_c$.

We define conjuguate domains

$$I(\boldsymbol{\theta}) \underset{FT}{\rightleftarrows} \tilde{I}(\boldsymbol{f})$$

and it is equivalent in terms of information content to know the image or its Fourier transform. The information can be sampled in either two-dimensional spaces. For a general discussion of information in images, see [1].

1.1 A Real Image Detector

The **pixel** (picture element) is the periodic pattern, $a \times a$ in size. The sensitive area may be smaller. The **geometric coverage** is the ratio of sensitive area to pixel area a^2. The **format** is $N \times N$, for a square detector of N^2 pixels. The sensitivity profile $S(\theta)$ across the pixel is not necessarily uniform. Fig.1.1 illustrates some of these properties.

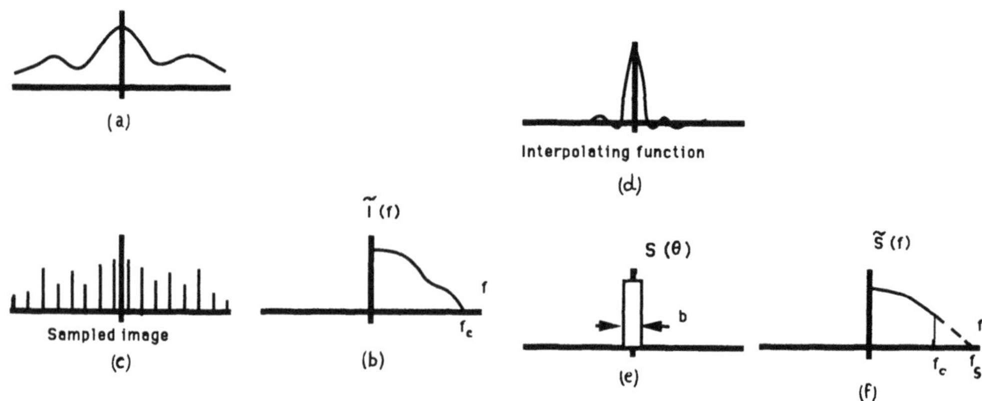

Fig. 1.1. Sampling of an image : (a) the input intensity profile; (b) the input spectrum, band-limited by the receiving system, with cut-off frequency f_c; (c) the required sampling of the image at intervals $\Delta\theta = (2f_c)^{-1}$; (d) the interpolating function sinc $(2f_c\theta)$; (e) the pixel sensitivity profile of $S(\theta)$ of width b ; (f) the filtering effect of the pixel finite size.

1.2 Digitization

Let represent $I(\theta)$ on a finite, discrete number of values $i_1, i_2, ..., i_k..., i_n$. These values can be equally spaced, or on log scale, or other : $i_{k+1} = i_k + \Delta i$, $\log i_{k+1} = \log\, i_k + \Delta i$...etc., Let define the dynamic range as the ratio i_N/i_1 and define the number of bits n : $n \geq \log_2$ (dynamic range). One often considers multi-dimension data storage and handling. For example data cubes $I(\theta_x, \theta_y, \mathrm{P})$ where P is a parameter such as wavelength, time...

1.3 Graphic Presentation

The graphic presentation of an image with appropriate display methods is essential, not only for adequate communication, but also for proper perception of the searched information by the eye. An image can be made more explicit by a number of ways :

- Iso $-$ I(θ) contours or iso $-$ log $I(\theta)$ contours.
- Gray levels : the dynamic range of the eye is 16 at most.
- Color levels : they offer a much larger dynamic range (10^3) due to the chromatic sensitivity of the eye.
- Three-dimensional plots.

...etc.

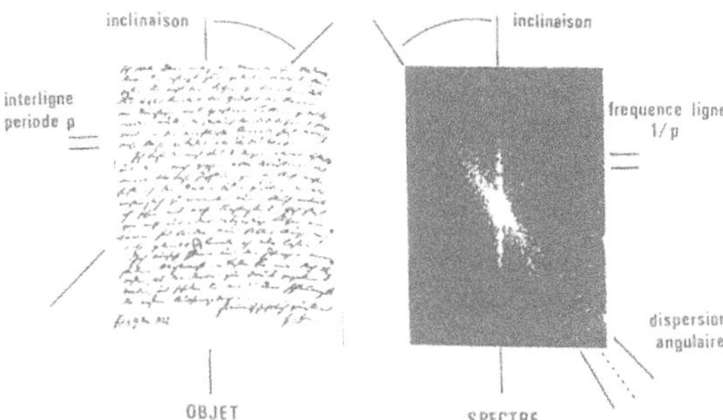

Fig. 1.2. An example of pattern analysis using the Fourier transform of an image . On left $I(\theta)$ is the writing of somebody. On right the modulus $|\tilde{I}(\boldsymbol{f})|$, clearly showing the global properties of the writing (mean spacing, inclination, angular dispersion...). [After Charraut et al. Courrier du CNRS, 66, 105 (1987)]

1.4 Image storage

The amount of information present in an image may be very large. Assume for example a Charge Coupled Device or CCD [format $N^2 = (1024)^2$], taking exposures every 10 secondes for one night of 10 hours on a 16 bits range. This is

$$16 \times N^2 \times 3.6\, 10^3 \sim 6.10^{10} \mathit{bits}$$

larger than the storage capacity of an optical disk ($\sim 10^9 \mathit{bits}$) ! Compare with the eye, which can absorb about 10^{10} bits per second !

1.5 Image Processing

All sorts of transformations may be a posteriori applied to an image $I(\theta)$. Examples will be found in radioastronomy such as filtering, smoothing, increasing the spectral coverage (i.e. creating new information taking in account physical constraints on the object...) or in optical astronomy as time-averaging, noise filtering...(Fig 1.2 and 1.3).

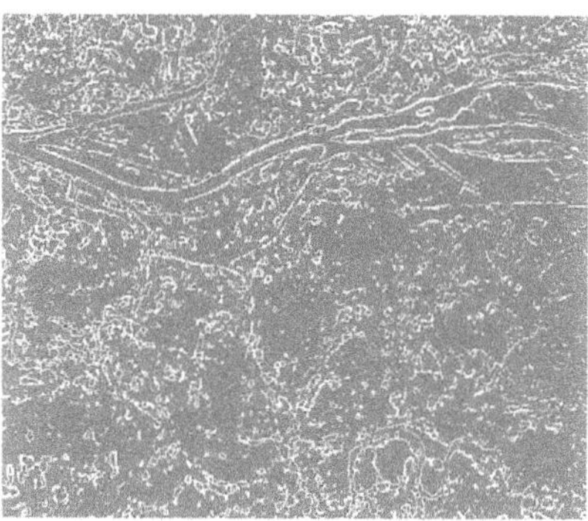

Fig. 1.3. An example of a posteriori image processing. The low spatial frequencies have been suppressed in this image, taken by the SPOT satellite on Strasbourg area. High spatial frequencies reveal the discontinuities (such as harbour, roads, the Rhine...). ©M.P. Stoll, ENSPS

1.6 Non-Optical Imaging

It may happen that it is impossible, for some radiation, to find suitable materials or systems (lenses, mirrors) to produce an image, or to achieve the required resolution. Interesting approaches have been developed to circumvent these cases.

The trivial case is the Camera obscura or stenope, which provides a quasi-image, with a very poor use of the available energy. The principle is impr oved in γ-ray cameras, where the input pinhole is replaced by a half-transparent, half-opaque mask, projecting its shadow on a multi-pixels detector. A suitable choice of the mask transparency function $P(\boldsymbol{r})$ at position $\boldsymbol{r}$ leads to a unique inversion of the convolution relation

$$S(\boldsymbol{r}) = P(\boldsymbol{r}) * I\left(\boldsymbol{\theta} = \frac{\boldsymbol{r}}{d}\right)$$

where $S(\boldsymbol{r})$ is the signal at pixel $\boldsymbol{r}$, d is the distance mask-detector, I the source intensity in direction $\boldsymbol{\theta}$.

Several γ-telescopes use this technique (Fig.1.4). The SIGMA telescope (France-USSR, launch 1989) has 53 x 49 pixels, and a 10 arc min angular resolution, to be compared with the best achieved to date, namely 5° with the COS-B satellite. The planned GRASP telescope (NASA-USA, launch ca.1993), has 360^2 pixels, with 1 arc min resolution and d = 4m.

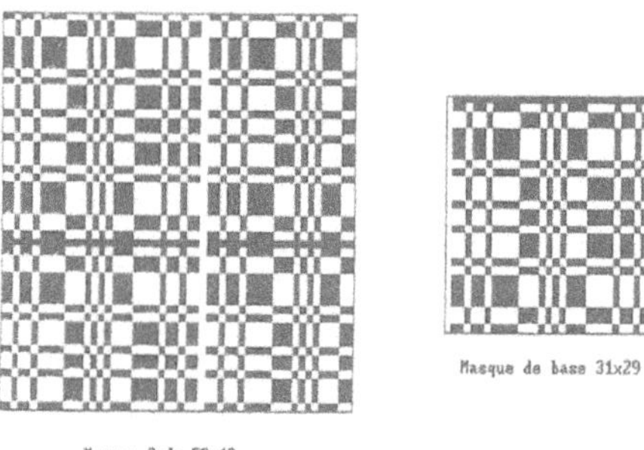

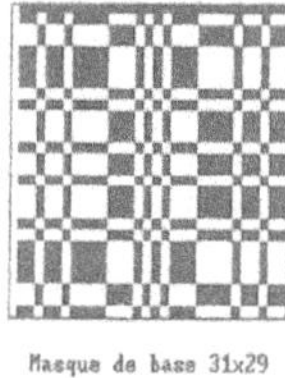

Fig. 1.4. Coded masks for the SIGMA telescope, in the γ-rays energy range 30 keV-1.5 MeV. The Field of View is 4.4 x 4.2 degrees.

1.7 Doppler Imaging

The goal here is to "image" the surface of a star having a diameter smaller than the angular resolution of an optical telescope. One uses the fact that the emission of a given spectral line varies from point to point on the star surface, and has a line-of-sight velocity, hence a Doppler effect on the line frequency, depending on the point, due to the stars rotation. The intensity and frequency of the line become periodic and time-dependent. Assuming a proper model of the emitting zone, latitude, longitude and extension of this zone (i.e. an "image") can be obtained. The sensitivity can be increased by using a set of lines and a spectral mask. The interesting fact about this method is that a unique solution may be found by inversion, even for rather complicated distributions of intensity on the star surface.

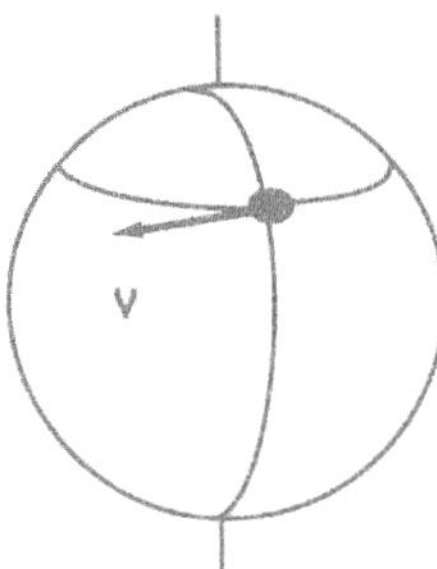

Fig. 1.5. Doppler Imaging. The spot (gray) gives one or several emission lines in the star spectrum; the wavelength and intensity of these are modulated by the star rotation. Velocity V, projected on the line of sight, depends on the spot location.

1.8 Aberrations

Let consider an optical system forming an image. When, from the point of view of geometrical optics, the image of a point is a point, the system is diffraction-limited and free of aberrations (perfect stigmatism). Then diffraction modifies this "geometric point" according to diffraction theory (Chap.2).

Consider the case of a paraboloïd mirror, perfectly stigmatic at its focus for rays parallel to axis. The osculating sphere to the paraboloïd has same focus, but the geometric image given by this spherical mirror is no longer a point : it presents **spherical** aberration [2].

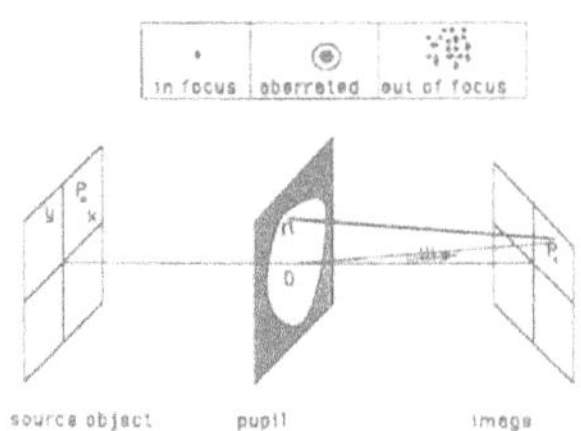

Fig. 1.6. P_0 is object, P_1 is the geometric (Gaussian) image of P_0 through system. The insert gives the traces of rays around P_1.

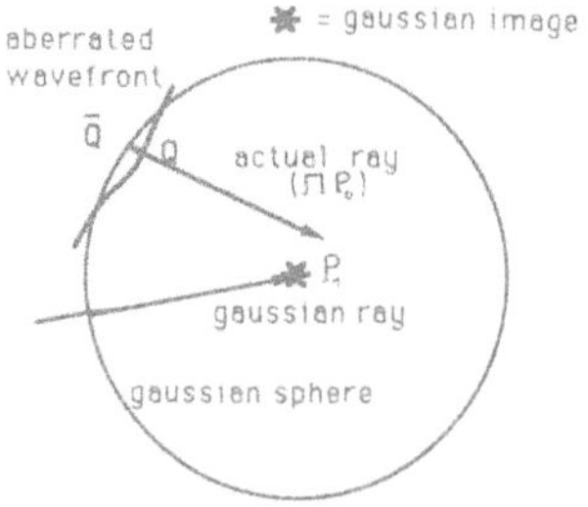

Fig. 1.7. The formation of the gaussian image P_1 (perfect spherical wave fronts) and the aberrated intensity distribution around P_1, produced by the aberrated wavefront. The optical path is $\Phi = \Phi(P_0; \Pi) = [P_0; \Pi] = [\bar{Q}Q]$. From Ref[2]

On Figures 1.6 and 1.7 is shown the process of formation of the aberrated intensity distribution around P_1. The optical path Φ between P_0 and P_1 can be expressed as a function of coordinates of P_0 (source) and of Π (part of the system used by the ray). If there are departures from axis of system, Φ can be developped in limited power expansion of P_0 and Π coordinates. Let assume the system to be rotationnally symetric, define position of Π by polar coordinates ρ, α, define position of P_0 by distance y_0 to axis, which is both convenient and not limiting the generality (Fig.1.8), then

$$\Phi^{(4)} = -\frac{1}{4}B\rho^4 - Cy_0^2\rho^2\cos^2\alpha - \frac{1}{2}Dy_0^2\rho^2 + Ey_0^3\rho\cos\alpha + Fy_0\rho^3\cos\alpha$$

This polynomial shows the lowest degree terms of aberrations, known as **the five primary SEIDEL aberrations.**

- all zero if $\rho = y_0 = 0$
- B, C, D, E, F are coefficients characterizing the system quality
- the values of these five terms may be as small as $\frac{\lambda}{100}$ or as large as 100λ depending on the system.

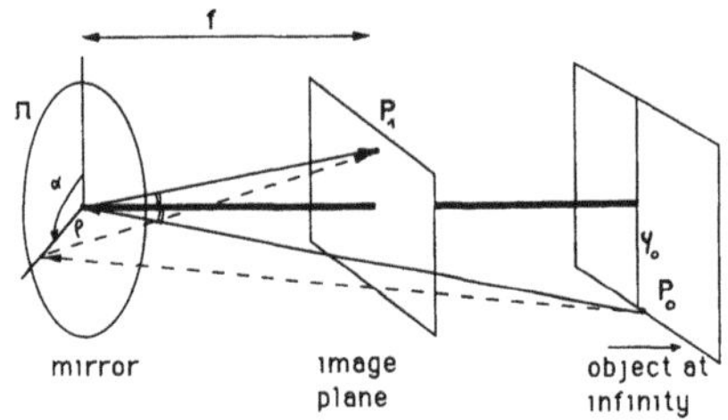

Fig. 1.8. This diagram shows the meaning of ρ, α, y_0 on the simple case where object is at infinity (and defined by angle $\theta_0 = lim\ y_0/\text{distance}$), and system reduces to a simple mirror (telescope primary). From Ref.[2].

Let characterize these aberrations in the following list :

- Spherical aberration $-\frac{1}{4}B\rho^4$. It affects the on-axis image ($y_0 = 0$) for a finite diameter D aperture. Hence, for given f, it varies as D^4.
- Coma $+y_0\rho^3\cos\alpha$. It affects the off-axis image since it is slightly y_0 dependent, but is strongly aperture dependent (ρ^3).
- Astigmatism $-Cy_0^2\rho^2\cos^2\alpha$.
- Field curvature $-\frac{1}{2}Dy_0^2\rho^2$. It affects more the off-axis image, and is not circularly symmetric for $C \neq 0$.
- Distorsion $Ey_0^3\rho\cos\alpha$. One shows that this leads to a stigmatic but distorted image.

Figure 1.9 shows the corresponding wavefront.

1.9 Overview of Images in Astronomy

We cannot describe here in detail the diversity of astronomical instruments, namely the telescopes, aimed at the production of images. They are described in references [3], [4], [5].

Fig.1.10 gives an overview of the angular resolution achieved to date in images formed in the spectral domain extending from 0.1 μm to 1 cm (ultraviolet to radio), limited either by diffraction or by the atmospheric turbulence (see Chap.3), obtained with the most powerful astronomical instruments. This graph clearly points on interferometric techniques as the tool to achieve high angular resolution (Chap.5).

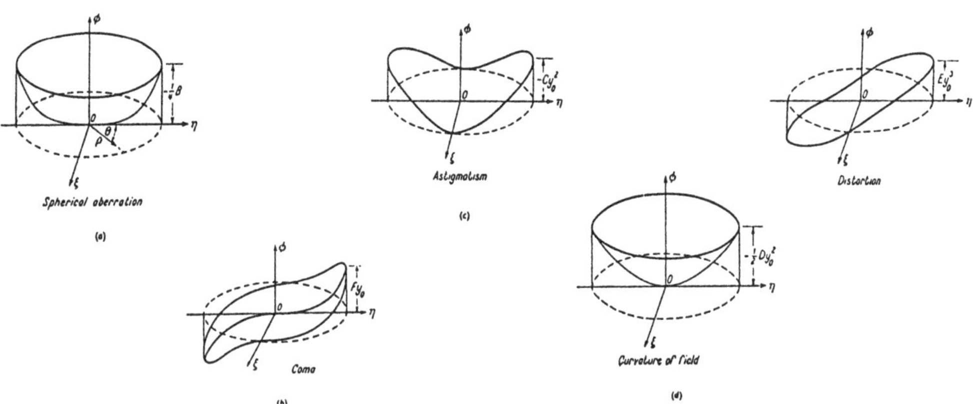

Fig. 1.9. The five optical aberrations. Φ is the departure of the wavefront from the gaussian sphere, ξ and η are the coordinates in the exit pupil of the system (from "Born & Wolf - *Principles in Optics*", Ref.[2])

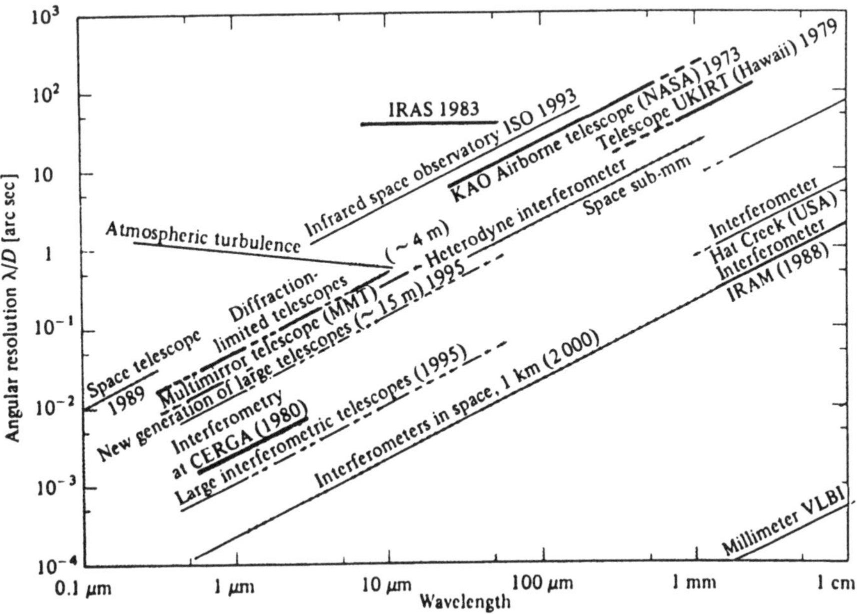

Fig. 1.10. Angular resolution in the wavelength range 100 nm – 1 cm obtainable in 1985 (*heavy lines*) and foreseen for the year 2000 (*lighter lines*). All lines with unit slope correspond to the fundamental diffraction limit. The gain in angular resolution and the extended spectral coverage for space-borne instruments is obvious. The grey areas show the absorption of the Earth's atmosphere (From Léna P., *Observational Astrophysics*, Springer 1988).

The format of astronomical images is heavily dependent on the format of the detectors themselves. Fig.1.11 gives an overview of the extreme variability of these across the electromagnetic spectrum. The outstanding position of the photographic plate emerges, even if it sensitivity is nowdays surpassed by other types of photoelectric detectors.

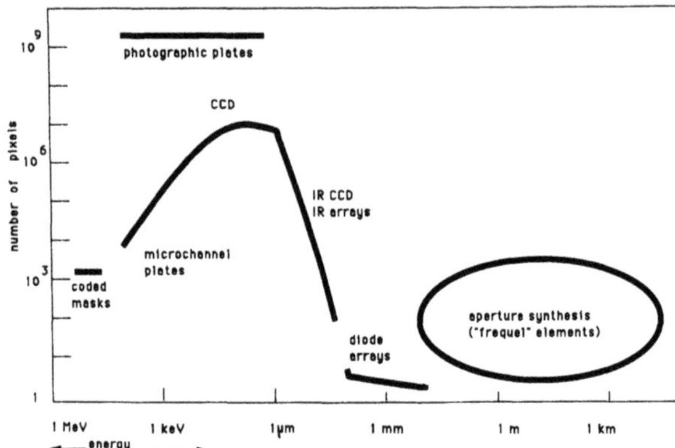

Fig. 1.11. The format of astronomical detectors, or astronomical images at radiowavelengths, across the electromagnetic spectrum.

2. Optical Imaging

The notion of **wave coherence** is central in imaging, especially in the interferometry techniques developed at radio and currently at optical wavelength (Chap.5). It is therefore useful to recall the main concepts of coherence.

2.1 Coherence of the Electromagnetic Field

Let $V(\boldsymbol{r},t)$ be the electric field, a random, stationary process. We define the complex degree of coherence

$$\gamma_{12}(\tau) = \frac{\langle V_1(t)V_2^*(t+\tau)\rangle^{1/2}}{\left\langle |V_1(t)|^2\right\rangle^{1/2} \left\langle |V_1(t)|^{1/2}\right\rangle}$$

γ_{12} describes simultaneously temporal (τ) and spatial coherence ($\boldsymbol{r}_2 - \boldsymbol{r}_1$). For quasi-monochromatic radiation of spectral width $\Delta\nu$, one describes its power spectrum

$S(\nu) = \exp - \frac{(\nu-\nu_O)^2}{\Delta\nu^2}$ as the average spectral density of $V(t)$and $R(\tau) = \exp - \left(\frac{\tau^2}{\tau_c^2}\right)$ its autocorrelation ,

where τ_c is the coherence time, Δ_ν the spectral width, $\tau_c\Delta_\nu \sim 1, c\tau_c$ the coherence length.

2.2 Zernicke-Van Cittert Theorem. Etendue of Coherence

Consider quasi-monochromatic radiation $\boldsymbol{k}$ and search for area $\perp$ to $\boldsymbol{k}$ and solid angle around $\boldsymbol{k}$ where coherence is maintained, i.e. $|\gamma| \sim 1$. The etendue or throughput is $a^2\theta^2 \sim \lambda^2$, as a first approximation of the throughput where spatial coherence is maintained.

Consider the illumination of a screen by a source of quasi monochromatic radiation. Points of the source are mutually incoherent (atoms). The final illumination is the interference of all signals reaching the screen. Let simplify the case : large distance between source and screen, small source and small screen (Fig.2.2).

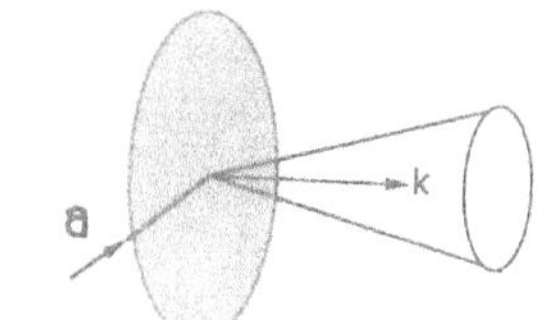

Fig. 2.1. Etendue of coherence

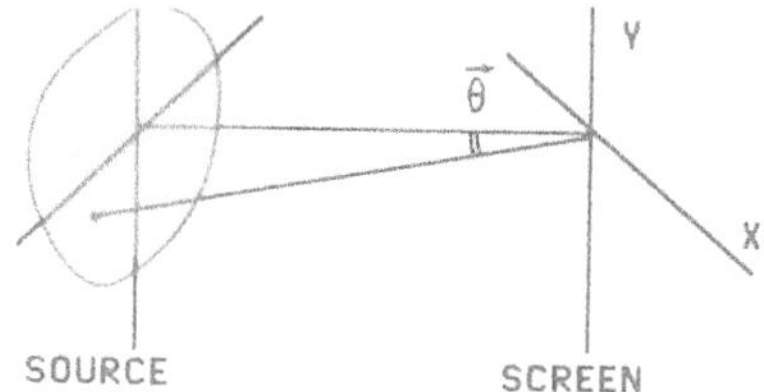

Fig. 2.2. Source and screen

$$\text{Then } |\gamma_{12}(0)| = \left| \frac{\iint I(\boldsymbol{\theta}) \exp\left\{ -2\pi i \left[\frac{X_2 - X_1}{\lambda}\theta_x + \frac{Y_2 - Y_1}{\lambda}\theta_y \right] \right\} d\boldsymbol{\theta}}{\int\int I(\boldsymbol{\theta}) d\boldsymbol{\theta}} \right|$$

$I(\boldsymbol{\theta})$ is the source intensity distribution $\gamma_{12}(0) \rightleftarrows I(\boldsymbol{\theta})$, with a normalization factor.
Example : Circular source of uniform brightness

$$|\gamma_{12}(0)| = \frac{|2J_1(u)|}{u},\ u = \frac{2\pi\theta_0\rho}{\lambda}\ ,\ J_1 \text{ is the Bessel function .}$$

If the source is at infinity, $\theta_0 \to 0$, $\gamma = 1$ over the whole screen.

If the source radius subtends angle θ_0 : $u \leq 2$, $\gamma \geq 0.577$, $E = \pi\rho^2\pi\theta_0^2 = \lambda^2$. This defines the etendue of coherence λ^2. Note that the value 0.577 is convenient, but arbitrary.

Application : Consider a star at 10 pc with radius $R_0 = 1.5\,10m$, no limb darkening (i.e. disc of uniform brightness) hence $\theta_0 = 100$milliarsec. At $\lambda = 0.5\ \mu m$, the radius of coherence is $\rho = \frac{\lambda}{\pi\theta_0} = 63.7cm$ and indeed wavelength dependent.

2.3 Image Formation

Any imaging system (telescope) only uses a limited fraction of the incoming wavefront from the source and diffraction will therefore perturb the ideal image.

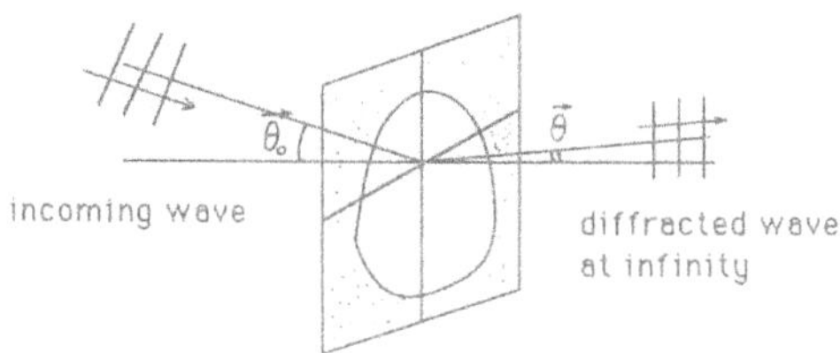

Fig. 2.3. Diffracted wave at infinity (Fraunhofer diffraction)

The amplitude of the diffracted field $V(t)$ is proportional to

$$\iint_{\text{screen}} G(\boldsymbol{r}) \exp\left[-2\pi i\,(\boldsymbol{\theta}_1 - \boldsymbol{\theta}_0) . \frac{\boldsymbol{r}}{\lambda}\right] \frac{d\boldsymbol{r}}{\lambda^2}$$

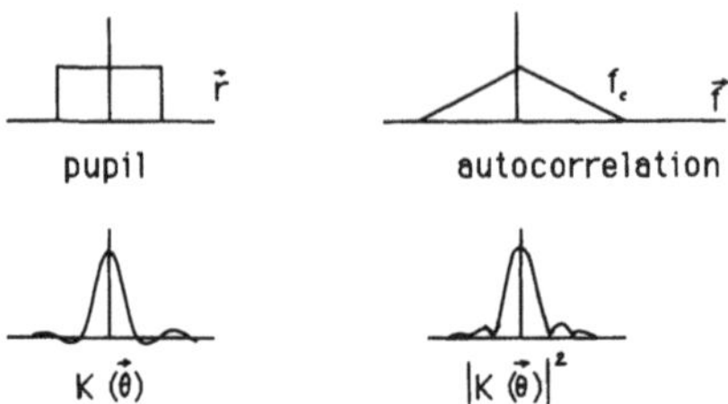

Fig. 2.4. A simple one-dimensional pupil and its filtering MTF. The impulse response $|K|^2$ is shown.

which again shows a pair of Fourier transforms :

$$\text{diffracted field} \underset{FT}{\rightleftarrows} \text{pupil function } G(\boldsymbol{r})$$

The pupil function $G(\boldsymbol{r})$ is defined as 0 or 1 depending on pupil transmission at point $\boldsymbol{r}$. The diffracted pattern at infinity, which is the interference of all diffracted waves, can be brought at finite distance with a lens in its focal plane.

Relation Source-Image. Let define $K(\boldsymbol{\theta}_0;\boldsymbol{\theta}_1)$ as the transmission of the system, i.e. the complex amplitude per solid angle unit around $\boldsymbol{\theta}_1$ as image of object of unit amplitude, zero phase at $\boldsymbol{\theta}_0$. If $V(\boldsymbol{\theta}_0) = \delta(\boldsymbol{\theta}_0 - \boldsymbol{\theta}_0')$, then $V(\boldsymbol{\theta}_1) = K(\boldsymbol{\theta}_0';\boldsymbol{\theta}_1)$. The image intensity is then $|K|^2$.

We define **isoplanicity** when there is translation invariance in a certain field of view

$$K(\boldsymbol{\theta}_0';\boldsymbol{\theta}_1) = K(\boldsymbol{\theta} B_1 - \boldsymbol{\theta}_0)$$

Then

$$V(\boldsymbol{\theta}_1) = \iint_{\text{source}} V_0(\boldsymbol{\theta}_0) K(\boldsymbol{\theta}_1 - \boldsymbol{\theta}_0) d\boldsymbol{\theta}_0$$

and

$$K(\theta) \propto \iint_{\text{pupil}} G\left(\frac{\boldsymbol{r}}{\lambda}\right) \exp\left(-2\pi i \frac{\boldsymbol{r}}{\lambda}.\boldsymbol{\theta}\right) \frac{d\boldsymbol{r}}{\lambda^2}$$

with a proportionality normalization factor.

These results lead to the incoherent illumination case giving the image I from the object O

$$I(\boldsymbol{\theta}_1) = \iint_{\text{source}} O(\boldsymbol{\theta}_0) \left|K(\boldsymbol{\theta}_1 - \boldsymbol{\theta}_0)\right|^2 d\boldsymbol{\theta}_0$$

which has the following consequences :

- V_{image} is a convolution : $V(\boldsymbol{\theta}) = V_O(\boldsymbol{\theta}) * K(\boldsymbol{\theta})$ easy to express in Fourier space as a multiplication $\tilde{V}(\boldsymbol{f}) = V_O(\boldsymbol{f}).\tilde{K}(\boldsymbol{f})$
- $\tilde{K}(\boldsymbol{f})$ is the amplitude transfer function of the pupil and acts as a spatial filter.
- I_{image} is a convolution : $I(\boldsymbol{\theta}) = O(\boldsymbol{\theta}) * |K(\boldsymbol{\theta})|^2$, or expressed in Fourier space

$$\tilde{I}(\boldsymbol{f}) = \tilde{O}_0(\boldsymbol{f}).\tilde{T}(\boldsymbol{f})$$

- $\tilde{T}(\boldsymbol{f}) = \dfrac{\iint G\left(\boldsymbol{f} + \frac{\boldsymbol{r}}{\lambda}\right) G^*\left(\frac{\boldsymbol{r}}{\lambda}\right) \frac{d\boldsymbol{r}}{\lambda^2}}{\iint G\left(\frac{\boldsymbol{r}}{\lambda}\right) G^*\left(\frac{\boldsymbol{r}}{\lambda}\right) \frac{d\boldsymbol{r}}{\lambda^2}}$ is the modulation transfer function (MTF) or spatial filter created by the pupil. Fig.2.4 gives a one dimensional example.

Point-Spread Function (PSF). Let assume a point source $\delta(\boldsymbol{\theta}-\boldsymbol{\theta}_0)$at $\boldsymbol{\theta}_0$. The image is $T(\boldsymbol{\theta}) = FT\left[\tilde{T}(\boldsymbol{f})\right]$. In the case of a circular pupil

$$I_1(\theta) = \left[\frac{2J_1\left(2\pi\frac{r_0}{\lambda}\theta\right)}{2\pi\frac{r_0}{\lambda}\theta}\right]$$

is the Airy function with its diffraction rings. Numerically, a 6 m diameter telescope (optical) at $\lambda = 0.5\mu m$ gives $w_c = 60\, arc\, sec^{-1}$, $\theta_c = \frac{1.22\lambda}{D} = 20\, milliarcsec$ while a 100 m radiotelescope at $\lambda = 18$ cm gives $f_c = 3.10^{-1}$, $\theta_c = 8\, arc\, sec.$

The Rayleigh criterion for resolution is

$$|\boldsymbol{\theta}_2 - \boldsymbol{\theta}_1| \geq 1.22\frac{\lambda}{D}$$

the first zero of point source 1 coinciding with the maximum of point source 2.

The MTF is a finer description of the system properties than the Rayleigh criterion.

Disconnected Pupils. It is possible to build the pupil function $G(\boldsymbol{r})$ as a multiple aperture, therefore extending arbitrarily the frequency cut-off, making $f_c = B/\lambda$ as large as needed.

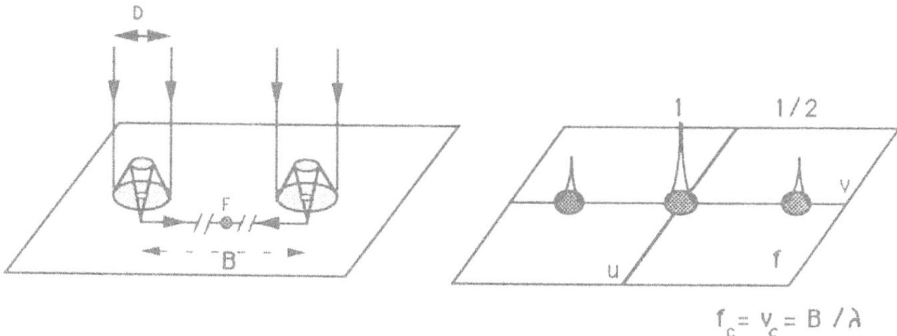

Fig. 2.5. Two sub-pupils of diameter D separated by a distance $\boldsymbol{B}$ form the pupil $G(\boldsymbol{r})$. The MTF $\tilde{T}(\boldsymbol{f})$ shows the band-pass around f_c. F is the common focus.

• The PSF contains only a limited range of spatial frequencies, i.e. the image appears as "fringes". The approximate correspondance between one object-point and one image-point is washed out by the spatial filtering. A single value of $\boldsymbol{B}$, distance between sub-pupils, is no longer adequate to get the image.

• A complete exploration of the $\boldsymbol{f}$ –domain for $|\boldsymbol{f}| \leq f_c$ allows the determination of $\tilde{I}(\boldsymbol{f})$ hence of $I(\boldsymbol{\theta})$ as if it where given by an optical system being diffraction-limited of diameter λf_c. A measurement made at a single frequency is a sample in Fourier space (it could, by analogy with *pixel*, be called *frequel*), of width $\delta f = \frac{2d}{\lambda}$. This is the basic principle of Aperture Synthesis, building up the image by successive measurements carried in the frequency domain.

The Normal Image Revisited. Consider the pupil $G\left(\frac{\mathbf{r}}{\lambda}\right)$ entirely made of N identical sub-pupils, $g_i(\mathbf{r}/\lambda)$, $i = 1, N$ (Fig.2.6). Linearity of field $V(t)$ superposition leads to

$$K(\boldsymbol{f}) = \sum_N k_i(\boldsymbol{f})$$

and the transfer function is, with designating convolution

$$T(\boldsymbol{f}) \propto K(\boldsymbol{f}) * K(\boldsymbol{f}) = \sum_N k_i * k_i(\boldsymbol{f}) + \sum_i \sum_j k_{i \neq j}(\boldsymbol{f}) * k_j(\boldsymbol{f})$$

Each pair of sub-pupils contribute to a spatial frequency in the image and the image is the sum of all Fourier components.

A pupil is considered as <u>redundant</u> if two sub-pupils contribute to the <u>same</u> frequency component.

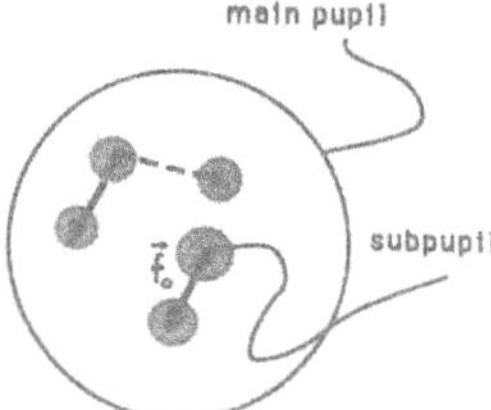

Fig. 2.6. Breakdown of a main pupil into several subpupils. Their size is arbitrary here. See Chap.3 for a practical example. A given spatial frequency $\boldsymbol{f}_0$ may be obtained from several pairs, as illustrated here (redundancy).

Strehl Ratio. The Strehl ratio is a simple criterion to evaluate the quality of the image given by a system as compared to the diffraction-limited image a perfect system would give of the same object.

$$S = \frac{I_{\max}(\boldsymbol{\theta})}{I_{\text{diffr}}(\boldsymbol{\theta})} = \frac{\text{Maximum brightness}}{\text{Maximum brightness of diffraction} - \text{limited image}}$$

One has the simple relation

$$I_{\max}(\boldsymbol{\theta}) = \iint_{\boldsymbol{f}} \tilde{I}(\boldsymbol{f}) d\boldsymbol{f},$$

where $\tilde{I}$ is the actual spectrum, and

$$I_{\text{diffr, max}}(\boldsymbol{\theta}) = \iint_{\boldsymbol{f}} \tilde{T}(\boldsymbol{f}) d\boldsymbol{f},$$

where $\tilde{T}$ is the diffraction-limited MTF.

Coherent and Incoherent detection.

• Coherent detectors measure average amplitude of field V_O

$$V(t) = V_O \exp(-2\pi i\nu\, t + \varphi)$$

This is obtained by a non-linear element (diode) mixing the input field V_{LO} with a local field (Local Oscillator or LO).

$$|V(t) + V_{LO}(t)|^2 = V(t)V^*(t) + V_{LO}(t)V^*_{LO}(t) + 2Re\,[V(t)V^*_{LO}(t)]$$

The mixed term contains the intermediate frequency (IF)

$$V(t)V^*_{LO}(t) = V_0 V_{LO} \cos\,[2\pi(\nu - \nu_0)t + \varphi]$$

$|\nu - \nu_0| \leq 1GHz$ can be measured in **amplitude** and **phase**. Hence $V(t)$ is fully determined by V_O, φ, ν. Coherent detection is possible in any frequency domain where a mixer can be obtained by technological means, and where the method is not paying a noise penalty.

In astronomy, mixers are built for $\lambda \geq 0.5\,\mathrm{mm}$, (radioastronomy) and $\lambda \gtrsim 10\,\mu\mathrm{m}$ (submillimetric and far IR astronomy) but the noise analysis shows that coherent detection is only adequate when thermal noise dominates, i.e. when $h\nu/kT_{\mathrm{source}} \ll 1$.

As a consequence, in coherent detection, if the detector (mixer) accepts more throughput than a single coherence etendue, interference is destructive and detection inefficient. Hence $s\omega \sim \lambda^2$. Fundamentally, a coherent detector is limited to *one* pixel of the image.

Radio-images are made by raster-scanning the telescope, since arrays of diodes are not yet available.

• Incoherent detectors measure the intensity of the field, i.e. the average quantity

$$< V(t)V^*(t) >_{\text{time constant of the detector}}$$

Time-constant varies from ms to ns. Incoherent detectors (quadratic) are : eye, photomultipliers and all quantum detectors such as scintillators, spark chambers... Images are easily obtained by mosaïcs of incoherent detectors. Fig.1.11 gave an overview of their format.

Methods of Aperture Synthesis. Aperture synthesis is a two-step process to increase the angular resolution in an image : a) combine coherently the light of two or more separate telescopes over a long baseline (Fig.2.5); b) vary the baseline configuration to explore the maximum number of spatial frequencies and therefore **synthetize** the full aperture.

Several possibilities exist to achieve this coherent combination :

• Combine directly the beams from each subpupil in a common focus on a quadratic, incoherent detector and therefore measure γ. This is the Michelson (1920) and Labeyrie (1976) method (Chap.5).

• Detect locally the electric fields $V_1(t)$ and $V_2(t)$ at subpupils 1 and 2, build the intermediate frequency (IF) signals by beating them with a local oscillator and carry the IF to a correlator which will measure the requested quantity, i.e. the field coherence

$$\gamma_{12}(\tau) = \langle E_0 E_1(t) \exp\,[2\pi i(\nu - \nu_0)t] \times E_0 E_2(t+\tau) \exp\,[-2\pi i(\nu - \nu_0)(t+\tau)] \rangle_{\text{time}}$$

This is the method in radio interferometry and also in the mid-infrared (10.6 μm) where coherent detection is satisfactory.

• Measure the time-correlation-of photocurrents at pupils 1 and 2, each signal being detected by a quantum detector integrating over a finite time Δt.

$$i_k(t) = \langle E_k(t')E^*_k(t') \rangle_{(t,t+\Delta t)} \quad k = \text{resp.}\, 1,2$$

$$\gamma_{12}(\tau) = \langle i_1(t) i_2(t+\tau) \rangle_t$$

This is called **intensity interferometry** (Hanbury-Brown and Twiss, 1950) but one shows that it is fundamentally limited to very bright sources like $m_V \sim 1$ stars.

3. Imaging through a Turbulent Medium

Fluctuating index of refraction causes optical path variations and phase fluctuations over a wavefront, hence a degradation of the image. Fluctuations may be due plasma inhomogeneities (interstellar and ionospheric effects on radio waves), H_2O concentration fluctuations (tropospheric effects on milimetric waves), air temperature fluctuations (tropospheric and stratospheric effects on optical waves). A similar formalism may be used in all these cases; to be specific, it is presented here for atmospheric temperature fluctuations. For additional discussion, see [3], [6], [7], [8].

3.1 Temperature Fluctuations

Let $T(\boldsymbol{r})$ be the temperature, $\Theta(\boldsymbol{r}) = T(\boldsymbol{r}) - < T(\boldsymbol{r}) >$, $\Phi_\Theta(K)$ its power-spectrum. Kolmogorov (isotropic) turbulence gives

$$\Phi_\Theta(\boldsymbol{K}) \propto \boldsymbol{K}^{-5/3}$$

between an inner scale ($K_{max}^{-1} \sim 1mm$ at STP, fixed by viscosity) and an outer scale (K_{min}^{-1} = meter to kilometer). Define the Fourier transform of $\Phi(\boldsymbol{K})$, the covariance $B_\Theta(\boldsymbol{\rho}) =< \Theta(\boldsymbol{r})\Theta(\boldsymbol{r}+\boldsymbol{\rho}) >$, and the structure function, useful to avoid diverging integrals

$$D_\Theta(\boldsymbol{\rho}) =< |\Theta(\boldsymbol{\rho}+\boldsymbol{r}) - \Theta(\boldsymbol{r})|^2 >= C_T^2 \rho^{2/3}$$

C_T^2, the structure constant of temperature fluctuations, sets the average amplitude by scaling the spectrum $\Phi_\Theta(\boldsymbol{K}) = 0.033 C_T^2 K^{-11/3} = \frac{\Phi_\Theta(K)}{4\pi K^2}$ and is directly related to the index of refraction ($C_n = 8\,10^{-5} T^{-2} P_{mb} C_T$). Orders of magnitude of tropospheric fluctuation are, for a layer of thickness $\Delta h : C_n^2 \Delta h \sim 1$ to $10.10^{13}\, cm^{1/3}$, which immediatly gives for example the rms differential fluctuation $[D_\Phi(\rho = 1m)]^{1/2} \sim 0.009$ Kelvin for $\Delta h = 100m$.

Time dependence. One generally assumes a "frozen" turbulence $\Phi_\Theta(K)$, carried by a wind V. Fluctuation become time-dependent with temporal frequency $f = VK$, and a cut-off $f_c \sim V K_{\max} \sim kilohertz$.

3.2 Propagation of the Wavefront

Fig.3.1 shows the incoming unperturbed wavefront, the turbulent layer, (altitude $h, h + \Delta h$), the wavefront at exit and at the telescope input. Assume quasi monochromatic wave, geometric optics approximation, frozen turbulence, average values taken over space $\boldsymbol{x}$.

The perturbed wavefront is

$$\psi_h(\boldsymbol{x}) = \exp\left[i\varphi_h(\boldsymbol{x})\right]$$

with the phase $\varphi_h(\boldsymbol{x}) = k \int_{\Delta h} n(\boldsymbol{x}, h) dh$, $k = 2\pi/\lambda$, the phase being a random variable of $\boldsymbol{x}$ at time t, assumed to be gaussian if $\Delta h \gg K_{min}^{-1}$ by addition of independent fluctuations.

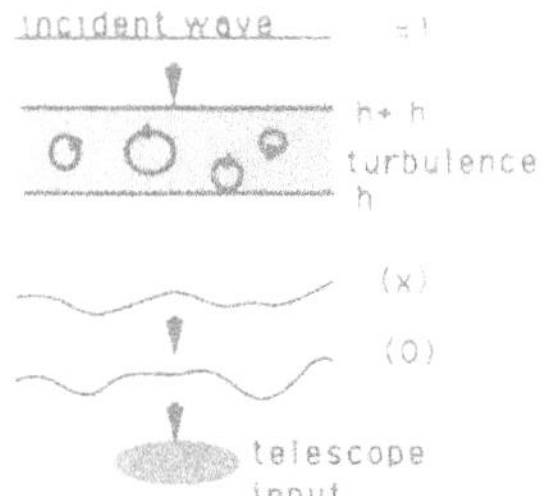

Fig. 3.1. The propagation of the wavefront across the atmosphere

The phase structure function is $D_{\varphi,h}(\boldsymbol{x}) = \left\langle |\varphi_h(\boldsymbol{x}+\boldsymbol{\xi}) - \varphi_h(\boldsymbol{\xi})|^2 \right\rangle$, leading to the covariance of the wavefront

$$B_{\psi,h}(\boldsymbol{x}) = \langle \psi_n(\boldsymbol{x}+\boldsymbol{\xi})\psi_h^*(\boldsymbol{\xi})\rangle = \exp\left\{-1.45k^2(C_n^2\Delta h)x^{5/3}\right\}.$$

Since covariance is invariant by Fresnel diffraction, $B_{\psi,O}(\boldsymbol{x}) = B_{\psi,h}$.

Correlation Length. This quasi-gaussian (5/3 instead of 6/3) function leads to a log-decrement and characteristic correlation distance $x_c \sim \left(1.45k^2C_n^2\Delta h\right)^{-5/3}$. Typically for $C_n^2\Delta h = 4.10^{-13}$, $\lambda = 0.5\mu m$, we find $x_c = 17cm$, a value strongly wave-length dependent as $\lambda^{6/5}$: the wavefront is correlated on longer distances in the infrared. Below, we discuss the quantity r_o, close to x_c, but directly related to the appearance of the image.

Power-Spectrum of the Phase. This spectrum $S_\varphi(\boldsymbol{f})$ is the Fourier transform of $B_\varphi(\boldsymbol{x})$ in the spatial frequency two-dimensional space $\boldsymbol{f}$. Hence

$$S_\varphi(\boldsymbol{f}) = 9.7\,10^{-3}k^2 f^{-11/3}\left[C_n^2(h)\Delta h\right]$$

and the spectrum of optical path fluctuations $z = \lambda\varphi/2\pi = \varphi/k$

$$S_z(\boldsymbol{f}) = 9.7\,10^{-3} f^{-11/3}\left[C_n^2(h)\Delta h\right]$$

<u>which is wavelength independent</u>. We deduce the variance of the phase difference between two points $D_\varphi(\boldsymbol{x}) = \left\langle |\varphi_o(\boldsymbol{x}+\boldsymbol{\xi}) - \varphi_o(\boldsymbol{\xi})|^2 \right\rangle = 1.1\,10^2\lambda^{-2}(C_n^2\Delta h)x^{5/3}$, or expressed in optical path difference

$$\sigma_z(x) = 1.7(Cn^2\Delta h)x^{5/6}$$

Numerical application shows $\sigma_2(x) \sim$ microns for $x \sim$ meters (Fig.3.2).

Angle-of-arrival fluctuations. The ray, normal to wavefront, makes with the vertical the angles $\alpha(\boldsymbol{r}) = -\frac{\partial}{\partial x}z(\boldsymbol{r}), \beta(\boldsymbol{r}) = -\frac{\partial}{\partial y}z(\boldsymbol{r})$ with power spectra $S_{\alpha(\text{resp.}\beta)}(\boldsymbol{f}) = \lambda^2u^2S_{\varphi,o}(\boldsymbol{f})$, (resp. v^2). The variance of this angle is $\sigma_m^2 = \langle\alpha^2\rangle + \langle\beta^2\rangle = \lambda^2\int f^2 S_{\varphi,o}(\boldsymbol{f})d\boldsymbol{f}$. [6]

3.3 Image Formation

One defines a new pupil $G(\boldsymbol{r})\psi_o(\boldsymbol{r},t)$ which "includes" the instantaneous phase perturbation. Using the standard image formation (Chap.1), this stochastic pupil creates a MTF random in space and time. In the simple case of a circular pupil (diameter$D$), with $D \gg x_c$, the <u>time-averaged</u> MTF is

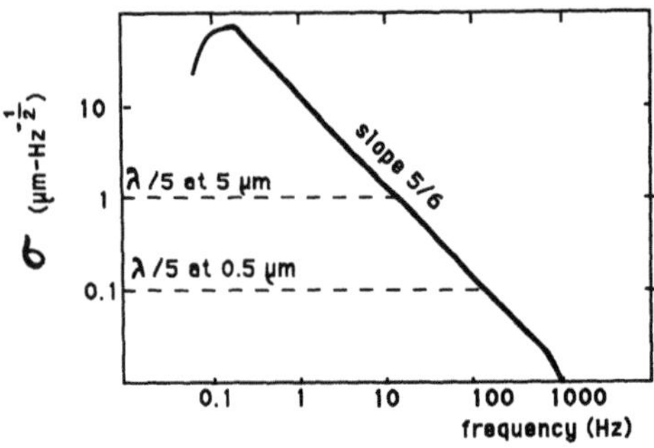

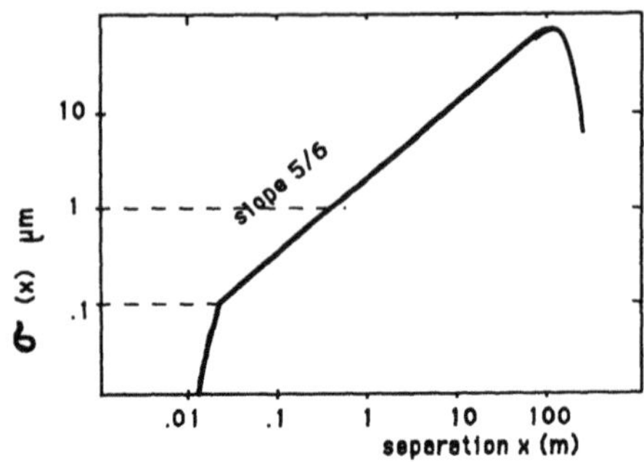

Fig. 3.2. Examples of temporal and spatial rms fluctuations of the turbulence-induced path error δz (t) [upper] or δz(X) for two points distant of X [lower]. High-and-low-cutoffs are due to the scales K_{min} and K_{max} of the turbulence.

$$\left\langle \tilde{T}(f) \right\rangle \simeq B_o(f) = \exp\left[-1.45k^2\left(C_n^2 \Delta h\right)(\lambda f)^{5/3}\right] = \exp\left[-\frac{\lambda f}{x_c}\right]^{5/3}$$

and the time-averaged point spread function (PSF) becomes its Fourier transform. $< \tilde{T} >$ being "almost gaussian", the size of the PSF, or image, is an halfwidth of order of λ / x_c. This is the seeing disc size.

Fried's Parameter $r_o(\lambda)$. This parameter is the diameter r_o of a circular aperture leading to the same central intensity as in a pure diffraction-limited imaging. Using the Strehl criterion

$$\iint \tilde{T}_{Airy,r_o}(f)df = \iint \tilde{T}_{turbulence}(f)df$$

and solving for r_o gives $r_o(\lambda) = 0.185\lambda^{6/5}\left(C_n^2 \Delta h\right)^{-3/5}$. Although more correctly defined r_o is close to and has the same physical meaning as x_c. Numerically, with the same values as above, $r_o = 35cm \sim x_c$. We can understand r_0 as a length over which, in the statistical sense, the phase varies by less than one radian, or the optical path by less than $\lambda/2\pi$.

Instantaneous MTF. At any given time, $\tilde{T}(f)$ is complex, with random phase in time, for any $f \gtrsim r_o/\lambda$. Only its time-averaged modulus may be non-zero. In the simple case considered here, we get $< \left|\tilde{T}(f)\right|^2 >= \frac{\sigma}{A}\tilde{T}_{Airy}(f)$, with $\sigma = \int B_o^2(u)du = 0.34\left(\frac{r_o}{\lambda}\right)^2$. σ is called the Fried's coherence area, a central notion in interferometric imaging.

$A = \frac{\pi}{4}\left(\frac{D}{\lambda}\right)^2$ is the pupil area in the same units. The quantity $< \left|\tilde{T}(f)\right|^2 >$ acts as a low-pass spatial filter with an attenuated but non-zero high frequency transmission. This non-zero transmission is the basis of speckle interferometry, allowing to restore the object high frequencies despite the atmospheric phase corrugation, since these high spatial frequencies are partly transmitted.

$$< \left|\tilde{I}(f)\right|^2 > = \quad < \left|\tilde{T}(f)\right|^2 > \quad \cdot \left|\tilde{O}(f)\right|^2$$

measured — Known if r_o is known or measured on a reference — Object

The instantaneous MTF contains also phase-distorted information, therefore methods of image $\tilde{O}(f)$ retrieval exist, leading not only to $\left|\tilde{O}(f)\right|$ but also to arg $\tilde{O}(f)$, hence to an estimate of $O(x)$ from its complete complex spectrum in the frequency interval $(0, D/\lambda)$.

3.4. Image Motion.

Since the image is a random distribution of intensity, its center of gravity will fluctuate with time. To compute it, one integrates the angle-of-arrival fluctuation over the frequencies from the largest K_{max} to the smallest D^{-1}(frequencies$f < D^{-1}$ are averaged out). The *rms* angle-of-arrival fluctuation σ_m is

$$\sigma_m^2 = \lambda^2 \int_{K_m}^{D^{-1}} S_{\varphi_o}(f) df$$

with $K_{max} \ll D^{-1}$, and $K_{max} \sim 0$ (not always true)

$$\sigma_m^2 = 6(C_n^2 \Delta h)/D^{1/3} = 0.14 \text{ arcsec for } D = 1m$$

3.5. Conclusion.

The basic understanding of the phenomena described in this chapter led to the possibility of crossing the classical seeing limit (angle $\lambda/r_0(\lambda)$) imposed on large ground based telescopes and to operate in a diffraction-limited resolution (λ/D). Speckle interferometry at visible and infrared wavelengths has developed from these premises : the recovery of the phase led to many developments, often inspired by similar problems encountered at radio wavelengths. The understanding of phase perturbation is also crucial for multitelescope optical interferometry (Chap.5) on decametric or hectometric baselines. Finally, the real-time correction of these perturbations can also be considered by using adaptive optics, as developed in the next Chapter.

3.6. Summary.

It is useful to summarize in a Table the various quantities related to the atmospheric perturbations, as introduced in this Chapter. Whenever possible, we give two expressions, using either the turbulence intensity due to a layer $\Delta h(C_n^2 \Delta h)$, or the Fried's parameter r_o.

- covariance $B_f(x) = \langle f(x+\xi) f(\xi) \rangle$
- structure function $D_f(x) = \left\langle |f(x+\xi) - f(\xi)|^2 \right\rangle$
- Temperature fluctuations structure function

$$D_\Theta(\rho) = C_T^2 \rho^{2/3}$$

C_T structure constant of temperature

C_n structure constant of refraction index at pressure P and temperature T.

$$C_n = 8.10^{-5} P(mb) T^{-2}(K) C_T$$

- Phase-structure function

$$D_\varphi = 2.9 k^2 \left(C_n^2 \Delta h\right) x^{5/3}$$

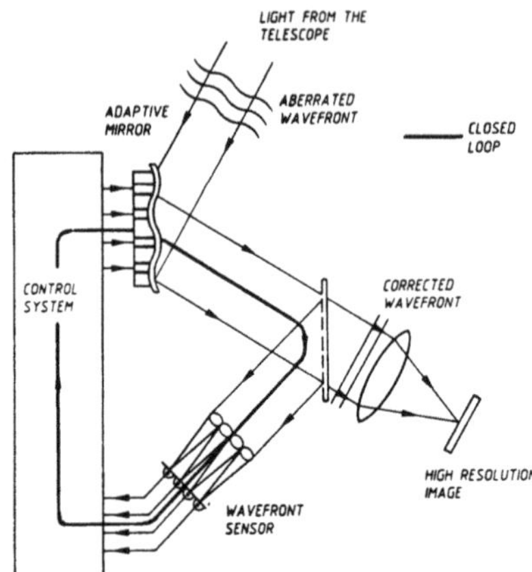

Fig. 4.1. The basic principles of an adaptive optics system, showing the incoming perturbed wavefront, the correcting element (deformable mirror), the wavefront sensor and the final image formed on an image detector.

$$D_\varphi = 6.88 \left(\frac{x}{r_o}\right)^{5/3}$$

- Phase power-spectrum (spatial)

$$S_\varphi(f) = 9.7\,10^{-3} k^2 \left(C_n^2 \Delta h\right) f^{-11/3}$$

$$S_\varphi(f) = 0.023\, r_o^{-5/3} f^{-11/3}$$

- Phase excursion (rms) between two points distant of x.

$$\sigma_\varphi(x) = [D_\varphi(x)]^{1/2} = 1.7k \left(C_n^2 \Delta h\right)^{1/2} x^{5/6}$$

$$\sigma_\varphi(x) = 2.62 \left(\frac{x}{r_o}\right)^{5/6}$$

- Optical path difference

$$z = \frac{\lambda}{2\pi}\varphi = \frac{\varphi}{k}$$

- Fried's parameter

$$r_o = 0.185\, \lambda^{6/5} \left(C_n^2 \Delta h\right)^{-3/5}$$

coherence area $\sigma = 0.32 r_o^2$

4. Adaptive Optics

We discuss a new method to restore images degraded by atmospheric turbulence. Instead of an a-posteriori treatment of the distorted image (speckle interferometry), costly in computing time and in sensitivity, and subject to systematic errors, the goal is to restore the original wavefront by creating artificial local phase delays at any instant, to maximize the spatial coherence of the incident wavefront. Paradoxically, the light of the distorted signal itself is used for this restoration [9], [10], [13], [16].

The incoming wavefront is described by the local optical path distorsion at point $\boldsymbol{x}$ of the wavefront and instant t $\Delta z(x, y, t) = \Delta z(\boldsymbol{x}, t)$. The statistics of Δz was discussed in Chap.3.

4.1. Why the use of adaptive optics ?

Its main purpose is to restore the capability of diffraction-limited imaging for large apertures, despite the effects of the atmospheric turbulence.

Moreover, it can improve the signal-to-noise in the image, when this ratio is set by other limitations than pure signal photon noise. This can be illustrated by a few examples:

• **At infrared wavelengths** ($\lambda \gtrsim 2.5\mu m$), the background thermal noise is dominant and varies with beam throughput. For a source close in size to λ/D, i.e. barely resolved by the telescope of diameter D

$$\text{signal/noise} = \frac{N_S S \omega_S}{\sqrt{N_B S \omega_P}} \propto \frac{\omega_S}{\sqrt{\omega_P}}$$

where N_S, N_B are intensities of source and background, ω_S and ω_P the source and pixel solid angles, S the telescope area. Comparing $\omega_P = \omega_s$ (seeing case) to $\omega_P = \left(\frac{\lambda}{2D}\right)^2 \simeq \omega_s$ in the diffraction limited case gives a considerable gain in signal-to-noise for large D values. Compare for instance with $D = 3.6m$, $\lambda = 3.6\mu m$, $\frac{\lambda}{2D} = 0.1$ arc sec, the gain for a seeing of 1 arc sec is $(\omega_s/\omega_P)^{1/2} = \sqrt{10}$.

• **At visible wavelengths**, the sky brightness (airglow) is of the order of 22 to 23 magnitude-arc sec^{-1} and the quantum noise of this background effectively limits the sensitivity of ground-based telescopes for deep surveys of faint objects. Any method to reduce the image size of a resolved object (i.e. close to λ/D) will enhance the contrast, hence provide an easier detection.

Let compare now with speckle imaging. Assume a source just at the size λ/D, providing n photons per coherence area (r_0^2) per coherence time of the atmosphere (τ_c), with $n > 1$.

Signal-to-noise ratio (SNR) in the image spectrum $\tilde{I}(\boldsymbol{f})$ is unity per exposure during τ_c in speckle mode [12]: this result is a basic conclusion of speckle imaging, hence $SNR = 1$. If adaptive mode is used, $N = (D/R_0)^2$ being the number of "phase cells" over the pupil, one gets $SNR = \sqrt{N} = D/r_0 \gg 1$. This simple reasoning shows again that direct imaging is better, if feasible, than speckle imaging.

4.2. Wavefront Analysis

To describe the perturbed wave over a circular aperture one decomposes it in Zernike polynomials, forming a complete set of orthogonal functions inside a circle:

$$V_n^l(X,Y) \Longrightarrow V_n^l(\rho \sin \Theta, \rho \cos \Theta) = R_n^l(\rho) e^{il\Theta}$$

with $l \gtrless 0$, $n \geq 0$, $n \geq l$, $n - |l|$ even, n, l integers.

The orthogonality is

$$\iint_{\rho^2 \leq 1} V_n^{l*}(X,Y) V_{n'}^{l'}(X,Y) dX\,dY = \frac{\pi}{n+1} \delta_{ll'} \delta_{nn'} \text{ (δ is Kronecker symbol)}$$

$R_n^l(l)$ are polynomials in ρ, $n \leq$ degree $\leq l$, with terms $n, n-2, ..., |l|$. For a real perturbation, one uses $Re\left[V_n^l\right]$. High l values correspond to high spatial frequencies.

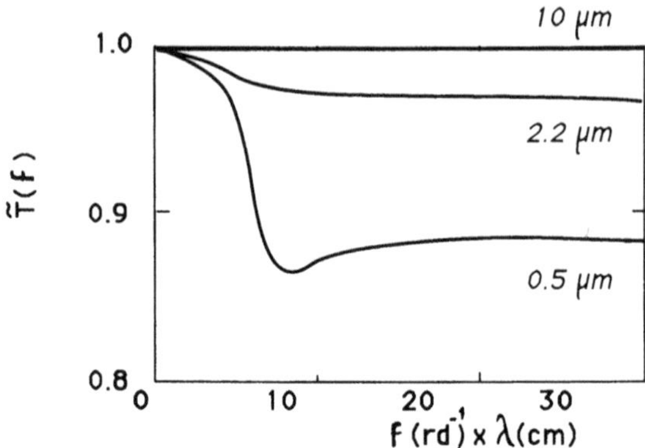

Fig. 4.2.. The result of scintillation on the MTF $\tilde{T}(\boldsymbol{f})$ after full correction of the phase errors by an adaptive system. Each curve refers to a wavelength of operation. Computations are made for $r_0(0.5\,\mu m) = 10\,cm$, i.e. a 1 arc sec seeing, and a turbulent layer located at h = 10 Km. In abcissa, the scale is spatial frequency f multiplied by wavelength (λ), hence is different for every curve

Wavefront sensors (WFS). The goal is to measure the local path error $\Delta z(\boldsymbol{x},t)$. It is fairly obvious that this quantity must be sampled at time intervals shorter than the atmospheric coherence time τ_c, and at spatial distances of the order of r_0 over the wavefront.

• **Local tilt** (α,β) can be obtained at the focus of a small lens (diameter $d \lesssim r_0$) sampling the wave-front. Local values of $\frac{\partial \Delta z}{\partial x}, \frac{\partial \Delta z}{\partial y}$ provide then, by integration, the quantity $\Delta z(\boldsymbol{x},t)$. All measurements must be done within τ_c. The knowledge of the statistics of Δz allows to estimate the error. Known as the **Shack-Hartman** sensing, this has the remarkable property to give a measure of the distorsion even on a object resolved by the aperture of diameter D, as long as it is unresolved by the subpupils (r_0).

• **Curvature sensors** measure the total wavefront curvature or Laplacian

$$\nabla^2(\Delta z) = \frac{\partial^2}{\partial x^2}(\Delta z) + \frac{\partial^2}{\partial y^2}(\Delta z) = r(\boldsymbol{x},t)$$

The knowledge of $r(\boldsymbol{x})$ everywhere on the wavefront <u>and</u> of $\left[\frac{\partial \Delta z}{\partial \rho}\right]_{\rho=1}$ at the edge allows to solve the Poisson equation and to compute the wanted quantity $\Delta z(\boldsymbol{x},t)$.

• **Shear measurements** make the wavefront interfere with itself after some geometric modifications. The resulting intensity is then modulated by the phase difference between the two interfering points $\varphi(\boldsymbol{x}_1) - \varphi(\boldsymbol{x}_2)$. Shearing can be done by translation or rotation of the splitted wavefront.

Amplitude errors. In Chap.3, we neglected scintillation, which is an amplitude fluctuation of the wavefront $\psi_0(\boldsymbol{x},t)$ due to variable interference. In a more general case, one would write for the perturbed MTF :

$$\tilde{T}(\boldsymbol{f}) = \exp - \frac{1}{2}\left[D_\varphi(\boldsymbol{f}) + D_\chi(\boldsymbol{f})\right],$$

introducing both phase <u>and</u> amplitude structure fonctions (cf.3.3). If only phase errors $\varphi = 2\pi\,\Delta z/\lambda$ are corrected, the residual pupil appears as a screen with variable transmission on a characteristic scale $\sim r_0$. It can be shown that this residual effect will create a halo in the image, of size typically r_0/λ, around the diffraction-limited central peak, with a few percents of the energy in the halo. An analogous effect appears in Fig.4.6.

Flux Requirements for Wavefront Sensing. Assuming the only limitation to be the signal noise itself (perfect photon counting mode), the requirement for sensing is "one photon per coherence area per coherence time".

If extra noise is present (detector noise, thermal noise, read-out noise) one must compute the equivalent photon number N_{ph} equal to this noise during the coherence time τ_c.

Referring to the first case, the relation between photon and magnitude is

$$p \simeq \tau\sigma\,\Delta\lambda\,10^{8-0.4m_v}$$

p number of photons
$\Delta\lambda$ bandwidth in nm
σ area in m^2
$m_V = V$ magnitude (at 550 nm)
τ time in s

Taking $r_0 = 0.1m$, $\sigma = 0.34r_0^2$, $\frac{\Delta\lambda}{\lambda} = 0.1$, $\tau = 10ms$, $p = 10$ gives $m_v = 10.6$ if the total transmission and detection quantum efficiency were unity and with some margin in signal-to-noise ($p > 1$). This requirement is relaxed if the correction is searched for at longer wavelengths, since, as shown on Chap.3,

$$r_0^2\tau \propto \lambda^3$$

hence going from $0.5\mu m$ to $5\mu m$ represents 7.5 magnitude gain. The correction at longer wavelengths relaxes thoroughly the limiting magnitude of the reference object required by the wavefront sensor.

Numerically assuming a 10% detection efficiency, reference objects at $0.5\mu m$ must be brighter than $m_V \simeq 8$ (about 1 star per square degree average), but at $5\mu m$ they only need to be brighter than $m_V = 15.5$ (about 1000 stars per square degree average). This must be understood as using a reference source at $\lambda = 0.5\,\mu$m, but only deducing from it the low-frequency (spatial) terms of $\Delta z(\boldsymbol{x})$ slowly varying in time.

Field of view. How correlated are wavefronts coming from two directions $\boldsymbol{\theta}_1$ and $\boldsymbol{\theta}_2$ making an angle $\boldsymbol{\theta} = \boldsymbol{\theta}_1 - \boldsymbol{\theta}_2$ between themselves ? This correlation will determine how a correction can be made on an object in direction $\boldsymbol{\theta}_1$, using a reference in direction $\boldsymbol{\theta}_2$. [12]

Let define the ratio

$$A_{\boldsymbol{\theta}}(\boldsymbol{f}) = < \frac{MTF(\boldsymbol{f}, \boldsymbol{\theta}_2, t)}{MTF(\boldsymbol{f}, \boldsymbol{\theta}_1, t)} > \quad , \quad < > \text{ over time}$$

The problem is not isotropic around $\boldsymbol{\theta}_1$, so let define the conjuguate space $\boldsymbol{f}(u,v)$ with u parallel to $\boldsymbol{\theta}_1 - \boldsymbol{\theta}_2 = \boldsymbol{\theta}$ and v perpendicular. Let γ be the zenithal distance, h the height of the turbulent layer (supposed unique to simplify), and $P(h)$ the normalized turbulence profile

$$P(h) = \frac{C_n^2(h)}{\int_{\text{atmosphere}} C_n^2(h)dh}$$

One shows [12] that $A_{\boldsymbol{\theta}}(\boldsymbol{f}) =$

$$\exp\left\{-.024r_0^{-5/3}(\cos\gamma)^{-11/6}\lambda^{5/6}\int_{atm} h^{5/6}P(h) \times I_3\left[\boldsymbol{f}\left(\frac{\cos\gamma}{\lambda h}\right), \boldsymbol{\theta}\left(\frac{h}{\lambda\cos\gamma}\right)^{1/2}\right]dh\right\}$$

where

$$I_3(\boldsymbol{f},\boldsymbol{s}) = \iint_{-\infty}^{+\infty} (x^2+y^2)^{-11/6}(1-\cos 2\pi \boldsymbol{r}.\boldsymbol{f}) \times \left[1+\cos^2 \pi r^2(1-2\cos^2 \pi s x)\right] d\boldsymbol{r}$$

with $\boldsymbol{f}(u,v), \boldsymbol{r}(x,y)$.

This rather heavy expression can be approximated for a single turbulent layer at mean altitude h. Looking for $\boldsymbol{\theta}$ such as

$$A_{\boldsymbol{\theta}}(\boldsymbol{f}) = 0.5 A_0(\boldsymbol{f}) = 0.5$$

one obtains the approximate expression

$$\theta_{\rm ISO} \simeq 0.31 \frac{r_0(\lambda)}{h}$$

which has a trivial physical explanation: rays issued at altitude from within the same Fried's area behave identically. This angle $\theta_{\rm ISO}$ is called the atmospheric isoplanatic angle or patch. It varies, as r_0, with $\lambda^{6/5}$. Numerical values of typically $r_0(500nm) = 10cm$, $h = 4000m$, lead to $\theta_{\rm ISO}(500nm) \simeq 2\,\text{arc sec}$, a rather small angle, obviously increasing to over 30 arc sec at $\lambda = 5\mu m$.

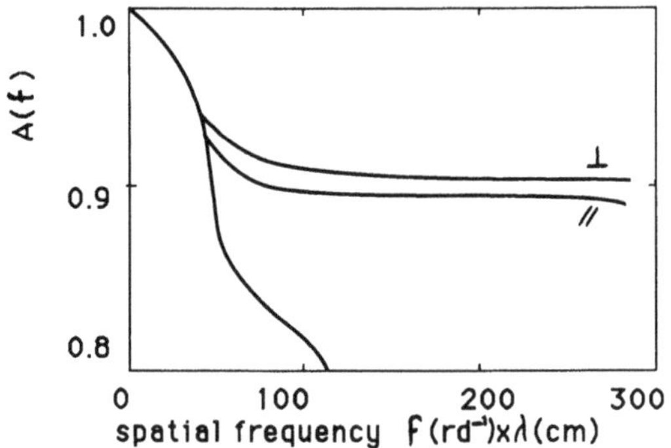

Fig. 4.3. The average of two MTFs distant of angle θ, versus spatial frequency, for $\theta = 10''$. The two curves refer respectively to $\boldsymbol{f} \perp \boldsymbol{\theta}$ and $\boldsymbol{f} \parallel \boldsymbol{\theta}$.$r_0 = 10cm$, $h = 4km$. It is clear that $\theta > \theta_{\rm ISO}$ makes impossible a full correction, since then $A(\boldsymbol{f}) \ll 1$.

4.3 Correcting the Wavefront

We discuss here how to restore the unperturbed wavefront, assuming that its perturbation has been measured as described above.

Tilt correction. Let consider the average slope of the wavefront over the pupil, i.e. the term $aR_1^1(\rho)\cos\theta = a\rho\cos\theta$, where the coefficient a is proportional to the angle-of-tilt $\boldsymbol{\xi}(\alpha, \beta)$. This tilt angle is the displacement of the image center-of-gravity with respect to the optical axis of the system.

A two-axis tilt mirror can be servoed on this center of gravity and recenter continuously the image, proviso the servo temporal band-pass Δf is such that $\Delta f \gg \tau_c^{-1}$, where τ_c is the atmosphere coherence time.

The required amplitude of correction is easily deduced from the expected image motion (Sec. 3.3)

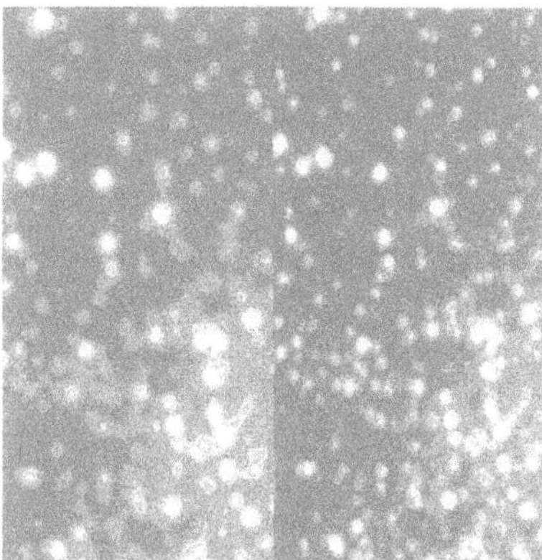

Fig. 4.4. Two exposures of the object 47 Tuc made without (right) and with (left) image stabilisation. Exposure time 45s, red filter (668nm). Full width half maximum of images is 1.2" at left, 0.9" at right. Thanks to the superior light concentration, the stabilized image reveals more details and reaches fainter magnitudes (ESO Disco system. F. Maaswinkel et al., Messenger **51**, 41, 1988).

$$\sigma_m^2 \simeq 0.36 \left(\frac{\lambda}{D}\right)^{1/3} \left(\frac{\lambda}{r_0}\right)^{5/3}$$

which is independent of λ, and reaches a few tenths of arc sec.

This tilt correction is most efficient when σ_m becomes comparable to or smaller than the seeing blur angle.

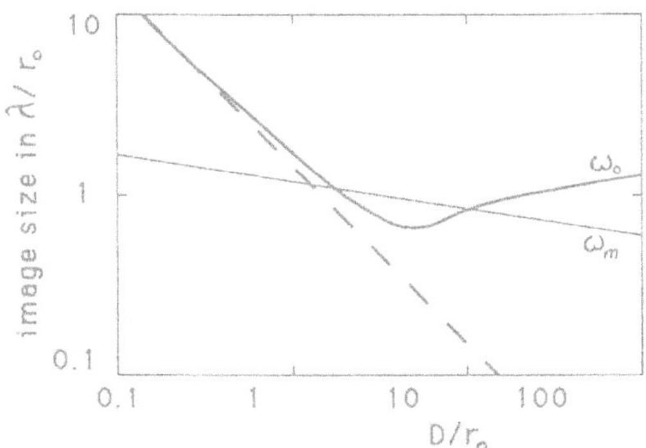

Fig. 4.5. Image size (in units of λ/r_0 seeing disc) versus D/r_0, measuring the number of speckles in the image. ω_0 is the total image blur, ω_m the image motion, the dashed line is the diffraction effect. Asymptotic value of ω_0 is $1.27\lambda/r_0$. This graph identifies a range $D/r_0 \sim 3$ where the image motion correction achieves maximum sharpening of the final image. [6]

Pupil Segmentation. It is clear from above that splitting the pupil D in sub-pupils of size 2 to $3r_0$, centering each sub-images, then coadding them will lead to a better tilt correction than averaged over the full pupil. This method is applied on the CFHT telescope. With good seeing, $r_0 \sim 30cm$, and with $n = 5$ to 10 segments a significant increase in quality is obtained. The condition remains that for $t < \tau_c$, a sufficient number of photons be present on each subpupil for the value of the local tilt $\boldsymbol{\xi}_n$ to be determined.

Multi-Elements Correction. Assume a flexible membrane, elastic, stretched over a circle, and the image of the entrance pupil being formed on it.

Let $z(\boldsymbol{x},t)$ be the wave-front deformation and $Z(\boldsymbol{x},t)$ the membrane displacement.

If $Z = -\frac{z}{2}$ and the membrane is used as a mirror, the reflected wave-front is fully corrected. In practice the system will have limitations such as:

- finite temporal bandwidth ;
- finite membrane flexibility ;
- finite number of actuators displacing the membrane...etc

and the quantity $2Z(\boldsymbol{x},t) - z(\boldsymbol{x},t)$ will not vanish. There will be a residual error

$$\epsilon(\boldsymbol{x},t) = 2Z(\boldsymbol{x},t) - z(\boldsymbol{x},t)$$

The problem is to minimize this error in a certain sense (e.g. the best MTF at low spatial frequencies as well as high ones...) with a statistical analysis properties, starting from the statistical properties of $z(\boldsymbol{x},t)$.

Intuitively, one may assume that if the number of actuators is $N \sim \left(\frac{D}{r_0}\right)^2$, i.e. the number of coherence areas over the pupil and the band pass of the system is larger than τ_c^{-1}, adequate correction may be achieved. Nevertheless, it is important to compute and/or simulate the properties of the residual MTF.

Choice of Amplitude for the Reflecting Surface. Take the rms excursion between two points distant of x on the wavefront, as derived in Chap.3

$$\sigma_z(x) = \frac{\lambda}{2\pi}\sigma_\varphi(x) = 0.42\lambda \left(\frac{x}{r_0}\right)^{5/6},$$

$\lambda = 0.5\mu m$, $r_0 = 10\,cm$, $D = 4m$, $\sigma_z = 4.5\mu m$. $5\sigma_z$ gives a correction for over 99% of the cases, since the wavefront statistics of Δz is gaussian.

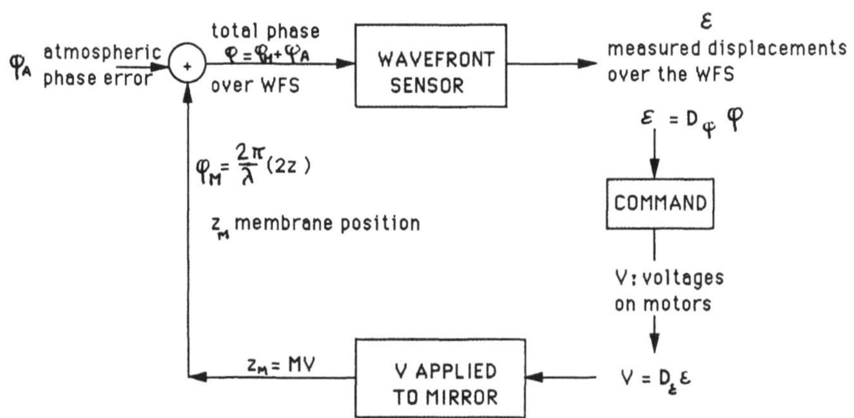

Scheme of Correction.

φ is a continuous function over the surface, discretized in N points;

ϵ is the output of the WFS, 2N points (2 values per phase point);

V are the voltages over the m motors;

z_M are the positions of the m motors;

φ , vector, dimensions $\geq$ N;

ϵ , vector, 2 N dimensions;

V, Z_M , vectors, m dimensions;

$\boldsymbol{D}_\varphi$, matrix 2N $\times$ N , connects WFS measurement and phase over the wavefront;

$\boldsymbol{D}_\epsilon$, matrix 2N $\times$ m , connects WFS measurements and voltages to apply to each motor;

$\boldsymbol{M}$, matrix N $\times$ m , connects voltages and displacements (could be diagonal if no cross-talk).

Note that

$$\epsilon = \boldsymbol{D}_\varphi \varphi = \boldsymbol{D}_\varphi \frac{4\pi}{\lambda} z_M = \frac{4\pi}{\lambda} (\boldsymbol{D}_\varphi \boldsymbol{M}) V.$$

$$\boldsymbol{D} = \boldsymbol{D}_\varphi \boldsymbol{M}$$

is the interaction matrix $2N \times m$ in size which characterizes the displacements obtained for a set of voltages, with $\varphi_A = 0$. This matrix forms the calibration of the system. We note this theory to apply only if linearity applies.

Let examine the matrix $\boldsymbol{D}_\epsilon$. $\boldsymbol{D}_\epsilon$ would be diagonal if there were no cross-talk. Each phase correction would be produced by a motor located at the same point (zonal control), having no effect on any other point. In practice, the diagonal terms are much superior to non diagonal terms and residual errors occur due to the finite accuracy of the computation.

The solution is to establish a new base over which to decompose φ. One may choose as base the natural deformation modes of a membrane, which are orthogonal.

The wavefront phase φ becomes $\boldsymbol{\Phi}$ over this new base, and one computes the matrix $\boldsymbol{D}$ such as $V = \boldsymbol{D}\boldsymbol{\Phi}$ or directly $V = \boldsymbol{D}_\epsilon \epsilon$. Terms in $\boldsymbol{D}$ are both diagonal and non-diagonal, but this time of similar magnitude.

Deconvolution. A special case would be to use the wavefront values ϵ to obtain φ, but then to deconvolve the image by the following procedure. The phase $\varphi(\boldsymbol{r})$ gives the complex amplitude $\psi(\boldsymbol{r}) = \exp i\varphi(\boldsymbol{r})$. We deduce the instantaneous MTF (pupil and atmosphere)

$$\tilde{T}(\boldsymbol{f},t) = \frac{1}{A} \int\int \psi\left(\frac{\boldsymbol{\rho}}{\lambda}\right) \psi^*\left(\boldsymbol{f} + \frac{\boldsymbol{\rho}}{\lambda}\right) G\left(\frac{\boldsymbol{\rho}}{\lambda}\right) G^*\left(\boldsymbol{f} + \frac{\boldsymbol{\rho}}{\lambda}\right) \frac{d\boldsymbol{\rho}}{\lambda^2}$$

and compute the restored object spectrum

$$\tilde{O}(\boldsymbol{f}) = \frac{\tilde{I}(\boldsymbol{f},t)}{\tilde{T}(\boldsymbol{f},t)}$$

In this case, it is convenient to represent φ on a different base, the Zernicke polynomials hence

$$\varphi \longrightarrow \boldsymbol{\Phi}_Z$$

whose higher terms are directly related to classical aberrations. Such deconvolution only requires a wavefront sensor, no actual active mirror is needed. If the noise is pure signal photon noise, this process is strictly equivalent to the process of using a correcting deformable mirror. It would be ideal at visible wavelengths under two conditions:

a) no background noise (i.e. magnitude of object brighter than about 20).

b) no detector noise despite the fast read-out rate ($\sim \tau_c$).

Effect of Partial Correction. One may assume the number N of correcting elements adequate for a given wavelength λ_0 i.e. $N = \frac{D}{(r_0(\lambda_0))^2}$, but too small for $\lambda < \lambda_0$. At λ, there will only be a partial correction, the highest spatial frequencies in the wavefront phase error remaining uncorrected. These high frequencies errors produce a phase screen which will scatter energy outside the Airy core of the image, hence a halo of size fixed by $\lambda/r_0(\lambda)$.

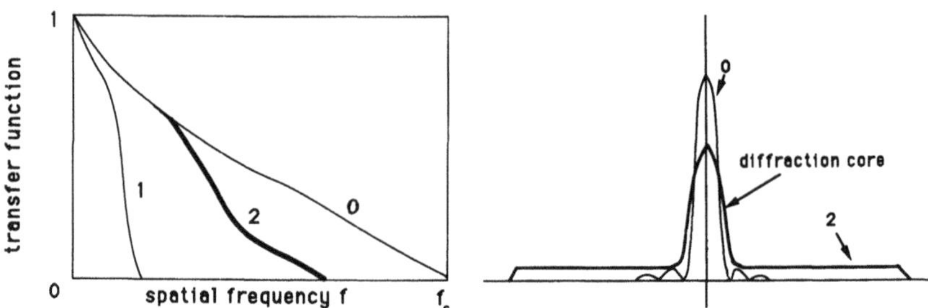

Fig. 4.6. Left: The telescope MTF (0), degraded by atmosphere (1) and partially corrected (2) at scale $r_0(\lambda_0)$. Right: the ideal Airy image (0) and the partially corrected image (2).

As example of application, let assume a 4m telescope, at $\lambda = 0.8\mu m$, where $\lambda/D =$ 40 milliarcsec. Assume a 1 arcsec seeing and a partial correction, with 10% of the total energy going in the central Airy disk, 90% remaining in the halo. The image brightness in the central core is increased in the ratio $(10/0.04^2)/(100/1^2) = 66$, a very large factor to feed the energy in the slit of a high-resolution spectrograph or to enhance the contrast with a sky background.

4.4 Astronomical Imaging with Adaptive Optics

We describe briefly a complete system currently under design for the 3.6 m ESO telescope. The Table 4.1 gives numerical values of the various parameters mentioned above, which provide useful dimensions for the system, and show that reasonably small values of N can achieve full correction of large telescopes in the near infrared. [Kern et al., in 10]

Table 4.1.

λ (μm)	0.5		2.2		3.85		10	
$r_0(\lambda) \propto \lambda^{1.2}$	$0.10m$	$0.20m$	$0.60m$	$1.20m$	$1.15m$	$2.30m$	$3.60m$	$7.20m$
$N(\lambda) \approx (D/r_0)^2$	1600	400	45	12	12	3	2	1
$\tau(\lambda) \approx (r_0/\bar{v})$	$10ms$	$20ms$	$60ms$	$120ms$	$115ms$	$230ms$	$360ms$	$720ms$
$\Theta(\lambda) \approx \left(r_0/\bar{h}\right)$	3"	6"	20"	40"	40"	80"	125"	250"

The system itself is shown on Fig.4.7, where the various command functions are clearly identified.

An example of isoplanatic field is given on Fig.4.8. The relaxation of the magnitude of the reference object (Sec.4.2 above) and the increase of the isoplanatic angle θ (Table 4.1) with increasing wavelength lead to a good coverage of the sky with reference objects, each one being surrounded by its isoplanatic field.

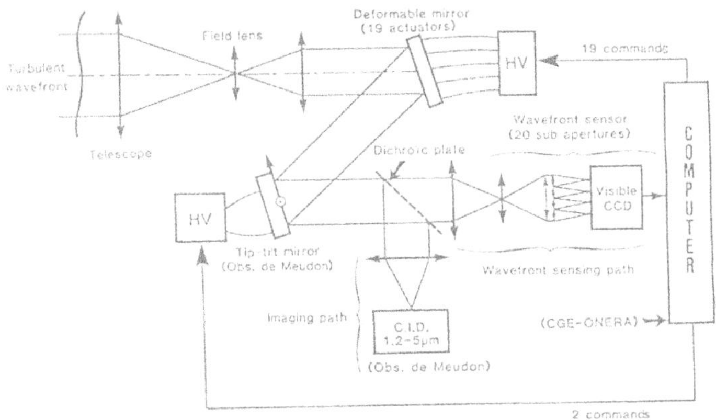

Fig. 4.7. An overview of the COME-ON adaptive system, under construction for the 3.6 m ESO telescope. The deformable mirror (N = 19) is able to correct down to $\lambda_0 = 3.5\mu m$ and partially at shorter wavelength. A Charge Injection Device (CID Camera) is used for diffraction-limited imaging.

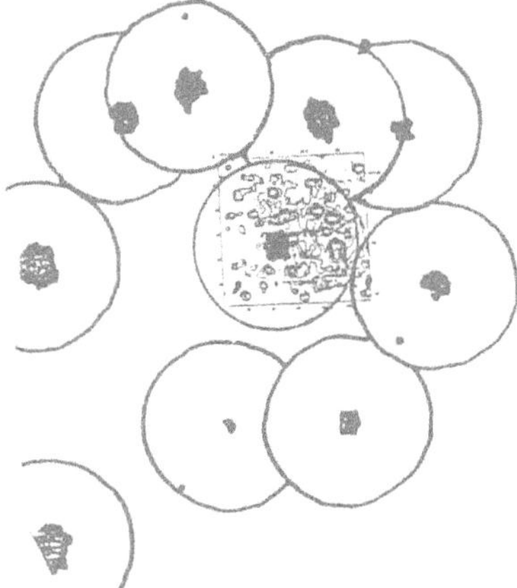

Fig. 4.8. The area of the Galactic Center: The square is a contour map at a wavelength of 2.2 μm. The isoplanatic circle is drawn around each reference star sufficiently bright ($m_v \lesssim 15$) for wavefront sensing. Note the large fraction of the sky covered, i.e. the area where an instantaneous MTF could be derived and used for correction. [Mourard and Mercouroff 1987. Priv. Comm.]

5. Multi Telescope Imaging

Radio-interferometry with several independent telescope is a well established technique, extensively described in the litterature [D. Downes, in this Volume], [4] and capable of remarkable achievements in terms of angular resolution (currently 50 micro arc sec at 2.6 mm wavelength) and image reconstruction. The challenging field of today is the application of the same basic physical principles, described in Chap.4, to optical wavelengths, i.e. from 0.3 to 20 μm (on the ground) or extended to the ultraviolet in space. This extension requires to transpose the technical solutions found at radio wavelengths, in terms of photon detection, beam recombination... etc, taking also in account the significant differences of the atmospheric properties (Chap.3). Since the development of optical interferometry shall be very fast in the coming years, we intend to give here some basic points of reference to understand the principles [13], [14] follow the expected progress and their application to real astronomical images of unprecedented resolution (i.e. in the range 0.1 to 10 milliarsec) at these wavelengths [15]. Note that similar angular resolution may be reached with baselines of a few hundreds of meters in length, and are yet comparable to the resolution obtained with radio baselines of the order of Earth's diameter ! [13]

5.1 Status of Optical Interferometry

Since the first results obtained with two independent telescopes by Labeyrie (1976), a number of optical interferometers have been brought in operation, and even more are in the planning stage (Fig.5.1). All the instruments having actually measured spatial coherence to date are made of $N = 2$ telescopes. New instruments aim to $N \lesssim 10$, improving the aperture synthesis capability. Other parameters appear in the instrument design and performances:

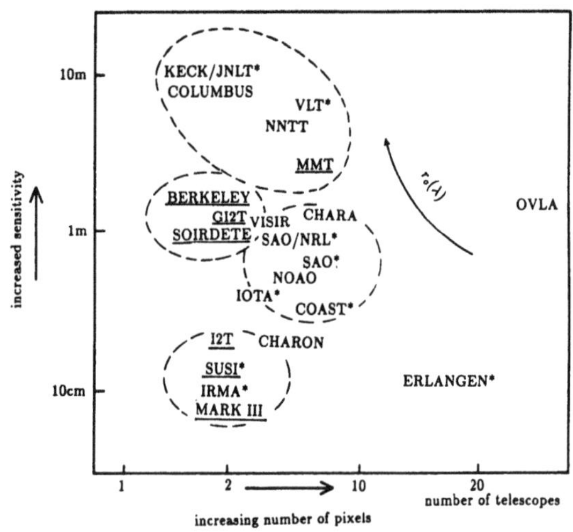

Fig. 5.1. Existing and currently planned interferometers or large telescopes. The diameter of individual telescopes in ordinate, the number of telescopes in the interferometer in abcissa. Existing instruments are underlined. Dashed curves are arbitrary but group types of instruments (see text). New programs are: IRMA (University of Wyoming), COAST (Cambridge, UK), SAO (Smithsonian)/ NRL (Naval Research Laboratory), VISIR (Cerga/Lyon), OVLA (Cerga), CHARON (Cerga), KECK (California), VLT (Very Large Telescope), JNLT (Japan), COLUMBUS (Arizona/Ohio/Illinois/Italy), NOAO (Tucson NSF), NNTT (United States). Funded programs are marked with asterisk*. Source: I.A.U. Transactions, General Assembly 1988, Joint Session on Interferometry, 1988, Reidel, in press.

• the wavelength of operation, which has to be compared to the telescope diameter D. If $D < r_0(\lambda)$, each telescope input is fully phased, and the measurement of the coherence γ is more straigtforward than in the opposite case, encountered with large ($D > 20cm$) apertures at visible wavelengths.

• the baseline $\boldsymbol{B}$ orientation and value: some interferometers operate on fixed baselines, others on variable ones, either continuously or by fixed increments.

• the coherence measuring technique: all interferometers, but one (Berkeley) use the broad-band, direct Michelson combination; heterodyne combination at $\lambda = 10.6\mu m$ provides some experimental convenience, essentially an easy detection of the beam combination, at the expense of a reduced sensitivity.

Table 5.1 . Optical Interferometers (*)

Name	Location	N	D(cm)	Baseline(m)	Array Shape	Position strategy	Results (science)
I2T	Cerga(France)	2	26	144	NS+EW	continuous	Star diameters[1] Star envelopes[2]
GI2T	Cerga(France)	2	150	70	NS linear	continuous	In progress
MMT	Mt Hopkins (Arizona)	6	180	6.8	Hexagon	monolithic (single mount)	-
Mark III	Mt Wilson (California)	2	7.5	12	NS	discrete steps	Astrometry[3] positions & diameters
SUSI Prototype	Univ.Sydney (Australia)	2	10	15	EW	fixed telescopes	Vega diameter[4]
SOIRDETE	Cerga(France)	2	100	15	EW	fixed telescopes	In progress
Berkeley	Mt Wilson (California)	2	150	28	NS	discrete steps	In progress

[1] di BENEDETTO G.P., RABBIA Y. (1987) Astron.Astrophys. **188**, 114
[2] THOM C., GRANES P., VAKILI F. (1986) Astron.Astrophys. **165**, L13
[3] MOZURKEWITCH D. et al. (1988) Astron.J. **95**, 1269
[4] DAVIS J., TANGO W.J. (1986) Nature **323**, 234
(*) All the interferometers figuring in this Table have to date produced quantitative visibility data, i.e. $|\tilde{O}(\mathbf{f})|$ for one or several frequencies **f**.

Table 5.1 details these different parameters and lists the main astronomical results.

Radio interferometry is a classical subject (see D. Downes, in this Volume), which is discussed in may textbooks and references. It is therefore useful to pinpoint similarities and differences with optical interferometry, since the physical and conceptual approach is basically identical. This is schematically proposed in Table 5.2. Some of the points mentionned in this table are not discussed in this lecture. Detailed discussion may be found in Ref.[12]

5.2 Beam Combination

We consider the simple case of a two-telescopes (or two-pupils, or two-apertures) interferometer (N=2), illuminated by a point source, quasimonochromatic and on-axis. Fig. 5.2a shows the **image-plane combination**: fringes are obtained in the common focal plane, and the focal arrangment is perfectly homothetic of the telescope itself as seen from the astronomical object, if the condition $B'/B = D'/D$, where B is the baseline. The focal plane intensity distribution $I(\boldsymbol{\theta})$ is the system PSF, its Fourier transform the system MTF $\tilde{T}_0(\boldsymbol{f})$. Measuring the fringes visibility in amplitude and phase is therefore a direct measure of the object spectrum at spatial frequency $\boldsymbol{B}/\lambda$, where $\boldsymbol{B}$ is the projected baseline on the sky. For a resolved, extended source, the classical Fourier relation holds

$$\tilde{I}(\boldsymbol{f}) = \tilde{O}(\boldsymbol{f}).\tilde{T}_0(\boldsymbol{f})$$

and the measure of $\tilde{I}(\boldsymbol{f})$determines $\tilde{O}(\boldsymbol{f})$ at frequency $\boldsymbol{f} = \frac{\boldsymbol{B}}{\lambda}$.

Table 5.2.. Comparison of Radio and Optical Interferometry

Radio	Optical
Uniform phase over individual apertures	Many phase cells over the pupil [1]
Unknown phase differences between apertures	Unknown phase differences between apertures [2]
Time integration	Short atmospheric coherence time [3]
Phase stability of delay lines	Mechanical stability of telescopes [4]
Electric delay lines	Optical delays [5]
Beam separation and mixing at IF	Optical beam recombination (pupil or image plane
No field-of-view, throughput = λ^2	Possible field-of-view (~1 as) [7]
Limited accuracy on amplitude and phase estimates, limited u-v coverage [8]	

Possible remedies

(1) Pupil reconfiguration, adaptive optics, multipixels detectors, telescope size limited to $D \sim r_o$.
(2) Phase closure, speckle masking, holography (self-calibration).
(3) Phase tracking with reference source.
(4) Active stability control.
(5) Movable telescopes, variable optical delay lines.
(6) Complex beam combining optics if $N \geq 3$.
(7) Pupil reconfiguration and matching.
(8) All restoration methods (Maximum Entropy, Clean ...).

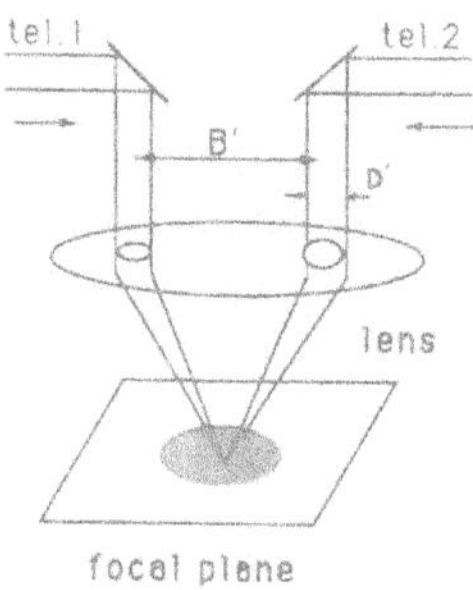

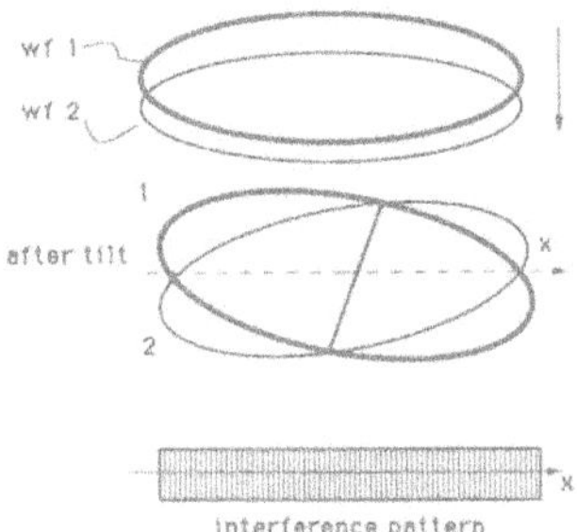

Fig. 5.2. (a) Beams from telescopes 1 and 2 recombine in the focal plane, where $\boldsymbol{\theta}$ is the local coordinate. (b) Wavefront recombination in the common pupil plane.

Alternatively, it is possible to superimpose the pupils of telescopes 1 and 2, after reimaging them (Fig.5.2.b). This makes the wavefront directly interfering on a detector. To measure γ, one can tilt one wavefront with respect to the other: fringes are obtained, their amplitude and phase gives $\tilde{T}_0(\boldsymbol{f})$. Or introducing an optical path modulation between the two beams modulates the output signal and produces the same result. This is called **pupil plane recombination.**

Both methods are strictly equivalent from the point-of-view of signal-to-noise ratio. Either may be a matter of experimental convenience, depending on the detectors, the wavelength of operation...etc.

Source with an Extended Spectrum [1]. In the case $N = 2$, the information in the focal plane

[1] "Spectrum" means here the distribution of intensity with wavelength in the case of a non monochromatic

(Fig.5.2a) is along the direction $\boldsymbol{B}$ only. Compression of the information in this direction is possible, and the perpendicular direction may be used to recover the spectral information on the object $O(\boldsymbol{\theta}, \lambda)$ in the image plane. This is illustrated on Fig.5.3 where are shown the successive steps of image compression, image dispersion and fringes measurements $\tilde{I}(\boldsymbol{f}, \lambda)$ at any wavelength in the dispersed spectrum.

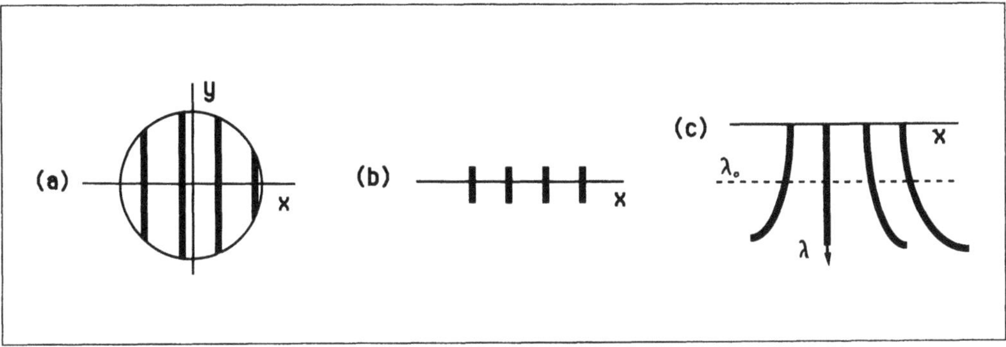

Fig. 5.3. (a) Intensity distribution in the $\boldsymbol{\theta}(x, y)$ image plane (b) Compressed image (e.g. with a cylindrical lines) (c) Dispersed image. Fringe spacing varies proportionaly to λ. Fourier transform may be taken at any wavelength λ_0 to measure $\tilde{O}\left(\boldsymbol{f} = \frac{\boldsymbol{B}}{\lambda_0}, \lambda_0\right)$.

Again, in the case $N = 2$, an interesting method, illustrated in Fig.5.4 is proposed for sources having an extended spectrum. Beams from telescopes 1 and 2 are fed in a classical Fourier Transform Spectrometer (FTS) which, at its output, superimposes the wavefronts and the pupils. When the FTS optical path difference is δ, the output intensity is, within an additive constant

$$I_{\text{out}}(\delta) = \int_{-\infty}^{+\infty} \tilde{O}(\boldsymbol{f} = \sigma \boldsymbol{B}, \sigma) \cos 2\pi\sigma\delta \; d\sigma$$

where $\sigma = 1/\lambda$ is the wave number. Inverse Fourier transform of $I_{\text{out}}(\delta)$ provides the "double" spectrum $\tilde{O}$ of the source in terms of the variables $\boldsymbol{f}$ and $\lambda = \sigma^{-1}$. This method is a multiplex one, with its classical advantages, since it mixes on a single detector the spatial and all the spectral information.

Operating optical interferometers achieve beam combination with $N = 2$ using these methods. An additional complexity comes from the need of adding to one of the beams a time-variable optical delay to compensate for the continuous variation with time of the projected baseline $\boldsymbol{B}(t)$ due to Earth's rotation. In principle, this delay can be accurately compute, computer-controlled and inserted in the beam with no reduction on the γ measurement accuracy.

When $N > 2$, as it will be the case with the new generation of optical interferometers, the experimental set-up and the γ measurements will become slightly more complex: each optical path from each telescope to the common focal plane must be kept constant, either by a continuous motion of each telescope, or by a delay line; the beam recombination does not allow simple schemes as above, and the simultaneous measurement of the spatial and temporal

source. "Spatial spectrum" will always be used to designate the complex Fourier transform of the source intensity distribution in angular direction $\boldsymbol{\theta}$. Both informations indeed are of astrophysical value.

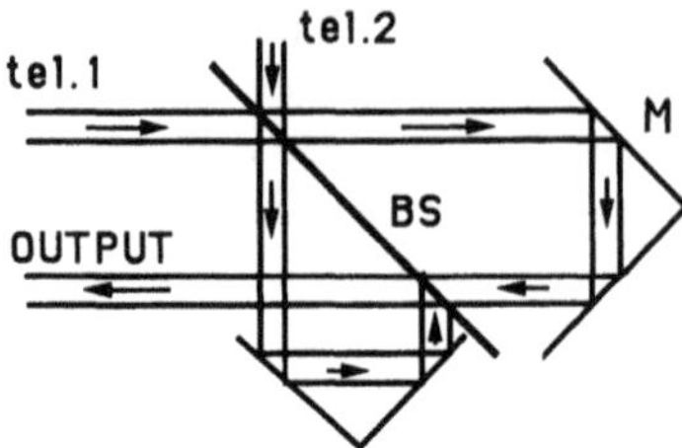

Fig. 5.4. A Fourier transform spectrometer fed by telescopes 1 and 2, with a beamsplitter BS and a movable mirror M introducing an optical delay δ between the two arms. A detector will measure the output intensity $I_{out}(\delta)$.

spectrum is more difficult. These points will undoubtly progress when these interferometers come to operation.

Comparing with radio wavelengths, it is clear that this complexity is related to the very basic fact that in the counting photon mode, characteristic of optical wavelengths, photons can not be split as waves are in a radio interferometer.

5.3 The Phase Problem: What does the Interferometer measure ?

In principle, the interferometer measures $\tilde{I}(\boldsymbol{f})$ at $\boldsymbol{f} = \frac{\boldsymbol{B}}{\lambda}$ as a complex quantity. In practice, atmospheric phase errors will perturb this measurement in two ways: random phase errors on each telescope pupil, and random phase difference between telescopes. Let assume the former ones to be negligeable or corrected (Sec. 5.4) and study the effect of the latter called **external** phase errors, in the simple case where $N = 2$.

Additional phase errors may be generated within the interferometer itself, due for instance to mechanical vibration, destroying the exact optical path equality from telescopes 1 and 2 to the common focus. Let call these errors **internal** ones. They have been the main practical difficulty encountered by Michelson in 1920, and are still fought against by modern optical interferometers. In principle, it is possible to internally monitor the interferometer components, by accurate distance measurements and active servo-control, in order to maintain absolute stability. This servo-control is achieved with the Berkeley and the Mark III interferometers, at respective wavelengths of 10.6 and 0.5 μm. If the internal path difference $\Delta z_{int}(t)$ is maintained zero during operation, the interferometer is called **absolute**: it is equivalent then to a perfectly rigid telescope.

If it were not for $\Delta z_{ext}(t)$, the interferometer could "lock" on zero path difference (ZPD) for an object anywhere in the sky, since all other optical delays can be computed from the geometry of observation: position of the interferometer, the Earth and the object.

Let briefly examine the effect of atmospheric path drift $\Delta z_{ext}(t)$. As shown in Chap. 3, this quantity is baseline dependent, with *rms* value

$$\sigma_{\Delta z} = 0.42\lambda \left(\frac{B}{r_0}\right)^{5/6} .$$

Δz can be much larger than λ, as illustrated by Fig.5.5.

These phase fluctuations have several detrimental effects, also known in radio interferometry although with a lesser impact. First, with $N = 2$, the only possible measurement is $\left|\tilde{I}(\boldsymbol{f})\right|$, since $\arg\tilde{I}(\boldsymbol{f})$ is now random. Measurements cited in Table 5.1, all obtained with $N = 2$, have only measured $\left|\tilde{I}(\boldsymbol{f})\right|$, which does not indeed provide by Fourier inversion the object, even with several frequencies $\boldsymbol{f} = \boldsymbol{B}/\lambda$ obtained at different baselines.

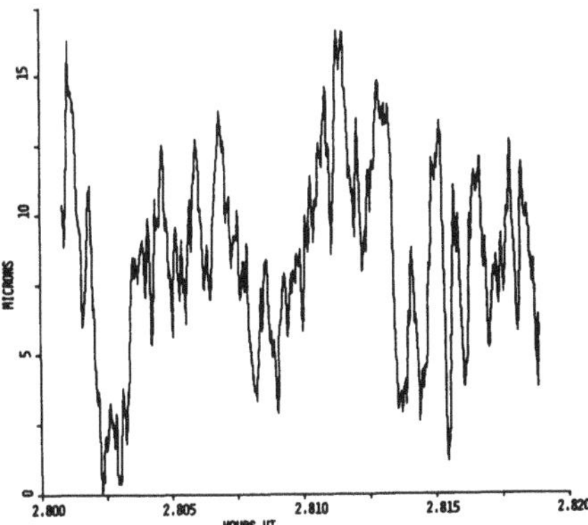

Fig. 5.5. Phase drift $\Delta z_{\rm ext}(t)$, measured over $B = 12m$, by the Mark III interferometer at $\lambda = 0.5$ μm. (Colavita et al. 1988, Appl. Opt.).

Second, even the measurement of $\left|\tilde{I}(\boldsymbol{f})\right|$ must be achieved during a time shorter than τ_c, the atmospheric coherence time. This has no detrimental effect on the signal-to-noise ratio S/N in the pure signal photon noise, since then, for a total observing time $T = M\tau$ of the M cumulated exposures of duration τ, with n photon per exposure (also called **frame**), the total S/N is $\sqrt{nM\tau}$, equivalent to the S/N obtained with a single exposure of duration T. On the other hand, if detector read out noise is present, with rms value σ_R, $(S/N)_T = nM/\sigma_R$, while $(S/N) = nM/\sigma_R\sqrt{M}$ when coadding M frames. Since $\tau \sim 10 - 100$ ms, while T may be several hours, M can be very large $(10^4 - 10^6)$ and the loss in S/N considerable.

All optical interferometers to date have measured $\left|\tilde{I}(\boldsymbol{f})\right|$ in the so-called "snapshot" mode, similar to the one used in speckle interferometry. In the future, high sensitivities and long time integration will become possible with the same approach as in adaptive optics: using either the photons from the source itself, or the ones from an adjacent reference, one compensates actively for $\Delta z_{\rm ext}(t)$ by a continuously variable, servo-controlled delay. This is indeed not sufficient to measure $\arg\tilde{I}(\boldsymbol{f})$, lost in the atmosphere.

To really obtain images with an optical interferometer –the ultimate goal indeed– one must circumvent the effect of $\Delta z_{\rm ext}(t)$. This can be achieved by **phase closure**, a method well-developed in radio-interferometry, but only applicable for $N \geq 3$, and preferably $N \gg 2$. It relies on the simple following principle: if $\Delta\varphi_i = 2\pi\Delta z_i/\lambda$, $(i = 1, 2..N)$, are the phase errors over the N apertures, then in the sums such as

$$\arg\tilde{I}(\boldsymbol{f}_{12}) + \arg\tilde{I}(\boldsymbol{f}_{23}) + ... + \arg\tilde{I}(\boldsymbol{f}_{31})$$

$$= \arg\tilde{O}(\boldsymbol{f}_{12}) + \arg\tilde{O}(\boldsymbol{f}_{23}) + \arg\tilde{O}(\boldsymbol{f}_{31}) + [\Delta\varphi_{12} + \Delta\varphi_{23} + \Delta\varphi_{31}],$$

the brackett cancels out. A sufficient number of such relations determines the quantities $\tilde{O}(\boldsymbol{f}_{ij})$ over the different baselines ij, and this determination may even be redundant, improving therefore the S/N ratio.

Although optical interferometers with $N > 2$ do not operate yet, the application of phase closure methods to optical wavelengths has been demonstrated by the following method: take N "pieces" of a wavefront, each smaller than r_0 in dimension, over the full pupil of a large telescope (e.g. the 5 m Palomar) and recombine them as if they were coming from separate telescopes. Pairs of pupils form $N(N-1)/2$ baselines and phase closure relations are derived, allowing to cancel the atmospheric effects [15].

5.4 Pupil Phasing and Sensitivity

As demonstrated in Chap.3, the phase is not uniform on a single telescope of diameter D illuminated by a quasi-monochromatic point source, as soon as $D > r_0(\lambda)$. Above discussion was restricted to the cases where this place is uniform or the pupil is phased[2].

We shall not discuss here in detail why it is so advantageous to work with phased pupils, but the basic reasons are fairly simple to understand. If each image contains numerous speckles, when $D > r_0$, the interference, even in the simple case $N = 2$, is not uniform in the full image field. Measurement of phase and amplitude of $\tilde{I}(\boldsymbol{f})$ must be done within each speckle, the complex quantity $\tilde{I}(\boldsymbol{f})$ becomes a random variable, and average values must be taken, as in speckle interferometry.

In the signal photon noise limited case ($\lambda \lesssim 1\mu m$), this process creates additional noise. In the detector- or background-limited noise case, it becomes necessary to measure the image with a many-pixels detector (of the order of the speckle number D^2/r_0^2), to get a single final value $\tilde{I}(\boldsymbol{f} = \boldsymbol{B}/\lambda)$, and this again cumulates noise. Fig.5.6 shows the S/N ratio per exposure of duration τ_c in the signal noise limited mode with phased and unphased apertures. The gain due to a phased aperture is obvious.

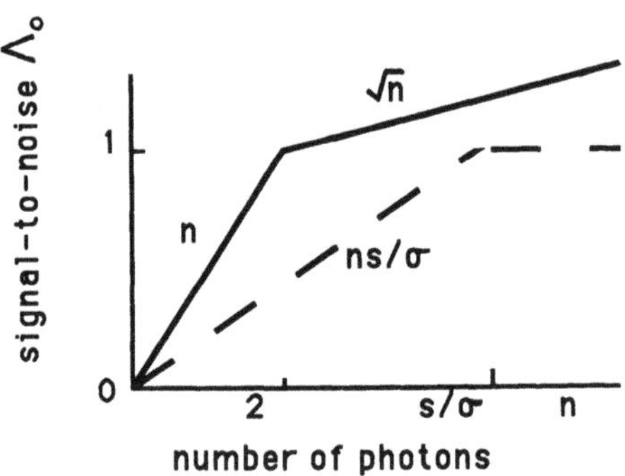

Fig. 5.6. Signal-to-noise ratio Λ_0 obtained with a phased pupil (full) and a multi-phases (dashed), versus the total number of photons available per frame (i.e. during τ_c). σ is the Fried's coherence area $0.3r_0^2$, $s = \pi D^2/4$ the telescope area. The slopes are indicated. This is valid only in the signal photon-noise limited case [12].

Phased apertures may be encountered in a number of situations:

• telescope diameter $D \lesssim r_0(\lambda)$, i.e. a few centimeters at $\lambda = 500nm$, to a few meters at $10\mu m$.

• telescope diameter $D \gg r_0(\lambda)$, but phase uniformity over D restored by adaptive optics (Chap.4) at the operating wavelength.

• telescope in space with any aperture D, where phase fluctuation across the wavefront will disappear. This latter enventuality is not considered in a near future, but will present remarkable opportunities in the long term.

How bright a source can be measured in the phased case ? A very simple computation gives the answer. A source of magnitude m_V gives a number p of signal photoelectrons, in the pure photon counting mode,

$$p = q\tau\, s\Delta\lambda\, 10^{(8-0.4m_V)}$$

[2] Optical interferometry with unphased pupils is discussed by G. Weigelt in this Volume. Most of the fundamental notions introduced here also apply to this case, but the analysis of a speckled image and the phase closure become less straightforward.

q is the detector quantum efficiency (0.05 to 5), τ(second) the measurement time ($\leq \tau_c$), s the telescope area (m^2), $\Delta\lambda$ the spectral bandwidth (in nm). This is of the order of a few nm to preserve temporal coherence across the diameter D despite the rms path difference $\Delta z(D)$. The S/N ratio on $\tilde{I}(\boldsymbol{f})$ is simply $\Lambda = p^{1/2}$. With D = 1m, $\tau = 10ms$, $q = 0.1$, $\Delta\lambda = 10nm$, one gets for $\Lambda = 10$ the limit $m_V = 10$. This is an indication of the sensitivity of an interferometer for a short exposure. Long exposures are indeed conditioned by phase stabilization as briefly discussed above.

A similar computation may be done for the near-infrared ($\lambda \geq 3.5\mu m$) in the background noise limited case. Assuming $T_{\mathrm{back}} = 300\mathrm{K}$, the emissivity $\epsilon_{\mathrm{back}} = 0.1$, the beam throughput being λ^2 (coherence throuput for a phased aperture, independent of diameter), the noise signal is, in photons,

$$p_{\mathrm{noise}} = \left(\epsilon_{\mathrm{back}} B_\lambda(T_{\mathrm{back}}).\Delta\lambda.\lambda^3 \tau_c/hc\right)^{1/2}$$

and the minimum detectable flux with $\Lambda = 10$ as above

$$F = \left(\frac{4\Lambda p\, h\nu}{\pi D^2\, \Delta\nu}\right) \quad \text{in } W\, m^{-2}\, Hz^{-1}$$

The bandwidth $\Delta\lambda$ required for coherence, as above, varies as λ^2, hence one may adopt $\Delta\lambda(@3.5\mu m) = \Delta\lambda(@0.5\mu m) \times \lambda^2 \sim 1\mu m$, in practice reduced to 0.1μm by the width of the atmospheric transmission band. Putting numbers, with D = 1 m, leads to $p \sim 500$, $F \sim 3$ millijansky.

These numbers are indicative of the short exposure sensitivity of perfect interferometric instruments in the phased mode. When apertures become unphased, the sensitivity computation becomes slightly more complex, involving the speckle statistics as well as the signal statistics [12].

5.5 Image Reconstruction

Astronomers work with real images, i.e. an intensity distribution $I(\boldsymbol{\theta}, \lambda)$. This has to be recovered from a certain set of quantities $\tilde{I}(\boldsymbol{f}, \lambda) = \tilde{I}(\boldsymbol{B}/\lambda, \lambda)$ obtained from a set of bases $\boldsymbol{B}$. At this point, the problem becomes completely similar to the one fully covered at radio-wavelength (D. Downes, this Volume) and does not deserve a special treatment. In practice, since data with only $N = 2$ were obtained to date, optical interferometrists have not been yet confronted to the use of such methods. On the other hand, numerous simulations are made, and the method of phase closure itself has been tested, by simulating $N > 2$ telescopes with a mask over a large telescope mirror.

5.6 Summary and conclusion

We have discussed the main steps leading to the reconstruction of an astronomical object $0(\boldsymbol{\theta})$ from measurements of its spatial frequency spectrum $\tilde{O}(\boldsymbol{f})$, a complex function of spatial frequency $\boldsymbol{f}$, determined over a set of baselines $B = \boldsymbol{f}\lambda$, sampling a certain domain in the $\boldsymbol{f}-$ frequency space. These steps are summarized in Table 5.3, which also indicates what has been achieved to date with $N = 2$.

As was apparent in Table 5.2, optical interferometry develops its specific methods quite in parallel to radio interferometry. The current science achievements are rather limited, since the use of $N = 2$ telescope does not provide the complex quantity $\tilde{O}(\boldsymbol{f})$, but only its modulus. This is the reason, apparent in Table 5.1, for the astronomical results to have only covered very simple cases, like measuring a diameter of a star assumed to be a disc of uniform brightness or with a simple center-to-limb variation. In the same time, the physical understanding of the various factors appearing in Table 5.2 has greatly progressed, essentially over the last ten years: properties of the atmospheric turbulence and of astronomical sites, beam combination, pupil phasing, active control of path differences... This is the reason why astrophysical results are now expected.

Table 5.3. Overview of optical interferometry. Successive steps discussed in the text are shown schematically here to emphasize how each step contributes either to improved sensitivity, or to an improved image by better frequency coverage. The heavy line path shows current operation of existing instruments; the reconstructed $O(\theta)$ can be of high accuracy (S/N ratio of a few percents) but far from an image in the usual sense because of the very limited coverage in the frequency domain, and of the knowledge of $|\tilde{O}(f)|$ in modulus only.

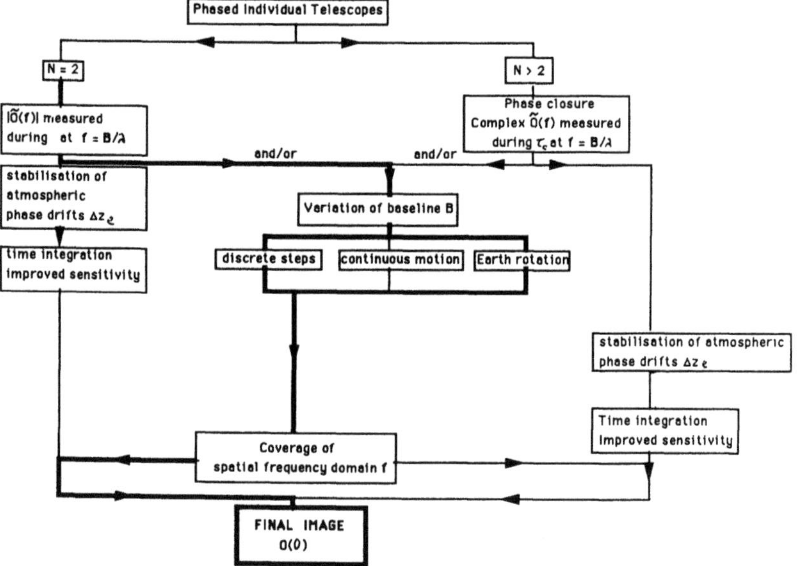

Appendix to Chap.5

We extract from a recent publication [Gay and McKarnia, in 15] Figs.5.7, 5.8, 5.9, which well illustrate the points made in this Chapter. They describe the operation and results of the Cerga SOIRDETE interferometer (Table 5.1) at infrared wavelengths.

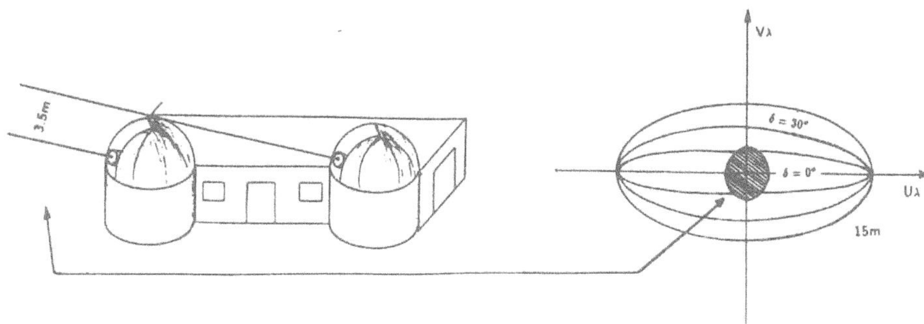

Fig. 5.7. The two 1m-telescopes on a fixed E-W 15 m baseline. The frequency (u,v) coverage is on the right for sources of different declinations δ. The blind area is due to the domes.

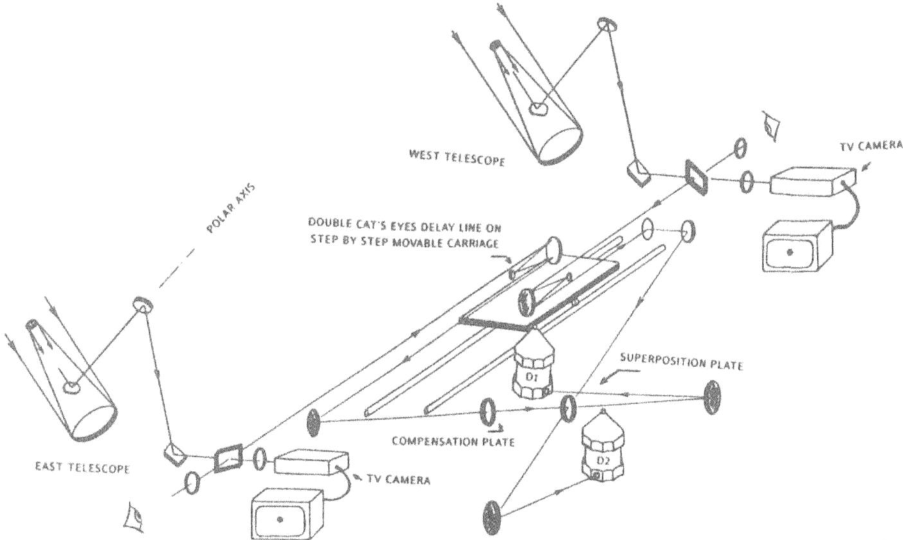

Fig. 5.8. The optical scheme of the SOIRDETE optical interferometer, working at infrared wavelengths. Detectors in the $\lambda = 1 - 10\mu m$ range are located in cryostats D1 and D2.

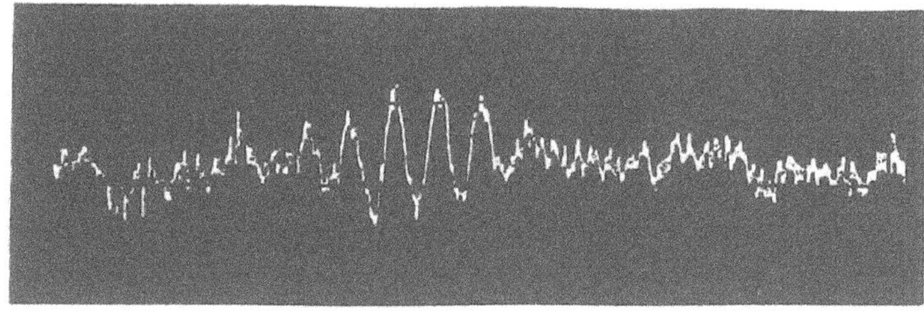

Fig. 5.9. Interference fringes obtained on α *Ori*, with a broad band interval $\lambda = 8$ to $11.5\mu m$. Maximum modulation corresponds to zero path difference. In abcissa, optical delay between beams 1 and 2.

References

Basic Imaging Theory

1 Dainty J.C., Shaw R. (1974) Image Science (Academic, New York)

2 Born M., Wolf E. (1980). Principles of Optics, 6th ed. (Pergamon, Oxford)

Imaging in Astronomy

3 Léna P. (1988). Observational Astrophysics (Springer, Heidelberg)

4 Rohlfs K. (1986) Tools of Radio Astronomy (Springer, Heidelberg)

Imaging through turbulence

5 Kitchin C.R. (1984), Astrophysical Techniques (Hilger, Bristol)

6 Roddier F. (1981) : "The Effects of Atmospheric Turbulence is Optical Astronomy" in Progress in Optics Vol XIX, 281

7 Tatarski, V.I. (1961) : Wave Propagation in a Turbulent Medium (McGraw-Hill, New York)

8 Woolf N. (1982) : "High resolution imaging from the ground" Ann. Rev. Astr. Ap. 20, 367

Adaptive Optics

9 Rousset G., Fontanella J.C., Primot V., Sève A. (1987). "Imagerie Optique à travers la Turbulence Atmosphérique." La Recherche Aérospatiale, 5, 47.

10 Ulrich M.H., Kjar K. Ed. (1989). Very Large Telescopes and their Instrumentation. (European Southern Observatory, Garching)

Speckle Interferometry

11 Perrier C. (1987) : "Apport de l'interférométrie des tavelures à l'étude des sources infrarouges". Thèse de Doctorat d'Etat, Université Paris 7.

12 Roddier F. (1988) : "Interferometric Imaging in Astronomy" Phys. Reports (in press).

Interferometry (multi telescopes)

13 Alloin D., Mariotti J.M. Ed. (1989). Diffraction Limited Imaging with Large Telescopes. Cargèse Summer School 1988. (Kluwer, Amsterdam)

14 Labeyrie A. (1978) Stellar Interferometry Methods. Ann. Rev. Astr. Ap. 16, 77

15 Merkle F. Ed. (1988) High resolution Imaging by Interferometry. Vol I and II. (European Southern Observatory, Garching).

This article was processed by the author using the TEX Macropackage from Springer-Verlag.

Interferometric Imaging in Optical Astronomy

Gerd Weigelt
Max-Planck-Institut für Radioastronomie, D-5300 Bonn, FRG

Summary

The atmosphere of the earth restricts the resolution of conventional astronomical imaging to about 1". Much higher resolution can be obtained by speckle methods. Speckle interferometry, the Knox-Thompson method and the speckle masking method (bispectrum or triple correlation processing) yield diffraction-limited resolution in spite of image degradation by the atmosphere and telescope aberrations. For example, with the ESO 3.6-m telescope a resolution of 0.028" is attained at a wavelength of 400 nm. The limiting magnitude is about 18. We will outline the theory behind the three methods. High-resolution images and *simultaneously* the spectrum of each resolution element can be obtained by the objective prism speckle spectroscopy and projection speckle spectroscopy methods. Finally, we will discuss the application of speckle masking to coherent arrays of telescopes. A very interesting example is the 4x8-m ESO VLT, which should yield the fantastic angular resolution of about 2 milli-arcsec.

1 Speckle Interferometry

Speckle interferometry has been invented in 1970 by Antoine Labeyrie (1970). In speckle interferometry the high-resolution autocorrelation of the object is reconstructed from a large number of speckle interferograms. Speckle interferograms are short-exposure photographs of the object recorded with an exposure time of about 0.02 sec and an interference filter with a bandwidth of about 30 nm. The angular diameter of speckle interferograms is typically about 1" (see Figure 1). The size of individual speckles (interference maxima) is about 0.03" for a 3.6-m telescope and $\lambda \sim 500$ nm. In speckle interferograms high-resolution object information is encoded in decoded form. Figure 2 a shows a typical speckle interferogram. The intensity distribution $I_n(\mathbf{r})$ of the n-th recorded speckle interferogram can be described by the incoherent, space-invariant imaging equation

$$I_n(\mathbf{r}) = O(\mathbf{r}) * S_n(\mathbf{r}) \qquad n = 1, 2, 3, \ldots, N \; (N \sim 10^3 \text{ to } 10^5), \tag{1}$$

where $O(\mathbf{r})$ denotes the object intensity distribution and $S_n(\mathbf{r})$ the intensity point spread function of atmosphere/telescope during the n-th exposure. $\mathbf{r} = (x,y)$ is a 2-dimensional space vector and $*$ denotes the convolution operator. In the following text the subscript n will be omitted. $I(\mathbf{r})$ and $S(\mathbf{r})$ denote random functions of a random process. In speckle interferometry many different speckle interferograms $I(\mathbf{r})$ are processed to the ensemble average power spectrum of all speckle interferograms. With the convolution theorem, the Fourier transform of Equation (1) becomes

$$\hat{I}(\mathbf{f}) = \hat{O}(\mathbf{f})\, \hat{S}(\mathbf{f}) \tag{2}$$

where the hat ^ denotes a Fourier transform, i.e., $\hat{I}(\mathbf{f}) \equiv \int I(\mathbf{r}) \exp(-2\pi i \mathbf{r} \cdot \mathbf{f})\, d\mathbf{r}$. $\mathbf{f} = (f_x, f_y)$ is a 2-dimensional coordinate vector in Fourier space. From Equation (2) follows for the ensemble average power spectrum $\langle |\hat{I}(\mathbf{f})|^2 \rangle$ of all $I(\mathbf{r})$:

$$\langle |\hat{I}(\mathbf{f})|^2 \rangle = |\hat{O}(\mathbf{f})|^2 \, \langle |\hat{S}(\mathbf{f})|^2 \rangle \,, \tag{3}$$

where the brackets $\langle \ldots \rangle$ denote ensemble average over all speckle interferograms. $\langle |\hat{S}(\mathbf{f})|^2 \rangle$ is called the speckle interferometry transfer function (SITF). The SITF can be determined by calculating the average power spectrum of speckle interferograms of an

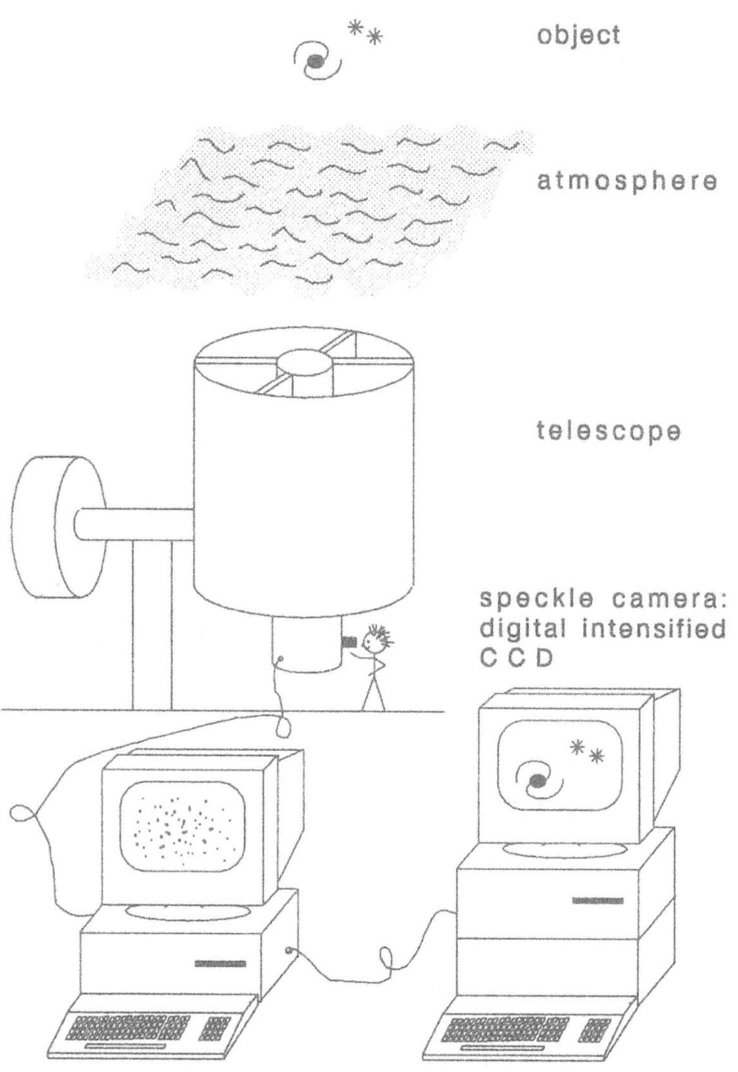

Fig. 1. Data recording and processing in speckle imaging

unresolvable star. The speckle interferograms of an unresolvable star are recorded before or after the observation of the object. The SITF is larger than zero up to the diffraction cut-off frequency. Therefore we can divide Equation (3) by the SITF and obtain the object power spectrum

$$|\hat{O}(\mathbf{f})|^2 = \langle|\hat{I}(\mathbf{f})|^2\rangle \,/\, \langle|\hat{S}(\mathbf{f})|^2\rangle \tag{4}$$

for frequencies $\mathbf{f}$ up to the diffraction cut-off frequency. From the object power spectrum the diffraction-limited object autocorrelation $AC[O(\mathbf{r})]$ can be obtained by an inverse Fourier transformation (autocorrelation theorem).

$$AC[O(\mathbf{r})] = \int O(\mathbf{r}')\, O(\mathbf{r}'+\mathbf{r})\, d\mathbf{r}'. \tag{5}$$

The autocorrelation is obtained instead of a true image since the phase of the object Fourier transform is not reconstructed by speckle interferometry. This is the so-called phase problem of speckle interferometry.

However, a true image can be reconstructed by speckle interferometry if there is an unresolvable object in the same isoplanatic patch as the object (i.e., closer than about 5"). The application of such an unresolvable object for image reconstruction is called *holographic speckle interferometry* (Liu and Lohmann, 1973; Bates et al., 1973). It can easily be shown that in this case true images can be reconstructed. We assume that a total object $O(\mathbf{r})$ consists of a point source $\delta(\mathbf{r}-\mathbf{r}_R)$ and an object $O'(\mathbf{r})$:

$$\text{total object } O(\mathbf{r}) = O'(\mathbf{r}) + \delta(\mathbf{r}-\mathbf{r}_R). \tag{6}$$

Speckle interferometry can reconstruct the autocorrelation $AC[O(\mathbf{r})]$:

$$\begin{aligned} AC[O(\mathbf{r})] &= AC[O'(\mathbf{r}) + \delta(\mathbf{r}-\mathbf{r}_R)] \\ &= \int [O'(\mathbf{r}') + \delta(\mathbf{r}'-\mathbf{r}_R)]\; [O'(\mathbf{r}'+\mathbf{r}) + \delta(\mathbf{r}'-\mathbf{r}_R+\mathbf{r})]\, d\mathbf{r}' \\ &= \int \delta(\mathbf{r}'-\mathbf{r}_R)\, \delta(\mathbf{r}'-\mathbf{r}_R)\, d\mathbf{r}' + \int O'(\mathbf{r}')\, O'(\mathbf{r}'+\mathbf{r})\, d\mathbf{r}' + \int O'(\mathbf{r}')\, \delta(\mathbf{r}'-\mathbf{r}_R+\mathbf{r})\, d\mathbf{r}' \\ &\qquad + \int \delta(\mathbf{r}'-\mathbf{r}_R)\, O'(\mathbf{r}'+\mathbf{r})\, d\mathbf{r}' \\ &= \delta(\mathbf{r}) + AC[O'(\mathbf{r})] + O'(-\mathbf{r}+\mathbf{r}_R) + O'(\mathbf{r}+\mathbf{r}_R). \end{aligned} \tag{7}$$

Here we can see that the autocorrelation of $O(\mathbf{r})$ contains an image of $O'(\mathbf{r})$. $O'(\mathbf{r})$ is separated from all other terms if the distance $\mathbf{r}_R$ of the reference star from $O'(\mathbf{r})$ is larger than 3/2 object diameter. This is the same condition found in Fourier holography.

2 Knox-Thompson Method

The Knox-Thompson (Knox and Thompson, 1974) and speckle masking methods have the advantage that they *do not require* an unresolvable object in the isoplanatic patch. The raw data for both methods are speckle interferograms as in speckle interferometry. The first processing step of the Knox-Thompson method is the calculation of the cross-spectrum

$$\langle \hat{I}(\mathbf{f})\, \hat{I}^*(\mathbf{f}+\Delta\mathbf{f})\rangle = \hat{O}(\mathbf{f})\, \hat{O}^*(\mathbf{f}+\Delta\mathbf{f})\, \langle \hat{S}(\mathbf{f})\, \hat{S}^*(\mathbf{f}+\Delta\mathbf{f})\rangle, \tag{8}$$

where $\Delta\mathbf{f}$ is a small shift vector with $|\Delta\mathbf{f}| \sim 0.2\ r_o/\lambda$ (see Dainty, 1984). r_o is the Fried parameter ($\sim$10 to 30 cm at optical wavelengths) and λ is the wavelength of light. If we divide Equation (8) by the Knox-Thompson transfer function (KTTF), we obtain the Knox-Thompson spectrum of the object:

$$\hat{O}(\mathbf{f})\,\hat{O}^*(\mathbf{f}+\Delta\mathbf{f}) = \langle \hat{I}(\mathbf{f})\,\hat{I}^*(\mathbf{f}+\Delta\mathbf{f})\rangle \,/\, \langle \hat{S}(\mathbf{f})\,\hat{S}^*(\mathbf{f}+\Delta\mathbf{f})\rangle. \tag{9}$$

The KTTF is derived from the speckle interferograms of an unresolvable object, which is observed before or after the object. For the Knox-Thompson spectrum of the object we can write

$$\hat{O}(\mathbf{f})\ \hat{O}^*(\mathbf{f}+\Delta\mathbf{f}) = |\hat{O}(\mathbf{f})|\ \exp[i\varphi(\mathbf{f})]\ |\hat{O}(\mathbf{f}+\Delta\mathbf{f})|\ \exp[-i\varphi(\mathbf{f}+\Delta\mathbf{f})], \tag{10}$$

where $\varphi(\mathbf{f})$ denotes the desired phase of the object Fourier transform $\hat{O}(\mathbf{f})$. If we take the phase terms of Equation (10), we see that we have

$\exp[i\varphi(\mathbf{f})]\ \exp[-i\varphi(\mathbf{f}+\Delta\mathbf{f})] = \exp\{i[\varphi(\mathbf{f}) - \varphi(\mathbf{f}+\Delta\mathbf{f})]\}$ or the

$$\text{phase difference } \Delta\varphi(\mathbf{f}) \equiv \varphi(\mathbf{f}+\Delta\mathbf{f}) - \varphi(\mathbf{f}) \tag{11}$$

between coordinate $\mathbf{f}+\Delta\mathbf{f}$ and $\mathbf{f}$. In other words, we have a recursive equation for calculating the desired phase of the object Fourier transform:

$$\varphi(\mathbf{f}+\Delta\mathbf{f}) = \varphi(\mathbf{f}) + \Delta\varphi(\mathbf{f}). \tag{12}$$

From the object Fourier phase measured by the Knox-Thompson method and the Fourier modulus a diffraction-limited image of the object can be reconstructed.

3 Speckle Masking

Speckle masking (Weigelt, 1977; Weigelt and Wirnitzer, 1983; Lohmann, Weigelt, Wirnitzer, 1983) is, as the Knox-Thompson method, a method for reconstructing true images from speckle interferograms. The additional advantage of speckle masking is that it can be applied to *diluted arrays of telescopes*, as discussed in Section 6. In speckle masking the same speckle interferograms $I(\mathbf{r})$ are reduced as in speckle interferometry and in Knox-Thompson processing. Speckle masking consists of the following processing steps:

STEP 1: *calculation of the ensemble average triple correlation*

$$C(\mathbf{r}_1,\mathbf{r}_2) = \langle \int I(\mathbf{r})\, I(\mathbf{r}+\mathbf{r}_1)\, I(\mathbf{r}+\mathbf{r}_2)\, d\mathbf{r}\rangle \tag{13}$$

or the ensemble average bispectrum

$$\hat{C}(\mathbf{f}_1,\mathbf{f}_2) = \langle \hat{I}(\mathbf{f}_1)\,\hat{I}(\mathbf{f}_2)\,\hat{I}^*(\mathbf{f}_1+\mathbf{f}_2)\rangle, \tag{14}$$

where $\hat{I}(\mathbf{f}_1)$, $\hat{I}(\mathbf{f}_2)$, and $\hat{I}^*(\mathbf{f}_1+\mathbf{f}_2)$ denote the Fourier transforms of $I(\mathbf{r})$, i.e.

$$\hat{I}(\mathbf{f}_1) = \int I(\mathbf{r})\exp(-2\pi i\mathbf{f}_1\cdot\mathbf{r})\, d\mathbf{r},$$

$$\hat{I}(\mathbf{f}_2) = \int I(\mathbf{r})\exp(-2\pi i\mathbf{f}_2\cdot\mathbf{r})\, d\mathbf{r},$$

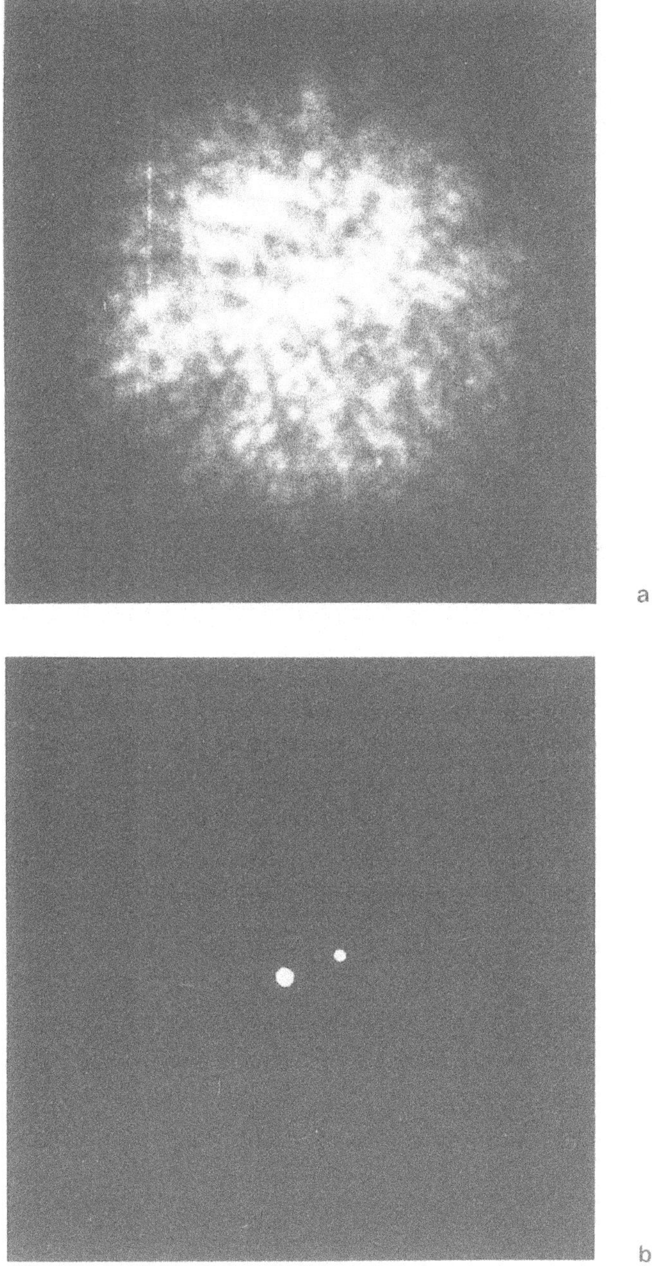

Fig. 2. Speckle masking observation of the spectroscopic double star Psi SGR.
a: One of 300 speckle interferograms recorded with the ESO 3.6-m telescope.
b: Reconstructed diffraction-limited image of Psi SGR (same scale as Figure 2a). The separation of the double star is 0.184" (epoch 1982.378).

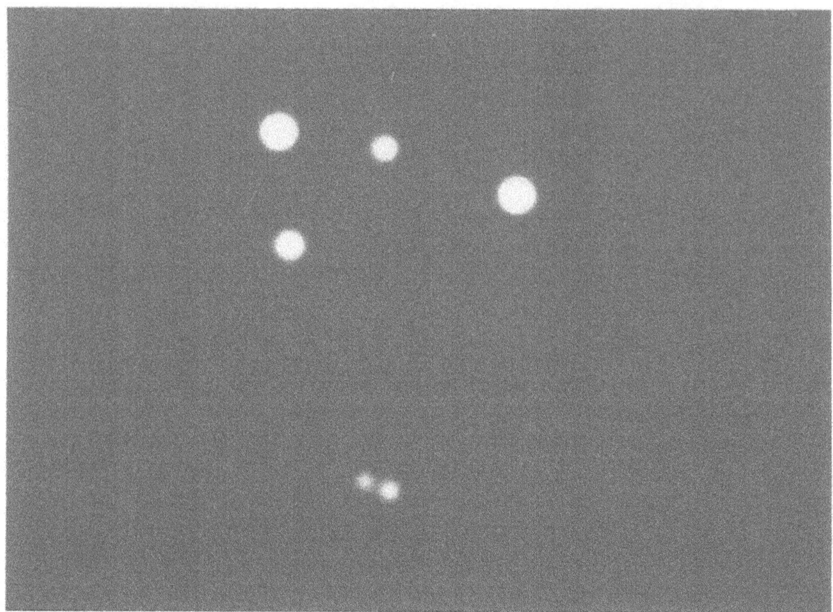

Fig. 3. Speckle masking observation of the central object in the giant H II region NGC 3603. The diffraction-limited image was reconstructed from 300 speckle interferograms recorded with the 2.2-m ESO/MPG telescope. The separation of the close double star at the bottom is about 0.08".

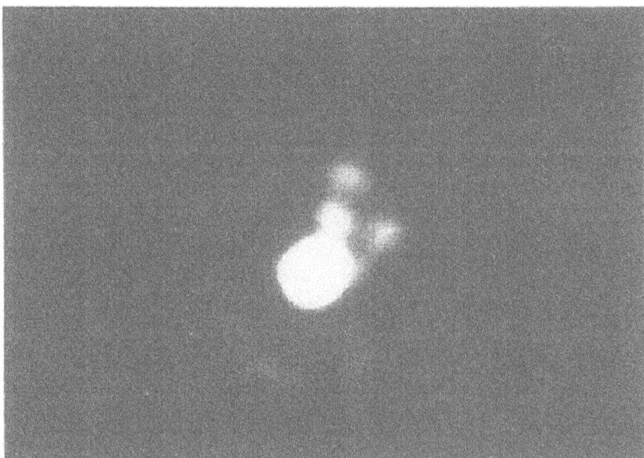

Fig. 4. Speckle masking observation of Eta Carinae at $\lambda \sim 850$ nm. The diffraction-limited image was reconstructed from 300 speckle interferograms recorded with the 2.2-m ESO/MPG telescope. The separations of the three faint star-like objects are 0.11", 0.18", and 0.21".

$$\hat{I}^*(\mathbf{f}_1+\mathbf{f}_2) = \int I(\mathbf{r}) \exp[2\pi i(\mathbf{f}_1+\mathbf{f}_2)\cdot\mathbf{r}]\, d\mathbf{r}. \tag{15}$$

$\langle ... \rangle$ denotes ensemble average over many speckle interferograms and the hat ^ denotes Fourier transforms. $\mathbf{f}_1$ and $\mathbf{f}_2$ denote 2-dimensional vectors in Fourier space. The bispectrum is the Fourier transform of the triple correlation (see Appendix A).

In most applications it is advantageous to use the bispectrum processing. The advantage of the triple correlation is that it can be easily visualized and that it can be used for photon-counting triple correlation techniques (Schertl et al., 1987). In the text below we will discuss the theory of bispectrum processing.

STEP 2: *compensation of the photon bias in the ensemble average bispectrum* (Wirnitzer, 1985; Hofmann and Weigelt, 1987).

STEP 3: *compensation of the speckle masking transfer function:*
From $I = O * S$ follows $\hat{I} = \hat{O}\,\hat{S}$ (convolution theorem). If we insert $\hat{I} = \hat{O}\,\hat{S}$ into Equation (14), we obtain

$$\begin{aligned} \hat{C}(\mathbf{f}_1,\mathbf{f}_2) &= \langle \hat{O}(\mathbf{f}_1)\,\hat{S}(\mathbf{f}_1)\,\hat{O}(\mathbf{f}_2)\,\hat{S}(\mathbf{f}_2)\,\hat{O}^*(\mathbf{f}_1+\mathbf{f}_2)\,\hat{S}^*(\mathbf{f}_1+\mathbf{f}_2)\rangle \\ &= \hat{O}(\mathbf{f}_1)\,\hat{O}(\mathbf{f}_2)\,\hat{O}^*(\mathbf{f}_1+\mathbf{f}_2)\ \langle \hat{S}(\mathbf{f}_1)\,\hat{S}(\mathbf{f}_2)\,\hat{S}^*(\mathbf{f}_1+\mathbf{f}_2)\rangle \quad \text{or} \end{aligned} \tag{16}$$

object bispectrum $\hat{C}_O(\mathbf{f}_1,\mathbf{f}_2) =$

$$\hat{O}(\mathbf{f}_1)\,\hat{O}(\mathbf{f}_2)\,\hat{O}^*(\mathbf{f}_1+\mathbf{f}_2) = \langle \hat{I}(\mathbf{f}_1)\,\hat{I}(\mathbf{f}_2)\,\hat{I}^*(\mathbf{f}_1+\mathbf{f}_2)\rangle \,/\, \langle \hat{S}(\mathbf{f}_1)\,\hat{S}(\mathbf{f}_2)\,\hat{S}^*(\mathbf{f}_1+\mathbf{f}_2)\rangle. \tag{17}$$

STEP 4: *derivation of modulus* $|O(\mathbf{f})|$ *and phase* $\varphi(\mathbf{f})$ *of the object Fourier transform* $\hat{O}(\mathbf{f})$ *from the object bispectrum* $\hat{C}_O(\mathbf{f}_1,\mathbf{f}_2)$:

We denote the phase of the object Fourier transform by φ and the phase of the object bispectrum by β, i.e.,

$$\hat{O}(\mathbf{f}) = |\hat{O}(\mathbf{f})|\ \exp[i\varphi(\mathbf{f})] \text{ and} \tag{18}$$

$$\hat{C}_O(\mathbf{f}_1,\mathbf{f}_2) = |\hat{C}_O(\mathbf{f}_1,\mathbf{f}_2)|\ \exp[i\beta(\mathbf{f}_1,\mathbf{f}_2)]. \tag{19}$$

Inserting Equations (18) and (19) into Equation (17) yields

$$\begin{aligned} \hat{C}_O(\mathbf{f}_1,\mathbf{f}_2) &= |\hat{C}_O(\mathbf{f}_1,\mathbf{f}_2)|\ \exp[i\beta(\mathbf{f}_1,\mathbf{f}_2)] \\ &= |\hat{O}(\mathbf{f}_1)|\ \exp[i\varphi(\mathbf{f}_1)]\ |\hat{O}(\mathbf{f}_2)|\ \exp[i\varphi(\mathbf{f}_2)]\ |\hat{O}(\mathbf{f}_1+\mathbf{f}_2)|\ \exp[-i\varphi(\mathbf{f}_1+\mathbf{f}_2)] \ \longrightarrow \end{aligned} \tag{20}$$

$$\exp[i\beta(\mathbf{f}_1,\mathbf{f}_2)] = \exp[i\varphi(\mathbf{f}_1)]\ \exp[i\varphi(\mathbf{f}_2)]\ \exp[-i\varphi(\mathbf{f}_1+\mathbf{f}_2)] \quad \longrightarrow \tag{21}$$

$$\beta(\mathbf{f}_1,\mathbf{f}_2) = \varphi(\mathbf{f}_1) + \varphi(\mathbf{f}_2) - \varphi(\mathbf{f}_1+\mathbf{f}_2) \longrightarrow \tag{22}$$

$$\varphi(\mathbf{f}_1+\mathbf{f}_2) \equiv \varphi(\mathbf{f}) = \varphi(\mathbf{f}_1) + \varphi(\mathbf{f}_2) - \beta(\mathbf{f}_1,\mathbf{f}_2). \tag{23}$$

Equation (23) is a recursive equation for calculating the phase of the object Fourier transform at coordinate $\mathbf{f} = \mathbf{f}_1+\mathbf{f}_2$. For the recursive calculation of the object Fourier phase $\varphi(\mathbf{f}) \equiv \varphi(\mathbf{f}_1+\mathbf{f}_2)$ we need in addition to the bispectrum phase $\beta(\mathbf{f}_1,\mathbf{f}_2)$ the starting values $\varphi(0,0)$, $\varphi(0,1)$, and $\varphi(1,0)$.

Since $O(\mathbf{r})$ is real, $\hat{O}$ is hermitian. Therefore, $\hat{O}(\mathbf{f}) = \hat{O}^*(-\mathbf{f})$, $\hat{O}(0) = \hat{O}^*(0)$, and $\varphi(0) = 0$.

$\varphi(0,1)$ and $\varphi(1,0)$ can be set to zero since we are not interested in the absolute position of the reconstructed image.

With $\varphi(0,1)=0$ and $\varphi(1,0)=0$ we have the required starting values. We obtain, for example:

$$\begin{aligned}
&\varphi(0,2) = \varphi(0,1) + \varphi(0,1) - \beta[(0,1),(0,1)]\\
&\varphi(0,3) = \varphi(0,2) + \varphi(0,1) - \beta[(0,2),(0,1)]\\
&\varphi(0,4) = \varphi(0,3) + \varphi(0,1) - \beta[(0,3),(0,1)]\\
&\ldots\\
&\varphi(2,0) = \varphi(1,0) + \varphi(1,0) - \beta[(1,0),(1,0)]\\
&\varphi(3,0) = \varphi(2,0) + \varphi(1,0) - \beta[(2,0),(1,0)]\\
&\varphi(4,0) = \varphi(3,0) + \varphi(1,0) - \beta[(3,0),(1,0)]\\
&\ldots\\
&\varphi(1,1) = \varphi(1,0) + \varphi(0,1) - \beta[(1,0),(0,1)]\\
&\varphi(2,1) = \varphi(2,0) + \varphi(0,1) - \beta[(2,0),(0,1)]\\
&\varphi(3,1) = \varphi(3,0) + \varphi(0,1) - \beta[(3,0),(0,1)]\\
&\ldots
\end{aligned} \tag{24}$$

The advantage of this recursive phase calculation is the fact that for each element of the object Fourier phase $\varphi(\mathbf{f})$ there are many different recursion paths and that it is possible to average over all $\varphi(\mathbf{f})$-values to improve the signal-to-noise ratio. For example, for the element (3,2) there are 8 useful recursion paths, for (6,4) there are 32 paths. Averaging over all paths yields

$$\varphi(\mathbf{f}) = \text{const.} \sum_{0<\mathbf{f}_1\cdot\mathbf{f}/f\le f/2} \varphi(\mathbf{f}_1) + \varphi(\mathbf{f}-\mathbf{f}_1) - \beta(\mathbf{f}_1,\mathbf{f}-\mathbf{f}_1). \tag{25}$$

However, in actual applications, the phase calculation is performed with complex exponential functions:

$$\exp[i\varphi(\mathbf{f})] = \text{const.} \sum_{0<\mathbf{f}_1\cdot\mathbf{f}/f\le f/2} \exp[i\varphi(\mathbf{f}_1)]\,\exp[i\varphi(\mathbf{f}-\mathbf{f}_1)]\,\exp[-i\beta(\mathbf{f}_1,\mathbf{f}-\mathbf{f}_1)]. \tag{26}$$

Since not all recursion paths for the same $\varphi(\mathbf{f})$-value lead to the same SNR, different weighting factors have to be chosen for different paths.

The modulus $|\hat{O}(\mathbf{f})|$ can be derived from the object bispectrum in two different ways. From Equation (17) follows for $\mathbf{f}_1=\mathbf{0}$

$$\hat{C}_o(\mathbf{0},\mathbf{f}_2) = \hat{O}(\mathbf{0})\,\hat{O}(\mathbf{f}_2)\,\hat{O}^*(\mathbf{0}+\mathbf{f}_2) = \text{const.}\,|\hat{O}(\mathbf{f}_2)|^2. \tag{27}$$

The second way is the *recursive* calculation of $|\hat{O}(\mathbf{f})|$. From Equation (20) follows

$$|\hat{C}_o(\mathbf{f}_1,\mathbf{f}_2)| = |\hat{O}(\mathbf{f}_1)|\,|\hat{O}(\mathbf{f}_2)|\,|\hat{O}(\mathbf{f}_1+\mathbf{f}_2)| \quad \text{or} \tag{28}$$

$$|\hat{O}(\mathbf{f}_1+\mathbf{f}_2)| = |\hat{C}_o(\mathbf{f}_1,\mathbf{f}_2)| \,/\, [\,|\hat{O}(\mathbf{f}_1)|\,|\hat{O}(\mathbf{f}_2)|\,] \quad \text{for } \mathbf{f}_1, \mathbf{f}_2 \text{ with } |\hat{O}(\mathbf{f}_1)|\,|\hat{O}(\mathbf{f}_2)| \neq 0. \tag{29}$$

As in the case of the recursive calculation of the phase, it is very important to use different weighting functions for different recursion paths. Christou et al. (1987) and Weghorn (1988) reported that the recursive derivation of the modulus $|\hat{O}|$ from the object bispectrum yields higher SNR than the standard power spectrum analysis.

Modifications of speckle masking are photon-counting triple correlation processing (Schertl et al. 1987), cross-triple correlation processing (Hofmann and Weigelt, 1987), and tomographic speckle masking (Schertl et al., 1987).

Pupil plane methods were not discussed in this paper (see Hofmann and Weigelt, 1986 a; Roddier and Roddier, 1986; Ribak, 1987).

Figures 2 to 4 show the application of speckle masking to the spectroscopic double star Psi SGR (Weigelt and Wirnitzer, 1983), the central object in the giant H II region NGC 3603 (Hofmann and Weigelt, 1986 b; Baier et al., 1988) and to Eta Carinae (Hofmann and Weigelt, 1988).

4 Objective Prism Speckle Spectroscopy

Objective prism speckle spectroscopy (Weigelt, 1981; Stork and Weigelt, 1984; Weigelt et al., 1986; Grieger et al., 1988) yields objective prism spectra with diffraction-limited angular resolution. The raw data for this technique are objective prism speckle spectrograms, which are obtained by inserting a prism or grating into a pupil plane (in the speckle camera). In this case each speckle is dispersed in a linear spectrum. The intensity distribution $K(\mathbf{r})$ of an instantaneous objective prism speckle spectrogram can be described by

$$K(\mathbf{r}) = \sum_m O_m(\mathbf{r}-\mathbf{r}_m) * G_m(\mathbf{r}) * S(\mathbf{r}) \,, \tag{30}$$

where $O_m(\mathbf{r}-\mathbf{r}_m)$ denotes the m-th object pixel, $G_m(\mathbf{r})$ is the spectrum of the m-th pixel, and $S(\mathbf{r})$ denotes the point spread function of telescope/atmosphere. $S(\mathbf{r})$ is wavelength independent in narrow wavelength bands (typically less than 30 nm).

From the speckle spectrograms $K(\mathbf{r})$ the

$$\text{objective prism spectrum } \sum_m O_m(\mathbf{r}-\mathbf{r}_m) * G_m(\mathbf{r}) \tag{31}$$

can be reconstructed. A laboratory simulation of objective prism speckle spectroscopy with speckle masking image reconstruction was described by Weigelt et al. (1986) and Grieger et al. (1988).

5 Wideband Projection Speckle Spectroscopy

A disadvantage of objective prism speckle spectroscopy is that it cannot be applied to general objects since in many cases the spectra of different object parts will overlap, as in the case of ordinary objective prism spectroscopy. This disadvantage can be overcome by the projection speckle spectroscopy technique (Grieger et al. 1988). The principle of this technique is summarized in Figure 5. The drawings show from top to bottom:

Image 1: 2-dimensional object, a triple star.

Image 2: 2-dimensional speckle interferogram $I(\mathbf{r}) = I(x,y)$ of the object.

Image 3: 1-dimensional projection $J(x)$ of the 2-dimensional speckle interferogram $I(\mathbf{r})$. The projection can be performed by an anamorphic imaging system of two cylinder lenses (Cooke, 1956; Labeyrie, 1981; Kingslake, 1983).

Image 4: Spectrally dispersed image $D(x,\lambda)$ of the 1-dimensional speckle interferogram $J(x)$. The spectral dispersion is, for example, performed by a non-deviating prism or by a grating.

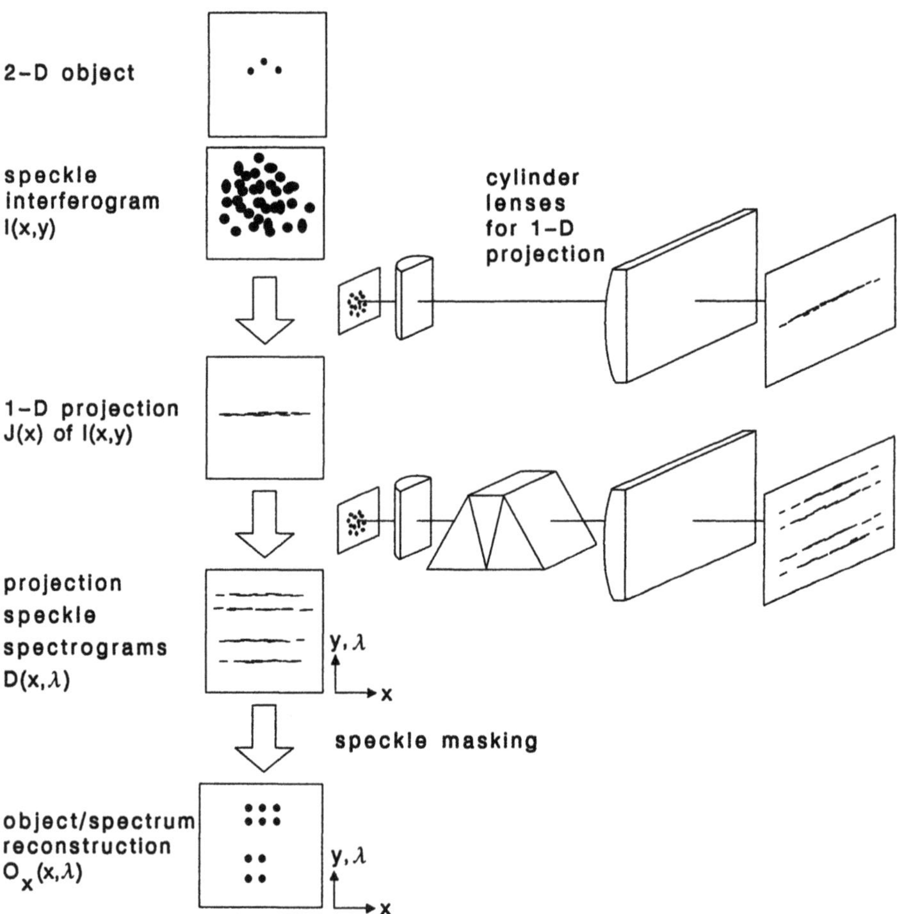

Fig. 5. Principle of projection speckle spectroscopy

The same technique was used by A. Labeyrie (1981) to obtain spectrally dispersed 1D-projections of Michelson interferograms. The spectrograms $D(x,\lambda)$ are the raw data for projection speckle spectroscopy. High-resolution spatial information is contained in the x-direction. The spectral information is found in the y-direction. The spectrograms $D(x,\lambda)$ have many advantages:
- speckle interferograms of different wavelengths do not overlap,
- the speckle interferograms of each line can separately be reduced by 1D-speckle masking,
- the wavelength dependence of the atmospheric point spread function is not problematic since each λ-line is reduced separately,
- general objects can be studied from 350 nm to 900 nm simultaneously (to 5 μ with two different detectors).

The intensity distribution $J(x;\lambda_0)$ of each individual line of $D(x,\lambda)$ can be described by

$$J(x;\lambda_0) = \mathrm{PROJ}[O(\mathbf{r};\lambda_0)] * \mathrm{PROJ}[S(\mathbf{r};\lambda_0)], \quad (32)$$

where $J(x;\lambda_0)$ is a single line of $D(x,\lambda)$ at wavelength λ_0. PROJ denotes the projection operator (projection parallel to the y-axis). $O(\mathbf{r};\lambda_0)$ is the object intensity distribution at wavelength λ_0. $S(\mathbf{r};\lambda_0)$ is the point spread function of telescope/atmosphere at wavelength λ_0 and $*$ denotes the convolution operator (Grieger et al., 1988).

Image 5: object-versus-spectrum reconstruction $O_x(x,\lambda)$, which can be reconstructed from the spectrograms $D(x,\lambda)$ by speckle masking. $O_x(x,\lambda)$ is a spectrally dispersed version of a 1-dimensional projection of the 2-dimensional object $O(\mathbf{r})$. 3-dimensional data cubes $O'(x,y,\lambda)$ can be obtained if many 2-dimensional object/spectrum reconstructions are measured with different projection and dispersion directions. Then $O'(x,y,\lambda)$ can be reconstructed using tomographic techniques.

A laboratory simulation of the $O_x(x,\lambda)$-projection speckle spectroscopy method was described by Grieger et al. (1988). The spectrum of the simulated stars consisted of emission lines at 467 nm, 480 nm, 509 nm, 546 nm, 578 nm, and 644 nm. In the case of emission line objects the limiting magnitude of projection speckle spectroscopy is about the same as for speckle imaging.

6 Optical Long-baseline Interferometry and Aperture Synthesis

The great advantage of optical long-baseline interferometry is the fact that it can yield images and spectra with fantastic angular resolution. For example, at $\lambda \sim 600$ nm and with 75 m baseline a resolution of 0.002" can be obtained. Possible image reconstruction methods are the phase closure method (Jennison, 1958; Rhodes and Goodman, 1973; Baldwin et al., 1986) and the speckle masking method.

Various computer simulations of speckle masking with coherent optical arrays were described by Hofmann and Weigelt (1986 c), Reinheimer and Weigelt (1987), and Reinheimer et al. (1988). The dependence of the signal-to-noise ratio in the reconstructed image on the following parameters has been studied in these experiments:
- number of photon events per interferogram
- optical transfer function (MMT, SMT, diluted array)
- redundancy of the pupil function
- number of r_0-subpupils per telescope (i.e., "multi-speckle mode"; r_0 = Fried parameter)
- number of rings in 2-dimensional ring-shaped arrays.

The experiments show that speckle masking can reconstruct true images in spite of large gaps in the optical transfer function. The Knox-Thompson method cannot be applied in this case. Speckle masking works since it can measure all closure phases.

Speckle masking can easily be applied to arrays of *large telescopes* (diameter larger than Fried parameter r_0; "multi-speckle mode"). In this case phase closure imaging has the disadvantage that it requires

- a non-redundant mask of many r_0-holes in front of each large telescope, or
- a non-redundant pupil reconfiguration (redundant input pupil → non-redundant exit pupil), or
- arrays of large numbers of beam splitters and beam combining mirrors or fibers or similarily difficult techniques.

For very large telescopes, as the ESO VLT, and optical wavelengths it is extremely difficult to apply these techniques and phase closure imaging. For example for $r_0 = 16$ cm, the 4 x 8-m VLT has to be regarded as a highly redundant array of 4 x 50 x 50 = 10 000 "16-cm telescopes" or r_0-subapertures with 10 000 different phase errors. In this case a non-redundant mask has the disadvantage that only a small fraction of the light can be used. For example, at magnitude 19 the speckle interferograms of the 4 x 8-m VLT will consist of about 10 photon events. If a non-redundant mask with 1% light transmission is inserted into the pupil for phase closure measurements, then each non-redundant mask interferogram will consist on average of only 0.1 photon events. Each individual Michelson interferogram would consist of 10/5000 ~ 0.002 photon events on the average. Even more difficult than non-redundant mask techniques is a non-redundant pupil reconfiguration or beam splitting techniques since non-redundant mirror arrays of 10 000 mirrors or arrays of 10 000 beam splitters would be required. These devices would introduce aberrations and cause light loss. Speckle masking can avoid the described difficult pupil manipulations since it is a phase closure method that can easily be applied to *redundant arrays* (a single-dish telescope is a highly redundant array). Of course applications of speckle masking to the VLT require detectors with large numbers of pixels. For a baseline of 75 m and $\lambda \sim 600$ nm the speckle or fringe diameter in VLT interferograms will be $1.22\,\lambda$ / baseline ~0.002". Therefore a photon-counting detector of ~1000 x 1000 pixels or better 2000 x 2000 pixels is required for 1" seeing. Such detectors exist already, obvious computers for the reduction of the interferograms are transputer arrays.

Acknowledgements. We thank ESO for observing time. The results shown in Figures 2 to 4 are based on data collected at the European Southern Observatory, La Silla, Chile.

Appendix A

We will show that the Fourier transform of the triple correlation $\int I(\mathbf{r})\, I(\mathbf{r}+\mathbf{r}_1)\, I(\mathbf{r}+\mathbf{r}_2)\, d\mathbf{r}$ is the bispectrum $\hat{I}(\mathbf{f}_1)\hat{I}(\mathbf{f}_2)\hat{I}^*(\mathbf{f}_1+\mathbf{f}_2)$:

Fourier transform of $\int I(\mathbf{r})\, I(\mathbf{r}+\mathbf{r}_1)\, I(\mathbf{r}+\mathbf{r}_2)\, d\mathbf{r}$

$$= \iiint I(\mathbf{r})\, I(\mathbf{r}+\mathbf{r}_1)\, I(\mathbf{r}+\mathbf{r}_2) \exp[-2\pi i(\mathbf{f}_1\cdot\mathbf{r}_1 + \mathbf{f}_2\cdot\mathbf{r}_2)]\, d\mathbf{r}\, d\mathbf{r}_1\, d\mathbf{r}_2$$

$$= \iint I(\mathbf{r})\, I(\mathbf{r}+\mathbf{r}_1) \exp[-2\pi i\mathbf{f}_1\cdot\mathbf{r}_1]\, d\mathbf{r}_1 \quad \left\{\int I(\mathbf{r}+\mathbf{r}_2) \exp(-2\pi i\mathbf{f}_2\cdot\mathbf{r}_2)\, d\mathbf{r}_2\right\} d\mathbf{r}$$

$$= \int\int I(\mathbf{r})\, I(\mathbf{r}+\mathbf{r}_1) \exp[-2\pi i \mathbf{f}_1 \cdot \mathbf{r}_1]\, d\mathbf{r}_1 \;\; \hat{I}(\mathbf{f}_2) \exp(2\pi i \mathbf{f}_2 \cdot \mathbf{r})\, d\mathbf{r} \quad \text{(shift theorem)}$$

$$= \int I(\mathbf{r}) \;\; \left\{ \int I(\mathbf{r}+\mathbf{r}_1) \exp(-2\pi i \mathbf{f}_1 \cdot \mathbf{r}_1)\, d\mathbf{r}_1 \right\} \;\; \hat{I}(\mathbf{f}_2) \exp(2\pi i \mathbf{f}_2 \cdot \mathbf{r})\, d\mathbf{r}$$

$$= \int I(\mathbf{r}) \;\; \hat{I}(\mathbf{f}_1) \exp(2\pi i \mathbf{f}_1 \cdot \mathbf{r}) \;\; \hat{I}(\mathbf{f}_2) \exp(2\pi i \mathbf{f}_2 \cdot \mathbf{r})\, d\mathbf{r} \quad \text{(shift theorem)}$$

$$= \int I(\mathbf{r}) \exp[2\pi i \mathbf{r} \cdot (\mathbf{f}_1 + \mathbf{f}_2)]\, d\mathbf{r} \;\; \hat{I}(\mathbf{f}_1) \;\; \hat{I}(\mathbf{f}_2)$$

$$= \hat{I}^*(\mathbf{f}_1+\mathbf{f}_2) \; \hat{I}(\mathbf{f}_2) \; \hat{I}(\mathbf{f}_2)$$

References

Baldwin, J.E., Haniff, C.A., Mackay, C.D., Warner, P.J.: 1986, *Nature* **320**, 595

Baier, G., Eckert, J., Hofmann, K.-H., Mauder, W., Schertl, D., Weghorn, H., Weigelt, G.: 1988, *The Messenger (ESO)* **52**, 11

Bates, R.H.T., Gough, P.T., Napier, P.J.: 1973, *Astron. Astrophys.* **22**, 319

Christou, J.C., Freeman, J.D., Roddier, F., McCarthy Jr., D.W., Cobb, M.L., Shaklan, S.B.: 1987, *Soc. Photo-Opt. Instr. Eng.* **808**, 32

Cooke, G.H.: 1956, *J. Soc. Motion Pict. Telev. Eng.* **65**, 151

Dainty, J.C., in *Laser Speckle and Related Phenomena*, J.C. Dainty, Ed. (Springer, Berlin, 1984)

Grieger, F., Fleischmann, F., Weigelt, G.: 1988, 'Objective Prism Speckle Spectroscopy and Wideband Projection Speckle Spectroscopy' in: Proc. High-resolution Imaging by Interferometry, ed. F. Merkle, ESO Conf., Garching, FRG, 15-18 March 1988, p. 225

Hofmann, K.-H., Weigelt, G.: 1986 a, *Appl. Opt.* **25**, 4280

Hofmann, K.-H., Weigelt, G.: 1986 b, *Astron. Astrophys.* **167**, L 15

Hofmann, K.-H., Weigelt, G.: 1986 c, *J. Opt. Soc. Am. A* **3**, 1908

Hofmann, K.-H., Weigelt, G.: 1987, *Appl. Opt.* **26**, 2011

Hofmann, K.-H., Weigelt, G.: 1988, *Astron. Astrophys.* **203**, L 21

Jennison, R.C.: 1958, *Mon. Not. Roy. Astr. Soc.* **118**, 276

Kingslake, R.: 1983, *Optical System Design*, Academic Press, p. 80

Knox, K.T., Thompson, B.J.: 1974, *Astrophys. J. Lett.* **193**, L 45

Labeyrie, A.: 1970, *Astron. Astrophys.* **6**, 85

Labeyrie, A.: 1981, 'Multiple Telescope Interferometry', in: Proc. of the ESO Conf. on Scientific Importance of High Angular Resolution at Infrared and Optical Wavelengths, eds. M.H. Ulrich and K. Kjär, Garching, 24-27 March 1981, p. 225

Liu, C.Y.C., Lohmann, A.W.: 1973, *Opt. Commun.* **8**, 372

Lohmann, A.W., Weigelt, G., Wirnitzer, B.: 1983, *Appl. Opt.* **22**, 4028

Reinheimer, T., Weigelt, G.: 1987, *Astron. Astrophys.* **176**, L 17

Reinheimer, T., Fleischmann, F., Grieger, F., Weigelt, G.: 1988, 'Speckle Masking with Coherent Arrays', in: Proc. High-resolution Imaging by Interferometry, ed. F. Merkle, ESO Conf., 15-18 March 1988, Garching, FRG, p. 581

Rhodes, W.T., Goodman, J.W.: 1973, *J. Opt. Soc. Am.* **63**, 647

Ribak, E.: 1987, *Appl. Opt.* **26**, 197

Roddier, F., Roddier, C.: 1986, *Opt. Commun.* **66**, 350

Schertl, D., Fleischmann, F., Hofmann, K.-H., Weigelt, G.: 1987, *Soc. Photo-Opt. Instr. Eng.* **808**, 38

Stork, W., Weigelt, G.: 1984, 'Speckle Spectroscopy', in: Conf. Proc. of the 13th Congress of the International Commission for Optics, ICO-13, ed. H. Ohzu, Sapporo, Japan, 20-24 Aug. 1984, p. 624

Weghorn, H.: 1988, 'Untersuchungen zur Speckle Masking-Übertragungsfunktion', Diplomarbeit, Universität Erlangen-Nürnberg

Weigelt, G.: 1977, *Optics Commun.* **21**, 55

Weigelt, G.: 1981, 'Speckle Interferometry, Speckle Holography, Speckle Spectroscopy, and the Reconstruction of High-Resolution Images from Space Telescope Data', in: Proc. Scientific Importance of High-angular Resolution at Infrared and Optical Wavelengths, eds. M.H. Ulrich, K. Kjär, ESO Conf., Garching 1981, p. 95

Weigelt, G., Wirnitzer, B.: 1983, *Optics Lett.* **8**, 389

Weigelt, G., Baier, G., Ebersberger, J., Fleischmann, F., Hofmann, K.-H., Ladebeck, R.: 1986, *Opt. Engineering* **25**, 706

Wirnitzer, B.: 1985, *J. Opt. Soc. Am. A* **2**, 14

Detectors and Receivers

Immo Appenzeller
Landessternwarte, D 6900 Heidelberg, FRG

PN diode
PIN diodes
FET

1 Fundamentals

1.1 Basic Principles, Definitions, and Notations

Practically all our knowledge about astrophysical objects is based on the the analysis of the electromagnetic radiation which reaches us from space. Because of the enormous distances of astronomical sources the observed energy fluxes usually are very low. Hence, the observing of astronomical objects means the detection of very weak electromagnetic waves. The principle instrumental set-up of such measurements is outlined in Figure 1.1. In most cases the observed radiation is collected and focused by a telescope which sorts the radiation according to its direction. Behind the telescope the radiation often passes another sorting device (labelled "analyzer" in Figure 1.1) which is sensitive to some intrinsic property of the radiation . Examples of such analyzing devices are filters, spectrographs, or polarimeters. Next, the radiation reaches the detector or receiver (the subject of this lecture) where the incident flux F_ν is absorbed and converted into a signal $S(F_\nu)$, which in most cases is an electrical current, voltage, or charge. After some on-line signal processing the signal is stored for a detailed analysis and the extraction of astrophysical information.

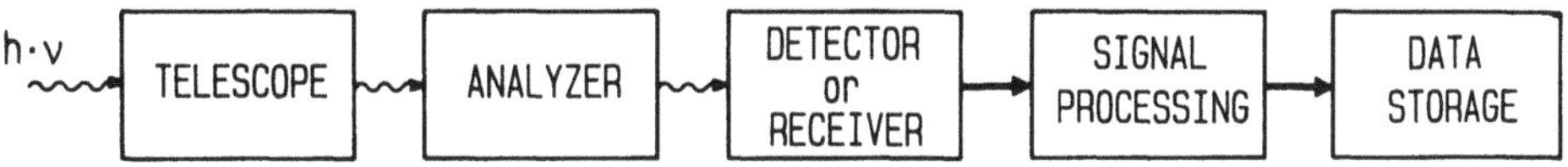

Fig. 1.1. The principle components of astronomical instrumentation

At present astronomers use three different basic detector types. In "coherent detectors", "radiometers" or "receivers" the incident electromagnetic waves are simply amplified and (sometimes after mixing with the signal of a local oscillator) rectified to produce a DC or low frequency electrical current. Such detectors are generally used at radio wavelengths. At higher frequencies the most common devices are "photon detectors" which make use of the interaction and energy exchange between the light quanta (photons) and electrons in the detector material. Finally, in "bolometers" the incident flux is determined indirectly from the heating of the detector material due to the absorbed radiation.

In practice a given detector may incorporate properties of these different classes in a single device. An example are the superconducting tunnel-effect diodes discussed in Section 3.3. These devices are clearly based on photon-electron interactions. But they are used as components of coherent receivers.

The quantity measured by an astronomical detector usually is a spectral flux density F_ν or F_λ integrated over a certain spectral range. Proper physical units for these quantities are, e.g., $\mathrm{Wm^{-2}Hz^{-1}}$ and $\mathrm{Wcm^{-2}\AA^{-1}}$. A more convenient measure of the small radiation fluxes from astronomical objects is the unit Jansky (Jy)($= 10^{-26}\mathrm{Wm^{-2}Hz^{-1}}$).

For historical reasons, in visual and infrared astronomy fluxes integrated over certain wavelength bands are also expressed in magnitudes m, where

$$m = -2.5 \log[\int g(\nu) F_\nu(\nu) d\nu] + C \tag{1.1}$$

The weighting functions g are usually realized by a combination of filters and the detector response functions and are defined implicitely by means of standard stars with fixed magnitudes. A visual magnitude of $m_V = 15$ in the UBV color system corresponds to a visual region spectral flux density of about 3 mJy.

In radio astronomy fluxes are often expressed as "antenna temperatures" T_A which are defined by

$$T_A = \frac{F_\nu A}{2k} \tag{1.2}$$

where A is the effective antenna surface and k the Boltzmann constant. This definition is motivated by the following fact. If an antenna observes an extended source (filling the whole antenna beam) of blackbody radiation of temperature T_{BB}, the observed flux is given by $F_\nu = \Omega B_\nu$, where B_ν is the Planck function and Ω the beam solid angle. At radio wavelengths where $h\nu \ll kT$ (h = Planck constant) B_ν can be approximated by the Rayleigh-Jeans law. Hence

$$F_\nu = 2\Omega(\nu/c)^2 kT_{BB} = 2\Omega\lambda^{-2} kT_{BB} \tag{1.3}$$

Furthermore, because of $\Omega A = \lambda^2$ (cf. P. Lena's contribution to this volume) we have $F_\nu = 2A^{-1} kT_{BB}$, or by inserting this expression into Equ. (1.2)

$$T_A = T_{BB} \tag{1.4}$$

Thus, for an antenna seeing only blackbody radiation, the antenna temperature is equal to the blackbody temperature and independent of the wavelength.

1.2 Noise

Besides the signal S resulting from the incident radiation all detectors produce additional unwanted output components. If the additional output power is constant or predictable, it can be subtracted and fully removed. However, all output signals also contain stochastic noise components N which result in a statistical uncertainty of the measured signal. Precise measurements obviously require low values of N and high values of the signal-to-noise ratio S/N.

Unavoidable noise sources in astronomical measurements are the quantum fluctuations of the incident radiation itself and the thermodynamic fluctuations of the energy content of the detector. The quantum fluctuations of the radiation result from its composition of individual light quanta or photons with individual energies $h\nu$. According to classical thermodynamics the energy of a thermodynamic system (such as a detector) fluctuates in units of kT where k is the Boltzmann constant. Hence, if the individual photons have energies

$$h\nu > kT \tag{1.5}$$

or

$$\nu > (k/h)T \approx 2\,10^{10}(T/\mathrm{K})\mathrm{Hz} \tag{1.6}$$

it is (at least in principle) possible to detect or count the light quanta individually. In this case the quantum fluctuations of the light become the limiting noise source. If a detector receives from a constant source an average of r photons per second, the actual number of photons collected during an integration time interval Δt will be

$$n = r\Delta t + s \tag{1.7}$$

where s is the deviation from the mean caused by the quantum fluctuations. Normally the arrival times of the individual photons show a random distribution resulting in random fluctuations of s around zero. Hence, for the arithmetic mean values $\bar{s}$ and $\overline{s^2}$ derived from a large number of individual integrations we have $\bar{s} \rightarrow 0$ and

$$(\overline{s^2})^{1/2} = \sigma = n^{1/2} = (r\Delta t)^{1/2} \tag{1.8}$$

Consequently we obtain for the signal-to-noise ratio

$$S/N = n/\sqrt{n} = \sqrt{n} = \sqrt{r\Delta t} \tag{1.9}$$

As the quantum fluctuations are an unavoidable noise source , Equ. (1.9) gives an upper limit of the S/N for any kind of photometric measurement. Increasing Δt or the photon rate r will increase S/N, but only proportional to the square root of these quantities.

As an illustrative example of the consequences of Equ. (1.9) let us consider medium resolution spectroscopic observations of a point source of visual magnitude $m_V = 15$ (such as a bright quasar, a typical T Tauri star, or a solar-type star at a distance of 1 kpc). A useful and easily memorized approximation formula for the photon flux at visual wavelengths (in photons $s^{-1}cm^{-2}Å^{-1}$) is

$$F_{Phot} \approx 10^{3-0.4m_V} \tag{1.10}$$

Assuming $m_V = 15$, the collecting area of a 4-meter telescope, a spectral resolution of 1 Å, and an optimistic total instrumental efficiency of 10 percent, we obtain for each spectral element a photon rate of $r \approx 10$ photons s^{-1}. Hence, according to Equ. (1.9) we have to integrate for 10 s to reach S/N = 10, for 10^3 s to reach S/N = 100, and 10^5 s (about 24 hours) to reach S/N = 1000.

A comparison of the prediction by Equ. (1.9) with actual observational results is presented in Figure 2. Obviously, the observations confirm the predicted relation rather well. Significant deviations occur only for the very faint and the very bright stars, where S/N becomes limited by other noise sources.

As demonstrated by the above examples and by Figure 1.2 , the photon statistics forms a critical limitation for astronomical measurements in the visual spectral range. Because of the higher energies of the individual photons and the resulting lower photon rates for a given energy flux, this effect is even more important at higher frequencies. The most energetic photons which so far could be ascribed to astrophysical objects have $h\nu = 10^{15}$ eV (UHE Gamma rays). If the radiation flux assumed in the visual spectroscopy example discussed above would reach the detector pixels in the form of such energetic photons, we would have a photon rate r = $10^{-14}s^{-1}$, or about one photon in $3\,10^6$ years. Obviously, observations at such high energies are feasible only if very large collecting areas and very long integration times are used.

The above estimates also show that astronomers cannot afford to waste photons, as any loss will decrease n and thus S/N. Therefore, a good astronomical detector must record the incident photons as completely as possible. A measure of a detector's photon collecting performance is its quantum efficiency (QE) or, more precisely, its "responsive quantum efficiency" (RQE) , which is defined as the ratio

$$RQE = QE = \frac{number\ of\ recorded\ photons}{number\ of\ incident\ photons} \tag{1.11}$$

A related quantity is the "detective quantum efficiency" (DQE) defined as

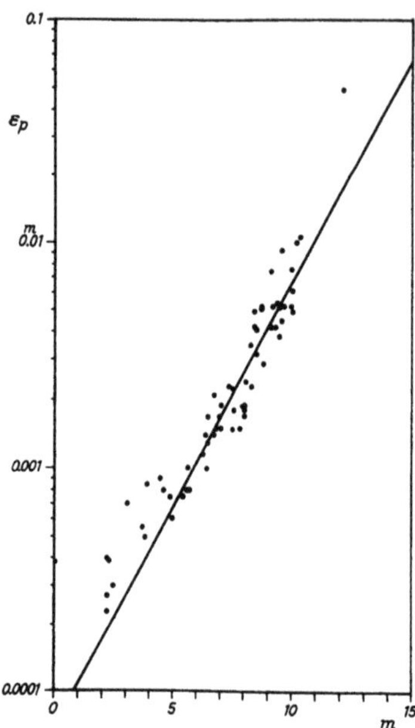

Fig. 1.2. Observational mean errors of stellar polarimetric observations as a function of the apparent stellar magnitudes. The solid line represents the relation predicted from the photon statistics according to Equ. (1.9)

$$DQE = \frac{(S/N)^2_{Output}}{(S/N)^2_{Input}} \tag{1.12}$$

From Equ. (1.9) and the above definitions it follows that in the case of pure photon noise we have DQE = RQE. If additional noise sources (such as fluctuations of detector properties) are present, we have $DQE < RQE$.

The additional noise produced by a detector itself is usually expressed by its noise equivalent power (NEP), which is defined as the power of the incident radiation at which the ratio of the signal and the detector noise reaches unity. In radio astronomy the intrinsic noise power of a receiver or a receiver component is customarily expressed (analogous to Equ. (1.2)) by a "noise temperature" T_N. $T_A + T_N = T_{Sys}$ is called the "system temperature". Measurements of T_{Sys} and thus of T_A are limited by the thermal fluctuations. If a receiver has a frequency bandwidth of $\Delta\nu$, per second an average of $\Delta\nu$ statistically independent noise contributions can be recorded. During an integration time Δt the number of noise contributions will be $n = \Delta\nu\Delta t$, with a statistical uncertainty of $\sqrt{n}$. Hence, the relative error of the temperature determination will be

$$\frac{\Delta T}{T_{Sys}} = \frac{1}{\sqrt{\Delta\nu\Delta t}} \tag{1.13}$$

The value $\Delta T = T_{Sys}(\Delta\nu\Delta t)^{-1/2}$ is often called "sensitivity limit".

1.3 Spectral Ranges

At present astronomical observations of electromagnetic radiation are carried out in the frequency interval $10^7 \leq \nu \leq 10^{30}$ Hz (about $10^2 \geq \lambda \geq 10^{-21}$ m or $10^{-8} \leq h\nu \leq 10^{15}$ eV). As outlined by Figure 3, for practical reasons this vast frequency range is subdivided into the Radio, Infrared, Visual, Ultraviolet, X-Ray, and Gamma regions. The Gamma-ray range is open-ended at high frequencies and by itself covers many orders of ten in photon energies, comprising a low-energy (LE) region ($10^5 - 10^8$ eV) , a high energy (HE) region ($10^8 - 10^{10}$ eV), a very high energy (VHE) region($10^{10} - 10^{13}$ eV), and an ultra-high energy (UHE) region($\geq 10^{14}$ eV).

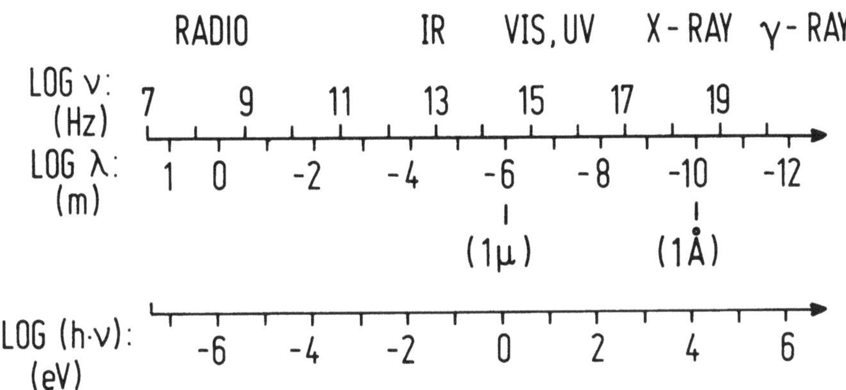

Fig. 1.3. Frequencies, wavelengths and photon energies of the electromagnetic frequency ranges used in astronomy

Electromagnetic radiation is by far our most important but not our only source of information about distant cosmic objects. Alternative techniques are e.g. neutrino astronomy and gravitational wave detection. Although applications of these methods have been very limited so far, they may become more important in the future. Hence, a brief discussion of the detector technology in these two fields will be presented at the end of this lecture. Astrophysical information can also be derived from cosmic ray observations. However, because of the cosmic magnetic fields the cosmic ray particles received on earth cannot be ascribed to individual astronomical objects. Therefore, cosmic ray observations cannot be regarded as an astronomical method.

2 Photon Detectors

2.1 Gas Ionization Devices

From the standpoint of the underlying physics, the least complex photon detectors are the gas ionization devices. The most common detector of this type is the proportional counter, which is widely used in X-ray astronomy. In its most simple form a proportional counter consists of a closed gas container with a transparent window and two electrodes (cf. Figure 2.1). Photons entering the gas interact with the bound electrons of the gas atoms. If the photon energy $h\nu$ exceeds the electron binding energy of the gas atoms, its absorption will result in the generation of a positive ion and a free photoelectron. In principle the ionization can occur from any permitted and populated electron energy level (Figure 2.2). But energetic photons most likely lead to the ejection of electrons from the lowest energy state. For atoms more massive than hydrogen a removal of the innermost electron leads to the generation of an excited ion. The ion can return to its groundstate either by radiative transitions or by using its excess energy to eject an additional (outer) electron (Auger effect). For light elements the Auger effect is the more likely de-excitation mechanism.

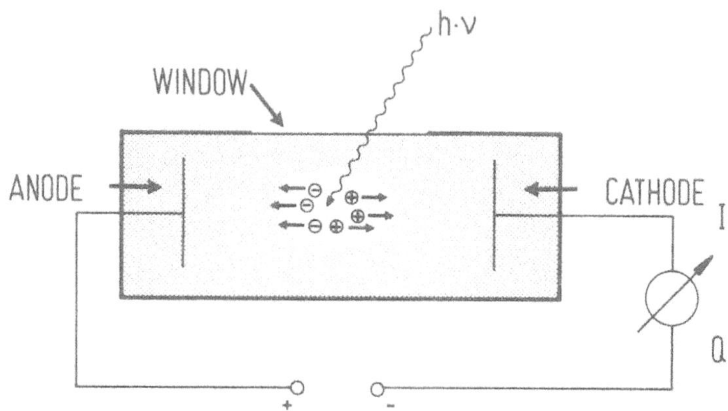

Fig. 2.1. Schematic arrangement of a proportional counter

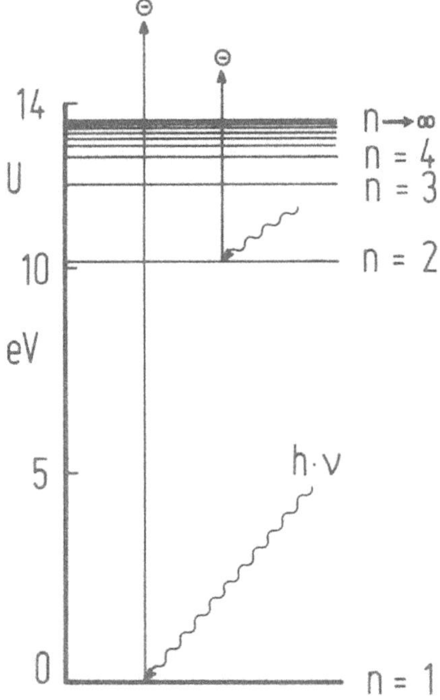

Fig. 2.2. Photoionization of the hydrogen atom. n denotes the principle quantum number of the different allowed electron energy states

If the photon energy exceeds the ionization energy, the excess energy will be transferred to the ejected electron. Energetic photons thus produce photoelectrons with high (kinetic) energies. When the electron energies exceed the gas ionization energy, their collisions with the gas atoms result in additional ionization events, until all the initial photon energy is spent in the form of ionization events. Hence, in the absence of other energy loss mechanisms the total number of free electrons (and positive ions) which can be produced in such a cascade is (approximately) proportional to the initial photon energy. If a voltage is applied to the electrodes of the proportional counter, the electrons will drift to the anode and the ions to the cathode. Normally the anode has the shape of a thin wire, resulting in a high local field gradient, resulting in a significant electrostatic acceleration on short distances. With an operating voltage of the order kilovolts, the energy gain between successive collisions with gas atoms can again exceed the gas ionization energy. The resulting secondary ionization events allow an amplification of the initial electron charges by factors of the order $10^3 - 10^5$. Therefore, each absorbed photon leads to an easily measurable current pulse. A photon flux can be measured by counting these pulses electronically. As the total charge of an individual current pulse is about proportional to initial photon energy $h\nu$, a proportional counter provides direct information on the photon energy. However, the stochastic character of some of the processes involved and energy losses (heating of the gas, radiative effects) limit the spectral resolution $E/\Delta E = \nu/\Delta\nu = \lambda/\Delta\lambda$ to values of the order 3 to 5.

By increasing the operating voltage, the charge amplification factor can be increased to values up to 10^8, where it saturates. In such high-gain "Geiger counters" the output pulses become independent of the photon energies and the spectral information is lost.Therefore, Geiger counters are rarely used in astrophysical applications.

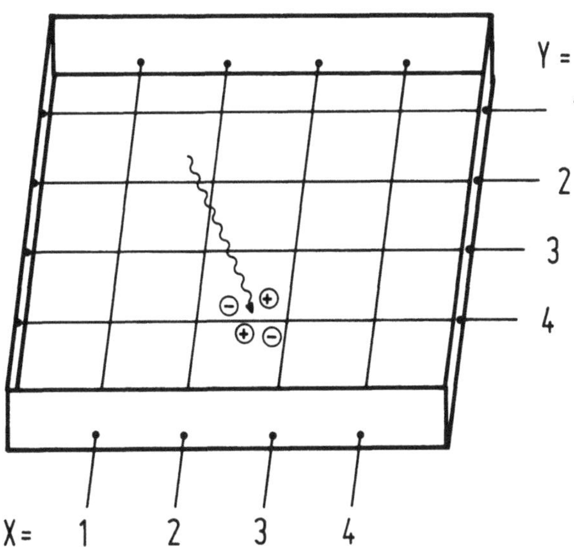

Fig. 2.3. Example of a position-sensitive proportional counter scheme

The use of multiple electrodes or resistive electrodes (with outputs at both ends) results in "imaging" or "position sensitive" proportional counters (PSPC). A schematic example is given in Figure 2.3. Here the electrodes form a rectangular grid of X and Y wires. A position sensitive signal is obtained by using the X-wires as anodes and the Y-wires as cathodes, each wire being connected to a separate electronic readout circuit. An event counted by a certain wire pair can be uniquely assigned to an X-Y coordinate position in the detector. A PSPC operated according to this principle is the main detector of the ROSAT X-ray satellite. With 80 anode wires and 160 cathode wires the ROSAT PSPC resolves nearly 10^5 pixels. Its spectral resolution is about 2.5.

Proportional counters are applied mainly in the photon energy range 0.1 - 20 keV. At lower (ionizing) energies no suitable window materials are available. At higher energies Compton scattering rather than ionization becomes the main photon- electron interaction. The filling gases of astronomical gas ionization devices are normally noble gases (e.g. Ne or Xe), to avoid losses due to the excitation of molecular vibration and rotation levels. The quantum efficiencies of proportional counters may reach peak values of 0.9 or more.

Closely related to the proportional counters are the gas scintillation counters (GSC). The GSCs are also based in the creation of electron-ion pairs by the absorption of energetic photons in a gas. However, instead of amplifying and measuring these charges, in a GSC the electrons are gently accelerated to energies where their collisions with the gas atoms result in an excitation and optical radiation of the atoms. The optical radiation pulses (which again are proportional to the photon energies) are recorded by means of photomultipliers. GSCs have QEs similar to those of proportional counters, but their spectral resolution can be significantly higher ($E/\Delta E \approx 15$).

2.2 Solid State Detectors

Solid state devices are by far the most common type of photon detectors. In order to understand their operating principles let us first recall a few basic physical properties of crystalline solids and of the photoeffect in semiconductors.

2.2.1 Energy Levels in Solids

As in the case of the gas ionization devices discussed above, solid state photon detectors are based on the interaction of photons with electrons which are bound in atoms. But in solids these atoms themselves are bound in the crystalline lattice. This has little influence on the innermost electrons and the corresponding lowest energy states. But the allowed energy states of the outermost electrons, which are characterized by overlapping wavefunctions and whose interaction with adjacent atoms is the basis for the formation and coherence of the lattice, are strongly modified. Their allowed energy states form broad energy bands which extend throughout the solid. As in the case of the discrete energy states of a free atom (Figure 2.2), electron energy levels between the allowed energy bands are forbidden and empty. The highest occupied bands, containing the electrons which are responsible for chemical effects, are called the valence bands. If an energy band is not fully occupied, electrons in this band can be lifted to slightly higher energies by acceleration in the electrostatic field of an outside voltage. The corresponding systematic motion of the electrons constitutes an electric current. Therefore, such incompletely occupied bands can carry an electric current and thus are called conduction bands.

As illustrated in Figure 2.4, the relative location of the valence and conduction bands in the energy diagram determine whether a solid is a metal, a semiconductor, or an insulator. At T = 0 K partially occupied valence bands (in metals) or an energy overlap of a fully occupied valence band and a conduction band (in semimetals) result in metalic properties. In semiconductors and insulators the valence bands are fully occupied and separated from the conduction bands by an energy gap E_G. Usually materials with $E_G < 5$eV are called semiconductors. With few exceptions only the semiconductors are of interest for the detection

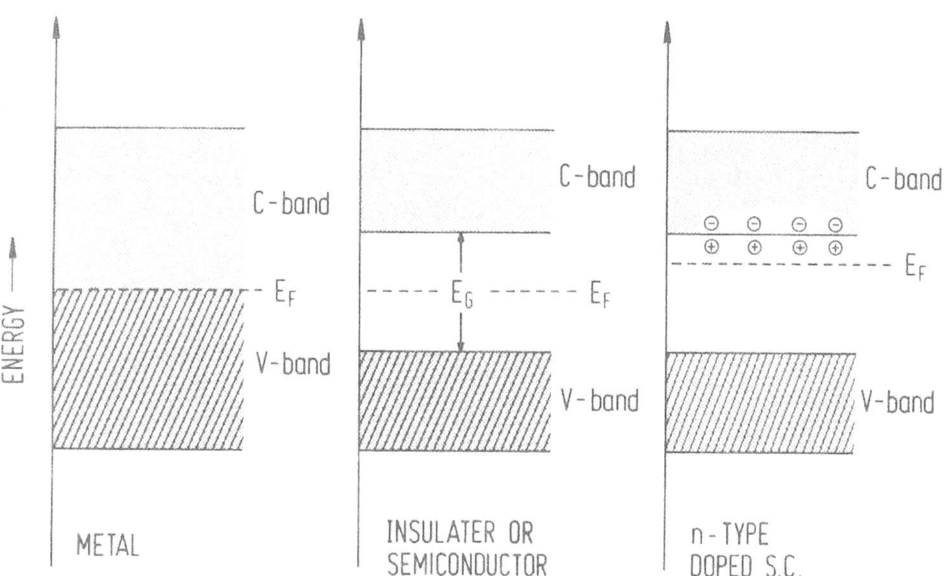

Fig. 2.4. Schematic energy diagram for different types of solids. E_F is the Fermi energy level. Allowed energy levels are shaded. Fully occupied bands are hatched

of photons (metals reflect most of the radiation, in insulators the photoeffect results in the buildup of charges). In all solids the probability of finding an electron at a permitted energy E is given by the Fermi distribution function

$$f = [\exp((E - E_F)/kT) + 1]^{-1} \tag{2.1}$$

where the Fermi energy E_F depends on the material. According to Equ.(2.1) E_F is the energy at which the probability of the corresponding electron state being occupied reaches 0.5. In a semiconductor or insulator at T = 0 K the Fermi energy level falls halfway between the lower boundary of the lowest conduction band and the upper boundary of the highest valence band energy. As also shown by Equ. (2.1), for $T > 0$ the probability of finding an electron in the conduction band of a semiconductor or insulator is never strictly zero. However, for a given temperature this probability rapidly decreases if the energy gap increases, becoming very low in typical insulators at low temperatures. On the other hand, from Equ. (2.1) it is clear that at sufficiently high temperatures any crystalline solid will show significant conductance. The conductivity of the insulator glas at T = 10^3 K is about equal to that of the typical semiconductor silicon at T = 10^2 K.

Modern microelectronics makes extensive use of the fact that the conductivity of semiconductors can be modified by implanting small concentrations of atoms with different valence electron numbers in the semiconductor lattice. Thus, it is possible to implant Sb atoms with five valence electrons per atom (Group V of the periodic system) into a crystal lattice of Group IV atoms (e.g. of the semiconductors Si or Ge) with four valence electrons. However, as only four of the five valence electrons are used in the Si or Ge lattice (cf. Figure 2.5), the extra electron of the Sb atom has a very low binding energy (< 50 meV) and a very large quantum-mechanical orbit. Hence, except at very low temperatures, the thermal energy of the lattice is sufficient to free these extra electrons into the conduction band, thereby increasing the conductivity significantly. A solid with conduction electrons produced in this way is called an n-type doped semiconductor. Because of the additional electrons in the conduction band the Fermi energy is increased relative to an intrinsic semiconductor of the same bulk material (cf. Figure 2.4). Similarly, group III atoms can become members of a group IV lattice by acquiring an additional outer electron from the valence band of the lattice, creating a positively charged "hole" in the valence-band electron energy distribution. Since now the valence band is no longer completely filled, it can carry a current which can be ascribed to the movement of positive holes. Obviously in such a positive or p-type doped semiconductor the Fermi energy level must be lower than in the corresponding pure material.

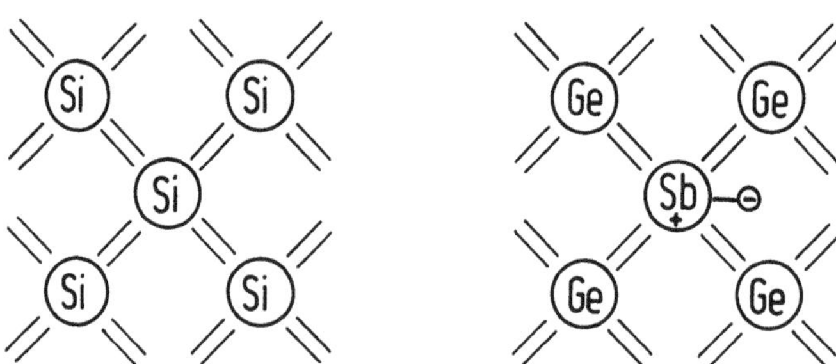

Fig. 2.5. Schematic representation of the lattice structure of an intrinsic and a doped (n-type) Group IV semiconductor. Each line represents an individual valence electron

As noted already, the loosely bound "extra electrons" of (n-type) doped semiconductors have very extended orbits which overlap even at small values of the impurity concentrations (down to 10^{-8}). As a result, doped semiconductors with sufficiently high impurity concentrations contain additional energy bands which are located between the valence and conduction bands of the corresponding pure solid. Because of the low binding energy of the corresponding electrons the energy gap between such an "impurity band" and the conduction band is always much smaller than E_G of the corresponding pure solid.

2.2.2 Photoconduction

As outlined by Figure 2.6, the absorption of a photon and the transfer of its energy to a valence band electron can lift the electron into the conduction band. Hence, in a semiconductor photon absorption increases the conductivity. If a voltage is applied to the semiconductor, the absorption results in a current which is proportional to the photon rate. Alternatively, if the photon energy is higher than the energy binding the electron to the solid, a photoelectron can be ejected from the lattice. The latter effect occurs in photocathodes. It will be discussed in Section 2.2.6.

The generation of conduction-band electrons by photon absorption in the valence band (which can take place in an intrinsic or a doped semiconductor) is called "intrinsic photoeffect". Obviously, it can occur only if the the photon energy exceeds the band gap energy E_G. Hence, all semiconducters have a cutoff frequency below which intrinsic photoabsorption and photoconduction are not possible. Above the cutoff frequency the intrinsic photoeffect can be very efficient, with RQEs sometimes exceeding 0.9. In a cooled, doped semiconductor where the temperature is low enough to prevent thermal excitation of the impurity band electrons, conduction band electrons can also be created by photoeffect from an impurity band. From Figure 2.6 it is clear that this "extrinsic" photoeffect, involving impurity band electrons, results in lower cutoff frequencies . Therefore, impurity band photoconductors are particularly useful for detecting low energy infrared photons.

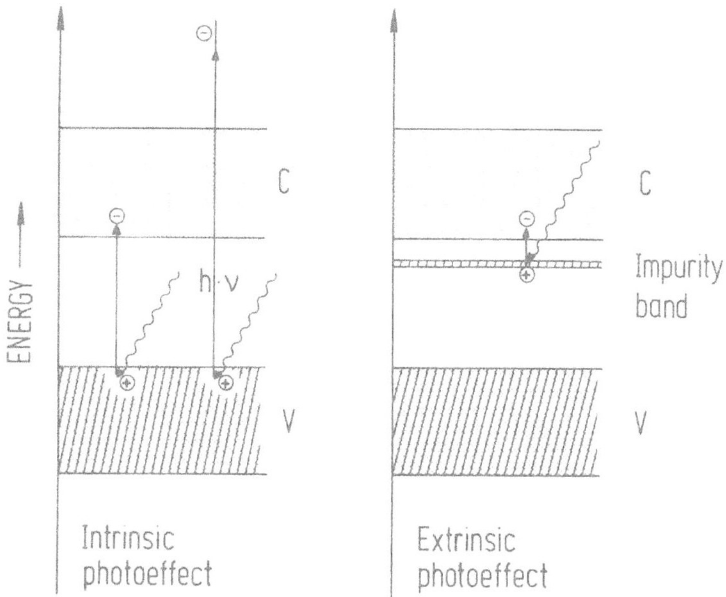

Fig. 2.6. Intrinsic and extrinsic photoeffect in photoconductors

Fig. 2.7. A stressed Ge:Ga detector with a cutoff wavelength of 240 μm. The semiconductor chip is the small black square in the lower left opening. It is mounted in vice which exerts the required stress. The device shown in this figure has been developed for use on the ISO infrared satellite (Courtesy D. Lemke, MPIA Heidelberg)

Even smaller effective band gaps can be achieved by subjecting impurity band conductors to a mechanical stress close to their elastic limit (cf. Figure 2.7). Small band gaps and correspondingly low cutoff frequencies can also be achieved in some intrinsic photoconductors where the band gap can be modified by changing the composition. An example is the alloy $Hg_{1-X}Cd_X Te$. For X = 0 this material has metallic properties (as the conduction and valence bands overlap slightly). For X = 0.2 the band gap becomes zero. For X = 1 the band gap reaches 1.6 eV. A listing of the band gaps and resulting cutoff wavelengths of various astronomically relevant photoconductors is given in Table 1. The relative response functions of some commercially available photoconductor materials are given in Figure 2.8.

Table 1. Band gaps E_G and cutoff wavelengths λ_{max} of astronomically relevant photoconductors

Composition:	E_G (eV):	λ max (μm):
	I. INTRINSIC PHOTOCONDUCTORS	
GaAs	1.35	0.92
Si	1.12	1.10
Ge	0.68	1.82
InAs	0.33	3.80
InSb	0.18	6.95
HgCdTe	adjustable	adjustable
	II. IMPURITY BAND PHOTOCONDUCTORS	
Si:In	0.16	8
Si:Ga	0.07	18
Si:Bi	0.06	18
Si:As	0.05	23
Si:P	0.05	28
Ge:As	0.01	95
Ge:Ga	0.01	120
Ge:Ga stressed	< 0.01	240

In astronomy photoconductors are used either as light sensitive resistors (as outlined above) or as photovoltaic diodes producing an illumination dependent voltage. Figure 2.9 outlines the principle of such a (PN) diode. It consists of a semiconductor crystal where in a narrow transition region the doping changes from p-type to n-type. In the absence of a voltage, the Fermi level must be constant throughout the lattice. On the other hand, as explained above, the location of the Fermi level relative to the conduction and valence bands depends on the doping type (and concentration). Hence, the constancy of the Fermi level results in a distortion of the conduction and valence bands. As the electrons tend to occupy the lowest allowed energy states, an electron (and hole) depletion zone forms at the junction of the p and n-type material. Consequently, in the absence of a bias voltage the diode has a higher resistance than its p and n type components. An outside voltage can deform the Fermi level. Depending on its polarity, the voltage can either compensate the band distortions and restore the conductivity or cause an even stronger depletion zone and higher resistance. If photons are absorbed in the depletion zone, the resulting photoelectrons and holes are seperated by the potential gradient accross the depletion zone (cf. Figure 2.9). If the two ends of the diode are connected by a load resistor, the absorbed radiation results in a current and a voltage proportional to the absorbed light flux.

Figure 2.10 shows the principle structure of two other diode types of astronomical interest. Closely related to the PN diodes are the PIN (= positive-intrinsic-negative) diodes, which contain a region of intrinsic material between the p- an n-type layers. As a result of

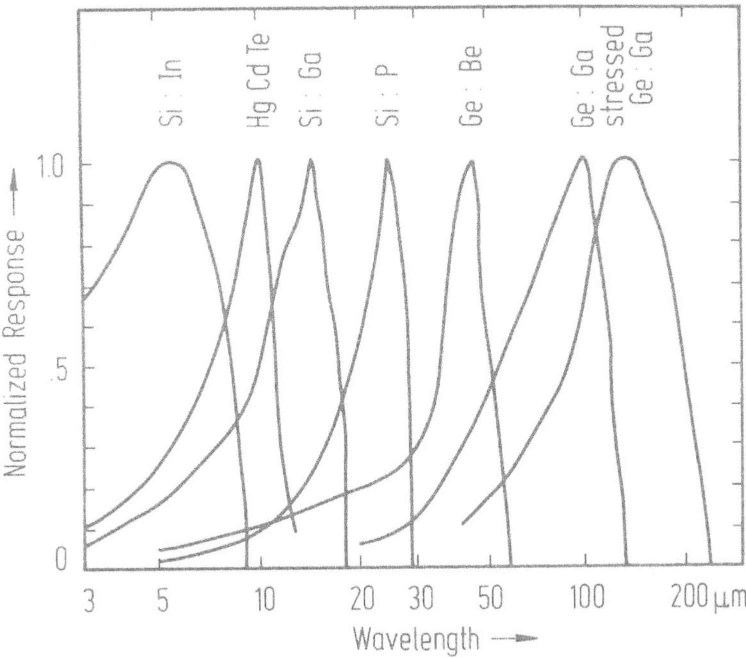

Fig. 2.8. Wavelength dependence of the relative response of various photoconducting materials

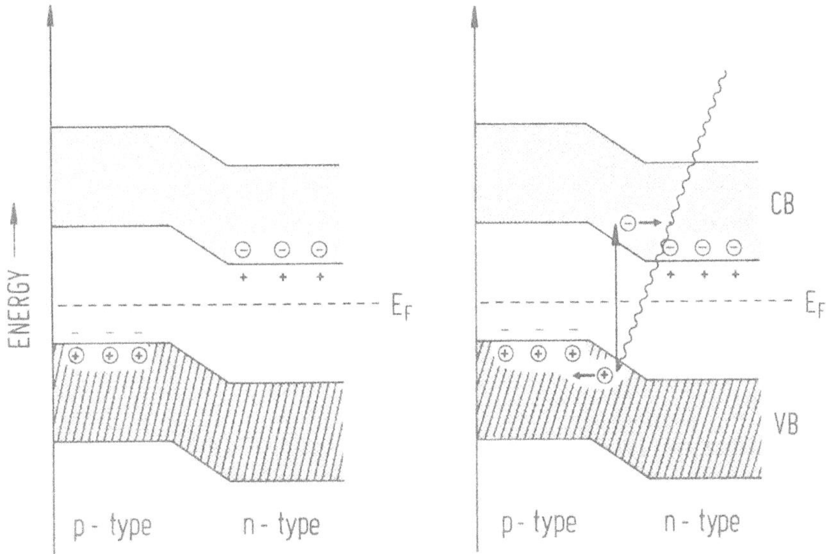

Fig. 2.9. Band distortion and photoeffect at PN junctions

this additional layer the extent of the depletion region is increased, giving a larger photosensitive volume. In the case of the"Schottky diodes" a metal and an n-type semiconductor are brought into contact . Charges trapped in the surface layers (or differential doping) cause the kind of band distortion shown in the figure. Thermal excitation or (at low temperatures) quantum-mechanical tunnelling allows electrons to cross the barrier formed by the depletion zone of the Schottky junctions in the direction of decreasing potential. As a result, a DC current can flow whithout an outside voltage. If a voltage is applied to a Schottky diode, its polarity determines whether the current is reduced or strongly enhanced. Hence, Schottky diodes exhibit a strongly nonlinear current-voltage relation.

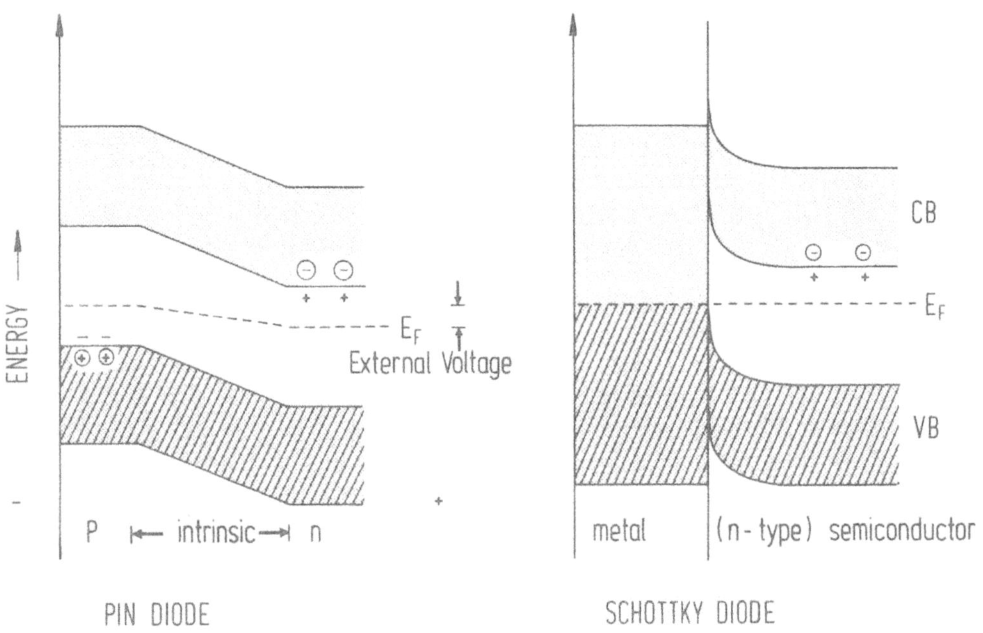

Fig. 2.10. Energy diagrams for PIN and Schottky diodes

Because of the small size of photoconducting detectors, imaging detector systems can be made by assembling arrays of individual detector elements, each one equipped with its own readout electronics. But, as these detectors are manufactured using the standard production methods for integrated electronic circuits, it is also possible to make one-chip integrated detector arrays, such as the Reticon arrays, the Charge-Coupled Devices (CCD), or the Charge Injection Devices (CID), which will be discussed in the following paragraphs.

2.2.3 Reticon

The Reticon device (named after the US firm which first developed this type of detector) is a silicon-based integrated circuit containing a large number(typically of the order 10^3) of photodiodes connected to FET switches and a shift register according to the scheme outlined by Figure 2.11. In most cases the diodes are arranged linearly (Figure 12.12) forming a long row. But other geometric arrangements are also possible and available. A Reticon is operated as follows: Before starting an exposure, temporarily a reverse voltage is applied to the diodes (and the associated capacitances) resulting in the storage of a charge of the order 10^6 electrons at each of the diodes. If the detector is cooled to $\approx$ 150 K the reverse conductivity of the diodes becomes very small and the charges can remain stored for hours. However, if the diodes are exposed to light, photoconduction will result in a discharge current proportional

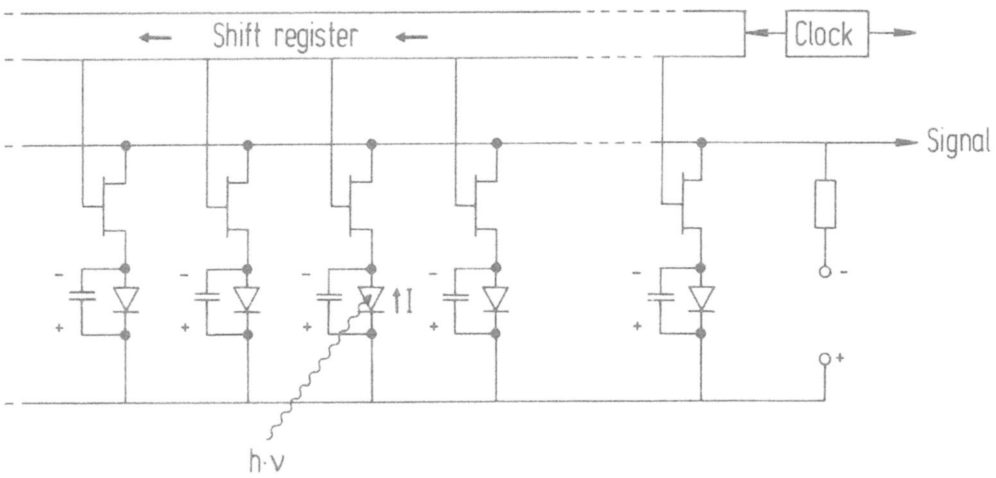

Fig. 2.11. Operating principle of the Reticon diode array

to the photon rate. After the end of the exposure the diodes are recharged individually by clocking a switching signal through the shift register. Each diode is sequentially connected to the output line, where an output signal is generated by the recharge current. Thus, the charge loss due to the illumination (and consequently the photon flux) is determined for each diode. The output signals are amplified, digitized and transferred sequentially to a computer where the recorded image can be restored from the known geometric arrangement of the diodes.

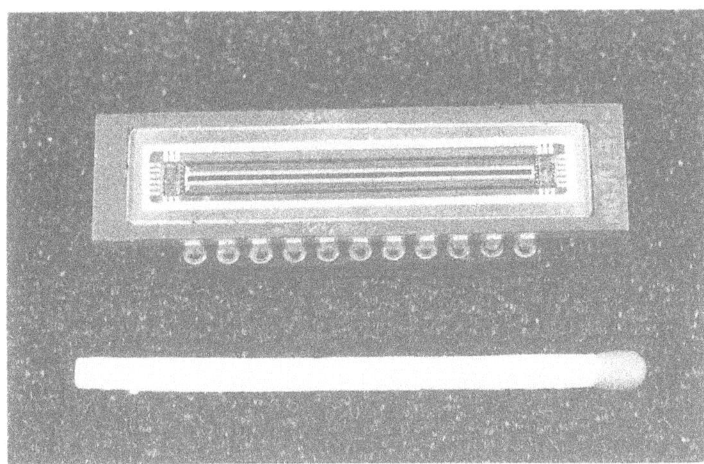

Fig. 2.12. A 1024 pixel linear Reticon array for astronomical spectroscopy. The central black strip is the photosensitive area

Silicon Reticons are most sensitive in the green to red spectral range but can be used at wavelengths up to 1.1 μm. Their peak quantum efficiencies reach values above .85. A disadvantage is the large capacitance of the Reticon output line and the associated charge. Electron fluctuations of this charge introduce a "readout noise" of the order 10^3 electrons r.m.s.. Hence, at each diode the signal must be at least of this oder to be measurable. On the other hand, up to 10^6 photoelectrons can be recorded at each pixel, allowing (according to Equ. 1.9) a $S/N \approx 1000$. Hence, a Reticon is an excellent device for precision measurements of relatively high radiation fluxes (number of recorded photons per diode $\gg 10^3$).

2.2.4 CCDs and CIDs

A CCD chip consists of a thin ($\lesssim$ 0.2mm), rectangular, mono-crystalline slab of photoconducting material. In most cases the main substrate is p-type silicon. Its surface is covered by very thin ($\lesssim 1\mu$m) strip-shaped conducting electrodes (cf. Figure 2.13) which are separated from the substrate by a $\approx 0.1\mu$ oxide insulating layer. The photosensitive part of the chip is a ($\approx 10\mu$) lightly doped or nearly intrinsic layer below the insulating layer where long narrow "channel stops" (consisting of heavily doped p-type material) are implanted by differential doping.

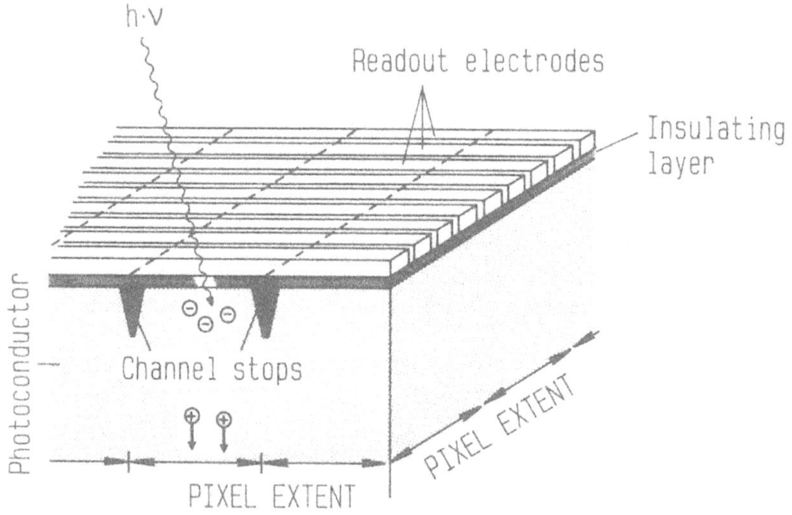

Fig. 2.13. Schematic structure of a CCD chip

Carrier depletion and the resulting negative charge efficiently prevents the movement of electrons across the channel stops. Hence, the channel stops define channels restricting the movement of the photoelectrons. These "read out channels" are oriented perpendicular to the electrode strips (cf. Figure 2.13).

At the edge of the chip the channel stops end in a "readout register" which has its own set of electrodes oriented perpendicular to the electrode strips in the detection section. For the most commonly used three-phase CCDs the whole electrode arrangement is outlined in Figure 2.14. As shown by this figure, in the 3-phase device the individual electrodes are connected alternatingly to three voltage supply lines. The size of the (square-shaped) pixels is defined by the channel width and three consecutive electrode strips.

When a CCD is exposed, at each pixel one electrode is kept at a (by a few volts) more positive voltage. If light is absorbed in the photoconductor the photoelectrons accumulate below this electrode while the holes are drawn into the substrate. After the exposure the chip shows a photoelectron charge distribution corresponding exactly to the distribution of the absorbed light. This two-dimensional charge distribution is read out by changing the electrode voltages according to the scheme indicated in Figure 2.15. Each full cycle of three voltage changes moves the whole charge distribution by one pixel along the readout channels. The pixel row adjacent to the readout register is moved through a transfer gate (omitted in Figure 2.14) into the register. By alternating the voltages of the register electrodes (following again the scheme of Figure 2.15) the content of the output register is then read out sequentially through an on-chip preamplifier to a signal line. This sequence of events is repeated untill the charge content of all pixels has been recorded, digitized, and stored in a computer.

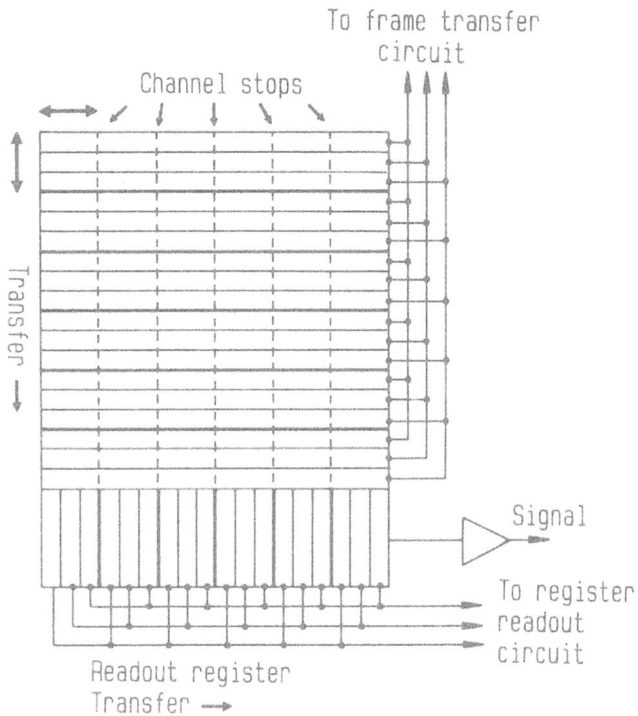

Fig. 2.14. Schematic electrode arrangement of a (6 × 6 pixel) three-phase CCD

The most serious problem in manufacturing CCDs are lattice defects which can trap electrons, resulting in an incomplete charge transfer during readout. As the charge of an individual pixel may experience up to 10^6 transfer steps, even a tiny fractional charge loss at each step will have disasterous consequences for the image. As lattice defects tend to result in the trapping of a fixed number of electrons, charge transfer deficiencies are particularly damaging to weakly exposed images containing few electrons per pixel. Surface or interface layers are particularly prone to contain lattice defects. Hence, high quality ("buried-channel") CCDs use a very thin ($\approx 0.2\mu$) n-type zone below the insulating layer to generate a minimum of the electrostatic potential slightly below the interface beteween the silicon and the insulating layer. The photoelectrons are accumulated in this potential minimum, where the lattice is more uniform and where the electrons can be transferred safely. Differential doping (or special electrode geometries) can also be used to define the charge transfer direction during readout. This allows to design CCDs with only two (2-phase) or even only one (virtual phase) electrode per pixel.

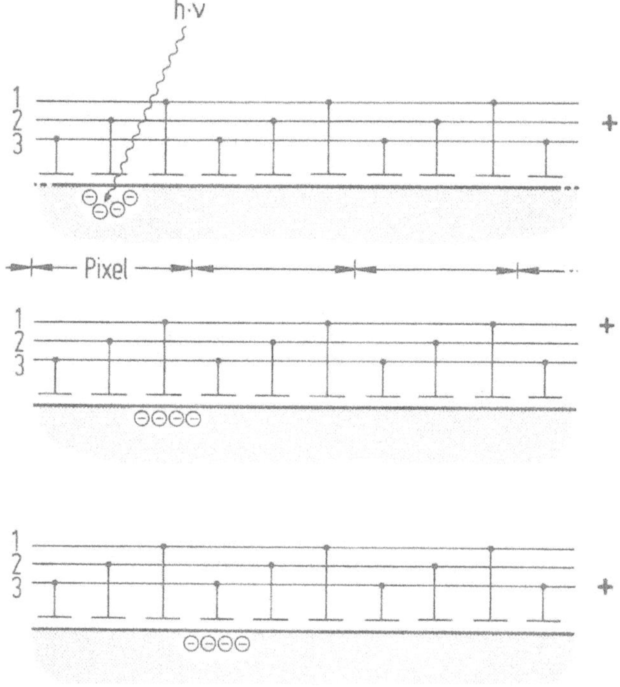

Fig. 2.15. Charge transfer in a three-phase CCD

A CCD chip can be illuminated either through the (transparent) electrodes or from the rear (substrate) side. Because of unavoidable light losses in the electrodes the latter solution is favoured in astronomical applications. However, since the absorption coefficient of silicon strongly increases with decreasing wavelength, most blue photons are absorbed near the surface of the silicon slab, before reaching the lightly doped photon conduction layer. Hence, such "back-illuminated" CCDs show a low sensitivity for blue light. A possible remedy is thinning the silicon slab (mechanically or chemically) to a thickness of $\approx 10\mu$. The dramatic improvement of the blue response achievable by this procedure is shown in Figure 2.16, where the measured RQE is plotted for two commercially available chips. However, even thinned, backside-illuminated CCDs normally show a steep sensitivity decrease in th UV. This behaviour is due to the very high absorption coefficient of silicon at short wavelengths, which results in an absorption of UV photons already in the surface layers where the photoelectrons tend to become trapped by lattice defects. Hence, CCDs for UV applications require specially treated surface layers. Alternatively, phosphorous paints can be applied which convert the UV photons into easily detected visual or red photons.

At present high quality CCDs suitable for astronomy are available with pixel sizes $\lesssim 30\mu$ and formats up to 1000×1000 pixels. Larger chips are under development. By cooling a silicon CCD to temperatures $\lesssim -100°$C (using e.g. liquid nitrogen, cf. Figures 2.17 and 2.18) the dark current due to thermally generated conduction electrons can be made negligible. Hence, as in the case of the Reticon, the sensitivity is limited mainly by readout noise. However, because of the absence of long signal lines, in CCDs readout noise values < 10 electrons (r.m.s.) can be achieved. Thus, CCDs are well suited for detecting weak light fluxes. On the other hand, because of the lower charge capacity of a CCD pixel (of the order 10^5 electrons) the maximum achievable S/N is lower than obtainable with a Reticon exposure.

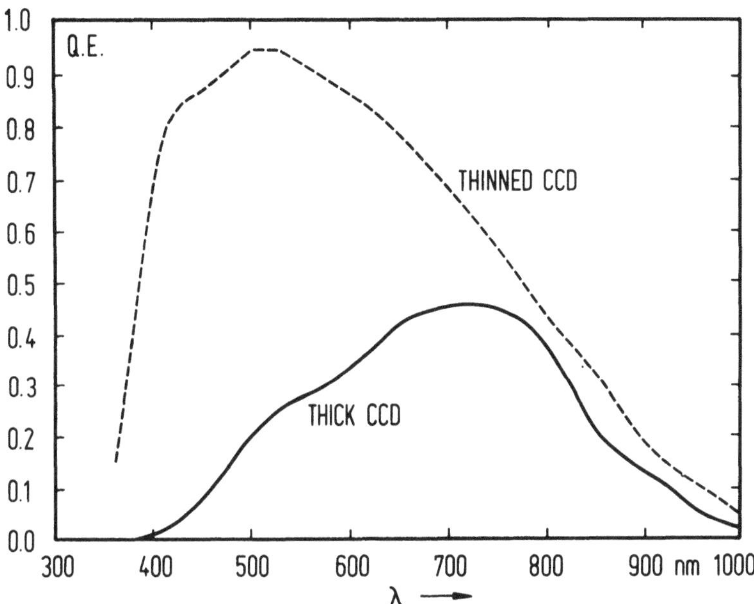

Fig. 2.16. Comparison of the measured responsive quantum efficiencies of a thick and a thinned back-illuminated CCD

As the conduction properties of the various layers of a CCD chip are based on the thermal excitation of the charge carriers, the CCD design cannot be directly applied to impurity band photoconductors. Therefore, for IR applications hybrid CCD detectors have been developed, where a matrix of impurity band detector elements is bonded to a silicon CCD. The hybrid design is also used in the commercially available HgCdTl infrared CCDs.

Closely related to the CCDs are the CID detectors. CIDs also consist of a photoconductor covered by a system of readout electrodes. However, each pixel has two individual electrodes which are connected to a rectangular grid of X and Y lines. Hence, an individual pixel corresponds to a distinct pair of these lines. As in the CCD, an exposure of the chip results in a "charge image". But, in contrast to the CCD, a CID can be read out without destroying the charge distribution. This is accomplished by temporarily varying the electrode voltages. If e.g. a charge has been accumulated at the pixel corresponding to the X-line i and the Y-line j, a voltage change at the X-line i will result in a charge dependent voltage variation at the corresponding Y-line (cf. Figure 2.19). A measurement of this voltage variation, gives the accumulated charge at the corresponding pixel.

To erase the exposure the photoelectrons are "injected"into the semiconductor substrate by means of a negative electrode voltage. The "non-destructive readout" obviously has the advantage that the growth of the image can be followed during the exposure. Furthermore, as lattice defects are much less critical than in CCDs, CIDs are easier to make and easily adoptable to different kinds of semiconductor materials. On the other hand, the length and the correspondingly high capacitances of the readout lines result in readout noise values which are comparable to those of Reticons and much higher than in the case of CCDs.

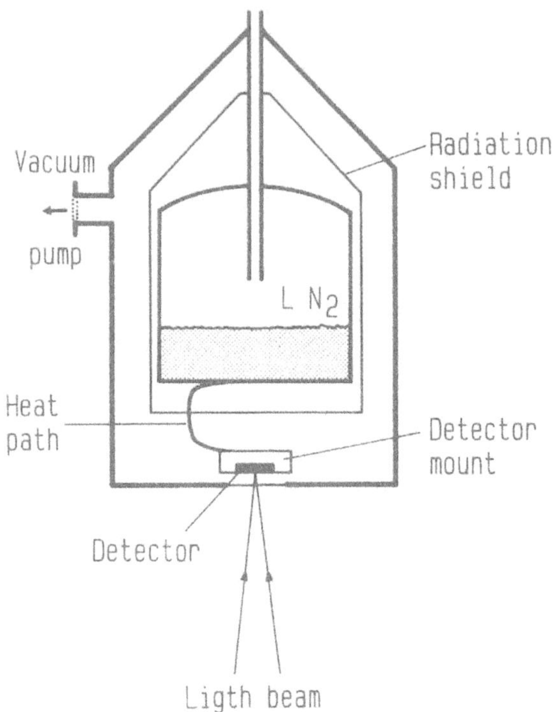

Fig. 2.17. Principle design of a liquid nitrogen cryostat for photoconducting detectors

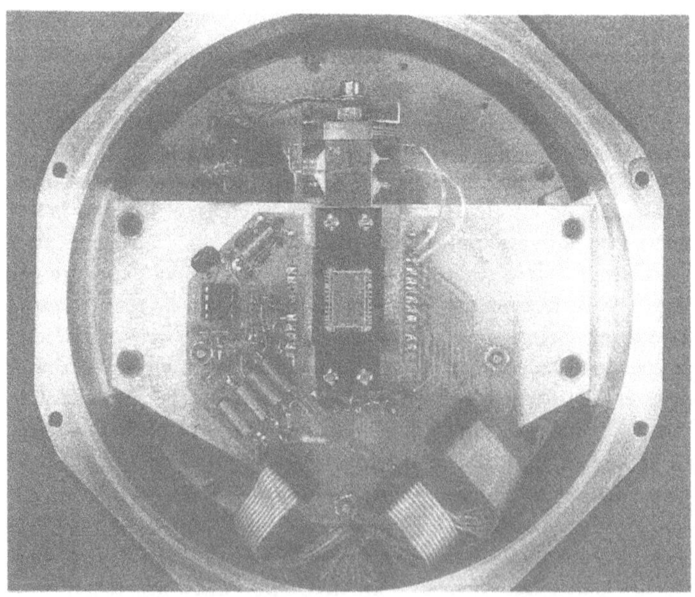

Fig. 2.18. A CCD chip (center) mounted in its (opened) cryostat. Also shown are the heat path and temperature control resistors (above) and part of the control electronics (to the left)

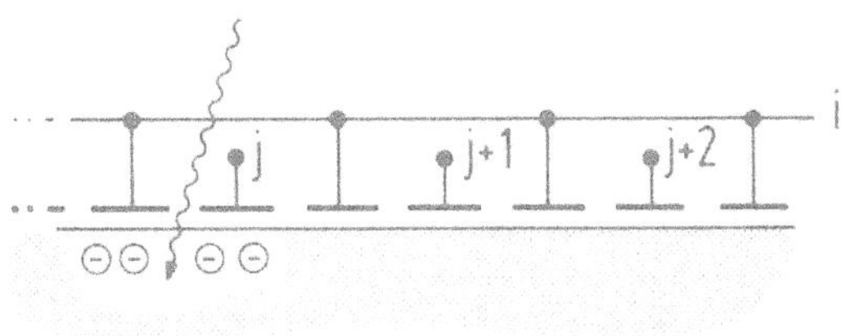

Fig. 2.19. Schematic electrode arrangement of a CID detector

2.2.5 Photography

Apart from the human eye, photographic plates are the oldest astronomical light detectors. Although for most astronomical applications photography has been replaced by CCDs and photocathode detectors, photographic plates are still being used for the imaging of large fields. Astronomical plates consist of small $(10 - 40\mu m)$ AgBr crystals (called grains) suspended in a thin dried gelatin layer. AgBr is a semiconductor with a band gap of about 2 eV. Hence pure AgBr absorbs blue light only. With impurities the sensitivity can be extended into the red and infrared spectral range. Impurities and lattice defects (including free Ag^+ and Br^- ions) are a prerequisite for reaching reasonable quantum efficiencies of photographic materials.

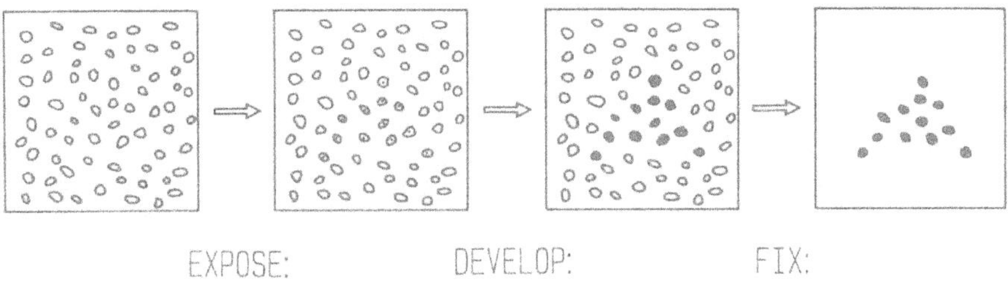

Fig. 2.20. The generation of a photographic image

As in the photoconductors discussed earlier, the absorption of a photon in an AgBr grain results in the generation of a conduction electron and a positive hole. Ions diffusing in from the gelatin tend to trap these holes , preventing a rapid recombination. (For this reason, the gelatin properties have a profound influence on the emulsion sensitivity). If a photolelectron diffusing through the lattice meets a free Ag^+ ion, a silver atom is formed. Because of its relatively small size, the Ag atom also can diffuse through the lattice. If it encounters a Br^- ion, AgBr is formed again. In this case the information about the photon absorption is lost. If an Ag atom meets a second Ag atom (produced by another photon reaching the same grain during the lifetime of the first silver atom), the two atoms form an Ag cluster (i.e. the beginning of a silver crystal). Clusters of only two silver atoms can still be destroyed again by Br^- ions. However, if the absorption of additional photons results in a further growth, the clusters become stable when they contain at least 3-5 atoms. Hence, after an exposure the grains which absorbed a sufficient numbers of photons contain tiny silver crystals. This "latent image" of tiny silver specks in part of the grains can be amplified and made visible by developing and fixing the plate (Figure 2.20). During the developing process the latent image is amplified (by a factor of the order 10^8) using a chemical process which converts AgBr into silver and a soluble bromium compound in those (and only in those) grains which already contain silver clusters or crystals. In the fixing process the remaining AgBr grains are removed chemically.

Since a grain contributes to the latent image only if several photons have been absorbed within the lifetimes and diffusion times of the silver atoms in the lattice, the photographic quantum efficiency depends on the incident flux, on the exposure time, on the size of the grains (larger grains obviously being more sensitive), and on the impurities and the diffusion time scale of the material. Astronomical applications, characterized by low photon rates, obviously require photographic emulsions with long diffusion times. Therefore, for astronomy special photographic plates are manufactured which contain highly imperfect microcrystals and high concentrations of impurities. The diffusion effects can be delayed further by cooling the plate during the exposure. As the lifetime and the survival chances of the silver atoms depend strongly on impurities, part of which are diffusing in from the gelatin and its surface, the plate sensitivity also depends on environmental characteristics such as the humidity. Removing selectively certain impurities (such as bromium ions and water molecules) by vacuum treatment, baking, or washing the plate in suitable chemicals before the exposure, may result in a sensitivity increase by large factors. Another method of increasing the sensitivity is a low light level uniform pre-exposure to have an initial supply of free Ag atoms in each grain when the object exposure is started. If the pre-exposure intensity is below the level where stable Ag clusters are formed, the background density of the plate remains unaffected. In order to be effective the time delay between the pre-exposure and the object exposure obviously must be less than the diffusion time scale of the emulsion.

In contrast to most other, solid state detectors, photographic plates use microcrystals. Hence, very large ($\approx m^2$) plates with $> 10^8$ pixels can be made. On the other hand, because of the competing physical effects in a grain during the exposure, the quantum efficiencies of photographic emulsions are generally rather low. Even for the best photographic plates the QE rarely exceeds 0.01. Furthermore, the nonlinearity of the photographic process and the many different parameters affecting the sensitivity make an accurate calibration of photographic images cumbersome and often impossible.

2.2.6 Photocathodes

Photocathodes are made of semiconductors with relatively low electron ejection energies, such as GaAs, Cs_3Sb, K_2CsSb, or(Cs)Na_2KSb. In a homogeneous semiconductor the photon energy required to eject a valence electron into the vacuum is always considerably higher than the band gap energy (cf. Figure 2.6). Therefore, modern photocathodes use thin inhomogeneous surface layers and the associated energy-band distortions (cf. Section 2.2.2) to lower the electron ejection energy. As illustrated by Figure 2.21, a substantial decrease of the ejection energy or "work function" can be achieved by n-type doping of the surface layers of an otherwise p-type (or intrinsic) semiconductor.

Using different doping concentrations the ejection energy and thus the spectral response can be tailored to specific needs. By means of such techniques photocathodes sensitive at wavelengths up to about 1.2 μm and with peak quantum efficiencies up to 0.6 have been developed. However, most practical photocathode devices have RQE values of $\approx$ 0.10 - 0.30 only. Furthermore, as good photocathodes depend on differential doping effects in very thin surface layers, they are very sensitive to overexposure, overheating, and chemical reactions with the rest gas of an incomplete vacuum. Dark emission of photocathodes can be made as small as 10^{-4} electrons s^{-s} per image element. If internal amplification methods are used (see below) photocathode devices are practically free of readout noise. Hence, in spite of their generally lower RQE, at very low light levels photocathode detectors may have higher DQEs than the photoconducting devices discussed above.

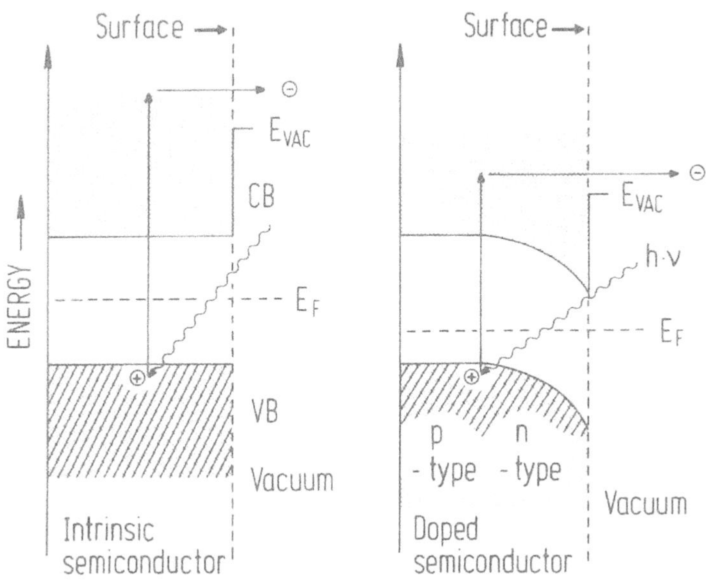

Fig. 2.21. Lowering of the ejection energy of photoelectrons in photocathodes by means of differential doping of the surface layers. As in Figure 2.9 the Fermi level remains constant while the permitted energy bands are bent at the p-n transition

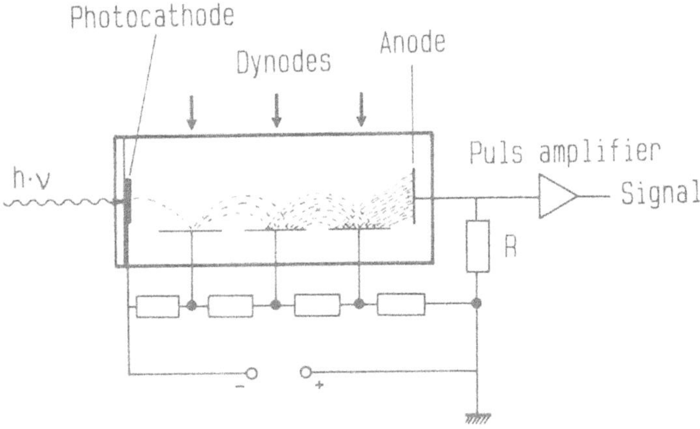

Fig. 2.22. Schematic arrangement of a photomultiplier tube

During the past decades a great variety of different phocathode detectors have been developed for astronomy. Only a few of these (photomultipliers, image intensifiers, intensified vidicons, and photon counting detectors) will be described here.

2.2.7 Photomultiplier Tubes

As shown schematically in Figure 2.22, a photomultiplier tube (PMT) consists of a photocathode, an anode, and several (up to about 15) secondary electrons emitting dynodes. The whole assembly is enclosed in a vacuum cylinder. Often the cathode is semitransparent and evaporated onto the inside of the tube entrance window. A photoelectron emitted by the cathode is accelerated by a high voltage ($\approx$ kV) electric field towards the first dynode. Since the energy gain of the electron is much larger than the electron ejection energy of the diode material (usually also a semiconductor) each primary photoelectron gives rise to the ejection of several (up to 50) secondary electrons. The secondary electrons are now accelerated towards the next dynode, where the charge is again multiplied by a factor larger than unity. As a result of the repeated charge amplification at the dynodes, each photoelectron from the cathode results in up to 10^8 electrons (i.e. an easily detectable charge pulse) at the cathode. Hence, like proportional counters, PMTs allow a direct detection and counting of the individual photoelectrons.

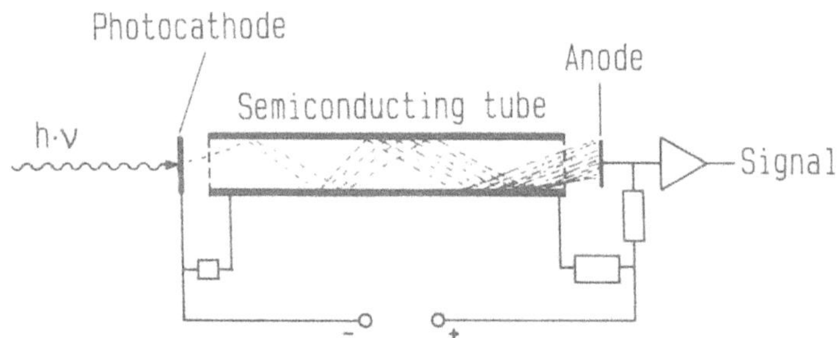

Fig. 2.23. Schematics of a channel electron multiplier

A variant of the PMT tubes are the channel electron multipliers (Figure 2.23) where the dynodes and their resistor chain are replaced by a tube with a semiconducting inner surface. By producing the photoelectrons in the entrance part of the tube, even a separate photocathode can be omitted. Channel multipliers obviously are very simple. Moreover, applying glas fiber production techniques, channel multipliers can be made rather small (down to diameters of about 10 μ). Bundles of large numbers of parallel channel multipliers fused together form Microchannel Plates (MCPs), which are used in imaging photocathode devices (see Figure 2.25).

2.2.8 Image Intensifier Tubes

Image intensifiers are vacuum tubes containing a semitransparent photocathode and an anode which consists of a thin metal film followed by a phosphor screen. The metal film is opaque to light but transparent to the energetic photelectrons which are accelerated by a high positive anode voltage to energies of the order 10 keV (cf. Figure 2.24).

Although part of the electron energy is lost in the metal film of the anode (which is needed to prevent an optical feedback), each accelerated electron still deposits several keV in the phosphor screen. This results in the emission of a large number (typically $\approx 10^2$) of secondary photons for each photoelectron. Using one of the techniques outlined in Figure 2.25, all photons emitted from a given point of the photocathode are focused to a corresponding point at the anode. Hence, on the phosphor screen we obtain an intensified image of the light distribution on the photocathode. This intensified image can be recorded by a less sensitive optical detector such as a photographic plate or a television camera. If the luminosity gain of a single image tube is not sufficient for this purpose, several intensifier stages can be used in series, resulting in a total amplification which is the product of the gains of the individual stages.

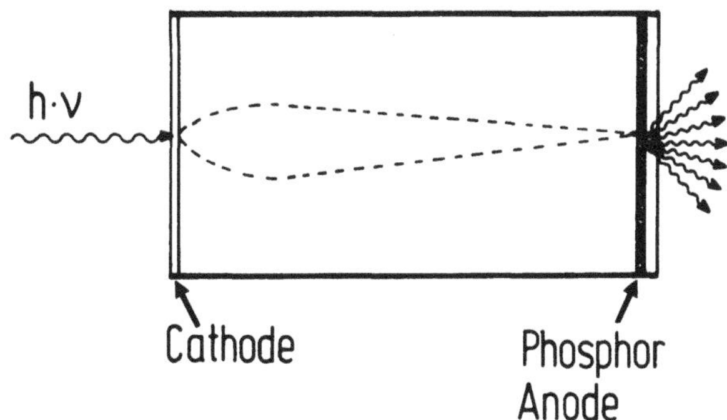

Fig. 2.24. Image intensifier principle

For image intensifier tubes a good vacuum is even more important than for other photocathode devices as the keV photoelectrons easily ionize the rest gas atoms and molecules. Ions produced in this way are accelerated towards the photocathode where they cause flashes of electron emission. Apart from increasing the noise level, the impact of the energetic ions also results in a slow destruction of the cathode material.

Present astronomical image intensifiers make use of four different electron imaging methods (cf. Figure 2.25). If the distance d between the photocathode and the anode is not more than about 10 - 50 times larger than the size of the required spatial resolution element, an image can be obtained without electron optics. Such "proximity focus" intensifiers (also called "proxifiers") are particularly simple and compact. However, since the spatial resolution is about proportional to d^{-1} while the luminosity gain is about proportional to the operating voltage V, and since the product Vd^{-1} is limited by the onset of field effect electron emission, high spatial resolution and high gain cannot be achieved simultaneously with proximity focus devices. For most astronomical measurements low spatial resolution is not acceptable. Therefore, in astronomy proxifiers find applications only as low gain devices for some special purposes.

Higher gains can be achieved when electrostatic lenses are used for focusing. If the lense electrode has a suitable geometry, a single lens at the anode potential produces an acceptable image. In this case the image quality is relatively insensitive to fluctuations of the operating voltage. A disadvantage of such simple electrostatically focused tubes is a substantial curvature of the object plane and the image plane. Hence the photocathode and the phosphor screen have to be curved. To compensate for this curvature the entrance and exit windows of electrostatic intensifiers normally consist of fiberoptic plates with a plane outside surface and a curved inner surface.

By far the best spatial resolution is achieved in magnetically focused intensifiers. Good images over an extended field are possible by matching the homogeneous magnetic field of a long coil to an approximately homogeneous electric field generated by means of a series of ring electrodes ("R" in Figure 2.25) at electric potentials intermediate between the cathode and anode. The achievable image quality is limited by the constancy of the coil current and accelerating voltage and on the efficiency of the shielding against outside magnetic fields (such as the terrestrial field). Examples of photographic spectrograms observed by means of magnetically focused image tubes are reproduced in Figure (2.26). Note the gain in exposure time *and* information content between the spectrograms (1) and (2). This difference is entirely due to the higher QE of the photocathode as compared to the unaided photographic plate. A further increase of the gain (spectrogram 3) reduces the time required to obtain a well exposed image (and allows the detection of single photoelectrons, made visible as individual small spots) but brings little additional information gain.

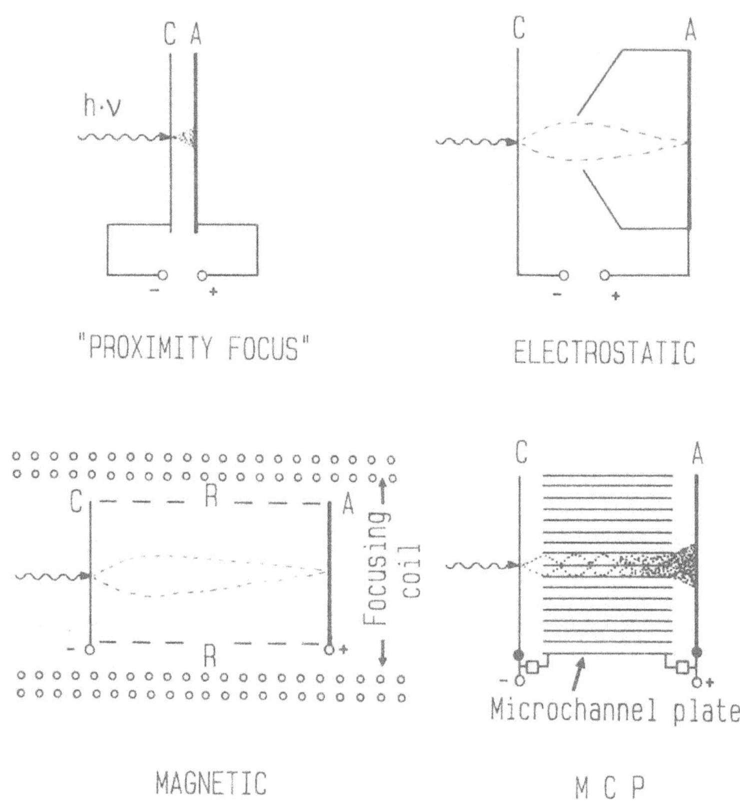

Fig. 2.25. Schematic representation of the different electron focusing methods used in image intensifiers

The principle of imaging by means of a microchannel plate has been mentioned already. As an important difference to other image intensifier types, in MCP devices the gain is achieved mainly by increasing the number of electrons rather than by the energy gain of single photoelectrons. Hence, MPC intensifiers produce very high gains (up to 10^6 in a single stage) at relatively low operating voltages. For this reason MCP intensifiers also are less sensitive to ion events. Moreover, the use of curved or V-shaped individual channels, efficiently prevents that ions produced and accelerated in the high field near the anode reach the sensitive photocathode.

2.2.9 Intensified Vidicons

The principle arrangement of vidicon and intensified vidicon tubes is outlined in Figure 2.27. Ordinary vidicons contain no photocathodes but use large scale (or matrix) photodiodes as primary detector surfaces.

Operationally a vidicon shows many similarities to the Reticon detector described above. As in a Reticon, the diode layer is charged uniformly at the beginning of the exposure. For this purpose an electron beam is scanned over the rear side of the diode layer. Incident light results

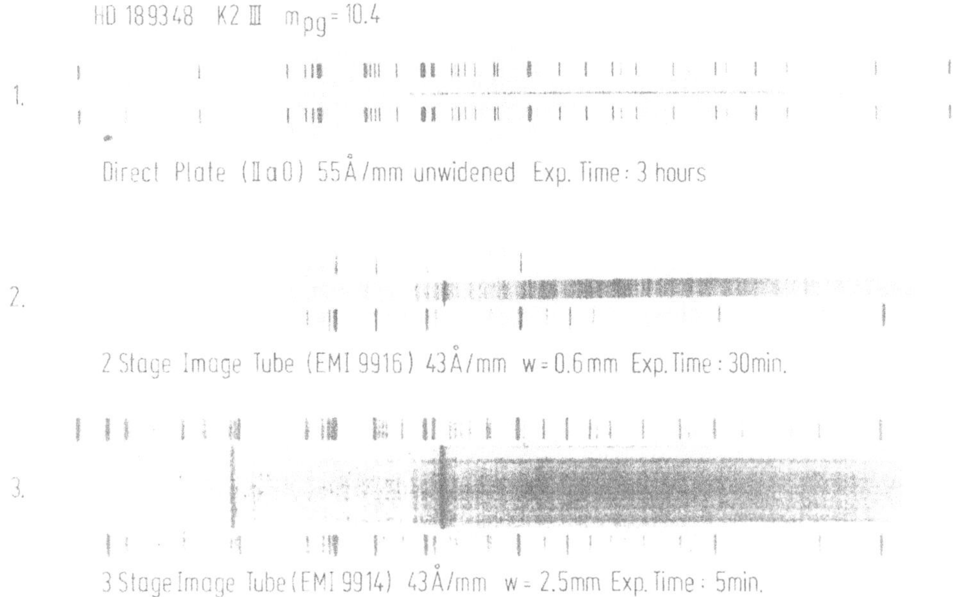

Fig. 2.26. Comparison of photographic spectrograms of the same object (1) obtained without an image intensifier, (2) with a medium gain intensifier, and (3) with a high gain intensifier recording individual photoelectrons

in photoconduction and thus in a charge reduction in the exposed areas. When the target is recharged again by the electron beam, a signal proportional to the absorbed light is produced by the recharge current. Intensified vidicons use as primary detector a photocathode. The photoelectrons are accelerated to about 10 keV and focused onto a vidicon target. Because of the energy gained in the high voltage field, each photoelectron produces a large number of secondary conduction electrons in the target. This again results in a discharge the target proportinal to the incident light intensity. The recharging of the target and the signal readout again occurs using an electron beam, as in the ordinary vidicon.

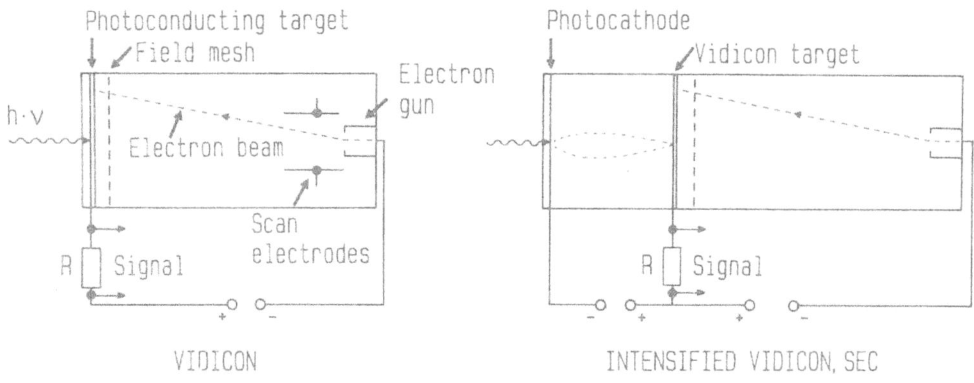

Fig. 2.27. Operating principle of vidicon and intensified vidicon tubes

Because of their high readout noise ordinary vidicon tubes (which are commonly used in TV cameras) are not suited for the low light levels encountered in astronomical measurements. On the other hand, in intensified vidicons the inherent high readout noise of the vidicon principle is overcome by the internal amplification of the photoelectrons. With KCl targets (which can be operated at room temperature) or silicon targets (at about - 100° C) exposure times of many hours between consecutive readouts are possible. Depending on the target material and the details of the target structure and readout procedures, the intensified vidicons are termed "Secondary Electron Conducting" (SEC), "Silicon Intensified Target" (SIT), or "Electron Bombarded Silicon" (EBS) tubes.

2.10 Photon counting Detectors

As illustrated in Figure 2.26, high gain image tubes allow detection of the individual photoelectrons even on photographic plates. Modern photon counting detectors combine the image tube principle with fast electronic readout schemes. During the past decade a great variety of arrangements based on this scheme have been developed for optical astronomy. All these systems use photocathodes as primary detectors and one of the focusing mechanisms discussed above for electron imaging. In the optically coupled devices the output screen of an image intensifier is imaged onto a secondary light detector by means of a transfer lens or a fiber plate. In the vacuum devices the electron image is recorded directly using an imaging electron detector or a position sensitive anode.

An important step towards the development of real imaging photon counting detectors were the image disector scanners (IDS). In an IDS the screen image of an image intensifier is read out by means of a position scanning PMT (called "image disector"). The decay time of the intensifier phosphor is used to store the image information over the scanning cycle. More efficient is the use of a (fast response) TV camera (based on a vidicon or a fast-scan CCD) behind an image intensifier. This combination is called Imaging Photon Counting System (IPCS). The image intensifier screen can also be imaged on a linear diode array) or a CCD. The latter combination is known as intensified CCD or ICCD. In this photon counting mode the CCD has to be read out on a timescale which is shorter than the inverse photon rate (per pixel). As there are upper limits to the readout frequency of a CCD, ICCDs are not suited for applications with high photon rates. A particular simple and fast intensifier readout scheme is used in the PAPA (= Precision Analog Photon Adress) detectors. In the PAPA device the output screen of a high gain image intensifier is imaged simultaneously on several coded masks in front of ordinary PMTs. The digital masks correspond to spatial frequencies increasing in increments of a factor of two. The position of a photon event on the screen can be determined uniquely from monitoring the PMTs and from determining through which of the masks the photon event has been seen. From the example of Figure 2.28 it is clear that with n such masks (and the associated PMTs) a total of 2^n pixels can be resolved in one dimension. Actual PAPA detectors use two (perpendicularly oriented) pairs of 9 masks and PMTs, resulting in a resolution of $2^9 \times 2^9 = 512 \times 512$ pixels. The spatial resolution can be increased by combining digital and analog masks.

Among the vacuum-coupled photon counting devices are tubes having a large number of individual anode sections which are connected to signal lines, again using binary coding schemes. An example of this detector type are the MAMA (=Multi-anode Microchannel Array) devices which are commercially available with formats of 10^6 or more pixels. Alternatively the anode of an image intensifier can be replaced by a diode array (such as a Reticon). In this case each accelerated electron produces a current pulse well in excess of the readout noise. Photon counting devices designed according to this principle (called Digicons) are the principal spectroscopic detectors of the Hubble Space Telescope. Another successful design

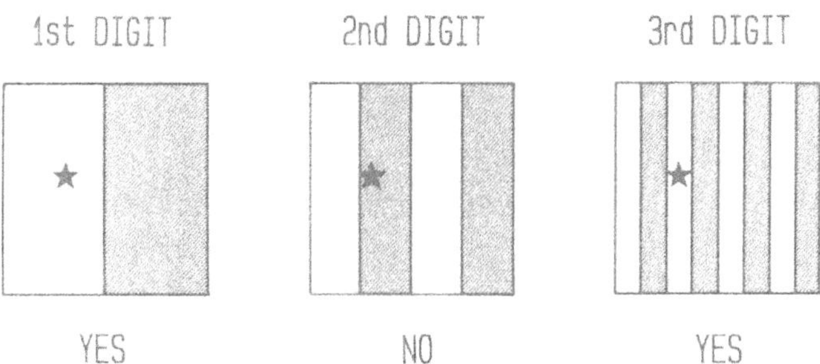

Fig. 2.28. Examples for binary coded masks used in PAPA detectors

principle is the combination of a microchannel plate and a uniform resistive anode. Using (e.g.) a square resistive anode with conections at each of the four corners, for each recorded photoelectron the impact position of the amplified electron cascade on the anode can be detemined from the voltage ratios at the four signal lines. Related are the Position Sensitive Anode (PSA) devices, which use an anode surface with several interlocking areas (cf. Figure 2.29) behind a MCP. Each primary photoelectron produces an unsharp electron beam reaching the anode. A measurement of the relative contributions of the charge avalanche to the different interlocking anode sections allows to determine the center of the unsharp impact region and thus of the position of the initial photoemission event on the cathode. The resolution of a PSA device depends on the accuracy by which the ratios of the output charges recorded by the different anode sections can be determined electronically.

2.3 High Energy Photon Detectors

As noted already, the gas ionization devices described in Section 2.1 find their main application in the detection of X-ray photons. However, energetic photons can also be recorded by means of the the solid state detectors discussed in the preceeding section. Historically, photographic plates were the first detectors used for X-ray observations of cosmic sources. Photocathode devices (such as MCPs) are still used in X-ray astronomy, notably for high angular resolution imaging. More recently CCDs optimized for X-ray detection have been developed. As $h\nu$ of X-ray photons is many times higher than the band gap energy of a semiconductor (or insulator), the absorption of an X-ray photon in a solid results in high energies of the photoelectrons. Most of this excess energy is lost by exciting additional electrons into the conduction band. Hence, each absorbed high-energy photon results in a large number of conduction electrons, with a total charge about proportional to the initial photon energy. Thus, at high photon energies, photoconductors (like proportional counters) allow a reliable detection of practically every incident photon (RQE ≈ 1), and provide information on the photon energies, i.e. the spectrum of the received radiation. The energy resolution increases with the photon energy and may exceed 10^2 for soft Gamma rays. An energy resolution of up to about 10^3 for X-rays can in principle be achieved by photon absorption in superconducting tunnel diodes (see Section 3.4 below) which may become available for future X-ray satellites.

At energies $\gtrsim$ 20 keV Compton scattering (rather than photoeffect) becomes the dominant interaction between photons and electrons bound to light atoms. The energetic Compton electrons can again be detected by their production of conduction electrons in solids, using large-volume PIN diodes.

However, most photon detectors based on Compton scattering make use of the scintillation effect in crystalline solids. The most commonly used scintillators are NaI and CsI

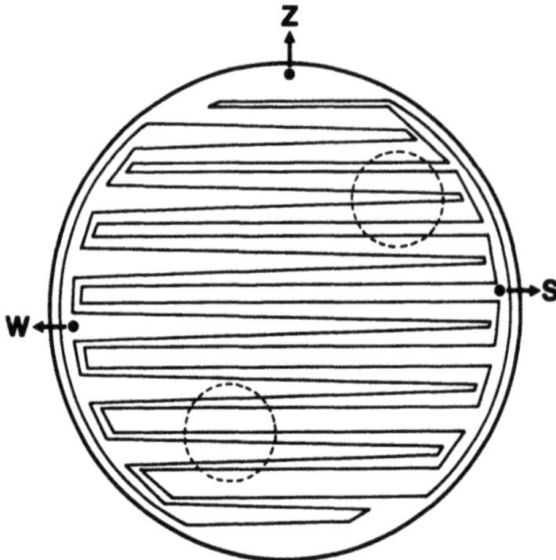

Fig. 2.29. Example of a position sensitive (wedge and strip) anode. Unsharp amplified electron images of individual photoelectrons (indicated by broken-line circles) result in relative contributions to the three anode areas (W,S,Z) which are unique functions of the impact position

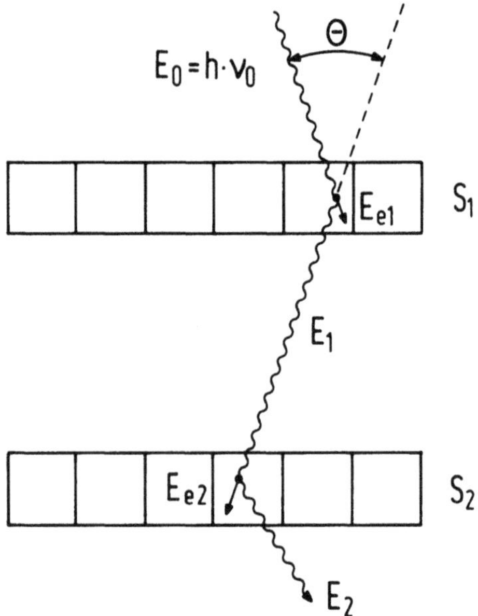

Fig. 2.30. Schematic arrangement of a Compton telescope

crystals doped with Tl or Na impurities. The valence-conduction band gaps of these materials are very large. Hence, a thermal excitation of conduction electrons is totally negligible. However, energetic Compton (or photo) electrons can easily lift valence electrons into the conduction band. The impurities mentioned above (Tl in NaI and CsI or Na in CsI) result in impurity bands which are only a few eV below the conduction bands. Hence, the de-excitation of conduction band electrons into the impurity band results in the emission of visual photons, for which the scintillator materials are transparent and which are detectable using optical detectors such as PMTs. The light flashes produced in this way by energetic photons again are approximately proportional to the initial photon energies ($E/\Delta E \lesssim 20$). Because of the absence of thermal excitation effects scintillation devices are well suited for the detection of low photon rates. On the other hand, scintillators are also sensitive to energetic charged particles originating from the cosmic radiation or from radioactive decays. To eliminate these spurious signals, scintillation photon detectors are "actively shielded" by enclosing the whole detector in thin organic plastic scintillators which are highly sensitive to energetic particles but interact only weakly with photons. The photon detectors are then operated in anticoincidence with the shield scintillators.

Compton detectors allow to derive directional information without optical components. The principle of such "Compton telescopes" is outlined in Figure 2.30. Basically a Compton telescope consists of two arrays (S_1 and S_2) of scintillation detetectors separated by (typically) a few meters. If a photon of initial energy E_0 is scattered in a detector element of S_1, it will change its direction by an angle θ. Furthermore, after the scattering it will have a lower energy E_1.The difference $E_{e1} = E_0 - E_1$ is deposited in the scintillator element. From the laws of energy and momentum conservation we obtain for these quantities the well known relation

$$\cos\theta = 1 + m_e c^2 (E_0^{-1} - E_1^{-1}) \tag{2.2}$$

If the scattered photon is completely absorbed in the second detector element S_2 ,i.e. $E_2 = 0$ and $E_0 = E_{e1} + E_{e2}$, we have

$$\cos\theta = 1 + m_e c^2 [(E_{e1} + E_{e2})^{-1} - E_{e2}^{-1}] \tag{2.3}$$

Hence, as E_{e1} and E_{e2} are measurable quantities, the angle θ can be determined.

For $h\nu > m_e c^2 = 1.02$ MeV (i.e. for Gamma-rays) electron-positron pair production becomes the most important photon interaction process. Again the resulting electrons can be recorded in semiconductor diodes or scintillation detectors. However, with increasing energies the mean free path of the photons become larger, requiring very large dimensions of such detectors. Hence for $h\nu \gtrsim 10^7$ eV spark chambers are more convenient devices. Like the corresponding devices used in high energy physics experiments, astronomical spark chambers consist of stacks of metal sheets which are immersed in a gas (usually Ne or Xe). The interaction of the incident Gamma-rays with the heavy atomic nuclei of the metal sheets results in energetic $e^- e^+$ pairs which ionize gas atoms along their path through the chamber. These ionization trails are made visible by applying a high voltage pulse to the metal electrodes which results in sparks along the trails. The length and orientation of the trails give the photon energy and the direction of the incident radiation. Furthermore, the plane defined by a trail pair provides information on the polarization of the radiation.

For $h\nu \gtrsim 10^{10}$ eV reasonable photon rates can be achieved only with very large collecting areas (cf. Section 1), and energy measurements require very large detector volumes. A solution to these problems is the use of the earth atmosphere as the detetor medium. A GeV Gamma-ray photon interacting with an atmospheric atomic nucleus produces a very energetic $e^- e^+$ pair. Interactions of these particles with the atmospheric atomic nuclei result (by means of Bremsstrahlung) in new energetic Gamma photons and in additional pairs. Hence, each primary photon can produce a large cascade of secondary photons and light charged particles (Figure 2.31). At photon energies $\gtrsim 10^{14}$ eV these "extensive air showers" (EAS) penetrate deeply into the atmosphere and are detectable with groundbased particle detectors (at least at high altitude sites). Air showers produced by cosmic ray particles (rather than by Gamma photons) can be discriminated by the muon content of the secondary particle spectrum. The source direction can be determined by measuring the arrival time of the shower as a function of

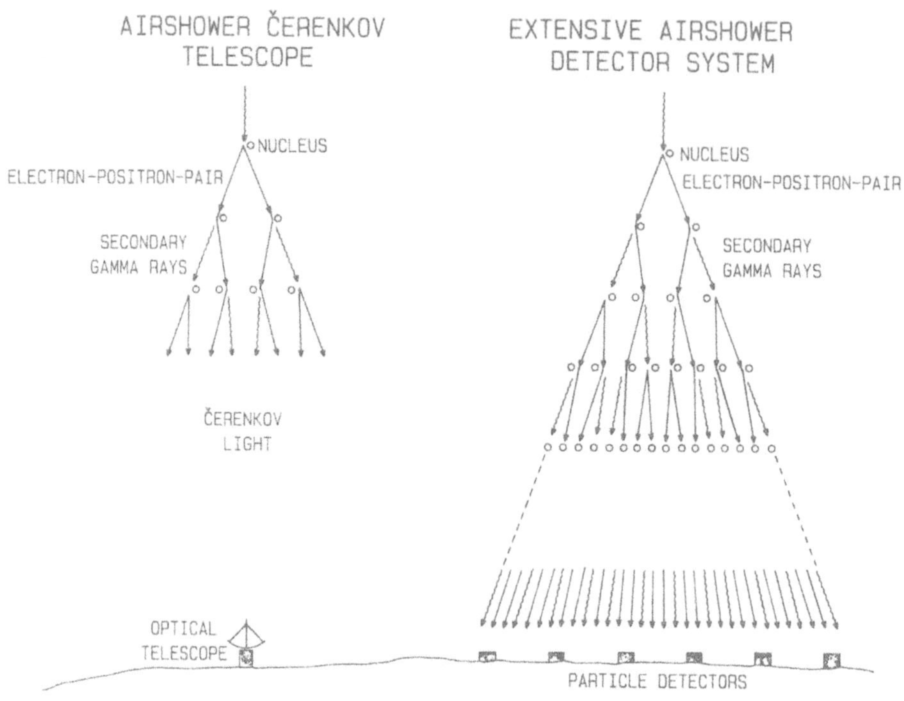

Fig. 2.31. Principle of Gamma-ray airshower detectors

the detector element position.The EAS arrays for Gamma-ray astronomy which are presently in use or under development have angular resolutions of about 2° and energy resolutions $\lesssim 2$.

For $h\nu \lesssim 10^{13}$ eV the air showers are absorbed already relatively high (> 10 km) in the atmosphere. However, it is still possible to measure such air showers by observing the Čerenkov radiation produced by the electrons and positrons moving faster than c/n (c being the vacuum velocity of light and n the atmospheric index of refraction). The "airshower Čerenkov telescopes" (ACT), operated according to this principle, consist of large low-quality optical light collectors (or arrays of such collectors) equipped with PMT tubes. In contrast to the EAS arrays, these light collectors have to be pointed towards the source position. Each VHE Gamma-photon results in a brief flash from the direction of the Gamma-ray source. By measuring the total amount of Čerenkov radiation and its angular distribution, the energy of the observed photon can be determined. However, normally only part of the radiation cone is recorded. Hence, ACTs usually provide only limited spectral information on the observed Gamma-rays.

3 Coherent Detectors

3.1 Design Principles

As noted already, coherent detection is the most straightforward way of recording electromagnetic radiation. However, with present technologies for $h\nu \gtrsim 10^{12}$ Hz coherent receivers either cannot be realized or are less efficient than photon detectors. Hence, in astronomy the use of coherent detectors so far has been restricted to radio wavelengths and (for special applications) to the IR ($\lambda \gtrsim 10\mu m$).

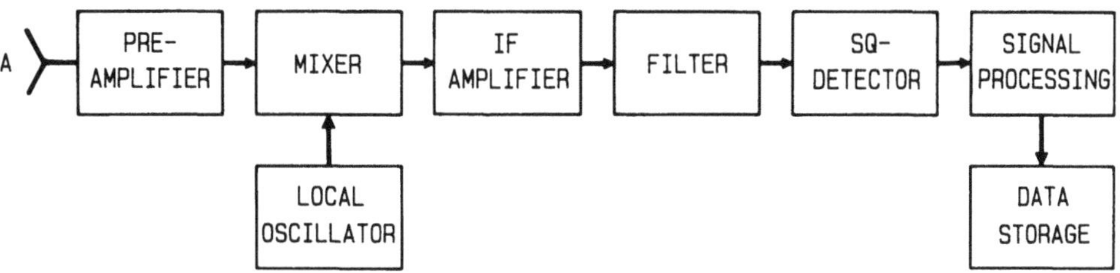

Fig. 3.1. Principle components of an astronomical receiver

Figure 3.1 shows the principle arrangement of a coherent detector system. As a rule astronomical receivers are designed according to the superheterodyne principle, where the received signal is (sometimes after preamplification) mixed with the signal of a local oscillator (LO) to produce an intermediate frequency (IF) signal. This IF signal is amplified, filtered, converted to a DC or low frequency current and then digitized and recorded by a data acquisition computer. Usually rectifiers generating a square of the IF signal ("square-law detectors") are used, as their output is directly proportional to the oscillation power (the square of the oscillation amplitude).

For astronomical applications the superheterodyne principle has two important advantages. Firstly, the very high amplification factors ($\approx 10^8$) required for the weak astronomical signals are difficult to achieve at a single frequency without feedback interferences. Secondly, the choice of a suitable LO allows the use of intermediate frequencies for which relatively inexpensive standard RF amplifiers and other commercial high frequency circuit components are available.

The mixer in a heterodyne receiver can be any non-linear electronic component, i.e. a device where the relation between the output voltage V_{OUT} and the input voltage V_{IN} can be approximated by

$$V_{OUT} = a_1 V_{IN} + a_2 V_{IN}^2 + a_3 V_{IN}^3 + ... \tag{3.1}$$

If $V_{IN} = V_1 + V_2$ is the sum of two harmonic contributions

$$V_1 = V_S \sin(\omega_S t + \phi_S)$$

and

$$V_2 = V_{LO} \sin(\omega_{LO} t + \phi_{LO})$$

the evaluation of Equ. (3.1) shows that V_{OUT} contains (among other components) terms proportional to

$$V_S V_{LO} \sin[(\omega_S + \omega_{LO})t + \phi]$$

and

$$V_S V_{LO} \sin[(\omega_S - \omega_{LO})t + \phi]$$

i.e. signal components oscillating with the sum frequency $\nu_S + \nu_{LO}$ and the difference frequency $\nu_S - \nu_{LO}$. One of these two frequencies is selected for (IF) amplification. However, each IF value corresponds to two different signal frequencies. If e.g. $\nu_S + \nu_{LO}$ is chosen for amplification, the signal frequency $\nu_M = \nu_S + 2\nu_{LO}$ is amplified, too, as its difference frequency results in the same IF frequency value. If the combined power of both ν_S and ν_M are recorded, a receiver is called a "double-sideband" device. In the case of spectral line observations the mirror frequency normally contains no (or unwanted) information and therefore is suppressed by prefiltering. In this case we have a "single-sideband receiver"

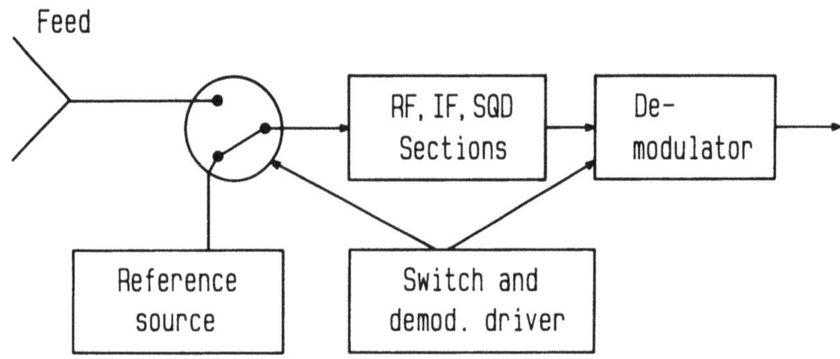

Fig. 3.2. Principle arrangement of a Dicke receiver

The high amplification factors of astronomical receivers often result in variations and drifts of the amplifier properties. In the presence of a strong background against which the signal has to be measured, amplifier variations directly limit the receiver sensitivity.Therefore, methods have been developed to eliminate the amplifier fluctuations from the output signal. As demonstrated first by the American astrophysicist R.H. Dicke, this can be accomplished by switching the receiver input (electronically) between the antenna and a reference source with a constant ouput, and measuring the difference signal by means of a synchronous demodulator (Figure 3.2). Suitable reference sources are low temperature blackbodies or temperarature-controlled Ohmic resistors which (due to thermal current fluctuations) emit an AC power which depends only on their temperature.

A disadvantage of the Dicke scheme is that half of the observing time is lost to the integrations on the reference source. This drawback is avoided in correlation receivers which observe the astronomical source and the reference source simultaneously. The two signals are mixed in two parallel channels using different phase shifts. An amplifier-drift free signal is then derived by correlating the output of the two receiver channels. Correlation techniques compensating instrumental variations are also used in receivers specialized for polarization measurements. For this purpose the signals of two antenna feeds which are sensitive to different Stokes parameters are correlated. Some other application of correlating receivers are described in the contribution of D.Downes to this volume.

3.2 Low Noise Amplifiers

As indicated in Figure 3.1, (except at very high frequencies) astronomical receiver systems usually are equipped with preamplifiers which are placed as close as possible to the antenna feed. The task of the preamplifiers is to increase the signal level to a value well above the noise contributions of the various following receiver components. In this case, the noise produced by the preamplifier determines the noise performance and sensitivity of the receiver. Hence, a low noise level is essential for a preamplifier. Noise powers (expressed as noise temperatures, cf. Section 1.2) achievable with different types of low-noise amplifiers at different frequencies are given in Figure 3.3. Also included in this figure are (very schematically) various background contributions (cosmic microwave background, radioemission of our galaxy, atmospheric background).

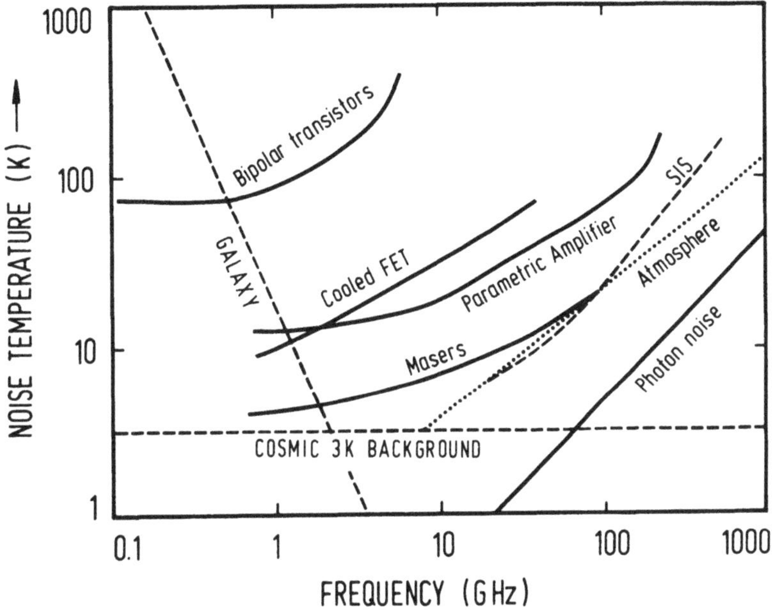

Fig. 3.3. Comparison of the minimum noise powers of different amplifier types

As shown by Figure 3.3, for $\nu \lesssim 1$ GHz the sensitivity of radio observations is normally limited by the galactic background. Hence, at such low frequencies ordinary (bipolar) transistor amplifiers are adequate. More demanding is the frequency range 1 - 10 GHz where the sensitivity is limited mainly by the cosmic microwave background. At present three different types of preamplifiers, namely cooled (to $\approx$ 20 K) field effect transistors (FET), parametric amplifiers, and masers, are used in this wavelength range. Their operating principles are described below. Parametric amplifiers and masers are also quite efficient in the adjacent 10 to 100 GHz region. Above 100 GHz (with present technologies) a better S/N can be achieved by feeding the antenna signal directly into a low-noise mixer.

3.2.1 FET Transistors

Field effect transistors are among the most common components of modern electronic circuits. In astronomical receivers usually the "metal-semiconductor" (MESFET) variety of the FETs is chosen. Its principle is outlined by Figure 3.4. On a thin ($\approx 1\mu$) conduction layer consisting of a doped semiconductor, three metal electrodes (source, gate, and drain) are deposited. By differential doping the source and drain metal-semiconductor interfaces are designed to have Ohmic properties. The gate-semiconductor interface forms a Schottky junction (cf. Figure 2.10). The extent of the associated depletion zone depends on the gate voltage. An increase of the depletion zone decreases the effective cross section of the conduction layer. Hence, by modifying the gate voltage it is possible to control the source-drain current. Consequently, a signal voltage at the gate results in an amplified voltage at the load resistor in the source-drain circuit.

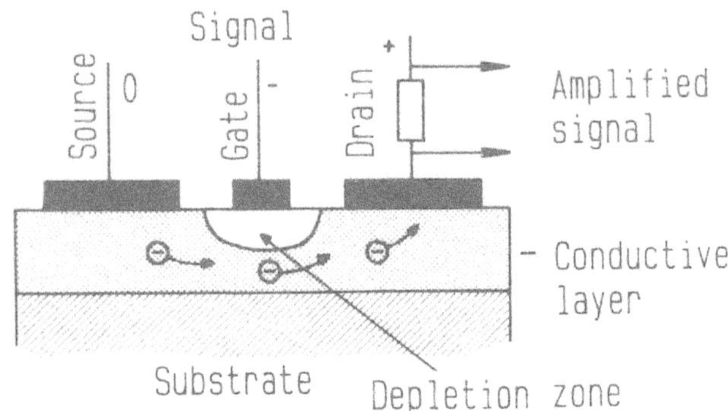

Fig. 3.4. Principle structure of a MESFET

For the amplification of signals in the GHz range an FET must have a low capacitance (to avoid shortening of the high frequency) and a fast response time. Therefore, for such high frequency applications semiconductors with a high carrier mobility (such as GaAs, cf. Figure 3.9) and small gate dimensions (down to 0.2μ) are used. Maximum amplification values are ≈ 10, maximum frequencies at present about 40 GHz. Special attention requires the impedance matching of the tiny FETs to standard waveguides. Compared to other amplifier types, FET transistors have the advantage of a broad bandwidth, good stability and reliability, and little operational constraints.

3.2.2 Parametric Amplifiers

Parametric amplifiers use electronic circuit components with voltage dependent parameters. The most common devices of this type are voltage dependent capacitors (called "varactors"). The principle of signal amplification by means of a variable capacitor is readily understood from the following experiment. In an oscillator circuit consisting of a capacitor and an induction coil connected in parallel (Figure 3.5), the oscillator energy is alternating periodically between the capacitor and the induction coil, resulting in a sinosoidal change of the voltage V at the capacitor.

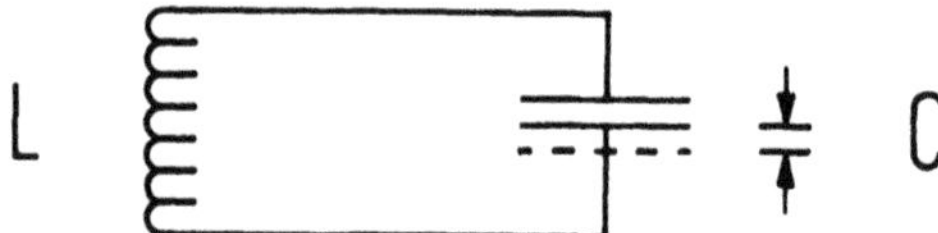

Fig. 3.5. Amplification by means of a variable capacitor

If the capacitance C is decreased at the phase when $|V|$ has a maximum, the voltage (and thus the electromagnetic energy) is increased, as the charge (which is given by $Q = CV$) remains constant. (If the capacitance decrease is accomplished by increasing the gap of a plate capacitor, the energy increase occurs at the expense of the mechanical work required for widening the gap). If the capacitor is returned to its original capacitance at the phase when the capacitor voltage passes through zero, the oscillation energy is not affected. Hence, with a properly timed change of the capacitance energy is fed into the electromagnetic oscillation. On the other hand, with a different timing, energy can also be extracted from an oscillation. Hence, in more general terms, a varactor can be used to exchange energy between oscillating signals.

Suitable variable capacitors are p-n or Schottky diodes (cf. Section 2.2.2). The capacitance of such diodes is determined by the width of the junction depletion zone. As noted in Section 2.2.2 the depletion zone width varies (nonlinearly) if a voltage is applied to the diode. The schematic circuit diagram of a varactor amplifier is given in Figure 3.6. The input signal and the signal from a local ("pump") oscillator are mixed by means of a varactor. Because of the nonlinearity of the varactor the mixing results in sum, difference, and overtone frequencies of the signal and the pump frequencies ν_s and ν_p. A detailed discussion shows that the values of the power P_{mn} flowing into the varactor at the frequencies $\pm(m\nu_s + n\nu_p)$ are related to one another by the Manley-Rowe relations

$$\sum_{m=-\infty}^{\infty} \sum_{n=0}^{\infty} \frac{mP_{mn}}{m\nu_s + n\nu_p} = 0 \tag{3.2}$$

and

$$\sum_{m=0}^{\infty} \sum_{n=-\infty}^{\infty} \frac{nP_{mn}}{m\nu_s + n\nu_p} = 0 \tag{3.3}$$

Using resonance techniques (e.g. by placing the varactor diode in a suitable cavity) it is possible to insure that only a few of the many possible frequencies are actually present at the diode, while all unwanted frequencies are suppressed.

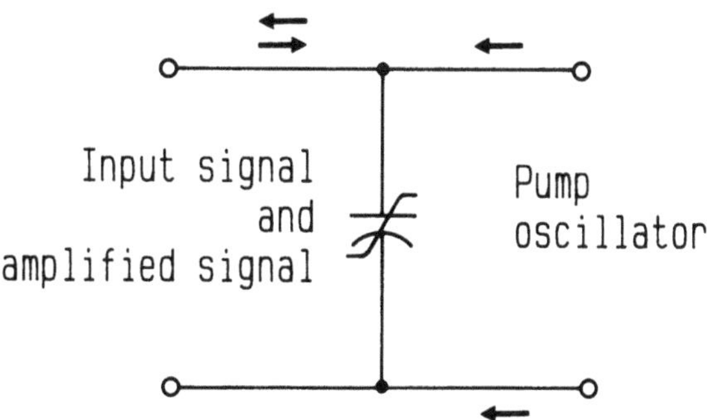

Fig. 3.6. Schematic circuit diagram of a varactor amplifier

In order to achieve amplification, in addition to ν_s and ν_p at least one further frequency must be present. In most varactor amplifiers the difference frequency $\nu_D = \nu_p - \nu_s$ is chosen as the third frequency. In this case the Manley-Rowe relations simplify to

$$\frac{P_s}{\nu_s} - \frac{P_D}{\nu_D} = \frac{P_p}{\nu_p} + \frac{P_D}{\nu_D} = 0 \tag{3.4}$$

As shown by this equation, P_p and P_D must have different signs, while P_s and P_D have the same sign. Hence, if power is transferred to the varactor at the pump frequency, this power can be withdrawn at the signal and at the difference frequency. If no power is withdrawn at the difference frequency, the signal itself is amplified while the difference frequency (in this case called "idler frequency") only acts as a temporary energy storage.

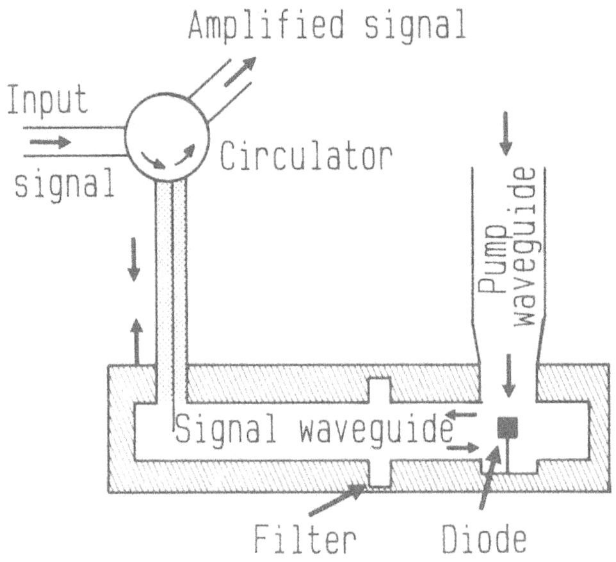

Fig. 3.7. Schematic layout of a parametric amplifier

The detailed theory of parametric amplifiers is rather complex, as many different operational modes and variants are possible. A possible technical realization of a PA is outlined (very schematically) in Figure 3.7. The signal and pump frequencies are combined at the varactor diode inside a resonance cavity. The same waveguide is used for the input and for the amplified signal. To separate these two signals the input waveguide is connected to a circulator (which sorts the two signals according to their propagation direction). With suitable pump oscillators parametric amplifiers can reach amplification factors up to about 10^2 in a single stage. Maximum bandwidths reach about 10 percent. The minimum noise temperatures of cooled parametric amplifiers are of the order 15 K.

3.2.3 MASERs

In a system of two atomic energy levels E_2 and E_3 with $E_3 > E_2$ and with radiative transitions between these levels (Figure 3.8) the radiation power absorbed at the frequency $\nu_{32} = \nu_{23} = (E_3 - E_2)/h$ is given by

$$P_{23} = h\nu_{23}(n_2 - n_3)W_{23} \tag{3.5}$$

where W_{23} is the corresponding transition probability and the n_i are the level populations. Obviously, for $n_3 < n_2$ radiation passing through the medium is absorbed, while for $n_3 > n_2$ the radiation is amplified. In thermal equilibrium the n_i are given by the Boltzmann distribution

$$n_i \sim \exp -(E_i/kT)$$

resulting in $n_3 < n_2$. Hence, to achieve amplification the thermal equlibrium must be replaced by a population inversion. In a modern MASER (= Microwave Amplification by Stimulated

Emission of Radiation) amplifier the population inversion is generated by exposing the system to a strong ("pump") radiation field at a third frequency ν_{13} involving a third level, as outlined in Figure 3.8. The absorption of the pump radiation results in an increase of n_3 and a decrease of n_1. If the temperature of the maser material is kept at a value of the order $\Delta E/k$, a rapid thermal readjustment of the level population can be prevented. Therefore, depending on the relative strength of the spontaneous transitions for ν_{32} and ν_{21} either n_3/n_2 or n_2/n_1 will become greater than unity, resulting in maser amplification of incident radiation at the corresponding frequency.

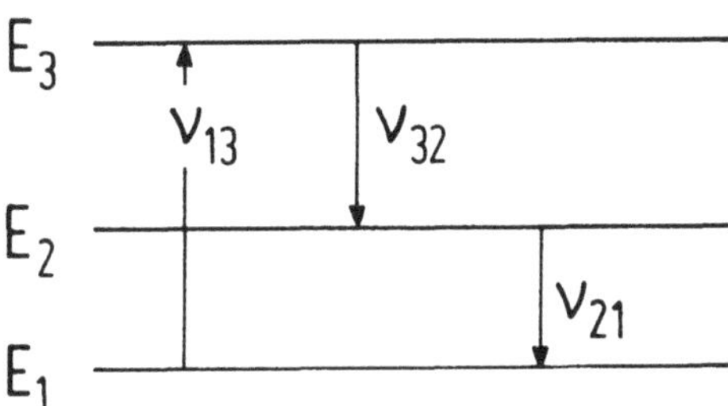

Fig. 3.8. Schematic energy diagram of a 3-level maser

The first masers used two-level transitions in gas molecules or atoms. Hydrogen masers are still used as high accuracy frequency standards in radio astronomy. In amplifiers normally solid state masers (with three or more levels) are chosen, as the higher atomic densities of solids allow higher amplification factors. Most of these devices make use of atomic transitions in the Fe or Cr impurity atoms of doped Al_2O_3 (ruby) or TiO_2 (rutil). Interactions with the lattice atoms result in a splitting of (twice) degenerate sharp energy levels of the impurity ions. The energy differences between these levels are in the range 10 - 100 GHz. A further splitting and thus additional energy levels can be produced by applying a strong (up to 1 Tesla) magnetic field. By adjusting the field strength it is possible to modify the level splitting and hence to tune the maser to a desired frequency band. Using a nonhomogeneous field the intrinsically very small band width of a maser amplifier can be extended to about $\Delta\nu/\nu = 0.01$.

The geometrical arrangement of a maser amplifier usually is similar to that sketched in Figure 3.7 for a parametric amplifier, except that the maser material replaces the diode. Alternatively, a maser can also be placed directly into the waveguide carrying the signal from the antenna. In this ("travelling wave") case the forward direction of the signal has to be insured by means of ferromagnetic isolators.

As spontaneous transitions are the only intrinsic noise source of the maser principle, maser amplifiers have low noise temperatures. A lower limit is given by the relation

$$T_N = |\frac{\Delta n_0}{\Delta n}| T \tag{3.6}$$

where T is the temperature of the material, Δn is the actual pupulation difference , and Δn_0 is the thermal equilibrium population difference of the two energy levels involved in the maser transition. With T = 4 K (boiling He) and $|\Delta n_0/\Delta n| \approx 0.4$ values as low as $T_N =$ 2 K can be reached. Typical noise temperatures of operational maser receivers are near 6 K. Amplification factors may be as high as 10^3. In spite of these impressive performance parameters, maser amplifiers have gained only limited popularity in radio astronomy, as the required low temperatures and high magnetic fields make maser amplifiers more complex and operationally more demanding than other types of amplifiers. Moreover, the relatively small band widths often are a disadvantage.

3.3 Local Oscillators

At present, basically three different types of local oscillators are used for astronomical radio receivers. At low frequencies ($\nu < 1$ GHz) conventional LC or RC oscillators are adequate. At higher frequencies (up to about 150 GHz) Gunn diodes provide a simple and compact solution to the oscillator problem. The active zones of these diodes are thin ($\approx 1\mu$m) low-conductivity GaAs (or GaP) layers. In vacuum and in most conducting materials (including the metals and the semiconductor silicon) the electron drift velocities increase with the electric field strength. However, as shown in Figure 3.9, in GaAs the electron velocities decrease again if the field strength is increased beyond a certain critical value. This unexpected behaviour is a consequence of the interaction between the moving electrons and the lattice. If a Gunn diode is operated with the field strength (across the low-conductivity layer) exceeding that corresponding to the maximum of the mobility function in Figure 3.9, the current flow will be unstable, as any space charge will become amplified. Hence, the current will start to oscillate. The oscillation frequency is given by the ratio between the (voltage dependent) electron drift velocity and the thickness of the low-conductivity layer. As the drift velocity is tunable only within about a factor 2 (cf. Figure 3.9), the dimensions of the active zone determine the frequency range at which the diodes can be used. Higher frequencies obviously require thinner active layers.

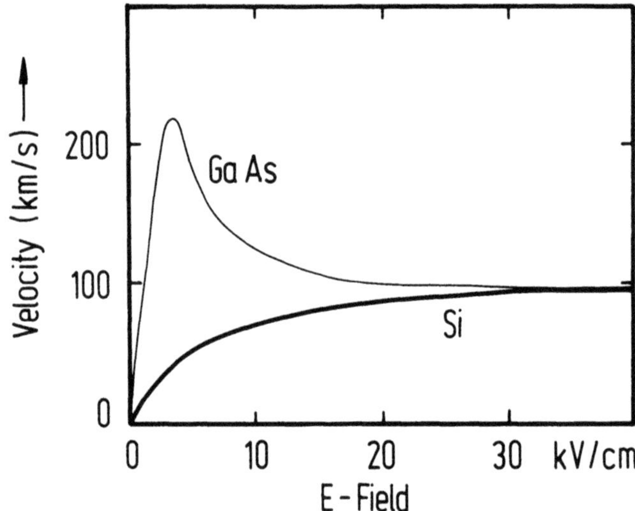

Fig. 3.9. Electron drift velocities as a function of the electric field in silicon and in GaAs

At frequencies where Gunn diodes would require unpracticable dimensions (or if a higher output power than achieveable with a Gunn diode is needed), electron-beam oscillators are used. These devices are based on the periodic modulation of an electron beam by means of a delayed feedback. Best known example of this type of oscillators is the return-beam klystron. Its operating principle is outlined in Figure 3.10. In these devices an electron beam is modulated when passing a resonance cavity. By reflecting the modulated beam back towards the cavity using a mirror electrode, a feedback is produced, which energizes the oscillation. The oscillation frequency follows from the electron travel times. A waveguide connected to the cavity is used to extract the oscillator power.

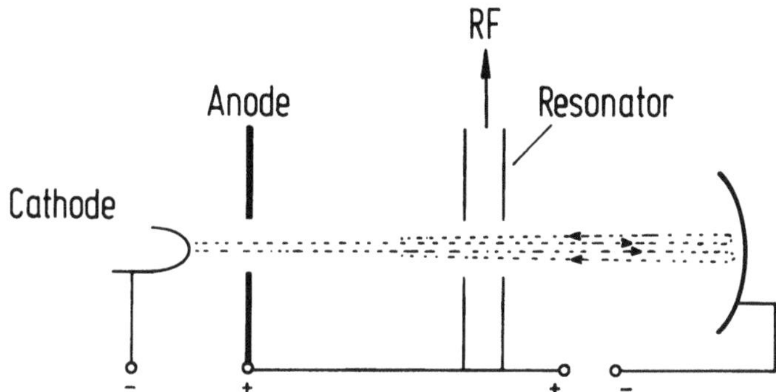

Fig. 3.10. Schematic arrangement of a return-beam klystron oscillator

Klystrons operating according to this scheme are available for frequencies up to about 250 GHz. Linear electron beam oscillators, where the delayed feedback originates from the velocity difference of the electron beam and the propagation speed of a guided wave, reach up to 1 THz. At even higher frequencies lasers are used, or the output signal of a conventional oscillator is converted to a higher frequency using a "frequency multiplier" (based on the generation of overtone frequencies in nonlinear mixers).

3.4 Mixers

In principle any nonlinear circuit component can be used as a mixer in a superheterodyne receiver. Examples of such nonlinear components are the PN semiconductor diodes discussed obove. Even better suited are Schottky diodes as their strong deviation from a linear current-voltage relation results in a higher amplitude of the IF signal. This is particularly important at high frequencies where no good preamplifiers are available and where consequently the system noise level is directly determined by the noise performance of the mixer.

In order to respond to high frequency radio signals a Schottky diode must have a low intrinsic capacitance, its dimensions must be small, and its conduction electrons must have a high mobility. Therefore, high frequency Schottky diodes consist of thin metallic wire tips ("whiskers") in contact with tiny GaAs semiconductor crystals. Figure 3.11 shows an example of a Schottky mixer for 460 GHz developed at the MPIfR Bonn. The diode is situated in the center of a metal block (which has been opened to take the photograph). The circular holes are screw holes for closing the block. The signals from the telescope and from a local oscillator (which are combined linearly by means of a diplexer) enter the block along the tapered wave guide ("feed") from the right. The diode is placed at the end of the waveguide and the whisker is pressed against the small (100μm) GaAs cube from below. The IF signal leaves the mixer along a standard 50 Ohm line at the top. Figure 3.12 shows an electron micrograph of the whisker, which consists of a chemically sharpened bronze wire.

At present, Schottky diode mixers are operated at frequencies up to 2.5 THz. However, the delicate point contacts require considerable skill for manufacturing and operating high

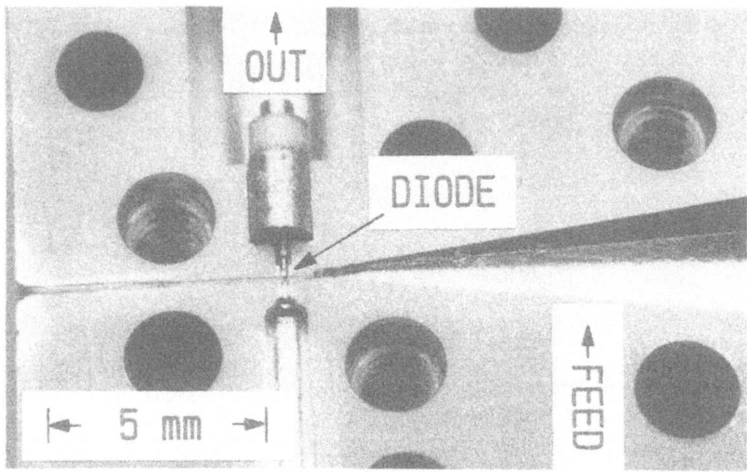

Fig. 3.11. Example of a high frequency (460 GHz) Schottky mixer (Courtesy A. Schulz, MPIfR Bonn)

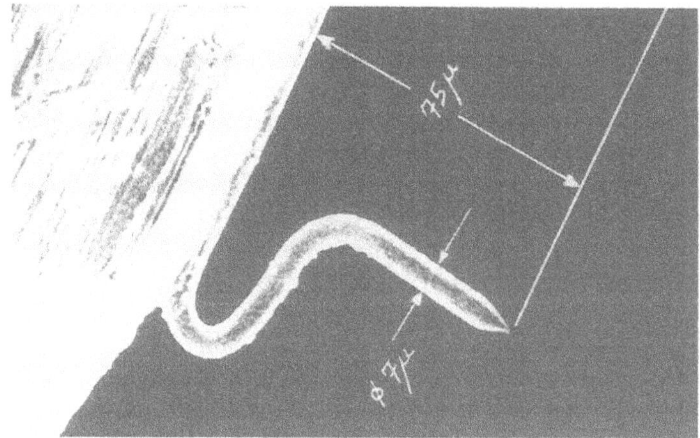

Fig. 3.12. Example of a metal whisker used for point contacts in high frequency Schottky mixers (Courtesy A. Schulz, MPIR Bonn)

frequency Schottky junctions. In the frequency range 100 - 700 GHz better stability and lower noise temperatures can be reached with SIS (= Superconductor-Insulator-Superconductor) and SIN (= Superconductor-Insulator-Normal) diode mixers. SIS diodes consist of two layers of superconducting material separated by a thin (10 - 20 Å) insulating layer. To understand its operating principle, let us first recall a few basic properties of superconductors. In Section 2.2 it was assumed that in a metal cooled to T = 0 K all electron energy states below the Fermi energy are filled and all energy states above the Fermi energy are empty. In superconducting metals this is not exactly true. In such materials, below the critical temperature a small fraction ($\lesssim 10^{-6}$) of the valence electrons combine to form Cooper pairs. In the absence of an electric field two electrons belonging to a Cooper pair have inverse momentum vectors and antiparallel spin. Hence, Cooper pairs behave like bosons. At T = 0 K, the Cooper pairs condense into a single energy state below the Fermi energy and above the energy states of the single electrons. Energy levels immediately above the Cooper pair groundstate are forbidden and form an energy gap symmetrical to the Fermi energy level (Figure 3.13). The resulting energy diagram resembles that of a semiconductor. But the energy difference of the gap (corresponding to the dissociation energy of the Cooper pairs, $\lesssim 3$ meV for conventional superconductors at T = 0) is much lower and comparable to the photon energies of (very) high frequency radio waves.

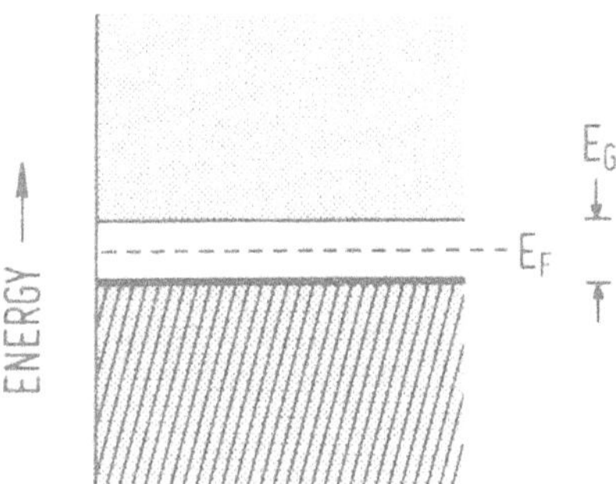

Fig. 3.13. The formation of a band gap in a superconducting metal

The presence of a band gap in the energy level diagram of superconductors results in a highly nonlinear voltage-current characteristic of the SIS junctions. Within the framework of classical physics the current should be exactly zero, as electrons cannot penetrate the insulating layer. However, since the thicknes of the insulating layer is not large compared to the extent of the electron wave functions, quantum mechanics allows the electrons to "tunnel" through the insulator from one superconductor to the other one.

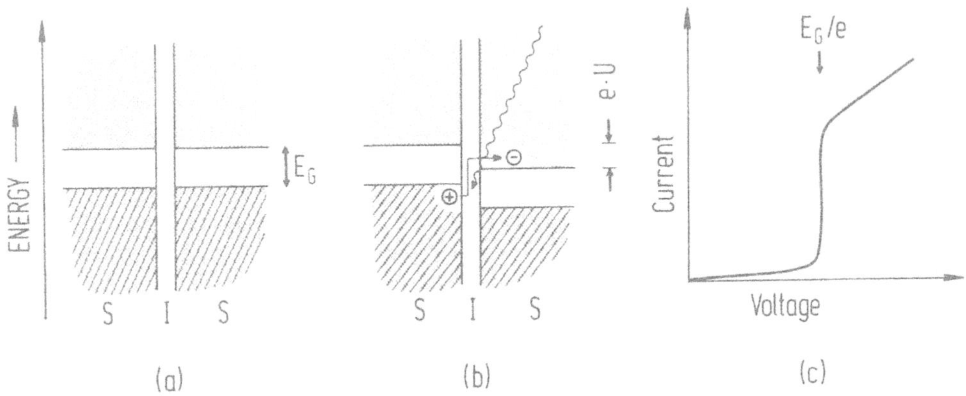

Fig. 3.14. Photoeffect and voltage-current characteristic of SIS junctions

If a voltage V is applied to the SIS junction, at small values of V the tunnel current will be very low, as the energy states below the gap are occupied while the conduction levels above the gap are almost empty at superconducting temperatures. However, an outside voltage also results in a difference of the Fermi level in the two superconductors (Figure 3.14). At the voltage where the bottom of the gap in the first superconductor reaches the energy level corresponding to the top of the gap in the second superconductor, electrons from the filled energy bands of the first superconductor start to tunnel into the empty allowed states of the second superconductor. Hence, at this voltage we experience a dramatic current increase. This effect results in the extremely nonlinear voltage-current characteristic of the SIS diodes (Figure 3.14c) and their suitability as mixer components. Furthermore, the absorption of microwave photons and the tunnelling of the photoelectrons, directly results in a photocurrent if the diode is exposed to radiation in the corresponding frequency range (Figure 3.14b). Hence, SIS junctions not only show a highly nonlinear voltage-current relation, but this function is directly influenced by incident microwave radiation.

In an SIN junction one of the two superconducting layers of a SIS diode is replaced by a normal metallic conductor. Hence, a band gap is present only on one side of the insulator. Otherwise a SIN junction operates according to the same principle. As the concentration of thermally excitated conduction electrons is very small, superconducting tunnel diodes have excellent noise properties. In laboratory experiments with SIS junctions noise temperatures close to the photon-statistics limit have been reached.

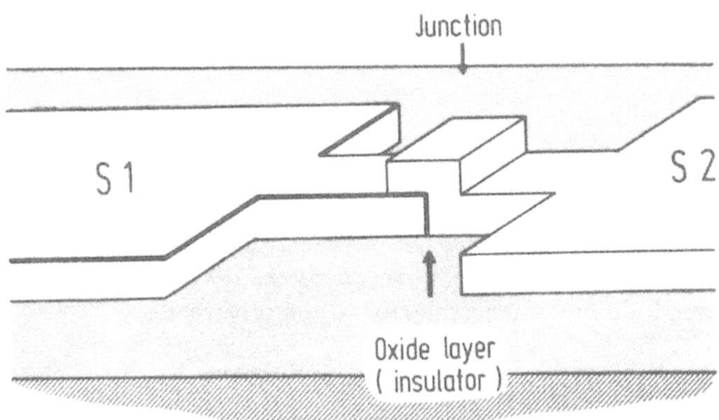

Fig. 3.15. Geometric structure of an SIS junction. Two thin layers (S_1 and S_2) of a superconductor metal are deposited on a substrate, forming a slight ($\approx 1\mu^2$) overlap region. S_1 and S_2 are insulated by a very thin ($\approx$ 10 Å) oxide layer covering the surface of S_1

Besides the single electrons, whole Cooper pairs can tunnel through SIS or superconducting point-contact junctions. The resulting complex voltage current behaviour is the basis of the Josephson effect mixers. However, although the Josephson effect has been known since many years, mixers based on this principle have been less successful than the single-electron tunnel effect SIS and SIN devices . Therefore, Josephson junction mixers will not be discussed here. In SIS junctions Cooper-pair tunnelling ("Josephson currents") constitute an unwelcome noise source. SIN junctions are safe from Cooper-pair tunnelling effects. But their less nonlinear voltage-current characteristics result in a somewhat poorer intrinsic noise performance.

Most superconducting tunnel-effect junctions are made of lead or niobium alloys which have realtively high critical temperatures. As insulators the natural oxydation layer forming on this metals in air can be used. To keep the capacitance small, geometrical arrangements resulting in junction surfaces of about $1\mu^2$ are chosen (Figure 3.15). Among the technical problems in designing SIS mixers is the need of matching waveguides for mm or longer wavelengths to such tiny devices. In praxis this can be achieved by depositing the junctions on a microantenna structure on the surface of a silicon semiconductor chip. In principle this technique also opens the possibility of producing "integrated arrays" of SIS junctions for imaging mm or sub-mm receiver systems.

4 Bolometers

The principle arrangement of a bolometer is shown schematically in Figure 4.1. The main component is a temperature-sensitive resistor R_T which is thermally connected with an absorber and with a heat sink of constant temperature (normally realized by a boiling liquid). If the resistor material has suitable optical and geometrical properties, the resistor itself can be the absorber. If no radiation is received by the absorber, the temperature of the resistor is equal to that of the heat sink. The absorption of radiation leads to a temperature increase, and thus in a change of R_T. This resistance variation is recorded by measuring the voltage change at the load resistor R_Ω.

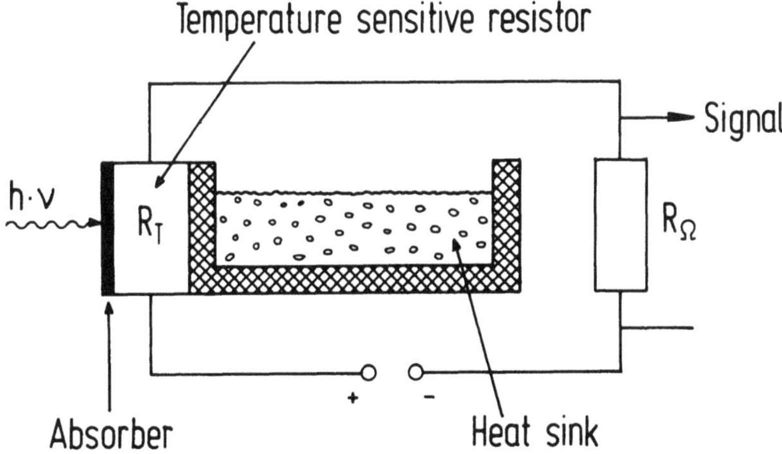

Fig. 4.1. Schematic arrangement of a bolometer

For a given radiation flux the strength of the signal depends on the slope of the R(T) relation of the resistor R_T, on the initial heat content (i.e. the product of the heat capacity and the temperature) of the resistor-absorber combination, and on the heat conduction between the resistor and the heat sink. The detection of very weak radiation fluxes obviously requires a small resistor size, a low operating temperature and a low heat conduction to the sink . Hence, bolometers are operated in a vaccum cryostat at the temperature of boiling ^{4}He (1.6 - 4.2 K, depending on the pressure) or ^{3}He (0.3 K). In most cases the small ($\lesssim$ mm-size) resistors are suspended by thin wires to keep the heat conductance small. On the other hand, in the presence of a strong radiation source (or background), a low thermal conductance may result in a saturation of the resistor or in a too slow response time of the bolometer. Therefore, the conductance has to be selected according to the particular application.

In most astronomical bolometers the temperature sensitive resistor consists of Ga-doped germanium. At liquid helium temperatures the conduction in Ge(Ga) results from the thermal excitation of impurity band electrons, which increases steeply with the temperature.

While the Ge(Ga) bolometer is based on the heating of the crystal lattice and the eventual transfer of the heat to the impurity band electrons, in "hot electron bolometers" (HEB) the absorbed radiation energy is tranferred selectively to conduction electrons. A suitable material is pure InSb. InSb HEBs (and mixers based on this effect) are used at FIR and mm wavelengths. In these bolometers the incident radiation does not produce additional conduction electrons. Instead, the (at liquid helium temperatures very few) existing conduction electrons are "heated" to higher energies, which changes their mobility and thus results in a change of the electrical conductance. As the energy exchange with the lattice is small, the energy input is largely restricted to the few conduction electrons. Hence, the HEBs have very low effective heat capacities and consequently short response times. This is the basis of their use as mixers in mm-wave receivers.

As any kind of absorbed radiation leads to a heating of the absorber, the bolometer principle can be used at all wavelengths (from X-ray to radio radiation), and (at least potentially) also for the detection of non-electromagnetic radiation such as neutrinos or graviational waves. At X-ray energies bolometers again allow the detection and (very accurate) energy determination of single photons. However, because of the required low operating temperatures, bolometers are relatively inconvenient for astronomical X-ray satellites. At present, bolometers are most commonly used for FIR and high-frequency radio astronomy. Here bolometers constitute efficient broad-band detectors. For narrow-band applications bolometers usually perform poorly, as their intrinsic broad response bands result in correspondingly large noise and instrumental background contributions.

5 Neutrino Detectors

Some types of astronomical objects (such as supernovae and stars with very hot cores) radiate a major fraction of their luminosity in the form of neutrinos. As these particles interact only through weak-force processes, their absorption cross section with matter is very low. Hence, neutrinos are particularly well suited for investigating the deep interior of stars and other dense astronomical objects. On the other hand, the low absorption cross section makes the detection of the weak neutrino fluxes from astronomical sources extremely difficult. In fact, so far reliable detections exist for two astronomical objects only. In the course of the past twenty years a total of several thousand neutrinos have been detected from the interior of our sun. Furthermore, in February 1987 at three different laboratories a total of about 24 neutrinos from the supernova 1987a in the LMC could be recorded.

At present, three different types of neutrino detection schemes are employed. All three methods are sensitive to electron neutrinos (or antineutrinos). The simplest process is the conversion of protons into neutrons by the capture of a (sufficiently energetic) antineutrino:

$$\overline{\nu_e} + p \rightarrow n + e^+$$

The reaction can be traced by means of the Čerenkov radiation emitted by the resulting relativistic positron. Hence, a neutrino detector based on this principle typically consists of a large volume of a transparent material with a high proton content, surrounded by arrays of light detectors. The largest existing detector of this type (the KAMIOKANDE detector in Japan) uses 6800 tons of very pure water. Other proton detectors use hydrocarbons (purified mineral oil). The existing proton neutrino detectors were initially designed to observe spontaneous proton decays. Their value for observing cosmic neutrinos was demonstrated by the simultaneous and clear detection of the SN1987a neutrino flux at three different geographic sites. Like all astronomical neutrino detectors, proton detectors have to be operated $\gtrsim 1$ km underground to avoid a too strong background of cosmic ray events.

For the detection of the low-energy electron neutrinos from the solar hydrogen-burning core radiochemical detectors are used. Those are based on the neutrino-induced conversion of a neutron into a proton in an atomic nucleus. Suitable reactions are e.g.

$$\nu_e +^{37}_{17} Cl \rightarrow^{37}_{18} Ar + e^-$$

$$\nu_e +^{71}_{31} Ga \rightarrow^{71}_{32} Ge + e^-$$

$$\nu_e +^{115}_{49} In \rightarrow^{115}_{50} Sn + e^-$$

$$\nu_e +^{7}_{3} Li \rightarrow^{7}_{4} Be + e^-$$

In all cases the reaction product is a radioactive isotope with a higher energy than the initial nucleus. As the energy difference must be supplied by the neutrino, each of these reactions has a characteristic cutoff energy. Historically, neutrino astronomy was initiated with the successful detection of solar neutrinos in a $^{37}Cl \rightarrow^{37} Ar$ detector constructed by R. Davis in 1968. The neutrino absorption cross section of $^{37}_{17}Cl$ is about 10^{-46} m^2. Hence, for neutrinos the mean free path in pure ^{37}Cl is just about one parsec. Using a well shielded tank with 610 tons of liquid C_2Cl_4 (containing about 2 10^{30} ^{37}Cl atoms), Davis was able to measure about 0.5 solar neutrinos per day. In order to determine the number of neutrino reactions in the C_2Cl_4 tank, the radioactive ^{37}Ar atoms are flushed out by percolating helium gas through the tank. After separating the argon from the helium (using a filter), the number of radioactive Ar nuclei is measured by counting the characteristic Auger electrons resulting from the decay of the Beta-active ($\tau \approx 35$ days) ^{37}Ar.

A disadvantage of ClAr detectors is that its relatively high cutoff energy (0.814 MeV) allows to detect only about 20 percent of the solar neutrino flux. The GaGe reaction listed above has a cutoff energy of 0.233 MeV only, covering most of the solar neutrino energy spectrum. A neutrino detector system based on this reaction has been developed by the international GALLEX cooperation in the Gran Sasso Underground Laboratory in Italy. This system uses about 30 tons of gallium contained in about 100 tons of a $GaCl_3$ solution. Again the characteristic Auger electrons of the decaying radioactive ^{71}Ge nuclei are used to trace the neutrino reactions.

In addition to the relatively well developed neutrino detection techniques described above, various other methods have been suggested for future astronomical neutrino detectors. Among those are superconducting bolometers which make use of the slight temperature increase caused by the scattering of neutrinos at heavy atomic nuclei. In a superconductor operated close to its critical temperature, the very small temperature change can in principle be detected from the change of the magnetic properties. Probably the most ambitious of the present neutrino detection schemes is the "Deep Underwater Muon and Neutrino Detector" (DUMAND) project. It envisages a very large Čerenkov radiation detector array (of about 10^3 PMTs) placed on a km-size section of the Pacific Ocean floor to detect energetic leptons resulting from neutrino reactions with the oxygen atoms of the seawater. If realized, the DUMAND project will be particularly useful for the detection of high-energy μ and τ neutrinos.

6 Gravitational Wave Detectors

According to General Relativity, time variations of the geometric distribution of masses result in the emission of gravitational waves. Physically such waves constitute a variation of the space geometry propagating with the velocity of light. As the geometry can be characterized by the metric tensor g_{ik} which connects the space-time line element ds with the coordinate differentials dx^i according to

$$ds^2 = \sum_{i,k} g_{ik} dx^i dx^k \tag{6.1}$$

the gravitational waves can be described as local variations of the components of g_{ik}. As the components of g_{ik} also determine three-dimensional distances between points at rest, a passing gravitational wave results in (differential) distance variations in planes parallel to the wave fronts.

In astrophysics ordered time variation of mass distributions result from orbital motions, pulsations, collapse and explosion, or rotation. According to Einstein's (GR) theory, waves are emitted if the variable object shows a changing quadrupole or higher momentum of its mass distribution. Furthermore, the radiation becomes significant only when the Schwarzschild radius of a system is not negligible compared to the system dimensions. Hence, gravitational wave detections of non-negligible intensity and a reasonable frequency of occurrence can be expected e.g. from very close or merging (galactic) binary systems and from supernova events in the Virgo cluster of galaxies. Unfortunately, in terrestrial detection systems such sources are expected to result in relative distance variations of typically about

$$\Delta x / x \approx 10^{-21} \tag{6.2}$$

only. Higher amplitudes are to be expected for SN events in our own galaxy or in nearby extragalactic sytems. However, because of the small SN rates ($\approx 10^{-2}$ per year in our galaxy) we may have to wait very long for such "nearby" SN events.

So far two types of gravitational wave detetors have been developed. In 1969 J. Weber used massive ($\approx$1 ton), acoustically well insulated aluminum cylinders of about 1 m length. If a gravitational wave hits such a "bar detector" the associated length change excites mechanical resonance oscillations, which can be monitored by means of a transducer belt of piezo-electric crystals around the circumference of the cylinder. In this way Weber could reach a sensitivity to relative geometrical distance variations of $\Delta x / x \approx 10^{-16}$ for oscillations at the resonance frequency (1.6 kHz) of his bars. (Note that for a 1-m bar this corresponds to measuring an absolute length variation of 10^{-6} Å, i.e. a fraction of an atomic nuclear radius). With this system Weber observed pulses which he ascribed to gravitational waves from the galactic center. However, other observers, using more sensitive cooled bar detectors with more sophisticated transducer schemes (reaching $\Delta x / x < 10^{-17}$), could not confirm the occurrence of such (unexpectedly strong) gravitational wave signals.

Because of unavoidable thermodynamic fluctuations and because of the serious mismatch of the propagation velocities of gravitational and acoustic waves, bar detectors are fundamentally inefficient. Therefore, during the past years, several groups developed gravitational wave detectors which rely on direct distance measurements using optical methods. A suitable arrangement, based on the Michelson interferometer principle, is outlined in Figure 6.1. A gravitational wave propagating perpendicularly to the plane defined by the two beams of the interferometer is expected to result in a differential geometrical distance change in the two branches. Obviously such a change will lead to a variation of the interference pattern on the optical detector D. This variation is recorded as the system's output signal. The

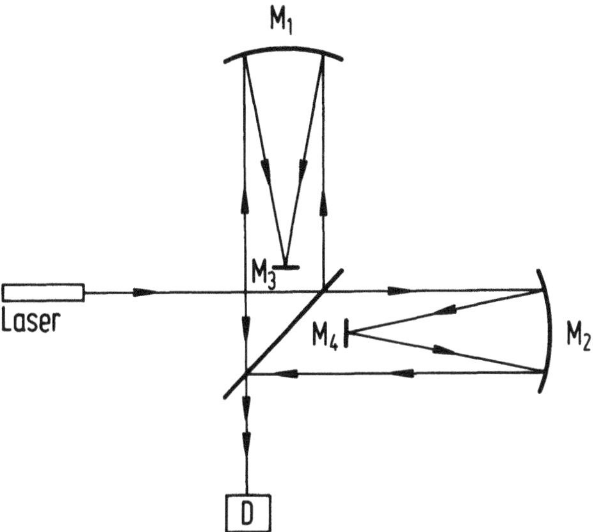

Fig. 6.1. Schematic arrangement of an interferometric graviational wave detector

sensitivity can be increased by adding additional reflections (up to 40 can be achieved with present optical components) between the mirrors M_1 and M_3, and M_2 and M_4, respectively. Presently existing interferometric systems (with arm legths $\leq$ 40 m) are still somewhat less sensitive than the best operational bar detectors. However, new interferometric gravitational wave detectors which presently are under development will have km-size interferometer arms. These systems are expected to reach sensitivities up to $\Delta x/x = 10^{-23}$. A comparison with Equ. (6.2) shows that with these detectors routine measurements of gravitational waves from astronomical objects should become possible. Such measurements will yield information on the geometrical structure of the sources that cannot be determined by any other astronomical observing technique.

Bibliography

General

Kitchin, C.R. Astrophysical Techniques (Adam Hilger, Bristol 1984)
Léna, P. Observational Astronomy (Springer, Berlin, Heidelberg, New York, 1988)

Solid State Physics

Kittel, C. Introduction to Solid State Physics, 6th ed. (Wiley, New York 1986)

Photon Detectors

Eccles, M.J., Sim, M.E., Tritton, K.P. Low Light Level Detectors in Astronomy (Cambridge University Press, 1983)
Mackay, C.D. Charge-Coupled Devices in Astronomy. In Ann. Rev. Astron. Astrophys. 24, 255 (1986)

Beynon, J.D.E., Lamb, D.R. eds. Charge-Coupled Devices and their Applications (McGraw-Hill, 1980)
Ramana Murthy P.V., Wolfendale A.W. Gamma-Ray Astronomy (Cambridge University Press 1986)
Robinson, L.B. ed. Instrumentation for Groundbased Optical Astronomy (Springer, Berlin, Heidelberg, New York, 1988
Walker, G., Astronomical Observations. An Optical Perspective (Cambridge University Press, 1987)

Radio Techniques

Kraus, J.D. Radioastronomy 2nd ed. Chapter 7 (Cygnus-Quasar Books, Powell, Ohio, 1986)
Meeks, M.L. ed. Methods of Experimental Physics Vol. 12B (Radio Telescopes) (Academic Press, New York 1976)
Philips T.G., Woody, D.P. Millimeter and Submillimeter Wave Receivers. In Ann. Rev. Astron. Astrophys. 20, 285 (1982)
Rolfs K. Tools of Radioastronomy (Springer, Berlin, Heidelberg, New York 1986)

Neutrino Detection

Barger V., Halzen F., Marshak M., Olive K. eds. Neutrino Masses and Neutrino Astrophysics (World Scientific Publ. Co., Singapore 1987)

Gravitational Wave Detection

Deruelle, N., Piran,T. eds. Rayonnement Gravitationnel (North-Holland, Amsterdam 1983)
Hawking S.W., Israel, W. eds. 300 Years of Gravitation, Chapter 9 (Cambridge University Press 1987)

This article was processed by the author using the TEX Macropackage from Springer-Verlag.

Radio Astronomy Techniques

D. Downes
Institut de Radio Astronomie Millimétrique, Grenoble

Single-Dish Radio Telescopes

Single-dish antennas are used to discover new spectral lines, to monitor time-variable objects, to study compact regions and to survey extended areas like the galactic plane or even the whole sky. Why are single dishes so useful? Because their full collecting area and diffraction-limited beam are available straightaway, and not after a synthesis lasting hours to months. They can be more easily re-equipped and converted from one frequency to another, than multi-element arrays, for which new receivers and correlators may be very costly. Since single dishes use their full aperture, there is no missing flux or zero-spacing problem as with arrays. Because of their flexibility, and because they are elements of synthesis arrays and VLBI networks, it is useful to understand the basic parameters of single dish antennas.

Fig. 1. The IRAM 30-m telescope on Pico Veleta. Photo by the author.

Atmospheric windows are limited by ionospheric electrons and tropospheric H_2O and O_2.

The domain of possible operation for ground-based radio telescopes extends from several MHz to 800 GHz. It is limited at low frequencies by the ionosphere and at high frequencies by the troposphere. The lower limit is due to the solar UV and X-rays creating a permanent, weakly ionized plasma around the earth, with plasma frequency, ν_p, and optical depth, τ_ν, of:

$$\nu_p = 9\sqrt{n_e} \qquad [\mathrm{Hz}] \tag{1}$$

$$\tau_\nu(\mathrm{day}) = 0.046\left[\frac{\nu}{100\ \mathrm{MHz}}\right]^{-2}, \quad \tau_\nu(\mathrm{night}) = 0.0046\left[\frac{\nu}{100\ \mathrm{MHz}}\right]^{-2} \tag{2}$$

where n_e is electron density [m^{-3}], typically $10^{12}\ \mathrm{m}^{-3}$. There is no propagation through the ionosphere below the plasma frequency, which is $\sim$ 2 to 10 MHz, depending on day/night conditions.

At higher frequencies, observations are limited by water vapour and oxygen lines in the troposphere, of which only the 22 GHz H_2O line is partially transparent. Table 1 gives values of atmospheric opacity, sky temperature and transmission for an excellent winter night on a mountain site, with a precipitable water vapour content of 1 mm, and an ambient temperature of -10°C. The values are taken from an atmospheric model (Cernicharo, 1988), and agree well with measurements on the site.

Table 1. Atmospheric absorption at Pico Veleta, Spain (2850 m altitude, winter night, 1 mm H_2O).

Freq. ν GHz	Zenith opacity τ_ν	Sky brightness temperature at zenith (K)	Zenith transmission $e^{-\tau_\nu}$	Remark
22	0.02	5	0.98	H_2O line
43	0.05	13	0.95	8 mm window
60	21.	263	0.00	O_2 (25 lines, 53 to 66 GHz)
90	0.05	13	0.95	3 mm window
115	0.25	58	0.78	CO (1–0) line in space
118	2.6	243	0.07	O_2 line
140	0.04	10	0.96	2 mm window
183	2.2	234	0.11	H_2O line
230	0.08	20	0.92	1 mm window
325	2.1	231	0.12	H_2O line
345	0.25	58	0.78	870 μm window
420	0.57	114	0.57	O_2
460	1.03	169	0.36	650 μm window
553	254.	263	0.00	H_2O
690	1.5	204	0.22	450 μm window
750	380.	263	0.00	H_2O and O_2
860	1.2	184	0.30	350 μm window

The antenna beam pattern is the Fourier transform of the illumination pattern.

Single-dish radio telescopes are usually used in prime-focus, Cassegrain, Gregorian or Nasmyth configurations. The radiation is focused to a feed whose electric field response to different annuli of the aperture is described by a grading function (apodising function in optics). As a function of the x, y coordinates of the aperture plane, the grading can be regarded as specifying the current distribution over the aperture, or electric field illumination pattern. The far-field voltage polar diagram, V, is the Fourier transform of the grading function, g, of the aperture:

$$V(l,m) \propto \mathcal{F}\{g(x,y)\} \tag{3}$$

where $\mathcal{F}\{\bullet\}$ is the two-dimensional Fourier transform operator, l, m = direction cosines relative to the x, y axes: ($l = \sin\theta \sin\phi$, $m = \sin\theta\cos\phi$, θ = angle of incidence relative to the normal to the x, y plane, ϕ = azimuth from y axis).

The power polar diagram, in the far field (the "beam pattern"), is proportional to the modulus squared of the voltage pattern:

$$P(l,m) \propto |\, V(l,m)\,|^2 \tag{4}$$

From eq (3), and the autocorrelation theorem of Fourier transforms (e.g., Goodman 1968), we have

$$\mathcal{F}\{P(l,m)\} = g \star g^* \equiv \int_{-\infty}^{+\infty}\int_{-\infty}^{+\infty} g(x,y)\cdot g^*(x-u,y-v)dx\,dy \equiv W(u,v) \tag{5}$$

where $W(u,v)$ is the instrumental transfer function, namely, the autocorrelation function of the grading, g^* is the complex conjugate of g, and the $\star$ symbol denotes autocorrelation, in two dimensions. Hence, the beam pattern, in the far field, is the Fourier transform of the transfer function.

For a uniformly illuminated circular aperture, as in an optical telescope, P is the Airy pattern. Radio and optical telescope illuminations differ in two ways. Firstly, for the low focal ratios common to radio telescopes, the image structure differs significantly from the Airy pattern (see Minnet and Thomas 1968, for the actual pattern). Secondly, radio feeds are usually monomode scalar horns whose lowest waveguide mode has a nearly gaussian variation of electric field strength across the aperture. Given the impossibility of building a monomode feed which uniformly illuminates an aperture, cutting off to zero exactly at the edge, a uniform illumination is undesirable for ground-based radio telescopes because of the sidelobe levels and the spillover. For a feed in the prime focus, the spillover around the edges of the aperture would pick up the 300 K ground radiation. For a feed in the secondary focus, the spillover around the edges of the secondary mirror would look at the sky; this is the classic advantage of secondary focus configurations, as the sky is colder than the ground at most radio wavelengths. However, in both the prime focus and the secondary focus, the broader the gaussian grading, the more uniform the aperture illumination, but the lower the aperture efficiency, which falls to zero for an infinitely broad grading function. (For communications or space antennas, higher sidelobes or more spillover may be acceptable, and non-gaussian illumination may be obtained with shaped sub-reflectors or lenses).

Table 2 gives, for the example of a circular aperture, the beam pattern formulae for uniform and for tapered illuminations. In the usual radio astronomy practice, one accepts a main lobe broader than that of the Airy pattern, in return for low sidelobes and low spillover. Typically, the feed horn has a gaussian taper to -10 to -14 dB at the edge of the dish. The far-field beam pattern is then, to a good approximation, a gaussian with full width to half power of

$$\theta_b \cong 1.2 \frac{\lambda}{D}, \tag{6}$$

and a main-beam solid angle of

$$\Omega_b \cong 1.133\, \theta_b{}^2, \tag{7a}$$

where θ_b is in radians, Ω_b in sterradians, λ = wavelength, D = dish diameter. For comparison, Table 2 also lists the beam pattern formulae for a gentler gaussian taper than that used for single dishes. This taper, to -6 dB, is used occasionally to weight the data from different baselines in aperture synthesis. As the array grading is a purely mathematical weighting to reduce sidelobes of the synthesized beam, the spillover considerations for a single-dish aperture are no longer relevant, and one can obtain a narrower beamwidth than for an equivalent single dish with diameter equal to the maximum array spacing.

As the Fourier transform of the truncated gaussian is not exactly gaussian, eq.(7a) gives the solid angle of an equivalent gaussian with the same halfwidth as the measured halfwidth. Usually eq.(7a) is good to 5 per cent. An even better approximation is the solid angle of an equivalent gaussian with the same full width to one-tenth-power, $\theta_{-10\,\mathrm{dB}}$, as the measured beam:

$$\Omega_b \cong 0.3411\, (\theta_{-10\,\mathrm{dB}})^2 \tag{7b}$$

Values obtained from eq.(7b) agree with those obtained from direct integration to better than one per cent (Goldsmith 1987).

If the beam is convolved with a gaussian source of halfwidth θ_s, the response pattern will also be gaussian, with halfwidth θ_r given by

$$\theta_r{}^2 = \theta_s{}^2 + \theta_b{}^2 \tag{8}$$

(in practice, this relation can be trusted only for $\theta_s \geq \theta_b/2$).

How does one do practical calculations of the power received by single-dish antennas, without knowing the details of the Fourier coverage, that is, the grading function, but nevertheless including the consequences of the aperture grading in global efficiency factors?

Antenna temperature is the temperature of an equivalent resistor; Brightness temperature is the temperature of an equivalent black body.

The radio telescope can be thought of as a watt meter, to measure the power coming from the sky, or more properly, the power per unit area per unit frequency, or *flux density*, S, in units of Janskys (1 Jy = 10^{-26} W m^{-2} Hz^{-1}). The antenna can be pointed toward a cosmic source and then away from it, the difference in the two outputs giving the contribution from the source alone, usually expressed in one of two ways. The first

Table 2. Examples of grading, transfer function and beam pattern for a circular aperture.

	Aperture (x,y) plane, units = wavelengths		Image (l,m) plane, angular units	Spatial frequency (u,v) plane, units = wavelengths		Image (l,m) plane, angular units	
Grading type	Grading function $g(x,y)$	$\rightleftharpoons$	Far-field voltage pattern $V(l,m)=\mathcal{F}\{g\}$	Transfer function $W(u,v)=g\star g^*$	$\rightleftharpoons$	Far-field power pattern $P(l,m)=\mathcal{F}\{W\}$	Beam (FWHP) θ_b
uniform	$\mathrm{circ}(r)$	$\rightleftharpoons$	$\frac{J_1(2\pi\theta)}{\theta}$	$\mathrm{circ}\star\mathrm{circ}$	$\rightleftharpoons$	$\left[\frac{2J_1(\theta)}{\theta}\right]^2$ 1.8 % sidelobes	$1.02\ \lambda/D$
gaussian to -10 dB at edge of dish.	$\exp[-\frac{\ln 0.1}{2}(r/R)^2]$ $\forall r\le R$, 0 otherwise.	$\rightleftharpoons$	$\exp[-\frac{\ln 2}{2}(2\theta/\theta_b)^2]$ (main lobe)	$\exp[-\ln 0.1\,(r/R)^2]$ $\forall r\le R$, 0 otherwise.	$\rightleftharpoons$	$\exp[-\ln 2\,(2\theta/\theta_b)^2]$ with $\le$ 1% sidelobes	$1.2\ \lambda/D$
gaussian to -6 dB at longest spacing.	$\exp[-\frac{\ln 0.25}{2}(r/R)^2]$ $\forall r\le R$, 0 otherwise.	$\rightleftharpoons$	$\exp[-\frac{\ln 2}{2}(2\theta/\theta_b)^2]$ (main lobe)	$\exp[-\ln 0.25\,(r/R)^2]$ $\forall r\le R$, 0 otherwise.	$\rightleftharpoons$	$\exp[-\ln 2\,(2\theta/\theta_b)^2]$ with 3 % sidelobes	$0.78\ \lambda/D$

Functions: circ = uniform circular aperture: $\mathrm{circ}(r) = 1\ \forall\ r \le R$, zero otherwise; J_1 = Bessel function, first kind, order one.

Operations: $\star$ = autocorrelation, $\mathcal{F}\{\bullet\}$ = Fourier transform operator, $\rightleftharpoons$ denotes Fourier transform pairs.

Symbols: D, R = aperture diameter, radius, $P \equiv$ power pattern $\equiv$ instrumental point spread function $\equiv$ beam (normalized to unity at peak), r = radius in aperture (x,y) plane and spatial frequency (u,v) plane, θ = radius in image (l,m) plane, θ_b = beamwidth (FWHP), λ = wavelength method is in terms of the *antenna temperature*, T_a, defined as the temperature of an equivalent resistor which would give the same power, as measured at the final output terminals, as the celestial source. The second method is in terms of the *brightness temperature*, T_b, defined as the Rayleigh-Jeans temperature of an equivalent black body which would give the same power per unit area per unit frequency and per unit solid angle as the celestial source. Both definitions yield convenient quantities which are *linearly* proportional to power. For this reason, antenna temperature and brightness temperature are not defined through the Planck formula, but only in the classical limit, even when $h\nu > kT$. As such, they are purely fictitious temperatures, useful for describing the outputs. In particular, the antenna temperature has nothing to do with the true physical temperature of the antenna itself, and the brightness temperature may also be a fictitious temperature which might vary with observing frequency, and be applied to objects like non-thermal sources for which there may be no unique physical temperature.

The relations among antenna temperature, brightness temperature and flux density are summarised for typical cases in Table 3.

Table 3. Equations for antenna temperature and brightness temperature.

	Antenna temperature, T'_a (temperature of equivalent resistor)	Brightness temperature, T_b (temperature of equivalent black body)
in general:	$S = \frac{2k}{A_e} \frac{\int T'_a \, d\Omega_r}{\int P \, d\Omega_b}$	$S = \frac{2k}{\lambda^2} \int T_{mb} \, d\Omega_r = \frac{2k}{\lambda^2} \int T_b \, d\Omega_s$
point source:	$S = \frac{2k}{A_e} T'_a$	$S = \frac{2k}{\lambda^2} T_{mb} \, \Omega_b$
gaussians:	$S = \frac{2k}{A_e} T'_a \frac{\theta_r^2}{\theta_b^2}$	$S = \frac{2k}{\lambda^2} T_{mb} \, 1.133 \, \theta_r{}^2$
formulae:	$\frac{S}{\mathrm{Jy}} = \frac{3516}{\epsilon_{ap}} \frac{D^{-2}}{\mathrm{m}^{-2}} \frac{T'_a}{\mathrm{K}}$	$\frac{S}{\mathrm{Jy}} = 2.64 \frac{\lambda^{-2}}{\mathrm{cm}^{-2}} \frac{T_{mb}}{\mathrm{K}} \frac{\theta_r^2}{\mathrm{arcmin}^2}$ (for gaussians)

k = Boltzmann constant = 1.38 10^{-23} J K^{-1}, λ = wavelength, A_e = effective collecting area,
D = dish diameter, ϵ_{ap} = aperture efficiency (eq.(11)), P = beam pattern (eq.(4)),
T_b = brightness temperature, T_{mb} = main-beam brightness temperature,
T'_a = antenna temperature outside the atmosphere, $T'_a = T_a \exp(\tau_o \sec z)$,
θ_b = beamwidth (FWHP), θ_r = response width (beam convolved with source),
Ω_b = main-beam solid angle; $d\Omega_r$, $d\Omega_b$, $d\Omega_s \Rightarrow$ integrate over response, beam or source, respectively.

Note that antenna temperatures are antenna-specific, and will vary from telescope to telescope. They are usually easy to measure, and are useful for understanding the system performance. In contrast, brightness temperatures are properties of the sources on the sky; provided the sources are resolved, different radio telescopes will measure the same value. The brightness temperatures are needed to do astrophysics. For example, in the equation for radiative transfer in a homogeneous medium, the direct proportionality of brightness temperature to intensity allows us to write

$$T_b = (T'_{ex} - T'_{bg})(1 - e^{-\tau_\nu}) \tag{9}$$

where the primes indicate equivalent Rayleigh-Jeans temperatures. Conversion to true excitation or background temperatures, T_{ex} or T_{bg}, is through the Planck formula, e.g.:

$$T'_{ex} = (h\nu/k)(\exp(h\nu/kT_{ex}) - 1)^{-1} \tag{10}$$

In eq.(9), the temperature is always the brightness temperature, never the antenna temperature.

The concepts of antenna and brightness temperatures allow us to calculate the power received by single dishes without knowing the details of the Fourier coverage of the aperture. The consequences of the grading are taken into account through global factors, the most important being aperture efficiency and beam efficiency.

The *aperture efficiency*, ϵ_{ap}, is defined as

$$\epsilon_{ap} \equiv A_e/A \tag{11}$$

where A_e is the effective collecting area, and A is the geometric area of the antenna.

The effective *beam efficiency*, B_{eff}, is most simply defined [1] by

$$B_{\text{eff}} \equiv T'_a/T_{mb} \tag{12}$$

where T'_a is the antenna temperature corrected for atmospheric extinction: $T'_a = T_a\, e^{\tau_\nu}$.

The aperture efficiency must be measured, but the beam efficiency may be calculated from the beam pattern.

From the definitions in eqs.(11, 12) and the relations in Table 3, we have

$$B_{\text{eff}} = \epsilon_{ap}\frac{A\Omega_b}{\lambda^2} \tag{13}$$

For a single dish of diameter D, $A = \pi D^2/4$, so eq.(13) and eq.(7a) yield

$$B_{\text{eff}} = 0.8899\,[\theta_b/(\lambda/D)]^2\,\epsilon_{ap} \tag{14a}$$

where θ_b = beamwidth (FWHP) in radians. The relevant information about the aperture illumination is in the ratio of beamwidth to λ/D. Alternatively, if there are accurate measurements of the beamwidth to one-tenth-power, $\theta_{-10\text{dB}}$, then eqs.(7b) and (13) yield

$$B_{\text{eff}} = 0.2679\,[\theta_{-10\text{dB}}/(\lambda/D)]^2\,\epsilon_{ap} \tag{14b}$$

Since this formula approximates the integral over the beam to an accuracy of 1% (cf. Goldsmith 1987), we can thus calculate main-beam efficiency to the same precision as we can measure aperture efficiency.

---

[1] This definition is easier to apply in practice than that in some other texts, where beam efficiency is given as Ω_b/Ω_a, the ratio of beam solid angle to the integral over the entire antenna pattern. See Appendix 1 for relations among quantities used here and those in some other references.

The aperture efficiency is usually measured via a celestial source of known flux density, whose antenna temperature is obtained by comparison with a noise standard, such as the difference between hot and cold loads. The beam efficiency is often estimated by observing a planet, whose true disk temperature is assumed, and whose angular diameter matches the beamwidth. The correction for the convolution of the beam with the planetary disk usually relies on an assumed gaussian beamshape, making this method mathematically equivalent to simply calculating the beam efficiency from a measurement of a point source in the first place, via eqs.(14a/b). As an illustrative example, Table 4 gives some typical values for these efficiencies for the IRAM 30-m telescope. Note that the illumination by different receiver feeds gives beamwidths varying from $\theta_b \sim 1.13\,\lambda/D$ to $\theta_b \sim 1.51\,\lambda/D$.

Table 4. Aperture and main-beam efficiencies of the IRAM 30-m telescope.

freq. ν GHz	receiver type	wave-length λ(mm)	beam width θ_b	illum-ination $\frac{\theta_b D}{\lambda}$	ratio $\frac{B_{\rm eff}}{\epsilon_{ap}}$	aperture effic. ϵ_{ap}	beam effic. $B_{\rm eff}$	S/T_{mb} Jy/K[a]
86	Schottky	3.49	27″	1.13	1.13	0.53	0.60	4.40
90	SIS	3.33	26″	1.14	1.15	0.47	0.54	4.47
106	Schottky	2.83	22″	1.13	1.14	0.50	0.57	4.44
111	Schottky	2.70	21″	1.13	1.14	0.49	0.56	4.43
140	SIS	2.14	17″	1.15	1.19	0.50	0.59	4.62
230	Schottky	1.30	13.5″	1.51	2.03	0.27	0.55	7.87
230	SIS	1.30	12.5″	1.39	1.74	0.27	0.47	6.74
250	bolometer	1.20	11.0″	1.33	1.58	0.28[b]	0.44[b]	6.17
265	SIS	1.13	10.8″	1.39	1.71	0.21	0.36	6.68
345	Schottky	0.87	8.5″	1.42	1.80	0.10[c]	0.18[c]	6.98

[a]For equivalent point source. [b]elevation 55°, [c]elevation 45°

The equipment behind the antenna feed is usually a low-noise pre-amplifier, followed by a mixer. In millimeter astronomy, because of the absence of good amplifiers at the RF frequency, the signal from the sky is fed directly to the mixer. The output of the mixing of the RF frequency with a local oscillator signal passes through one or several IF amplifier stages, and is then fed to a backend detector, whose DC output goes to a computer.

The noise of the radiometer is characterised by a *receiver temperature*, T_R, the temperature of an equivalent resistor with the same noise power, as given by the Nyquist formula, $P = k\,T\,\Delta\nu$. Normally, the receiver temperature consists of contributions from the mixer plus the IF stages, the latter being multiplied by the conversion loss, L, of the mixer, $T_R = T_M + L\,T_{IF}$.

The *system temperature* is the temperature of an equivalent resistor which would give the same noise power as the entire system, that is, the receiver, and the contributions received by the antenna from the sky and the ground:

$$T_{\rm sys} = T_R + T_{\rm sky} + T_{\rm ground} \tag{15}$$

where

$$T_{\rm sky} = F_{\rm eff}\, T_{\rm amb}(1 - e^{-\tau_\nu}) \tag{16}$$

and

$$T_{\rm ground} = (1 - F_{\rm eff})\, T_{\rm amb} \tag{17}$$

where $T_{\rm amb}$ is the ambient temperature (assumed to be the same for the air and the ground, in this example), and $F_{\rm eff}$ is the forward efficiency of the antenna.

The r.m.s *sensitivity*, ΔT_a, of the system is given by

$$\Delta T_a = T_{\rm sys}(\Delta\nu\tau)^{-0.5} \tag{18}$$

where $\Delta\nu$ = bandwidth per observing channel, and τ = integration time. This relation may be multiplied by an additional factor which depends on the observing mode; e.g., for on-off observations, with no digital clipping of the data, the right side of eq.(18) should be multiplied by $\sqrt{2}$.

In millimeter astronomy, the sky is used as a calibrator source.

Because the atmospheric extinction at millimeter wavelengths is significant (see Table 1), and varies with elevation, it is useful to take the sky itself as a calibrator source, as its signal also varies with elevation in the same way as the attenuation, so such a calibration yields the source temperature outside the atmosphere.

This calibration method, often called the chopper wheel method (Penzias and Burrus, 1973), defines the calibration signal to be the difference between an absorber at ambient temperature and the sky.

The output voltage from the ambient load is

$$V_{\rm amb} = G(T_{\rm amb} + T_R) \tag{19}$$

and from the sky and cabin, from eqs. (16, 17),

$$V_{\rm sky} \;=\; G\,\left[F_{\rm eff}\; T_{\rm sky} + (1 - F_{\rm eff})\; T_{\rm ground\ or\ cabin} + T_R\right] \tag{20a}$$

Or,

$$V_{\rm sky} \;=\; G\,\left[F_{\rm eff}\; T_{\rm amb}(1 - e^{-\tau_\nu}) + (1 - F_{\rm eff})\; T_{\rm amb} + T_R\right] \tag{20b}$$

where G is the varying gain factor to be calibrated out, $T_{\rm amb}$ = ambient temperature, T_R = receiver temperature.

The difference of these two outputs is the calibration voltage:

$$\Delta V_{\rm cal} \equiv V_{\rm amb} \;-\; V_{\rm sky+cabin} \;=\; G\; F_{\rm eff}\; T_{\rm amb}\; e^{-\tau_\nu} \tag{21}$$

The signal received from the radio source is

$$\Delta V_{\rm sig} \;=\; G\; T'_a\; e^{-\tau_\nu} \tag{22}$$

Hence, solving for T'_a,

$$T'_a \;=\; \frac{\Delta V_{\rm sig}}{\Delta V_{\rm cal}}\; F_{\rm eff}\, T_{\rm amb} \tag{23}$$

Here T'_a is the antenna temperature of the source outside the atmosphere, and $\tau_\nu = \tau_0$ sec z; τ_0 = zenith optical depth, z = zenith angle.

Or, defining $T_a^* \equiv T_a'/F_{\rm eff}$, we also have

$$T_a^* = \frac{\Delta V_{\rm sig}}{\Delta V_{\rm cal}} T_{\rm amb} \tag{24}$$

This is the reason why the chopper-wheel method is so convenient; one needs only to multiply the source to calibration ratio by the ambient temperature to obtain T_a^*, a temperature which is automatically corrected for atmospheric attenuation. Note that T_a^* is a fictional temperature convenient for the chopper-wheel method, and not as useful for other methods of calibration. T_a^* can be thought of as a "forward-beam brightness temperature". It is the brightness temperature of an equivalent source which fills the entire 2π sterradians of the forward beam pattern. Appendix 2 gives the interrelations among the various temperature scales defined here, in the context of the chopper- wheel method.

The equations presented here and in Appendix 2 are for the simplified case of single-sideband observing (image band suppressed) and $T_{\rm atm} \approx T_{\rm amb}$. In practice, the image sideband will have some gain factor g_i relative to that of the signal sideband ($g_s = 1$), and the average atmospheric temperature, $T_{\rm atm}$, will differ from the temperature of the chopper, here assumed to be at the ambient temperature, $T_{\rm amb}$. In this case, $T_a^* = (\Delta V_{\rm sig}/\Delta V_{\rm cal})\, T_{\rm cal}$, where

$$T_{\rm cal} = (T_{\rm amb} - T_{\rm atm})(1 + g_i)e^{\tau_s} + T_{\rm atm}(1 + g_i e^{\tau_s - \tau_i}) \tag{25}$$

where τ_s, τ_i = atmospheric opacities in the signal and image bands (zenith opacities x air mass) — see also eq.(4) of Davis and Vanden Bout (1973).

Note also that although the chopper wheel method corrects for atmospheric attenuation, it does **not** correct antenna temperature for all telescope losses. In particular it does not correct for the very significant telescope losses due to surface irregularities. For the IRAM 30-m telescope at 230 GHz, for example, this is half the power! For this telescope, from the values in Table 4, we have, for a source of main-beam brightness temperature T_{mb},

$$\text{power received} \propto T_a^* = \frac{B_{\rm eff}}{F_{\rm eff}} T_{mb} = \frac{0.45}{0.90}\, T_{mb} = \frac{1}{2}\, T_{mb} \quad \text{at 230 GHz,}$$

whereas the power lost due to the forward scattering by the surface irregularities is

$$\text{power lost to scattering} \propto (T_{mb} - T_a^*) = \left(1 - \frac{B_{\rm eff}}{F_{\rm eff}}\right) T_{mb} = \frac{1}{2}\, T_{mb} \quad \text{at 230 GHz.}$$

Surface irregularities scatter power into an error beam.

Reflector surface irregularities distort a plane wave into a wavefront with phase errors. A shallow paraboloid with average deviations Δz, normal to its surface will give deviations $2\,\Delta z$ to the wavefront after reflection, corresponding to phase changes of

$$\Delta\phi = 4\,\pi(\Delta z/\lambda) \tag{26}$$

(For deep reflectors, Δz is an effective reflector tolerance; for its relation to the actual physical deviation, see Ruze, 1966; for telescopes with f/D = 0.3, the deviations normal to the surface are $\sim 1.2\ \Delta z$, Greve and Hooghoudt, 1981).

The electric fields of the waves arriving at the focus will then have a phase error factor $e^{i\Delta\phi}$, which for small $\Delta\phi$ expands to $1+i\Delta\phi-\frac{1}{2}(\Delta\phi)^2$. The main-lobe *power* pattern, P_m, will then be reduced relative to its error-free value, P_o, by

$$\frac{P_m}{P_o} = 1 + \langle\Delta\phi\rangle^2 - \langle\Delta\phi^2\rangle \tag{27a}$$

If there are as many negative as positive irregularities, then by choosing a suitable aperture plane, one can always force $\langle\Delta\phi\rangle$ to be zero, so for $\Delta z < \lambda$,

$$\frac{P_m}{P_o} = 1 - \langle\Delta\phi^2\rangle \tag{27b}$$

which is valid for $\Delta\phi \ll 1$ radian (Väisälä 1922). [2] Alternatively, expanding $e^{i\Delta\phi}$ into cosine and sine components and squaring, the power reduction in the direction of maximum response (main lobe) due to the irregularities is

$$\frac{P_m}{P_o} = cos^2\Delta\phi \tag{27c}$$

which is an excellent approximation for $\Delta z \leq \lambda/20$ (see Kraus 1986, Fig. 6–51; Christianson and Högbom 1985, Fig. 3.5).

In general, if the errors are random, gaussian, and uniformly distributed over the aperture, and if they have correlation lengths, l_c, satisfying $\lambda \ll l_c \ll D$, then

$$\frac{P_m}{P_o} = e^{-\Delta\phi^2} = e^{-(4\pi\Delta z/\lambda)^2} \tag{28}$$

where $\Delta\phi$ is the r.m.s. phase error and Δz is the effective r.m.s surface tolerance. (Ruze, 1952; see also Bates, 1958, and the review by Ruze, 1966, and references therein).

For example, from the measured values of aperture efficiency for the IRAM 30-m telescope (Table 4), the r.m.s. phase error at 230 GHz due to surface irregularities is $[-\ln(\epsilon_{ap}(230\,\text{GHz})/\epsilon_{ap}(90\,\text{GHz}))]^{0.5} = [-\ln(0.27/0.50)]^{0.5} = 0.78$ radians, corresponding to an effective surface error of 80 μm r.m.s. (After correction for the under-illumination at 230 GHz, the value is 70 μm r.m.s.).

Statistically, the power which is removed from the main lobe is scattered into an "error" beam of width $\theta_e \approx \lambda/l_c$ and relative power

$$\frac{P_e}{P_o} = 1 - \frac{P_m}{P_o} = 1 - e^{-\Delta\phi^2} \tag{29}$$

For large phase errors, the main lobe disappears, and the error pattern approaches

$$P_e(\theta) = \left(\frac{\pi l_c}{\lambda}\right)^2 \frac{[1 - e^{-\Delta\phi^2}]}{\Delta\phi^2} \exp[-\left(\frac{l_c u}{8\Delta z}\right)^2] \tag{30}$$

[2] This reference was brought to my attention by A. Greve.

(Scheffler 1962), where $u = \sin\theta$. The width of the error beam (FWHP) in this case is $\theta_e = 13.3\ \Delta z/l_c$, and is independent of wavelength, as expected from geometric optics. Alternatively, if one wishes to define an angle analogous to the seeing disk, where the phase error $\Delta\phi = 1$ radian, then the beamwidth (FWHP) from eq.(30) is

$$\theta_e\,(\Delta\phi = 1) = 1.06\,\frac{\lambda}{l_c} \tag{31}$$

In this sense, the correlation length, l_c, is analogous to the Fried parameter, r_o, in optical seeing, and the error beam is analogous to the seeing disk.

The presence of an error pattern modifies the equations in Table 3, and Appendix 1 and 2 as follows. For source sizes comparable with the size of the error beam, the beam efficiency, $B_{\rm eff}$, must be interpreted as a *full-beam* efficiency (main lobe plus error pattern), rather than as a main-beam efficiency, and be evaluated appropriately. Similarly, T_{mb} becomes a *beam-averaged brightness temperature*, where the average now extends over the error pattern as well.

Radio "seeing" effects are caused by variations in the "wet" component of refractivity.

The refractive effect of the neutral atmosphere in the range 0 – 30 GHz, away from resonances, is characterised by the *refractivity*, N, given by Smith and Weintraub (1953), as:

$$N = (n-1)10^6 = \frac{77.6}{T_{\rm atm}}(P + \frac{4810\,e}{T_{\rm atm}}) \tag{32}$$

where $T_{\rm atm}$ = temperature [K] of the atmosphere ~280 K, P = total atmospheric pressure, in millibars (1 atmosphere = 1013 mb), e = partial pressure of water vapour [mb] ~10–30 mb at sea level, n = index of refraction. The first term is called the dry component of the refractivity, N_D, and the second term is called the wet part, N_W. The variation of N_D is exponential, with a scale height of 8 km, and can be predicted from the equation of hydrostatic equilibrium. For a typical value of N_D ~280, the excess path length, relative to the path of the rays in free space is $\Delta L_D = 10^6 \int N_D\, d\ell \cong 225$ cm.

The variation of the wet part, N_W, is also roughly exponential, with a scale height of ~2 km, but H_2O is not well mixed. For a value of N_W ~100, the excess path length due to the wet refractivity is $\Delta L_W = 10^6 \int N_W\, d\ell$ ~20 cm. For comparison, the excess path length due to refraction in the ionosphere is $\Delta L \sim 10\,{\rm cm}\,(\nu/10\,{\rm GHz})^{-2}$, or only 0.1 cm at a frequency of 100 GHz.

Anomalous Refraction

An atmospheric effect with stronger consequences for some mm telescopes and arrays than for cm telescopes, because of the smaller primary beams, is anomalous refraction. In these events, studied at 3 and 1.3 mm with the 30-m telescope on Pico Veleta and at 13 mm with the 100-m Effelsberg telescope (Altenhoff et al. 1987), radio sources appear to move away from their true positions on the sky by up to 40″ for periods up to 30 sec of time. The effect is stronger in the afternoon, and weaker in winter on cold sites. Anomalous refraction is caused by variations in the "wet" component of the refractive index, which change the electrical path length by ~0.5 mm on baselines of 30–100 m. The associated variations in the water vapour content are < 0.1 mm, so changes in atmospheric opacity or sky brightness are hardly noticeable. In mm and sub-mm interferometry, anomalous refraction will increase phase noise, broadening the synthesized beam, and will reduce fringe amplitude, as sources move out of the narrow primary beams. The electrical length changes seen with single dishes are consistent with the phase noise measurements made with interferometers (Armstrong and Sramek, 1982, Bieging et al. 1984, Kasuga et al. 1986), namely, $\Delta l \simeq 10^{-2} B^{0.8}$, where Δl = variation in electrical length, in mm, and B = baseline, in meters. However, the single-dish data suggest that one is seeing individual packets of moist air, rather than integrating over many cells at varying distances from the telescope.

Here we are concerned with variations, ΔN_W, in the wet part of the refractivity, arising from changes, Δe, in the partial pressure of water vapour. From eq.(32), we have $\Delta N_W \approx 5\,\Delta e$ [mbar]. At a temperature $T_{\rm atm}$ of 280 K, a relative humidity of 50 per cent, the water vapour partial pressure is e = 4.5 mbar. A 20 per cent fluctuation in relative humidity thus corresponds to $\Delta e \approx 1$ mbar, and $\Delta N_W \approx 5$. Over a thickness ΔL of 100 m, the typical variation in electrical pathlength will be $\Delta \ell \approx \Delta N_{wet} \cdot 10^{-6} \Delta L$ ≈ 0.5 mm.

As in the previous section on surface irregularities, the variation in electrical length across a single dish or an interferometer corresponds to a change in phase of the wavefront, with $\Delta\phi = 2\pi(\Delta\ell/\lambda)$, where one cycle, λ/D, of the Fourier component corresponding to the diameter, D, of the antenna (or separation of antennas, for an interferometer) is a phase change of 2π. Hence, the fluctuation in electrical path, $\Delta\ell$, corresponds to an apparent position shift, $\Delta\theta$, in the object being observed by $\Delta\theta = (\Delta\phi/2\pi)(\lambda/D) = \Delta\ell/D$. For D = 30 m, the expected position shifts are typically $\Delta\theta \sim 3'' - 5''$, as observed.

Single-dish imaging in the presence of atmospheric noise can be improved by beam switching.

We have seen in the previous discussion that the clear-sky anomalous refraction produces image motion, but the corresponding variation in the precipitable water vapour content is <0.1 mm, so there is a negligible effect on atmospheric opacity or sky brightness. With even stronger short-term variations in water vapour content, the sky brightness variations may exceed the system noise, strongly perturbing single-dish mapping programs. In this case, observations in the presence of atmospheric noise can be improved by using beam switching, with the beams having as small a separation as possible. For example, at a wavelength λ = 2 cm, with a 100-m telescope, the half-power beamwidth is $\lambda/D \approx 1'$. One might then consider beam switching over an angle of $8'$, which would correspond to a separation of $\sim$ 7 m in packets of water vapour located at a distance of 3 km from the telescope.

An important point is that the beams need not be well separated with respect to source structure. In the multi-beam data reduction technique developed by Emerson et al. (1979), a restoring function is defined, as follows: Let S be the source distribuion on the sky, and B the beam pattern, so that T = S * B is the true map, which would be obtained in the absence of atmospheric disturbances (* denotes convolution). If we now let C be the chopping, or beam switching, function, then the raw data is the observed map M, given by

$$\mathrm{M = S * B * C = T * C} \tag{33}$$

and

$$\hat{\mathrm{M}} = \hat{\mathrm{T}} \cdot \hat{\mathrm{C}} \tag{34}$$

where the ^ symbol denotes Fourier transform. Solving for the true map,

$$\hat{\mathrm{T}} = \hat{\mathrm{M}}/\hat{\mathrm{C}} \equiv \hat{\mathrm{R}} \cdot \hat{\mathrm{M}} \tag{35}$$

and hence,

$$\mathrm{T} = \mathrm{R} * \mathrm{M} \tag{36}$$

So the true map can be obtained by convolving the observed map with a restoring function, R, whose Fourier transform is given by $1/\hat{\mathrm{C}}$. By the sampling theorem, for mapping a region of size x, one need only sample the transform of the switching function $\hat{\mathrm{C}}$ at intervals $1/x$, so in practice, the restoring function is a comb with only a few positive- and negative-going elements. For field sizes up to four times the beam separation, observations with this beam-switching method should yield a better signal-to-noise ratio than Dicke switching against a load, even in the absence of atmospheric perturbations. The solution given by this method is unique, unlike the solutions given by algorithms like CLEAN. It applies to observing in which one can switch between beams faster than the atmospheric variations, and where the atmosphere, on short time scales, appears basically the same at the different switch positions. As such this method may be particularly useful for observing in the millimeter, sub-millimeter and near-infrared ranges.

Fig. 2. Antennas of the IRAM interferometer on Plateau de Bure. Photo A. Rambaud.

Radio Interferometers

Radio interferometer arrays are used to achieve high angular resolution. As interferometers sample the Fourier components of the brightness distribution of radio sources, it is only necessary to sample well enough to synthesise an aperture much larger than any physical antenna structure than could be built in practice, and to reconstruct the image of the radio source as seen by the synthetic aperture. As interferometers not only have higher resolution than single dishes, but also perform a filtering of the spatial frequencies, they can eliminate confusing extended radiation from their maps and provide better position measurements for compact sources than can single dishes. With the application of atomic frequency standards and video recorders, interferometry became able to span continents and oceans, and Very Long Baseline Interferometry now provides the highest possible angular resolution available in all of astronomy.

Interferometers measure the Fourier components of the source brightness.

The relation of the response, R, of a radio interferometer to the brightness distribution $b(x, y)$ of a radio source in the sky may be derived from the Van Cittert–Zernicke theorem (see Clark, 1986, for a concise exposition). The result may be expressed as a two-dimensional Fourier transform, in rectangular coordinates, as:

$$R(S_x, S_y) = \int\int b(x, y) e^{i2\pi(ux+vy)} dx\, dy \tag{37}$$

where

$$b(x, y) = \int\int R(S_x, S_y) e^{-i2\pi(ux+vy)} du\, dv \tag{38}$$

The function $b(x, y)$ is the source brightness distribution, a real, non-negative function, and u, v and x, y are baseline and sky coordinates respectively.

In actual radio astronomy practice, the voltage outputs, V_1 and V_2 of a pair of antennas, are multiplied to obtain the cross-correlated power, as a function of time, τ,

$$R(\tau) = \frac{1}{T} \int_0^T V_1(t) V_2^*(t - \tau) dt \tag{39}$$

As a one-dimensional example,

$$\begin{aligned} R(\tau) &= A \exp[i \frac{2\pi D}{\lambda} \cos \theta(\tau)] \\ &\equiv A\, e^{i\Phi(\tau)} \end{aligned}$$

where V_1, V_2 = outputs from antennas 1, 2, A = fringe amplitude, $\Phi(\tau)$ = fringe phase $\Phi(\tau) = \frac{2\pi D}{\lambda} \cos \theta(\tau)$, D = baseline, θ = angle to the source measured from the baseline direction. The response R has maxima when $\Phi(\tau) = n2\pi$. The spacing between maxima is the fringe angular spacing, or the resolution of the interferometer. This spacing is

$$\begin{aligned} \frac{d\Phi}{d\theta} &= \frac{d}{d\theta}\left(\frac{2\pi D \cos \theta}{\lambda}\right) \\ &= \frac{-2\pi D_p}{\lambda} \end{aligned} \tag{40}$$

where $D_p = D \sin \theta$ is the projected baseline. The separation of fringes is a phase change of $\Delta\Phi = 2\pi$ radians, or, an angular change of

$$\Delta\theta = \frac{\lambda}{D_p}$$

In terms of eqs. (37) and (38), we can say that an interferometer operating at wavelength λ and projected baseline D_p measures the flux in the Fourier components of the source brightness, at spatial frequency D_p/λ.
If the response R is normalized, we have

$$\mathsf{V}(u,v) = \frac{\int\int b(x,y)e^{i2\pi(ux+vy)}\,dx\,dy}{\int b(x,y)dx\,dy} \tag{41}$$

where V is called the *fringe visibility*, with $0 \leq |\mathsf{V}| \leq 1$.

Fringe stopping "tilts" the array antennas to a plane perpendicular to the source.

The normal radio astronomy practice is to lead the signals received at any pair of antennas in an array to two separate correlators. The branch to one of the correlators contains a delay of $\lambda/4$ relative to the branch to the other correlator. This $\pi/2$ phase shift means that one branch corresponds to the sine component and the other branch to the cosine component of the interferometer response. In both the sine and cosine branches, time delays are inserted electronically to compensate for the changes in the excess path to one of the antennas of an interferometer pair, due to the earth's rotation. The effect of this path compensation is equivalent to placing the individual antennas of an array on a giant, imaginary paraboloid, pointing at the position of the phase reference center. One may then think of the individual antennas as "panels" on this imaginary paraboloid. Alternatively, one may think of the delay lines as placing the array antennas on a plane perpendicular to the reference center in the sky, with equal-length cables thereafter to the correlators. In terms of eq. (39), where

$$R(\tau) = Ae^{i\Phi}, \quad \text{and} \tag{42}$$
$$\Phi = \frac{2\pi D}{\lambda} \cos\theta$$

the delay lines artifically set $\Phi = 0°$, $\theta = \theta' - \theta_0 = 90°$, $\cos\theta = 0$, and tilt the imaginary giant "dish" to the position θ_0, where θ_0 is the position of the phase reference center on the sky. The modified fringe phase will then be zero for a source component at position θ_0.

Hence, in radio astronomy usage, the "phase" usually refers to the residual fringe phase after the fringes have been stopped, and zero phase corresponds to the position of the phase reference center. If a source component is located at a small angle α from the reference position, it will have a residual phase, ϕ, given by

$$\begin{aligned} \phi &= \frac{2\pi D}{\lambda}\cos(\theta_0+\alpha) = \frac{2\pi D}{\lambda}\sin\alpha \\ &\cong \frac{2\pi D}{\lambda}\alpha \end{aligned} \tag{43}$$

The modulus of the response, $|R| = A = (R_{\cos}^2 + R_{\sin}^2)^{1/2}$, is the fringe amplitude, and $\phi = \tan^{-1}$ (Imag R/Real R) $\equiv \tan^{-1}(R_{\sin}/R_{\cos})$ is the residual "fringe phase" corresponding to position of the relevant source component.

With this idea of fringe amplitude and phase, the source brightness distribution, from eq. (38) can be re-written as

$$b(\alpha, \delta) = \int A_{(D)} e^{i\phi} \exp[i2\pi \frac{\boldsymbol{D} \cdot \boldsymbol{S}}{\lambda}] d(\frac{\boldsymbol{D}}{\lambda}) \tag{44}$$

where $\boldsymbol{D}/\lambda$ is the baseline vector, in units of wavelengths, and $\boldsymbol{S}$ is a unit vector in the direction of the source. For example (in one dimension), if $\phi = 0$, and if $A = 1$ Jy for all Fourier components (that is, at all baselines), then $\int e^{-i2\pi u\theta}\, du = \delta(\theta)$, that is, a 1–Jy point source located at the position of the phase reference center.

Delay tracking must be done to a fraction of the coherence length.

For an interferometer with a bandwidth $\Delta\nu$, the output, averaged over the frequency band, will be

$$R = \int_{\nu_1}^{\nu_2} V_1 V_2 \cos \frac{2\pi\nu\,\ell}{c}\, d\nu \tag{45}$$

where ℓ is the path difference. The output R will go to zero when $c/\Delta\nu = \ell$, for a square filter response. This value of $\ell = c/\Delta\nu \equiv \ell_c$ is called the *coherence length.* One may also think of the situation as follows. Radiation of bandwidth $\Delta\nu$ from a point source in the sky will have a coherence time $\Delta t = 1/\Delta\nu$. In this time, the wave will travel a distance $\ell_c = c\,\Delta t$. This is the length of the wave train, or the coherence length, and an interferometer pair can detect fringes as long as the cross correlation performed on the radiation arriving at the two antennas is done within the coherence length.

By the same token, an interferometer of baseline D will be sensitive to signals arriving from an angular range in the sky, provided the path difference corresponding to the angular displacement is less than the coherence length. This angular range is ℓ_c/D, and is often referred to as the "delay beam". Provided the delay beam is larger than the primary beams of the individual array antennas of size d, then sources within the field of view, λ/d, of these antennas can be detected by the interferometer. Hence in planning observations, one usually tries to have synthesized beam < field of view < delay beam, or

$$\frac{\lambda}{D} < \frac{\lambda}{d} < \frac{\ell_c}{D}$$

Table 5 gives some coherence lengths of the wave trains for bandwidths which might be considered for use in visible, infrared and radio interferometry. The last column of Table 5 gives the steps of delay tracking required for high accuracy work, as in astrometry, or accurate stellar diameter measurements, namely, $\sim \ell_c/16$. For lower-precision work, the $\ell_c/16$ criterion can be relaxed, and fringes can still be detected as long as the delay tracking is accurate to within the coherence length, ℓ_c.

Table 5. Typical coherence lengths for visible, infrared and radio interferometry.

Wavelength, λ or frequency, ν	Bandwidth $\Delta\lambda$ or $\Delta\nu$	Resolution $\frac{\lambda}{\Delta\lambda} = \frac{\nu}{\Delta\nu} = \frac{\text{Delay beam}}{\text{synthesized beam}}$	Coherence length, ℓ_c $\frac{\lambda^2}{\Delta\lambda} = \frac{c}{\Delta\nu}$	Delay tracking (resolution needed for 1% precision in amplitude measurements) $\frac{\ell_c}{16}$	$\Delta t = \frac{1}{16\Delta\nu}$
Visible					
5000 Å	5 Å	1000	500 μm	30 μm	100 fsec
	500 Å	10	5 μm	0.3 μm	1 fsec
Infrared					
5 μm	0.005 μm	1000	5 mm	300 μm	1 psec
	0.5 μm	10	50 μm	3 μm	10 fsec
10 μm	0.01 μm	1000	1 cm	600 μm	2 psec
	1 μm	10	100 μm	6 μm	20 fsec
20 μm	0.02 μm	1000	2 cm	1.2 mm	4 psec
	2 μm	10	200 μm	12 μm	40 fsec
Millimeter					
345 GHz	50 MHz	6900	6 m	37 cm	1 nsec
	1 MHz	345000	300 m	19 m	60 nsec
230 GHz	50 MHz	4600	6 m	37 cm	1 nsec
	1 MHz	230000	300 m	19 m	60 nsec
115 GHz	50 MHz	2300	6 m	37 cm	1 nsec
	1 MHz	115000	300 m	19 m	60 nsec
	100 kHz	$1.2\ 10^6$	3 km	190 m	0.6 μsec
Centimeter					
22 GHz	16 MHz	1400	18 m	1 m	4 nsec
	2 MHz	11000	150 m	9 m	30 nsec
	10 kHz	$2.2\ 10^6$	30 km	2 km	6 μsec
10.6 GHz	16 MHz	660	18 m	1 m	4 nsec
	2 MHz	5300	150 m	9 m	30 nsec
5 GHz	16 MHz	310	18 m	1 m	4 nsec
	2 MHz	2500	150 m	9 m	30 nsec
1.7 GHz	2 MHz	850	150 m	9 m	30 nsec
	10 kHz	$1.7\ 10^5$	30 km	2 km	6 μsec

Aperture synthesis

Consider a giant single dish with panels at positions 1, 2, ... n. The voltages excited across the aperture of this dish are V_1, V_2, ...,V_n. The output detector power is $(\Sigma V_i)^2 = \Sigma V^2 + \Sigma V_n V_m \, cos(\phi_n - \phi_m)$. Only the second term contains the high resolution information, through the cross products, which describe interference. Hence to *synthesize* a large aperture, we need only measure $V_n \, V_m cos(\phi_n - \phi_m)$ at all points within the aperture and sum. For an array with a large number of antennas distributed over a aperture of size D, one may synthesise an image on short time scales ("snapshot" mode). For an array with few antennas, one may sample the interferometer output at a number of positions within an aperture of size D, by displacing the antennas to different stations along baselines of maximum length D, and by using the earth's rotation to change the orientation of the baselines, thereby sweeping out rings or tracks in the aperture to be sampled. A classic example of this technique, called *earth rotation synthesis*, or *supersynthesis* was the map of Cas A made with the Cambridge One-Mile Telescope, with only three antennas, for which only one antenna was moved to different stations, and data were collected on only two baselines simultaneously. In some of the maps of Cas A, (e.g., that at 5 GHz by Rosenberg, 1970), the resulting image had about 10000 pixels and a resolution of 6 arc sec. Subsequent maps made at the VLA had many more pixels and higher resolution, but this example from the earlier days of aperture synthesis serves to remind us that extensive Fourier sampling, and hence, complex images, can also be made with interferometers with only a few elements.

Derivation of Sensitivity for an Aperture Synthesis Telescope.

The sensitivity of a radio synthesis array limited by system noise may be calculated as follows. The r.m.s. fluctuations in antenna temperature, ΔT_a, are given by

$$\Delta T_a = \frac{f \, T_{sys}}{\sqrt{t \Delta \nu}} \tag{46}$$

where T_{sys} = system temperature, t = integration time, $\Delta\nu$ = bandwidth, and f is a noise factor due to analog to digital conversion in the correlators. Table 6 gives examples of this factor for some radio interferometers.

The r.m.s. fluctuations in flux density, seen from a pair of antennas, will therefore be

$$\Delta S = \frac{2k \; \Delta T_a \; e^{\tau_\nu}}{A_e \; \sqrt{2}} \tag{47}$$

where τ_ν is the atmospheric opacity and $A_e = \epsilon_{ap} \, \pi D^2/4$ is the effective collecting area of a single dish of diameter D and aperture efficiency, ϵ_{ap}. Hence for an array of n identical dishes, with $N = n(n-1)/2$ baselines being observed simultaneously, the r.m.s. variation in flux density will be

$$\Delta S = \frac{2k \; f \; T'_{sys}}{A_e \; \sqrt{2 \, N \, t \, \Delta \nu}} \tag{48}$$

where $T'_{sys} = T_{sys} e^{\tau_\nu}$. If the array synthesizes a beam of solid angle Ω_b, then the (synthesized) main-beam brightness temperature, T_b, of a point source will be defined by

Table 6. Noise factors for digital correlators.

Correlator bits	Levels	Degradation	Noise factor f	Example
1	2	64 %	1.57	Cambridge 1/2-mile (line)
2	3	81 %	1.23	VLA line and continuum IRAM and Berkeley line systems.
2	4	88 %	1.12	VLBA
3	8	94%	1.07	Nobeyama spectral correlator
4	16	98 %	1.02	IRAM array, continuum

$$S \equiv \frac{2k}{\lambda^2}\, T_b\, \Omega_b \tag{49}$$

Hence, the r.m.s. variations in brightness temperature on aperture synthesis maps will be

$$\Delta T_b = \frac{\lambda^2\, f\, T'_{\rm sys}}{A_e\, \Omega_b\, \sqrt{2\, N\, t\, \Delta\nu}} \tag{50}$$

If the data from the array are weighted with a taper to –6 dB at the longest baseline length $L_{\rm max}$, then the full width to half-power of the synthesized beam will be

$$\theta_b = 0.7\, \frac{\lambda}{L_{\rm max}} \tag{51}$$

and if the synthesized beam is a circular gaussian,

$$\Omega_{mb} = 1.133\, (0.7\, \lambda / L_{\rm max})^2$$

For example, for a 2–bit, 3–level correlator, $f = 1.23$, so

$$\Delta T_b = \frac{2.0\, T'_{\rm sys}\, L^2_{\rm max}}{\epsilon_{ap}\, D^2\, \sqrt{N\, t\, \Delta\nu}} \tag{52}$$

Hence for the same integration time, dish size, and bandwidth, the noise, in brightness temperature, increases with the square of the maximum baseline of the array. If the data are also smoothed to keep the velocity resolution constant (e.g. in order to compare maps of spectral lines at different wavelengths), we then have

$$\Delta\nu\,[\mathrm{Hz}] = 10^6\, \frac{\Delta \mathrm{V}\ [\mathrm{km/s}]}{\lambda[\mathrm{mm}]} \tag{53}$$

so

$$\Delta T_b = \frac{42\,\lambda^{2.5}\,T'_{\rm sys}}{\epsilon_{ap}\,D^2\,\theta_b^2\,\sqrt{N\,t\,\Delta V}} \tag{54}$$

where ΔT_b, and $T'_{\rm sys}$ are in K, D in m, θ_b in arc sec, t in sec, λ in mm, and ΔV in km/s. Therefore, if maps at different frequencies are made with the same *angular* resolution, θ_b, and the same *velocity* resolution, ΔV, then the noise, in brightness temperature, improves by $\lambda^{2.5}$ as one goes to shorter wavelengths.

Other factors limiting sensitivity.

Often, however, the sensitivity will be limited not by system noise, but by the consequences of incomplete sampling in the Fourier transform plane, or u, v plane. Obvious contributers may be sidelobes (from the type of grading function used for the Fourier components), grating rings (from the Fourier response to a regular spacing of the antenna tracks), or aliasing (e.g., from interpolation of data onto a regular, rectangular grid, to allow use of a Fast Fourier Transform).

One procedure to overcome defects introduced by undersampling the (u, v) plane is the CLEAN algorithm (Högbom 1974), which has the following steps:

1. Start with the raw data ≡ "dirty" map (in α, δ plane).
2. Go to highest peak on map, subtract from it the "dirty" beam (the uncorrected response of the array to a point source). Remember the intensity and the position.
3. Go to the next highest peak on the new map, and subtract the dirty beam from this peak too.
4. Iterate, as desired.
5. When the noise level is reached, add back gaussians at the positions derived in the previous steps.

The result is a "cleaned" map, consistent with the observed data. (But not necessarily the true distribution on the sky).

The other popular algorithm for restoring data is the *maximum entropy method*, (MEM), in which the image brightness $b(x, y)$ is subdivided into M pixels, having brightness b_n. One then maximizes either the quantity $Q = -\Sigma \log b_n$, or $Q = -\Sigma b_n \log b_n$, where the sum is taken over the M pixels. Loosely speaking, CLEAN gives its best results when the image consists of compact sources, while the MEM methods are superior when the image has extended components.

Very Long Baseline Interferometry (VLBI).

The quest for ever higher angular resolution led radio astronomers to develop VLBI. With connected element interferometers, such as, for example, the Cambridge 5-km telescope, operating at a wavelength, λ, of 6 cm, and a maximum baseline, D, of 5 km, the resolution, λ/D, is ~2″. As many cosmic radio sources were still unresolved at arc second resolution, it became important to extend baselines as far as possible. In VLBI, the baselines are $\sim 10^3 - 10^4$ km, too large for cable or waveguide connections among the antennas, so at each individual station one must have:

1. **Amplification**, with preservation of phase;
2. **Mixing** with a very stable oscillator, to video frequencies;
3. **Recording** on video cassettes.

Days, weeks, or months later, the video recordings can be brought to a processing center and correlated. In order to preserve fringe phase, random jumps $\delta\nu$ in the frequencies of the local oscillators must be $\leq 1/2\pi$ cycles, or equivalently, the phase jumps must be $\leq$ 1 radian, during the integration time, $t_{\rm int}$. This means:

$$\delta\nu \; t_{\rm int} < \frac{1}{2\pi}$$

and hence the relative frequency stability of the local oscillators must be

$$\frac{\delta\nu}{\nu_o} < \frac{1}{2\pi\nu_o \, t_{\rm int}}$$

For example, for $t_{\rm int} \sim 10^2$ sec and $\nu_o = 10^{10}$ Hz, the requirement on frequency stability is:

$$\frac{\delta\nu}{\nu_o} < 2 \; 10^{-13}$$

Such a stability can be achieved with a hydrogen maser frequency standard, which has $\delta\nu/\nu_o \sim 10^{-14}$ over 100 sec, and drifts of < 1 μsec/year. An Allan-variance analysis of the stability of the hydrogen maser (Rogers and Moran 1981) shows that the upper limit in frequency to which this standard can be used for VLBI is about 1000 GHz.

In order to process the VLBI data on the astronomical sources, one must derive the coordinates of the baselines among the VLBI antennas. This baseline information is highly accurate, and leads to important geodetic applications of VLBI, as well as the astronomical ones. For example, if the VLBI baselines can be derived to an accuracy of ~30 cm, then observers can measure:

- time (UT1) to ±0.0002 sec,
- variations in the rate of the earth's rotation,
- polar motion to a precision of 30 cm.

As such, VLBI holds great promise in applications to the measurement of earth tides and continental drift.

The phase variations in the VLBI data are mainly due to the variations in the "wet" component of the refractivity, whereas in optical interferometry, phase variations are due to the "dry" component. Table 7 lists some typical phase errors in various types of interferometry, over typical baselines. There is, of course, a large variation depending on weather conditions.

Interferometric Imaging: the Closure Phase is a property of the source.

As is evident from Table 7, in both VLBI and optical interferometry, the wavefront phase is lost. However, if it is somehow possible to recover the Fourier phases of the source brightness distribution from the observations, then one can reconstruct the image of the source. Coarsely expressed, one may say that if fringes can be detected on a sufficiently large number of baselines, then the Fourier phases of the source can be recovered, and one can reconstruct images.

To do so, one makes use of the *closure phase*, Φ, which is the sum of the observed phases, ϕ, around any closed triangle of interferometer baselines (Jennison 1958):

$$\Phi_{123} = \phi_{23} + \phi_{12} + \phi_{31}$$

Table 7. Typical phase errors in various types of interferometry.

Wavelength range and baseline	r.m.s. phase errors (degrees)	r.m.s. path errors (wavelengths)	Time scale (sec)	Remark
Optical (100–m)	10^4	10^2	0.01	phase lost completely
mm VLBI (1000 km)	10^3	10	100	phase lost
mm Interferometer (300 m)	10^1–10^2	0.1–1	100	phase perturbed
cm Interferometer (1–10 km)	10	0.03	1000	phase retained

The observed phases, ϕ_{ij}, on the different baselines contain the phases of the source Fourier components, ψ_{ij}, and also error terms, ϕ_i, ϕ_j, introduced by errors at the individual antennas and by the atmospheric variations at each antenna, as illustrated in Table 8.

Table 8. Phase Closure

Antenna pairs	Observed phase	=	Source phase	+	Antenna or atmospheric errors
2, 3	ϕ_{23}	=	ψ_{23}	+	$\phi_2 - \phi_3$
1, 2	ϕ_{12}	=	ψ_{12}	+	$\phi_1 - \phi_2$
3, 1	ϕ_{31}	=	ψ_{31}	+	$\phi_3 - \phi_1$

$$\text{Sum} = \Phi_{123} = \phi_{23} + \phi_{12} + \phi_{31} = \psi_{23} + \psi_{12} + \psi_{31}$$

Hence, the closure phase, Φ, or the sum of observed phases around the triangle of baselines, is the sum of phases of the source Fourier components only. It is thus a property of the source, and the phase errors due to the atmosphere and the antennas cancel out.

For n antennas, there may be $n(n-1)/2$ independent baselines, but only $n-1$ unknown phase errors. This means that most of the phase information can be recovered when many telescopes take part in an observation. The information contained in closure phases may be thought of as the fraction of closure phases relative to the number of antennas (the number of unknown phase errors), or, $(n-2)/n$, where n is the number of antennas in the array. Table 9 gives some examples of this fraction for a arrays with 3, 4 and 27 antennas. The generalization of the phase closure method, to bi-spectrum analysis (Weigelt, this volume), allows us to regard the use of a 50 x 50 array camera on an optical telescope as an array of 2500 "antennas". Table 9 shows that arrays with a large number of antennas have nearly all the information needed. In fact, in such cases, it may not even be necessary to calibrate all the baselines, as calibration of a single baseline may suffice to determine all the rest.

Table 9. Information in Closure Phases.

Number of Antennas		Fraction of Phases
3		33 %
4		50 %
27	(VLA)	92 %
2500	($\sim$ 50 x 50 array, 1.5 m telescope)	100 %

Given the possiblity to recover the phase in radio and optical interferometry, we should distinguish between the image processing procedures such as CLEAN or Maximum entropy, which compensate for gaps in the Fourier sampling, and procedures which use techniques of phase recovery. Table 10 lists some of the current procedures in these two categories.

Table 10. Image processing procedures in radio interferometry.

Pure "Restoration" (Deconvolution) Methods:		
(1) — CLEAN	Compensate for poor sampling	
(2) — Maximum Entropy Method(s) (MEM)	of the aperture, or u, v plane	
Phase Recovery Methods:		
(3) — Phase closure		
(4) — Hybrid mapping	= (3) + (1)	VLBI
(5) — Self-calibration	= (4) + solve for antenna gains	VLA
(6) — Global fringe fitting	= (3) (τ, $d\phi/dt$) + (5)	VLBI

Hybrid mapping was one of the early methods of applying phase closure to VLBI data (Readhead et al. 1980), consisting of applying phase closure and interatively comparing with a source model, possibly aided by CLEAN, until the image converges.

Self-calibration (VLA).

One of the most often used phase-recovery and image-formation procedures, especially with VLA data, is *self calibration* (Schwab 1980, Cornwell and Wilkinson 1981; see also the review by Pearson and Readhead, 1984).

As we have seen, an interferometer with delay lines may be regarded as a giant paraboloid with many "panels". Each "panel" or element in the array, may be considered as contributing a "surface irregularity", or antenna gain factor, g_i, where $g_i = a_i e^{i\phi_i}$. The amplitude of this gain factor may be, for example, $a_i = e^{-\tau_\nu}$, the atmospheric attenuation at the i-th antenna, and the phase of this gain factor, $\phi_i = \Delta\phi_i$, the phase error caused by the atmosphere or by the instrumentation at the i-th antenna.

The self-calibration routine involves a doubly-nested iterative loop, where an initial model, e.g., a point source, is Fourier transformed to the visibility plane. The observed visibilities are then used by the self-calibration procedure to update the model visibilities, and the result is then Fourier transformed back to the image plane. In the image plane, the CLEAN (or MEM) algorithm is used to deconvolve the image and to derive a new "model" image. By successive iterations, one may solve for all of the antenna gains, using the phase closure information.

Global fringe fitting is a procedure devised by Schwab and Cotton (1983) which applies the closure relations to the time delays, τ, and fringe rates, $d\phi/dt$, in VLBI data. The search for these quantities is done on all baselines simultaneously. A source model is used to predict delays and fringe rates, and the visibilities so obtained are used to make an image with self-calibration. As such, global fringe fitting is a triply-nested iterative loop: the fitting procedure, the self-calibration, and the restoring routine (CLEAN or MEM) within the self-calibration.

Strategies for Fourier Coverage

In the radio domain, where one usually operates in the "single speckle" mode, the locations of array antennas have often been guided by the following considerations:

1.) Are the phenomena occurring on short time scales, like solar bursts? If so, then good "snapshot" coverage is required, and many baselines must be available simultaneously, to allow instantaneous imaging.

2.) Are there many different scale sizes of sources to be observed, or one dominant one? If there is only one, as in solar research, then a ring-shaped array, such as the Culgoora radio heliograph, may be appropriate. If there are many different scales, then the configuration should be flexible, allowing expansion and contraction of the array, like the VLA.

3.) Cost: If the configuration is to be flexible, can the array be designed to minimize the number of antenna stations, which are often expensive? Can the same stations be used in different configurations?

4.) Is the priority on imaging quality or sensitivity? If the size of the instrument is smaller than the Fried parameter, and if sensitivity is critical, then it may be advantageous to maximize the *size* of the individual array antennas. If high-quality images are desired, and if the angular sizes of the objects to be studied are comparable with or larger than the likely field of view of the array antennas, then it is more advantageous to maximize the *number* of array elements.

5.) Topography: What is possible? Should one choose a site which is adequate for dense, two-dimensional coverage of baselines in an imaging array, or a site which allows the longest possible baseline?

6.) Redundancy: Earlier treatments on array redundancy (Arsac 1955, Bracewell 1966, Moffet 1968) sought to obtain maximum resolution with the minimum number of redundant baselines. However, for arrays with a small number of antennas, it may be more advantageous to choose configurations which allow phases to be derived for all baselines, by use of the closure relations. In a new type of minimum-redundancy array consistent with phase closure (Morita and Ishiguro 1985), the baselines are redundant by a factor of 1.4.

7.) Holes in the u, v coverage: For imaging purposes, the main strategy is to have a reasonably "even" coverage of the u, v plane. As the image is the Fourier transform of the visibility samples in the u, v plane, an infinity of possible visibility functions can be "hidden" in the unsampled regions of the u, v plane. The larger the "holes" in the sampling, the greater the variety of possible images which can also be consistent with the observed data. One strategy to select the location of antennas to minimize the size of the "holes" in the Fourier coverage is described by Cornwell (1986, 1989). This method includes a "simulated annealing" algorithm (Kirkpatrick et al., 1983), which uses a thermodynamic analogy with a quantity corresponding to "temperature" and a function corresponding to "energy". The idea is to slowly lower the "temperature", so that the system has time to settle into its minimum "energy" configuration. The resulting antenna locations are regularly spaced, with symmetric patterns in the u, v plane, leading Cornwell to call them "crystalline arrays".

Table 11. Strategies for Fourier coverage when r_o < aperture size (after Roddier, 1987).

Array type	Condition	Strategy
Phased array: (fringes over identical baselines add in phase)	Bright reference source in the speckle isoplanatic patch	Configure to get best synthesized beamshape.
Coherent array: (bright sources) (fringes have random phases)	partially redundant arrays; integration time $\propto$ number of baselines	Phase closure; output pupil non-redundant
Coherent array: (faint sources)	Input and output pupils must be fully redundant	Phase is best recovered by triple correlation.

Fourier Coverage Strategies for the Speckle Domain: Roddier (1987) has elaborated the best strategies when the Fried seeing parameter, r_o, is smaller than the size of the telescopes, as in the visible or infrared. Table 11 gives a summary of his ideas. Roddier distinguishes between phased arrays and coherent arrays. In *phased arrays*, fringes over identical baselines add in phase, yielding images with a "beam" given by the diffraction pattern of the aperture. Radio arrays are usually phased arrays. In optical (visible and infrared) interferometry, arrays can be phased if there is a bright reference source in the speckle isoplanatic patch. In *coherent arrays*, interference fringes are observed but their phases are random, yielding a speckle pattern. For these interferometers, it may be of interest to consider moving the telescopes for some of the path compensation rather than relying entirely on delay lines. The Fourier coverage resulting from this type of strategy is explored by Vivekanand et al. (1988, 1989).

Acknowledgements. I thank J. Cernicharo, A. Greve, S. Guilloteau, D. Morris, S. Radford, C. Thum, and M. Vivekanand for helpful comments.

Bibliography

Bracewell, R.N., 1965, *The Fourier Transform and Its Applications*, McGraw-Hill, New York (2nd edition, 1978).
Christiansen, W.N., Högbom, J.A., 1985, *Radiotelescopes*, Cambridge University Press, Cambridge, 2nd edition.
Kraus, J.D., 1986, *Radio Astronomy*, 2nd edition, Cygnus-Quasar Books, Powell, Ohio.
Rohlfs, K., 1986,*Tools of Radio Astronomy*, Springer, Heidelberg.
Thompson, A.R., Moran, J.M., Swenson, G.W., 1986, *Interferometry and Synthesis in Radio Astronomy*, Wiley, New York.

References

Altenhoff, W.J., Baars, J.W.M., Downes, D., Wink, J.E., 1987, Astron. Astrophys. **184**, 381.
Armstrong, J.W., Sramek, R.A., 1982, Radio Science, **17**, 1579.
Arsac, J., 1955, Comptes Rendus Acad. Sci., **240**, 942.
Bates, R.H.T., 1959, IRE Trans. Ant. Prop., **AP–7**, 369.
Bieging, J.H., Morgan, J., Welch, W.J., Vogel, S.N., Wright, M.C.H., 1984, Radio Science, **19**, 1505.
Bracewell, R.N., 1966, in *Progress in Scientific Radio*, Pub. 1468, Nat. Acad. Sci., Washington DC, p. 243.
Cernicharo, J., 1988, Thèse d'Etat, Université de Paris VII.
Clark, B.G., 1986, in *Synthesis Imaging*, ed. R.A. Perley, F.R. Schwab, A.H. Bridle, Nat. Radio Astron. Obs., Green Bank, p. 1.
Cornwell, T.J., 1986, MMA Memo No. 38, NRAO; 1989, in preparation.
Cornwell, T.J., Wilkinson, P.N., 1981, Mon. Not. Roy. Astron. Soc., **196**, 1067.
Davis, J.H., Vanden Bout, P. 1973, Astrophys. Letters, **15**, 43.
Emerson, D.T., Klein, U., Haslam, C.G.T., 1979, Astron. Astrophys., **76**, 92.
Goldsmith, P.F., 1987, Int. Journ. Infrared & Millimeter Waves, **8**, 771.
Goodman, J.W., 1968, *Introduction to Fourier Optics*, McGraw-Hill, Appendix B, p. 278.
Greve, A., Hooghoudt, B.G., 1981, Astron. Astrophys., **93**, 76.
Högbom, J.A., 1974, Astron. Astrophys. Suppl., **15**, 417.
Jennison, R.C., 1958, Mon. Not. Roy. Astron. Soc., **118**, 276.
Kasuga, T., Ishiguro, M., Kawabe, R., 1986, IEEE Trans. Ant. Prop., **AP-34**, 797.
Kirkpatrick, S., Gelatt, C.D., Vecchi, M.P., 1983, Science, **220**, 671.
Kutner, M.L., Ulich, B.L., 1981, Astrophys. J., **250**, 341.
Minnett, H.C., Thomas, B.M., 1968, Proc. IEE, **115**, 1419; reprinted in *Reflector Antennas*, ed. A.W. Love, 1978, IEEE Press, New York, p. 56.
Moffet, A.T., 1968, IEEE Trans. Ant. Prop., **AP–16**, 172.
Morita, K.I., Ishiguro, M., 1985, in Proc. Int. Symp. Ant. Prop., ISAP–1985, Kyoto.
Pearson, T.J., Readhead, A.C.S., 1984, Ann. Rev. Astron. Astrophys., **22**, 97.
Penzias, A.A., Burrus, C.A., 1973, Ann. Rev. Astron. Astrophys., **11**, 51.
Readhead, A.C.S., Walker, R.C., Pearson, T.J., Cohen, M.H., 1980, Nature, **285**, 137.
Roddier, F., 1987, in *Interferometric Imaging in Astronomy*, ed. J.W. Goad, NOAO, Tucson, p.135.
Roddier, F., 1987, Journ. Opt. Soc. Am. **4**, 1396.
Rogers, A.E.E., Moran, J., 1981, IEEE Trans. Instrum. Meas., **IM–30**, 283.
Rosenberg, I., 1970, Mon. Not. Roy. Astron. Soc., **151**, 109.
Ruze, J., 1952, Suppl. al Nuovo Cimento, **9**, 364.
Ruze, J., 1966, Proc. IEEE, **54**, 663; reprinted in *Reflector Antennas*, ed. A.W. Love, 1978, IEEE Press, New York, p.56.
Scheffler, H., 1962, Zeitschrift für Astrophysik, **55**, 1.
Schwab, F.R., 1980, Soc. Photo-Opt. Inst. Eng., **231**, 18.
Schwab, F.R., Cotton, W.D., 1983, Astron. J., **88**, 688.
Smith, E.K., Weintraub, S., 1953, Proc. IRE, **41**, 1035.
Väisälä, J., 1922, Ann. Univ. Tennical Aboensis, Turku, Ser. A1, No.2.
Vivekanand, M., Morris, D., Downes, D., 1988, Astron. Astrophys., **203**, 195.
Vivekanand, M., Morris, D., Downes, D., 1989, Astron. Astrophys., in press.

Appendix 1. Correspondences with notation elsewhere.

The description of antenna efficiencies in the text follows the practice at the IRAM 30-m telescope, which simplifies the antenna parameters for observers to two quantities, the effective beam efficiency, $B_{\rm eff}$, and the forward efficiency, $F_{\rm eff}$. For readers wishing to relate the usage in this paper to that in more detailed analyses of antenna parameters, the following table gives correspondences among our notation and that of Kutner and Ulich (1981, Ap.J., **250**, 341) and Kraus (1986, *Radio Astronomy*, 2nd ed.).

Quantity	Notation used in text	Notation elsewhere: Kutner, Ulich	Kraus
forward efficiency	$F_{\rm eff}$	η_l	—
effective beam efficiency	$B_{\rm eff}$	$\eta_l \eta_{\rm fss} = \eta_s$	$k_o \dfrac{\Omega_M}{\Omega_A}$
forward scattering & spillover efficiency	$B_{\rm eff}/F_{\rm eff}$	$\eta_{\rm fss}$	—
antenna temperature, outside the atmosphere	$T'_a = B_{\rm eff} T_{mb} = F_{\rm eff} T^*_a$	$\eta_l T^*_a$	$T_A\, e^{\tau_\nu}$
T'_a, corrected for rear spillover & scattering and resistive losses	$T^*_a = (B_{\rm eff}/F_{\rm eff})\, T_{mb}$	$\eta_{\rm fss}\, T^*_R$	—
true source Rayleigh-Jeans brightness temperature	T_b	T_R	$T_b,\ T_s$
beam dilution factor	$(\theta_s/\theta_r)^2$ (for gaussians)	η_c	Ω_s/Ω_M
beam-averaged brightness temp.	$T_{mb} = (F_{\rm eff}/B_{\rm eff})\, T^*_a$ $T_{mb} = (\theta_s/\theta_r)^2\, T_b$ (for gaussians, and $\theta_s <$ error beam)	$T^*_R = \eta_c T_R$	T'_s

In the Kraus notation, k_o = ohmic losses in the antenna = η_r in the K-U notation. In general, $0 \leq k_o \leq 1$, and for a well-designed antenna, $k_o \approx 1$.

Note that in the Kutner and Ulich notation, $\eta_{\rm fss}$ is not a constant of the telescope, but a variable, which must be evaluated as a function of the diameter of the source to be observed. It is incorrect to take an $\eta_{\rm fss}$ evaluated on the moon, for example, to determine a T^*_R scale for observations of giant molecular clouds in galaxies. In general, if the sources are smaller than the error beam, one should use main-beam brightness temperature. In the K-U notation, this means choosing $\eta_{\rm fss} = B_{\rm eff}/F_{\rm eff}$, so that $T^*_R = T_{mb}$.

Appendix 2. Chopper-wheel temperature scales and interrelations (simplified example, $T_{\mathrm{atm}} \approx T_{\mathrm{amb}}$, image band suppressed; otherwise see eq.(25)).

Quantity:	Antenna temperature	Antenna temperature outside atmosphere	Forward-beam brightness temperature	Main-beam brightness temperature
Symbol :	T_a	T'_a	T^*_a	T_{mb}
Temperature of :	Equivalent resistor (Nyquist formula)	Equivalent resistor, outside atmosphere	Equivalent black body covering forward 2π	Equivalent black body in main beam (R-J formula)
Definition :	$T_a \equiv \dfrac{S\,A_e\,\mathrm{e}^{-\tau_\nu}}{2k}$	$T'_a \equiv \dfrac{S\,A_e}{2k} \equiv T_a\,\mathrm{e}^{\tau_\nu}$	$T^*_a \equiv \dfrac{S\,A_e}{2kF_{\mathrm{eff}}} \equiv \dfrac{T'_a}{F_{\mathrm{eff}}}$	$T_{mb} \equiv \dfrac{S\lambda^2}{2k\Omega_b} \equiv \dfrac{T'_a}{B_{\mathrm{eff}}} \equiv \dfrac{F_{\mathrm{eff}}}{B_{\mathrm{eff}}}T^*_a$
Calibration factor :	$T_1 = F_{\mathrm{eff}}T_{\mathrm{amb}}\mathrm{e}^{-\tau_\nu}$ $= T_{\mathrm{amb}} - T_{\mathrm{sky+cabin}}$	$T_2 = F_{\mathrm{eff}}T_{\mathrm{amb}}$ $= T_1\,\mathrm{e}^{\tau_\nu}$	$T_3 = T_{\mathrm{amb}}$ $= T_2/F_{\mathrm{eff}}$	$T_4 = T_{\mathrm{amb}}\dfrac{F_{\mathrm{eff}}}{B_{\mathrm{eff}}}$ $= T_2/B_{\mathrm{eff}}$
Measuring method :	$T_a \equiv \dfrac{\Delta V_{\mathrm{sig}}}{\Delta V_{\mathrm{cal}}}\,T_1$	$T'_a \equiv \dfrac{\Delta V_{\mathrm{sig}}}{\Delta V_{\mathrm{cal}}}\,T_2$	$T^*_a \equiv \dfrac{\Delta V_{\mathrm{sig}}}{\Delta V_{\mathrm{cal}}}\,T_3$	$T_{mb} \equiv \dfrac{\Delta V_{\mathrm{sig}}}{\Delta V_{\mathrm{cal}}}\,T_4$
Remarks :	Scale used for T_R, T_{sys}.	Scale used for T'_{sys}.	convenient for chopper wheel; less useful in other calibration methods.	OBSERVERS ARE STRONGLY URGED TO USE THIS SCALE!
R.m.s. noise :	$\Delta T_a = \dfrac{\sqrt{2}\,T_{\mathrm{sys}}}{\sqrt{\Delta\nu t}}$	$\Delta T'_a = \dfrac{\sqrt{2}\,T'_{\mathrm{sys}}}{\sqrt{\Delta\nu t}}$	$\Delta T^*_a = \dfrac{\sqrt{2}\,T'_{\mathrm{sys}}}{F_{\mathrm{eff}}\sqrt{\Delta\nu t}}$	$\Delta T_{mb} = \dfrac{\sqrt{2}\,T'_{\mathrm{sys}}}{B_{\mathrm{eff}}\sqrt{\Delta\nu t}}$
In reduction programs, to obtain temp. scale in :	T_a	T'_a	T^*_a	T_{mb}
set B_{eff} equal to:	e^{τ_ν}	1.0	F_{eff}	B_{eff} (measured value)

A_e = effective collecting area, λ = wavelength, k = Boltzmann constant, Ω_b = beam solid angle, τ_ν = atmospheric optical depth, $\Delta\nu$ = bandwidth/channel, t = on-source integration time, B_{eff} = beam efficiency, F_{eff} = forward efficiency, T_R = receiver temperature (units of antenna temp.). $T'_{\mathrm{sys}} = T_{\mathrm{sys}}\,\mathrm{e}^{\tau_\nu}$ = equivalent system temp. above atmosphere, (units of T'_a).

Principal Symbols. (Equation or Table numbers indicate location where first used).

A_e	effective collecting area	Table 3
$b(x,y)$	brightness distribution of radio source	eq.(37)
B	beam pattern	eq.(33)
B_{eff}	effective main-beam efficiency	eq.(12)
C, $\hat{\text{C}}$	chopping function and its transform	eqs.(33, 34)
D	antenna diameter	eq.(6)
D_p	projected baseline	eq.(40)
e	partial pressure of water vapour	eq.(32)
F_{eff}	forward efficiency	eq.(16)
$\mathcal{F}\{\bullet\}$	Fourier transform operator	eq.(3)
$g(x,y)$	aperture (current) grading function	eq.(3)
g_i, g_s	relative gain in image and signal bands	eq.(25)
G	gain factor	eq.(19)
k	Boltzmann constant = $1.38\ 10^{-23}$ J K^{-1}	Table 3
l, m	direction cosines relative to x, y axes, coordinates in image plane	eq.(3)
l_c ℓ_c	correlation length coherence length	eq.(28) eq.(45) ff.
$L_{\max}$	length of maximum baseline	eq.(53)
n	number of antennas in an array	eq.(48)
n_e	electron density	eq.(1)
M, $\hat{\text{M}}$	observed map and its transform	eqs. (33, 34)
N	refractivity, number of baselines in an array	eq. (32) eq. (50)
P	pressure	eq. (32)
$P(l,m)$	far-field power pattern	eq.(4)
P_m	power in main lobe	eq.(27)
P_o	main-lobe power, without surface errors	eq.(27)
r	radius in aperture (x,y) plane	Table 2
r_o	Fried parameter	eq.(31)ff
R, $\hat{\text{R}}$	restoring function and its transform	eqs.(35, 36)
R	aperture radius	Table 2
$R(u,v)$	interferometer response	eq.(37)
S	source distribution	eq.(33)
$\boldsymbol{S}$	source direction vector	eq.(44)
S	flux density	Table 3
T, $\hat{\text{T}}$	true map and its transform	eqs.(33, 34)
T_a	antenna temperature	Table 3
T_a'	antenna temperature outside atmosphere	Table 3
T_a^*	forward beam brightness temperature	eq.(24)
T_{amb}	ambient temperature (e.g., of receiver cabin)	eq.(16)
T_{atm}	mean temperature of atmosphere	eq. (19)
T_b	brightness temperature	Table 3
T_{mb}	main-beam brightness temperature	Table 3
T_R	receiver temperature	eq.(15)
T_R^*	beam-averaged brightness temperature	Appendix 1
T_{sys}	system temperature	eq.(15)

$\mathsf{V}(u,v)$	fringe visibility	eq. (41)
$V(l,m)$	voltage power pattern	eq.(3)
V_1, V_2	voltages from antenna 1, 2	eq.(39)
$W(u,v)$	instrumental transfer function	eq.(5)
u,v	coordinates in spatial frequency plane	eq.(5)
x,y	coordinates in aperture plane	eq.(3)
z	zenith angle	Table 3
ΔT	r.m.s. temperature sensitivity	eq.(18)
ΔV_{sig}, ΔV_{cal}	response to signal, calibration	eqs.(21, 22)
Δz	effective r.m.s. surface tolerance	eq.(26)
$\Delta\nu$	bandwidth (per frequency channel)	eq.(18)
$\Delta\phi$	r.m.s. phase error	eq.(26)
ϵ_{ap}	aperture efficiency	eq.(11)
θ	general angle, radius in image (l,m) plane, angle of incidence to aperture (x,y) plane	Table 2 eq.(3) ff.
θ_b	beamwidth (full width to half power)	eq.(6)
$\theta_{-10\mathrm{dB}}$	full width to one-tenth power	eq.(7b)
θ_e	width of error beam (FWHP)	eq.(31)
θ_r	response width (beam * source) (FWHP)	eq.(8)
θ_s	source width (FWHP)	eq.(8)
ν	frequency	eq.(1)
τ	integration time	eq.(18)
τ_ν	optical depth (opacity)	eq.(2)
τ_s, τ_i	atmospheric opacity in signal & image bands	eq.(25)
ϕ	phase angle, azimuth from y-axis in aperture (x,y) plane	 eq.(3) ff.
Ω_b	main-beam solid angle	eq.(7a,b)
Ω_r	response pattern solid angle (beam * source)	Table 3
Ω_s	source solid angle	Table 3

Superscripts:

*	complex conjugate	eq.(5)

Symbols:

$\star$	autocorrelation	eq.(5)
$*$	convolution	eq.(33)
$\rightleftharpoons$	"is the Fourier transform of"	Table 2
$\hat{\mathrm{F}}$	Fourier transform of the function F	eq.(34)

Special Functions:

circ	uniform circular aperture function	Table 2
J_1	Bessel function of first kind, order one	Table 2

INDEX

Lecture Notes in Mathematics

Vol. 1236: Stochastic Partial Differential Equations and Applications. Proceedings, 1985. Edited by G. Da Prato and L. Tubaro. V, 257 pages. 1987.

Vol. 1237: Rational Approximation and its Applications in Mathematics and Physics. Proceedings, 1985. Edited by J. Gilewicz, M. Pindor and W. Siemaszko. XII, 350 pages. 1987.

Vol. 1250: Stochastic Processes – Mathematics and Physics II. Proceedings 1985. Edited by S. Albeverio, Ph. Blanchard and L. Streit. VI, 359 pages. 1987.

Vol. 1251: Differential Geometric Methods in Mathematical Physics. Proceedings, 1985. Edited by P.L. García and A. Pérez-Rendón. VII, 300 pages. 1987.

Vol. 1255: Differential Geometry and Differential Equations. Proceedings, 1985. Edited by C. Gu, M. Berger and R.L. Bryant. XII, 243 pages. 1987.

Vol. 1256: Pseudo-Differential Operators. Proceedings, 1986. Edited by H.O. Cordes, B. Gramsch and H. Widom. X, 479 pages. 1987.

Vol. 1258: J. Weidmann, Spectral Theory of Ordinary Differential Operators. VI, 303 pages. 1987.

Vol. 1260: N.H. Pavel, Nonlinear Evolution Operators and Semigroups. VI, 285 pages. 1987.

Vol. 1263: V.L. Hansen (Ed.), Differential Geometry. Proceedings, 1985. XI, 288 pages. 1987.

Vol. 1265: W. Van Assche, Asymptotics for Orthogonal Polynomials. VI, 201 pages. 1987.

Vol. 1267: J. Lindenstrauss, V.D. Milman (Eds.), Geometrical Aspects of Functional Analysis. Seminar. VII, 212 pages. 1987.

Vol. 1269: M. Shiota, Nash Manifolds. VI, 223 pages. 1987.

Vol. 1270: C. Carasso, P.-A. Raviart, D. Serre (Eds.), Nonlinear Hyperbolic Problems. Proceedings, 1986. XV, 341 pages. 1987.

Vol. 1272: M.S. Livšic, L.L. Waksman, Commuting Nonselfadjoint Operators in Hilbert Space. III, 115 pages. 1987.

Vol. 1273: G.-M. Greuel, G. Trautmann (Eds.), Singularities, Representation of Algebras, and Vector Bundles. Proceedings, 1985. XIV, 383 pages. 1987.

Vol. 1275: C.A. Berenstein (Ed.), Complex Analysis I. Proceedings, 1985–86. XV, 331 pages. 1987.

Vol. 1276: C.A. Berenstein (Ed.), Complex Analysis II. Proceedings, 1985–86. IX, 320 pages. 1987.

Vol. 1277: C.A. Berenstein (Ed.), Complex Analysis III. Proceedings, 1985–86. X, 350 pages. 1987.

Vol. 1283: S. Mardešić, J. Segal (Eds.), Geometric Topology and Shape Theory. Proceedings, 1986. V, 261 pages. 1987.

Vol. 1285: I.W. Knowles, Y. Saitō (Eds.), Differential Equations and Mathematical Physics. Proceedings, 1986. XVI, 499 pages. 1987.

Vol. 1287: E.B. Saff (Ed.), Approximation Theory, Tampa. Proceedings, 1985–1986. V, 228 pages. 1987.

Vol. 1288: Yu. L. Rodin, Generalized Analytic Functions on Riemann Surfaces. V, 128 pages, 1987.

Vol. 1294: M. Queffélec, Substitution Dynamical Systems – Spectral Analysis. XIII, 240 pages. 1987.

Vol. 1299: S. Watanabe, Yu.V. Prokhorov (Eds.), Probability Theory and Mathematical Statistics. Proceedings, 1986. VIII, 589 pages. 1988.

Vol. 1300: G.B. Seligman, Constructions of Lie Algebras and their Modules. VI, 190 pages. 1988.

Vol. 1302: M. Cwikel, J. Peetre, Y. Sagher, H. Wallin (Eds.), Function Spaces and Applications. Proceedings, 1986. VI, 445 pages. 1988.

Vol. 1303: L. Accardi, W. von Waldenfels (Eds.), Quantum Probability and Applications III. Proceedings, 1987. VI, 373 pages. 1988.

Lecture Notes in Physics

Vol. 310: W.C. Seitter, H.W. Duerbeck, M. Tacke (Eds.), Large-Scale Structures in the Universe – Observational and Analytical Methods. Proceedings, 1987. II, 335 pages. 1988.

Vol. 311: P.J.M. Bongaarts, R. Martini (Eds.), Complex Differential Geometry and Supermanifolds in Strings and Fields. Proceedings, 1987. V, 252 pages. 1988.

Vol. 312: J.S. Feldman, Th.R. Hurd, L. Rosen, "QED: A Proof of Renormalizability." VII, 176 pages. 1988.

Vol. 313: H.-D. Doebner, T.D. Palev, J.D. Hennig (Eds.), Group Theoretical Methods in Physics. Proceedings, 1987. XI, 599 pages. 1988.

Vol. 314: L. Peliti, A. Vulpiani (Eds.), Measures of Complexity. Proceedings, 1987. VII, 150 pages. 1988.

Vol. 315: R.L. Dickman, R.L. Snell, J.S. Young (Eds.), Molecular Clouds in the Milky Way and External Galaxies. Proceedings, 1987. XVI, 475 pages. 1988.

Vol. 316: W. Kundt (Ed.), Supernova Shells and Their Birth Events. Proceedings, 1988. VIII, 253 pages. 1988.

Vol. 317: C. Signorini, S. Skorka, P. Spolaore, A. Vitturi (Eds.), Heavy Ion Interactions Around the Coulomb Barrier. Proceedings, 1988. X, 329 pages. 1988.

Vol. 318: B. Mercier, An Introduction to the Numerical Analysis of Spectral Methods. V, 154 pages. 1989.

Vol. 319: L. Garrido (Ed.), Far from Equilibrium Phase Transitions. Proceedings, 1988. VIII, 340 pages. 1988.

Vol. 320: D. Coles (Ed.), Perspectives in Fluid Mechanics. Proceedings, 1985. VII, 207 pages. 1988.

Vol. 321: J. Pitowsky, Quantum Probability – Ouantum Logic. IX, 209 pages. 1989.

Vol. 322: M. Schlichenmaier, An Introduction to Riemann Surfaces, Algebraic Curves and Moduli Spaces. XIII, 148 pages. 1989.

Vol. 323: D.L. Dwoyer, M.Y. Hussaini, R.G. Voigt (Eds.), 11th International Conference on Numerical Methods in Fluid Dynamics. XIII, 622 pages. 1989.

Vol. 324: P. Exner, P. Šeba (Eds.), Applications of Self-Adjoint Extensions in Quantum Physics. Proceedings, 1987. VIII, 273 pages. 1989.

Vol. 327: K. Meisenheimer, H.-J. Röser (Eds.), Hot Spots in Extragalactic Radio Source. Proceedings, 1988. XII, 301 pages. 1989.

Vol. 328: G. Wegner (Ed.), White Dwarfs. Proceedings, 1988. XIV, 524 pages. 1989.

Vol. 329: A. Heck, F. Murtagh (Eds.), Knowledge Based Systems in Astronomy. IV, 280 pages. 1989.

Vol. 330: J.M. Moran, J.N. Hewitt, K.Y. Lo (Eds.), Gravitational Lenses. Proceedings, 1988. XIV, 238 pages. 1989.

Vol. 331: G. Winnewisser, J.T. Armstrong (Eds.), The Physics and Chemistry of Interstellar Molecular Clouds mm and Sub-mm Observations in Astrophysics. Proceedings, 1988. XVIII, 463 pages. 1989.

Vol. 332: P. Flin, H.W. Duerbeck (Eds.), Morphological Cosmology. Proceedings, 1988. VII, 438 pages. 1989.

Vol. 333: I. Appenzeller, H.J. Habing, P. Léna (Eds.), Evolution of Galaxies – Astronomical Observations. Proceedings of the Astrophysics School I, 1988. X, 391 pages. 1989.

M. Stix, Kiepenheuer Institut für Sonnenphysik, Freiburg, FRG

The Sun

An Introduction

1989. XIII, 390 pp. 192 figs. (Astronomy and Astrophysics Library). ISBN 3-540-50081-2

Contents: Introduction: The Sun's Distance, Mass, Radius, Luminosity. - Internal Structure. - Tools for Solar Observation. - The Atmosphere. - Oscillations. - Convection. - Rotation. - Magnetism. - Chromosphere, Corona, and Solar Wind. - List of Symbols. - References. - Subject Index.

P. Léna, University of Paris, France

Observational Astrophysics

Translated from the French by A. R. King

1988. XII, 328 pp. 204 figs. (Astronomy and Astrophysics Library). ISBN 3-540-18433-3

Contents: Astrophysical Information. - The Earth's Atmosphere. - Photometry. - Measurement and Processing of Signals. - Detectors. - Images. - Spectral Analysis. - Appendix: Physical and Astronomical Constants. - General Bibliography. - Bibliography. - Subject Index.

M. Harwit, Smithsonian Institute, Washington, D.C., USA

Astrophysical Concepts

2nd ed. 1988. XV, 626 pp. 175 figs. (Astronomy and Astrophysics Library). ISBN 3-540-96683-8

(Originally published by John Wiley and Sons, Inc., 1973 and reprinted by Concepts, 1982)

Contents: An Approach to Astrophysics. - The Cosmic Distance Scale. - Dynamics and Masses of Astronomical Bodies. - Random Processes. - Photons and Fast Particles. - Electromagnetic Processes in Space. - Quantum Processes in Astrophysics. - Stars. - Cosmic Gas and Dust. - Structure of the Universe. - Life in the Universe. - Cosmic Origins. - Epilogue. - Appendix A: Astronomical Terminology. - Appendix B: Astrophysical Constants. - List of References. - Index.

Springer-Verlag
Berlin Heidelberg New York
London Paris Tokyo
Hong Kong